Bioorganometallics

Edited by
Gérard Jaouen

Further Titles of Interest

A. Messerschmidt (Ed.)

Handbook of Metalloproteins

3 Volumes

2004
ISBN 0-470-86981-X

M. J. Welch, C. S. Redvanly (Eds.)

Handbook of Radiopharmaceuticals

Radiochemistry and Applications

2003
ISBN 0-471-49560-3

K. Schwochau

Technetium

Chemistry and Radiopharmaceutical Applications

2000
ISBN 3-527-29496-1

Bioorganometallics

Biomolecules, Labeling, Medicine

Edited by
Gérard Jaouen

WILEY-VCH

WILEY-VCH Verlag GmbH & Co. KGaA

Editor

Prof. Gérard Jaouen
Ecole Nationale Supérieure de Chimie de Paris
Laboratoire de Chimie Organométallique
(UMR CNRS 7576)
11, rue Pierre et Marie Curie
75231 Paris Cedex 05
France

■ All books published by Wiley-VCH are
carefully produced. Nevertheless, editors,
authors, and publisher do not warrant the
information contained in these books,
including this book, to be free of errors.
Readers are advised to keep in mind that
statements, data, illustrations, procedural
details or other items may inadvertently
be inaccurate.

Library of Congress Card No.: applied for
A catalogue record for this book is available
from the British Library.

**Bibliographic information published by
Die Deutsche Bibliothek**
Die Deutsche Bibliothek lists this publication
in the Deutsche Nationalbibliografie;
detailed bibliographic data is available in the
internet at http://dnb.ddb.de.

© 2006 Wiley-VCH Verlag GmbH & Co.
KGaA, Weinheim

Printed in the Federal Republic of Germany
Printed on acid-free paper

Cover Design   4t Werbeagentur, Darmstadt
Typesetting   Manuela Treindl, Laaber
Printing   betz-druck GmbH, Darmstadt
Binding   J. Schäffer GmbH, Grünstadt

ISBN-13:   978-3-527-30990-0
ISBN-10:   3-527-30990-X

Preface

When Anne Vessières, Siden Top and I decided, in 1979, to explore the boundary between organometallic chemistry and biology, the situation was unpredictable, with some genuine but rare successes, a number of obstacles obscuring the way forward, and in general an outlook that did not appear bright. Dorothy Hodgkin had solved the X-ray crystal structure of vitamin B_{12}, John S. Thayer had studied the biological effects of σ-bond organometallics, such as the formation of certain poisons including alkyl tin and mercury compounds in nature (*J. Chem. Ed.*, 1971, 48, 806; *J. Organomet. Chem.*, 1974, 76, 265), while chemotherapy, which had been introduced by Ehrlich in the form of Salvarsan®, was abandoned in favor of antibiotics such as penicillin. However, progress was made when, thanks to Daniel Mansuy, the discovery and subsequent understanding of new metal–carbon bonds revealed the existence of carbenes of P-450 cytochromes; when the work of Petra Köpf-Maier, inspired by cisplatin, opened the way for therapeutic use of metallocenes such as $(C_5H_5)_2TiCl_2$; and finally, when the conceptual possibility was raised, by Michael Cais, of an analytical test called the Metalloimmunoassay. These were however only halting steps, taken at a time when organometallic catalysis was in full stride.

The discovery of the organometallic nature of an anaerobic methane-producing enzyme confirmed our belief that water was compatible with certain organometallics. Our field of study was defined by the expertise available within our small research group. Siden Top was an excellent synthetic organometallic chemist able to turn his hand to any challenge. Anne Vessières was fully qualified in two fields, having produced a thesis in each, one on the chemistry of iron–carbonyl complexes and the other in endocrinology. Accordingly, we chose to examine the potential of metal–carbonyl complexes in applications relating to endocrinology.

It transpired that we had misjudged the difficulty of the task, and it was five years before material progress was made. I am pleased to have this opportunity to recognize my Canadian friends, Michael J. McGlinchey (McMaster University) and Ian S. Butler (McGill University), whose tolerance for risk allowed us to navigate a difficult early period which might otherwise have proved fatal. From the start we also received concrete encouragement and valuable advice from Pierre Potier, F. Albert Cotton and the Lord (Jack) Lewis. Once we emerged from this start-up period, things became easier. Other adventurous and creative students, including Michèle Salmain, joined the original team. Other laboratories became

Bioorganometallics: Biomolecules, Labeling, Medicine. Edited by Gérard Jaouen
Copyright © 2006 Wiley-VCH Verlag GmbH & Co. KGaA, Weinheim
ISBN: 3-527-30990-X

interested in these multidisciplinary approaches. Organometallic enzymes revealed their full importance, as did radiopharmaceuticals, novel drugs, and even organometallic bioanalyses. The field has expanded to such a degree that the need for a comprehensive review, as presented in this book, is pressing. One of the exciting features of bioorganometallic chemistry at this stage of its development is that expanding horizons can still be glimpsed beyond what has been discovered. We hope that the prospect will stimulate the imagination of the reader.

Paris, October 2005   *Gérard Jaouen*

Contents

Bioorganometallics: Biomolecules, Labeling, Medicine. Edited by Gérard Jaouen
Copyright © 2006 Wiley-VCH Verlag GmbH & Co. KGaA, Weinheim
ISBN: 3-527-30990-X

Some authors of this book
From left to right: J. C. Fontecilla-Camps, G. Jaouen, S. Takenaka, T. B. Rauchfuss, M. Salmain, R. H. Fish, A. Vessières, N. Metzler-Nolte, R. Alberto, P. J. Sadler

List of Contributors

Roger Alberto
University of Zürich
Institute of Inorganic Chemistry
Winterthurerstrasse 190
8057 Zürich
Switzerland

Wolfgang Beck
Universität München
Department Chemie
Butenandstrasse 5–13
81377 München
Germany

Richard H. Fish
University of California
Laurence Berkeley National Institute
Berkeley, CA 94720
USA

Juan C. Fontecilla-Camps
Laboratoire de Cristallographie et
Cristallogenèse des Protéines
Institut de Biologie Structurale
"Jean-Pierre Ebel"
41, rue Jules Horowitz
38027 Grenoble Cedex 1
France

Gérard Jaouen
Ecole Nationale Supérieure de
Chimie de Paris
Laboratoire de Chimie
Organométallique
(UMR CNRS 7576)
11, rue Pierre et Marie Curie
75231 Paris Cedex 05
France

Rachel C. Linck
University of Illinois at
Urbana-Champaign
School of Chemical Sciences
106 Noyes Lab
505 S. Mathews
Urbana, IL 61801
USA

Michael J. McGlinchey
Department of Chemistry
Centre for Synthesis and Chemical
Biology
University College of Dublin
Belfield
Dublin 4
Ireland

Michael Melchart
University of Edinburgh
Department of Chemistry
West Mains Road,
Edinburgh EH9 3JJ
UK

Bioorganometallics: Biomolecules, Labeling, Medicine. Edited by Gérard Jaouen
Copyright © 2006 Wiley-VCH Verlag GmbH & Co. KGaA, Weinheim
ISBN: 3-527-30990-X

Nils Metzler-Nolte
Universität Heidelberg
Institut für Pharmazie und
Molekulare Biotechnologie
Im Neuenheimer Feld 364
69120 Heidelberg
Germany

Thomas B. Rauchfuss
University of Illinois at
Urbana-Champaign
School of Chemical Sciences
106 Noyes Lab
505 S. Mathews
Urbana, IL 61801
USA

Peter J. Sadler
University of Edinburgh
Department of Chemistry
West Mains Road,
Edinburgh EH9 3JJ
UK

Michèle Salmain
Ecole Nationale Supérieure de
Chimie de Paris
Laboratoire de Chimie
Organométallique (UMR CNRS 7576)
11, rue Pierre et Marie Curie
75231 Paris Cedex 05
France

G. Richard Stephenson
Wolfson Materials and Catalysis
Centre
School of Chemical Sciences and
Pharmacy
University of East Anglia
Norwich, NR4 7TJ
UK

Shigeori Takenaka
Department of Materials Science
Faculty of Engineering
Kyushu Institute of Technology
1-1 Sensu-cyo
Tobata-hu, Kitakyushu-shi
Fukuoka 804-8550
Japan

Siden Top
Ecole Nationale Supérieure de
Chimie de Paris
Laboratoire de Chimie
Organométallique (UMR CNRS 7576)
11, rue Pierre et Marie Curie
75231 Paris Cedex 05
France

Anne Vessières
Ecole Nationale Supérieure de
Chimie de Paris
Laboratoire de Chimie
Organométallique (UMR CNRS 7576)
11, rue Pierre et Marie Curie
75231 Paris Cedex 05
France

1
A Novel Field of Research:
Bioorganometallic Chemistry, Origins, and Founding Principles

Gérard Jaouen, Wolfgang Beck, and Michael J. McGlinchey

The two founding fathers.
Top: *Organometallic chemistry*: the French L.-C. Cadet de Gassicourt
(1713–1799; courtesy of J. Flahaut). Bottom: *Organometallic drugs*:
the German P. Ehrlich (1854–1915; from a German postage stamp).

1.1
Introduction

In 1985 the term bioorganometallic chemistry was first applied to the synthesis
and study of organometallic species of biological and medical interest [1, 2]. It
became apparent at that time that a critical mass of separate but novel results had

Bioorganometallics: Biomolecules, Labeling, Medicine. Edited by Gérard Jaouen
Copyright © 2006 Wiley-VCH Verlag GmbH & Co. KGaA, Weinheim
ISBN: 3-527-30990-X

been reached that required a unifying term to identify this type of research as a whole. The term in fact covered complexes formed using classical organometallic ligands (e.g. CO, alkyls, π-bonded species) or biomolecules (steroids, amino acids, sugars, peptides, DNA, vitamins, enzymes, antibodies) attached by direct metal–carbon linkages, but all having in common their ability to play a role in biological processes. This by no means exhaustive list should give an idea of the potential scope of this field [3–15].

The idea of combining organometallic and biological components had of course occurred before to a number of researchers, with varying results. This opening chapter aims to address this period of induction and emergence, viewed from today's perspective, and to attempt to show the development of the founding principles of this field, which is multidisciplinary by definition and thus requires advances on several fronts for its own development. In this respect, it is remarkable that one of the factors leading to the explosion of the field was the enhanced understanding of the metal–carbon (M–C) bond, with the result that all studies up to and including the structure of vitamin B_{12} consider only σ bonds [16–18]. However, following the discovery of ferrocene and its sandwich structure [19–21], we see the establishment of a theoretical basis for the understanding of new types of M–C bonds (σ, π and δ) such as, for example, those of metallocenes, metal carbonyls or metal carbenes (Scheme 1.1).

This structural diversity provided a wealth of new reactivity for chemical researchers and gave rise to the development of transition–metal organometallic chemistry in general, and later to bioorganometallic chemistry as we know it today [22–24]. The positioning of bioorganometallic chemistry is summarized schematically in Scheme 1.2.

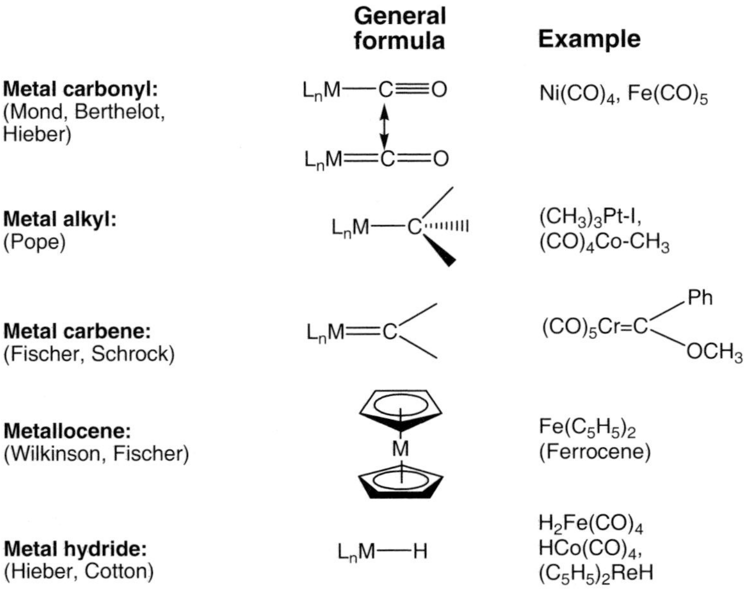

	General formula	**Example**
Metal carbonyl: (Mond, Berthelot, Hieber)	L_nM—C≡O ⇅ L_nM=C=O	$Ni(CO)_4$, $Fe(CO)_5$
Metal alkyl: (Pope)	L_nM—C	$(CH_3)_3Pt$-I, $(CO)_4Co$-CH_3
Metal carbene: (Fischer, Schrock)	L_nM=C	$(CO)_5Cr$=C Ph / OCH_3
Metallocene: (Wilkinson, Fischer)	M	$Fe(C_5H_5)_2$ (Ferrocene)
Metal hydride: (Hieber, Cotton)	L_nM—H	$H_2Fe(CO)_4$ $HCo(CO)_4$, $(C_5H_5)_2ReH$

Scheme 1.1 Transition metal organometallic compounds.

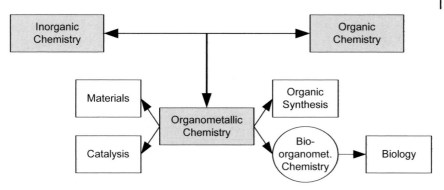

Scheme 1.2 Relationship of organometallic chemistry
(complexes with direct metal–carbon bonds) to the principal related disciplines.

With the perspective of time, it is easier to see how the profusion of novel functions, novel structures and thus novel properties led to the phenomenal success of transition–metal organometallic chemistry in the second half of the twentieth century. The most thoroughly explored fields of application were firstly in the area of catalysis, with ground-breaking reactions such as Ziegler–Natta polymerization, asymmetric synthesis (Sharpless, Noyori, Kagan, Knowles) and olefin metathesis (Grubbs, Schrock), as well as organic synthesis, with a number of advances such as those involving C–C coupling (Stille, Suzuki, Heck, Pauson), and access to new materials and polymers (Kaminski, Marks, Brintzinger). It was not until later, since it required the integration of two disciplines as well as the ability to control the behavior of organometallic complexes in water, that bioorgano-metallic chemistry, the subject of the current chapter, appeared on the scene. Recent expansion has been rapid and at least five major fields of activity can now be identified (Scheme 1.3). A brief history of each will be followed by a deeper discussion in the later chapters of the aspects of these areas that have already been elucidated.

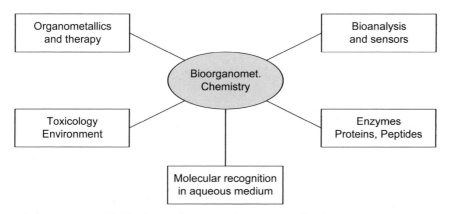

Scheme 1.3 Principal fields of research activity in bioorganometallic chemistry.

1.2
Organometallics and Therapy

1.2.1
The Founding Father

The pharmaceutical industry at the present time is in a relative lull in its ongoing search for truly innovative, therapeutically useful molecules. After witnessing the incredible success of antibiotics, and while awaiting the fulfilment of hopes placed in gene therapy, it is important to continue the search for innovative concepts. The study of the uses of organometallic complexes in medicine could be one of these innovative approaches, owing to the unusual reactivity of these entities. Seen from this perspective, organometallic chemistry had an appropriate genesis, arising from the mind of a distinguished pharmacist of the Enlightenment, a contemporary of Priestley and Lavoisier. This pharmacist was a member of the French *Académie Royale des Sciences* and had a surgery at 115, rue St Honoré in Paris.

Louis-Claude Cadet de Gassicourt (1713–1799) left a fascinating legacy; the son he recognized in 1769, Charles-Louis, was actually the illegitimate offspring of King Louis XV (and Marie-Thérèse Boisselet). On the other hand, he could never have known that he is now regarded as the father of organometallic chemistry. On August 27, 1758, he submitted a paper to the *Académie Royale des Sciences* which dealt with "Cadet's fuming liquid", and which was published in 1760 [25]. This heavy brown liquid is highly toxic, smells strongly of garlic, and spontaneously bursts into flame when exposed to air. It comprises mainly cacodyl oxide $(Me_2As)_2O$, but also contains other cacodyl compounds, such as dicacodyl $(Me_2As)_2$. For personal reasons, Cadet never went back to this seminal work. The unhappiness of his conjugal life led him to squander his money, among other things on the ladies of the nearby Palais Royal. He left practically no inheritance to his adopted son, other than a good education. This at least allowed the young man to become the pharmacist of the Emperor Napoleon and thus to participate in his campaigns in Europe [26]. The story is perhaps typical of many in those turbulent years of French history, when fates collided, and fortunes were made and unmade.

$$As_2O_3 + 4\ CH_3COOK \rightarrow \text{``Cadet's fuming liquid''} \qquad (1.1)$$

In Cadet's day there was no way to explain this reaction, obtained by distilling a mixture of arsenic oxide and "foliated earth of tartar" (potassium acetate), as part of a study on invisible inks. In fact, it was not until almost 80 years later, with the work of Bunsen, that the explanation was forthcoming: these reaction conditions led to the production of "cacodyls" [27, 28]. Cacodyl is derived from the Greek κακωδης *kakodes* meaning evil-smelling. In 1834, J. B. Dumas was forbidden by the administration of the *Ecole Polytechnique*, located in central Paris, to pursue his research on the fuming liquid. The University at Marburg was somewhat more accommodating. They allowed Bunsen (1811–1899) to use a thatch-covered hunting cabin in the middle of the woods to study these cacodyl species from 1837 to 1842, and so identify their organometallic nature, i.e. possession of a direct M–C bond [27, 29–31].

A number of cacodyl derivatives were subsequently discovered, notably various cacodylates used in the first half of the twentieth century as tonics and fortifying medicines for conditions associated with a lowering of hemoglobin levels (tuberculosis, multiple sclerosis, malaria, etc.) [28].

1.2.2
The First Significant Organometallic Drug

The negative effects of arsenic have been known throughout history (e.g. Nero, or the Borgias). In seventeenth-century France, white arsenic, As_2O_3, acquired the name of "poudre de succession" (inheritance powder). Advances in organometallic chemistry were made throughout the nineteenth century in Europe. For example, Bunsen and Frankland described alkylarsines and alkylmercury compounds, Löwig synthesized $(C_2H_5)_4Pb$, Friedel and Crafts investigated organochlorosilanes, and Schützenberger prepared $[Pt(CO)Cl_2]_2$. These reports led Ehrlich (1854–1915) and others to realize that organometallic compounds of arsenic could be less toxic, and easier to manipulate, than their inorganic counterparts. The idea was a novel one, but its implementation proved difficult, requiring the collaboration of a chemist, Bertheim, and a biologist, Sacachiro Hata. In turn, the systematic medical application of organoarsenicals as antiparasitic agents was due to the perseverance of Paul Ehrlich, who was awarded the Nobel Prize in 1908 [32–35].

Of all the compounds prepared before 1909, product No. 606 proved to be particularly effective against syphilis. It was marketed as Salvarsan® in Europe, and was known as Arsphenamine in the USA. It was originally thought that this compound was an arsenic analog of azobenzene; it was represented as **1**, a dimer with an –As=As– double bond and designated as 3,3'-diamino-4,4'-dihydroxy-arsenobenzene (Scheme 1.4).

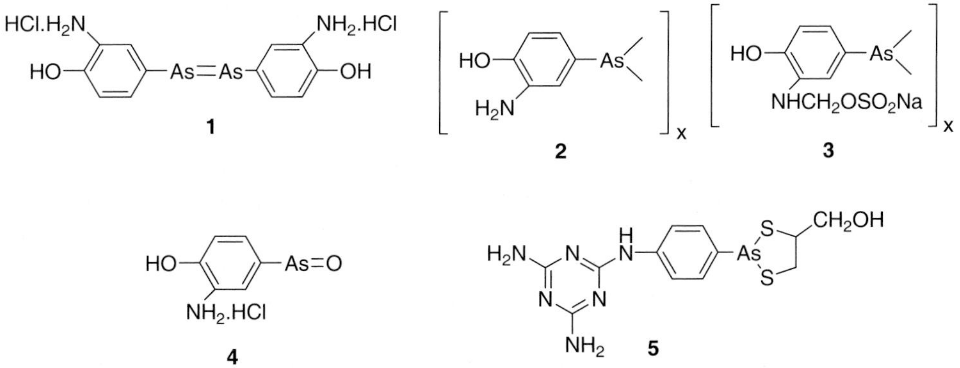

Scheme 1.4 Chemical structure of Salvarsan **2**, NeoSalvarsan **3**, Mapharsen **4** and Melarsoprol **5**.

Subsequent studies have shown that unsubstituted arsenobenzene is in fact a cyclic hexamer $[(C_6H_5As)_6]$ and it seems probable that Salvarsan **2**, is also polynuclear (Scheme 1.4) [30]. The 200DM banknote included an image of the structure in this configuration. Although Mapharsen®, **4**, is conventionally represented as an As(III) species with a double bond to oxygen (analogous to a nitroso-arene), this presumably cannot be correct and it is more likely to be an arsenious acid or a related cyclic structure. Unfortunately, X-ray crystallographic data do not appear to be available but the composition of Salvarsan has now been shown by mass spectroscopy (N. C. Lloyd, H. W. Morgan, B. K. Nicholson, R. S. Ronimus, *Angew. Chem. Int. Ed.* 2005, 44, 941–944).

Salvarsan rapidly became the drug of choice for the treatment of syphilis, whose ravages at that time may be compared to those of AIDS today. Salvarsan itself is not easily soluble in water and its hydrochloride form proved too toxic for medical use. Salvarsan was therefore dissolved in a basic solution and administered in that form [34]. Some time later, NeoSalvarsan® (No. 914; neoarsphenamine **3**), a water-soluble derivative, resolved this difficulty [34]. However, both Salvarsan and NeoSalvarsan oxidize in air, and must be stored in sealed ampoules under a nitrogen atmosphere.

1.2.3
Arsenic Compounds after Ehrlich

After the death of Ehrlich in 1915, research into the organoarsenicals continued, and it was discovered that, in the body, Salvarsan was oxidized to oxophenarsine hydrochloride **4** (marketed as Mapharsen®). Since this compound was stable in air, it gradually replaced Salvarsan for therapeutic use, to the point where it became the predominant organoarsenic medicine during the 1930s.

Mapharsen and other arsenic analogs remained in therapeutic use throughout the 1940s but were gradually replaced by penicillin. This does not mean that organoarsenic compounds have completely disappeared from the market. Indeed, they are still employed in veterinary medicine for their antiparasitic properties, and their antiviral activity has recently been studied.

There remain a number of open questions concerning Salvarsan and its relatives, not least its actual structure and its mode of action. H. Morgan of the University of Waikate has developed a research program into the treatment of antibiotic-resistant infectious agents, based on the work of Ehrlich. Certain bacteria, including some that cause gastroenteritis, syphilis, intra-uterine infection and lyme disease, share the ability to break down sugar using a unique enzyme, but this enzyme may be blocked by using compounds containing arsenic. It would thus be interesting to discover whether Salvarsan or other related molecules might be able to block this key enzyme. Another compound worthy of note is Melarsoprol **5** (Arsobal®), used in the treatment of African sleeping sickness [36, 37], while this and other compounds in the same series are effective in the treatment of refractory acute leukemia [38–40].

1.2.4
Organometallic Mercury Compounds

Mercury has been used in medicine since the time of Ancient Greeks, according to the account of Diascorides, and it is described by Avicenna in the early Middle Ages (980–1037) as a treatment for lice and scabies. Even in those times, the long-term toxicity of mercury was known [41], to the extent that the life expectancy of criminals used in the cinnabar mines of Spain by the Romans was as little as 3 years. The ideas of Paracelsus (1493–1541) expressed during his tenure at the University of Basel, are often quoted on the subject of the nature of toxicity: Paracelsus wrote, "*Alle Ding sind Gift und nichts ohn Gift; alein die Dosis macht das ein Ding kein Gift ist* [All things are poison and not without poison; only the dose makes a thing not a poison]. For example, all food and all drink is a poison if consumed in quantities beyond the normal usage." [41, 42]. On this basis, he re-commended the use of inorganic metal complexes for medical purposes, including compounds of arsenic and mercury. These compounds played a role in the treat-ment of syphilis for several centuries: see for example, J. J. Casanova de Seingalt (1725–1798) "Histoire de ma vie" (My life story). There is an interesting historical parallel between the compounds of mercury and those of arsenic, both in terms of the gradual change in approach from inorganic to organometallic chemistry, and the use of the resulting compounds **6** and **7** as medications (Scheme 1.5) or, in the case of the mercury or arsenic alkyls, as large-scale toxins [43].

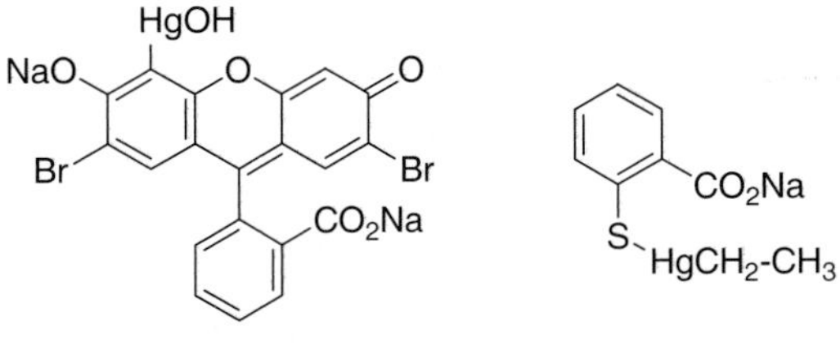

Scheme 1.5 Chemical structure of mercurochrome **6** and merthiolate **7**.

Organomercury compounds are considered of little therapeutic interest today, although mercurochrome and merthiolate are still used as mild local antiseptics. This underlines the common tendency for compounds of this type; they are normally prescribed as external antibiotics rather than for internal use. In fact most cases of organometallic poisonings are due to ingestion of methylmercury compounds either from marine organisms or from treated seed grain [44].

1.2.5
The Current Re-evaluation: Considerations of Efficacy, Toxicity and Selectivity

In addition to the explosion in synthetic potential given to organometallic chemistry by the wide range of new functional groups discovered in the second half of the twentieth century (Scheme 1.1), the other founding event that relaunched the field of metal-based drugs is directly attributable to Rosenberg. He discovered the antitumoral effects of the inorganic drug *cisplatin*, **8**, in particular against testicular cancer. The compound had been synthesized in 1844 by Peyrone [45] but it was not until 120 years later that Rosenberg revealed its properties [46, 47].

To date, around 20 000 articles on platinum complexes have been published; an excellent approach to the subject can be found in the book by Lippert [48], and overviews of metal antitumor agents can be found in the publications of Keppler [49], Sadler [50] or Clarke [51]. Despite this huge volume of work, the only inorganic complexes approved for clinical use as antitumor drugs are *cisplatin*, **8**, *carboplatin*, **9**, *nedaplatin*, **10**, and *oxaliplatin*, **11** (colorectal cancer) (Scheme 1.6). Several others are in clinical testing, and the mechanism of action has been examined [52].

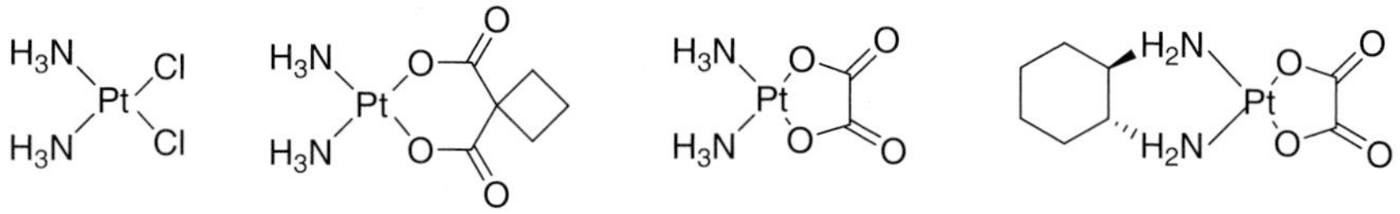

Scheme 1.6 Chemical structure of cisplatin **8**, carboplatin **9**, nedaplatin **10**, oxaliplatin **11**.

The impetus to search for anticancer complexes other than those of platinum stems from its various drawbacks, including cellular resistance to platinum which is sometimes encountered in the clinic; the toxic side effects of cisplatin **8**, which can be severe; and its limited spectrum of activity against other types of cancer [48]. In an attempt to overcome these difficulties, in particular those linked to general toxicity, the possibility of using inorganic complexes of other metals has been explored. For this purpose, complexes of ruthenium, a metal with properties similar to the naturally present element iron, seem particularly attractive. For example, a ruthenium(III) salt, **12** (NAMI-A), $[ImH]^+$ $[trans\text{-}RuCl_4(dmso)Im]^-$, imidazolium *trans*-imidazole(dimethylsulfoxide)tetrachlororuthenate (Scheme 1.7), shows interesting antimetastatic activity as well as possessing low toxicity *in vivo* [53, 54]. Phase I clinical trials indicate that it is well tolerated in human subjects [55].

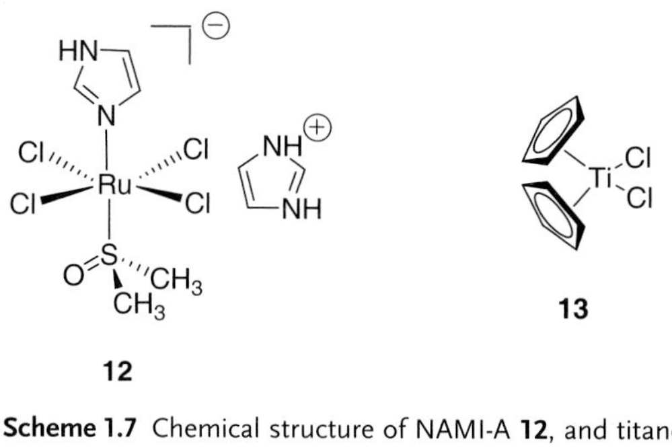

Scheme 1.7 Chemical structure of NAMI-A **12**, and titanocene dichloride **13**.

In parallel with the studies being carried out in coordination chemistry, organometallic complexes of the same type have also been the subject of considerable endeavor, particularly those in the metallocene series with metals such as Ti, Fe, Mo, V, Re, Co, Ru, [56–59]. Titanocene dichloride, **13**, is paradigmatic; the structural analogy between **13** and cisplatin **8** is immediately clear. This organometallic complex **13** has been in clinical trials and was believed to react with DNA in a similar manner to cisplatin. However this complex is difficult to formulate for administration because of its ready hydrolysis. The clinical response was not encouraging and trials have been abandoned. It should also be noted that the Ti(IV) products resulting from the hydrolysis process show a strong proliferative effect in hormone-dependent breast cancers [60]. The research continues, however, using titanium complexes of particularly robust ligands [61, 62]. A concurrent approach, with the aim of lowering the toxicity of these metallic drugs, is based on changing the central metal. The organometallic complexes of Ru are confirming their promising early results for this purpose. In particular, half-sandwich complexes, such as the (η^6-arene)Ru(II) systems **14** and **15** shown in Scheme 1.8, show only low general toxicity [63, 64].

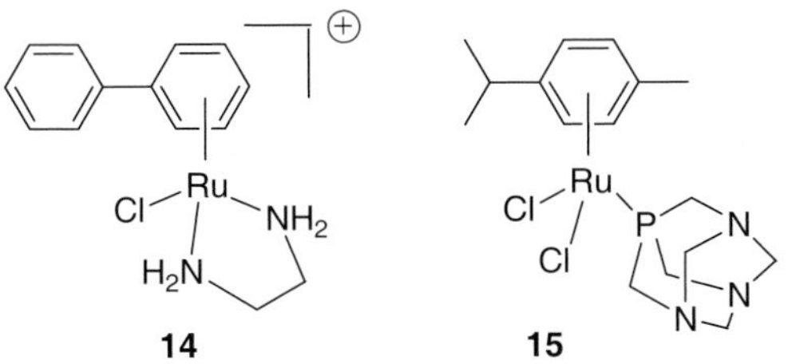

Scheme 1.8 Organometallic complexes of ruthenium **14** and **15**.

Complex **14** shows promising anticancer activity both *in vitro* and *in vivo*, including activity against cisplatin-resistant cell lines [63]. Such organometallic complexes offer much scope for optimization of biological activity by changing the coordinated arene, the chelated ligand, and the chloride leaving group. Molecule **15** shows antimicrobial properties and pH-dependent binding properties [65]. These results are discussed in Chapter 2.

The approach described below is different and promising: it is based on modification via organometallic chemistry of a known bioligand (tamoxifen, chloroquine) in such a way as to optimize its effects. This selective targeting aims to increase efficacy via the organometallic functionality, perhaps involving the formation of oxidized species proximate to the active site [66–73].

Scheme 1.9 summarizes two approaches that are most developed to date, to illustrate the targeting principle. The SERMs (Selective Estrogen Receptor Modulators), of which tamoxifen **16** (active metabolite hydroxytamoxifen, OH-Tam, **17**) is the prototype, represented a significant step forward in the treatment of hormone-dependent breast cancer (ER+). To address the problem of non-hormone-dependent (ER-) tumors, hydroxyferrocifen (OH-ferrocifen, **18**) was

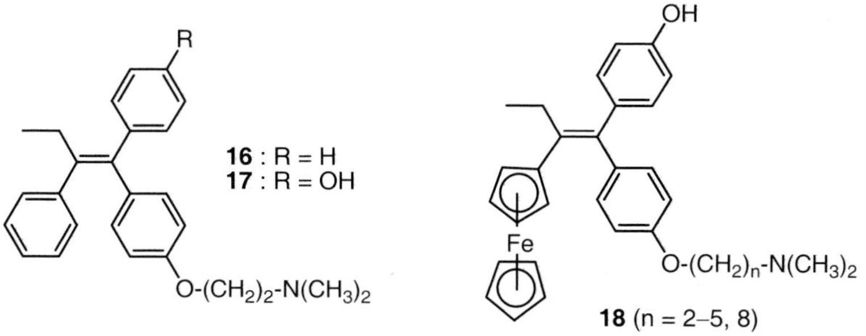

Scheme 1.9 The antiestrogen tamoxifen **16**, its active metabolite hydroxy tamoxifen **17** and hydroxy ferrocifens **18**.

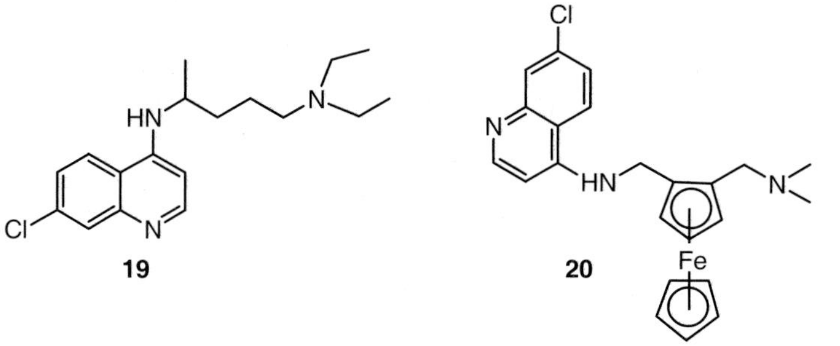

Scheme 1.10 Chemical structure of chloroquine **19** and ferroquine **20**.

prepared. This molecule shows antiproliferative effects on breast cancer cells classed either ER(+) or ER(−). This is due to the particular properties of ferrocifen in an oxidizing environment, which in this type of structure produces targeted apoptosis [67–69].

The other current example involves malaria. Chloroquine, **19**, one of the currently prescribed antimalarials, is subject to resistance phenomena, while ferroquine **20**, surprisingly, does not have this problem. The latter product is currently in clinical development at Sanofi-Aventis [73]. All of these recent developments are discussed in detail, with further examples, in Chapter 3.

1.3
Toxicology and the Environment

This book will not deal in detail with the toxicological and environmental problems associated with the use of organometallics since these subjects have been covered extensively in books by Thayer [44] and Craig [74], respectively. However, it is worth noting that the toxicity of organometallic compounds, e.g. $Ni(CO)_4$, was recognized almost as soon as they were discovered. For the record, and as an illustration, a few historical incidents are briefly outlined here.

Minamata disease was the name given to organomercury poisoning following the first reported large-scale event of this type, to which the inhabitants of the Minamata Bay area in Japan fell victim in 1953. The toxin was CH_3-Hg-SCH_3 [75], from the shellfish *Hormonya mutabilis*, but downstream of the effluents of a factory preparing acetaldehyde [76]. Owing to these serious toxicological problems, the biology of methyl mercuric chloride is the most intensely studied of all organometallics [44].

Organoarsenics were considered for use in poison gas attacks during the First World War. The best known of these gases is lewisite (ClCH=$CHAsCl_2$), which was proposed as a "blistering gas" but was never used [77]. Poison gas research led to the development of antidotes, of which the most widely used was 2,3-di-mercaptopropanol [$HSCH_2CH(SH)$-CH_2OH], known as mercaprol or BAL (British anti-lewisite). It works by binding to the arsenic atom to form a water-soluble complex that is excreted by the body [78, 79].

An early organometallic molecule, tetraethyl lead, which was first made in Zurich by Löwig [80], was widely used for a period of time as a fuel additive, until recognition of its effects on the environment and its implication in cases of environmental poisoning led to its prohibition [81]. The toxicology of (η^5-methyl-cyclopentadienyl) manganese tricarbonyl (MMT) has also been studied; it is a permissible fuel additive in some jurisdictions in North America.

The unfortunate story of "Stalinon" was very harmful to the biological development of tin-based organometallic compounds. Like Minamata disease, it severely delayed the development of therapeutic applications for organometallics. In 1954, 4 million capsules, prepared for the treatment of staphylococcal infections and believed to contain diethyltin diiodide, were distributed in France under the name of "Stalinon". This was a case of the cure being worse than the disease; 102 people died and as many were affected by various neurological disorders. It was subsequently shown that (C_2H_5)$_2$$SnI_2$, and even more so ($C_2H_5$)$_3$SnI (present as an impurity), were powerful neurotoxic agents [82, 83].

This reputation for toxicity clearly has not been advantageous for the development of therapeutic applications for organometallic compounds, although a more favorable tendency is in evidence for the derivatives of silicon, phosphorus or germanium [84–86]. It should be noted, however, that although attaching organic groups directly to the metal may increase toxicity, this is not always the case. It was noted earlier that organic compounds of arsenic are less toxic than their inorganic analogs and that this observation led to Ehrlich's pioneering work on chemotherapy.

We cannot conclude this brief historical outline of organometallics without mentioning Gosio gas. Starting in nineteenth-century Germany, cases were reported of poisoning occurring in rooms decorated with wallpaper that had been colored using arsenate salts. In dark, poorly ventilated, damp and musty conditions, the paper could be attacked to give off a volatile, evil-smelling compound that contained arsenic. Gosio proposed that this toxic compound was [(C_2H_5)$_2$AsH] [87]. But it was in fact Challenger who solved the problem [88, 89], identifying the gas as containing [(CH_3)$_3$As], trimethylarsine, and realized that it was caused by a biological process involving molds with *S*-adenosylmethionine, and coined the term "biological methylation" to describe this process which turned out to be a more general one. Such biological methylation processes to give toxic organometallics may also involve methylcobalamine. The same question has arisen for Hg and Sn, in the latter case concerning the organotins used as antifouling agents.

1.4
Bioanalytical Methods Based on Special Properties of Organometallic Complexes

The development of the radioimmunoassay (RIA), an indirect assay method using radioactive tracers and specific antibodies of the analyte, made great strides in the 1960s, following the work of Yalow and Berson [90, 91]. This development was decisive in the history of bioanalysis, since it increased assay sensitivity by a factor of at least 1000 compared to direct assay [92]. The possibility of using organometallics in this context was first raised by Gill and Mann [93]. These authors were interested in the possibility of labeling a ferrocene with ^{59}Fe, a radioactive isotope of iron, which is a mixed β and γ emitter with a half-life of 44.5 days. Injection of a synthetic polypeptide labeled with 2 or 16 ferrocene entities into rabbits produced anti-ferrocene specific antibodies, incidentally establishing the non-toxicity and stability of ferrocene *in vivo*. It was soon realized however that the use of radioisotopes was not without problems, including their limited shelf life, the relatively small number of usable isotopes, and cost. But it was the associated health risks that brought about draconian controls and led to their use being limited to a few specialized analytical laboratories. The early 1970s consequently witnessed the appearance of the non-isotopic approach, that is, the development of immunological tests not requiring the use of radioactive tracers. Of these, methods using enzymatic, fluorescent or luminescent probes received particular attention [94, 95]. The potential offered by metal probes, whether the metals were in the form of gold colloids, rare earth chelates with interesting fluorescence properties, or organometallic complexes, quickly became apparent [96–98]. Cais in particular introduced the term metallo immuno assay (MIA) as early as 1977 to refer to immunoassays using metal complexes as tracers [96]. Some of these in fact possess unusual physico-chemical properties that allow them to be analyzed in the picomole range, as required by immunoassay. In addition, the great versatility of organometallic synthesis permits relatively straightforward access to tracers corresponding to a large number of desired analytes (cf. Table 8.1). Since Chapter 8 provides a complete review of this subject, just a few examples of the pioneering work done in this field should suffice here.

The mid-1980s saw the development of the first three non-isotopic immunoassays to use organometallic tracers that actually came to fruition. They are based on three different analytical techniques: atomic absorption, electrochemistry, and Fourier-transform infrared spectrocopy (FT–IR) [99–101]. Cais himself did not bring to completion an actual immunoassay by atomic absorption, although he tested the technique with this view in mind [102]. Brossier was the first to validate

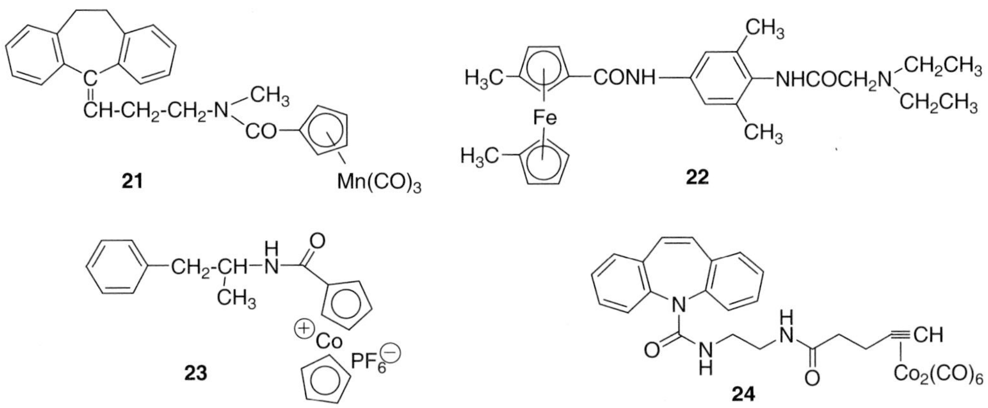

Scheme 1.11 Representative organometallic tracers used in metalloimmunoassays.
21 (cymantrenyl tracer of the nortryptiline),
22 (ferrocenyl tracer of lidocaine),
23 (cobaltocenium tracer of amphetamine),
24 (alkyne dicobalt hexacarbonyl tracer of carbamazepine).

this approach, in the case of antidepressants [103] by quantifying the manganese atom in the nortryptiline tracer **21**. This technique suffers, however, from a lack of sensitivity because of possible solvent contamination by trace metals, and has not been the subject of recent development.

Electrochemical analytical methods all possess high sensitivity, require only low-cost (and miniaturizable) instrumentation, and can be used in turbid conditions. In addition, the unusual electrochemical behavior of ferrocene, which possesses a reversible and stable redox response, was quickly recognized. McNeil accordingly published as early as 1986 an ingenious homogeneous ferrocene-mediated amperometric immunoassay of lidocaine, a drug used in the treatment of cardiac arrhythmias, by using the ferrocenyl tracer of lidocaine **22** [100]. Later, Degrand and Limoges developed another type of immunoassay based on electro-chemical detection of the electroactive cationic cobaltocenium derivative of amphetamine **23** combined with the specific properties of Nafion® [104]. Electrochemical detection is also the basis of DNA sensors or genosensors, a new family of biosensors for the detection of pathogens or specific genes associated with diseases (see Chapter 9). It is interesting to note that a biosensor of this type, the eSensor™, developed by Farkas [105–107] has reached the market. The principle of its activity is shown in Fig. 1.1.

Another commercially successful product is the "ExacTech Pen", manufactured by Medisense Inc., which is used to monitor glucose levels in people with diabetes. Glucose oxidase (GOD) is immobilized onto a pyrolitic graphite electrode, and the ferrocenyl center in **25** (Scheme 1.12) is used to shuttle electrons between the enzyme and the electrode. The ferricinium ion replaces oxygen as the cofactor for GOD. The formal potential of the ferrocene depends on the substituents on one or both rings, but the electron transfer reaction retains the advantageous qualities of rapidity and reversibility [100].

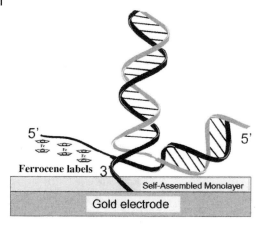

Fig. 1.1 Schematic representation of the eSensor™ DNA Detection System. A self-assembled monolayer is generated on a gold electrode. For electrochemical detection to occur across the monolayer, two hybridization events must occur: the first between the capture probe (shown in black perpendicular to the electrode) and the target (for example, human papillomavirus DNA, shown in gray), and the second between an adjacent region of the target and ferrocene labeled signal probe (shown in black parallel to the monolayer) [106].

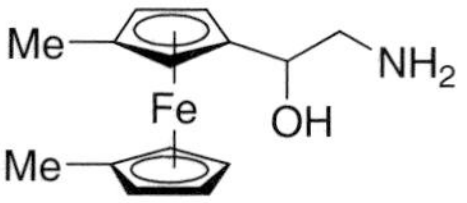

25

Scheme 1.12 Ferrocene complex used to monitor glucose in blood.

The third technique, infrared spectroscopy, was revolutionized at the beginning of the 1980s, when the old dispersive equipment was replaced by Fourier-transform instruments served by desktop computers. It was this technological revolution that permitted the development of infrared spectroscopy as a reliable and sensitive analytical technique [108, 109]. Metal carbonyls with an M–CO bond are one of the major families of organometallic compounds (see Scheme 1.1), at the basis of the first organometallic industrial applications (Mond, Roelen) as well as recently put to new uses (Fischer carbenes), but they are also well-known for their unique spectroscopic properties in infrared (see Chapter 7). In fact the very intense stretching vibrations of M–CO bonds resonate in infrared in the 2000 cm^{-1} region, a zone where practically no standard organic functions resonate at all. This is clearly visible in the spectrum of the chromium carbonyl complex, **26**, whereby the characteristic a_1 and e ν_{CO} stretching vibrations, typical of a C_{3v} symmetric moiety, appear at 1956 and 1876 cm^{-1} (Fig. 1.2). These bands are also 8–10 times more intense than any other band in the spectrum and appear in an absorption window for proteins (Fig. 1.3).

E_2 26-α 26-β

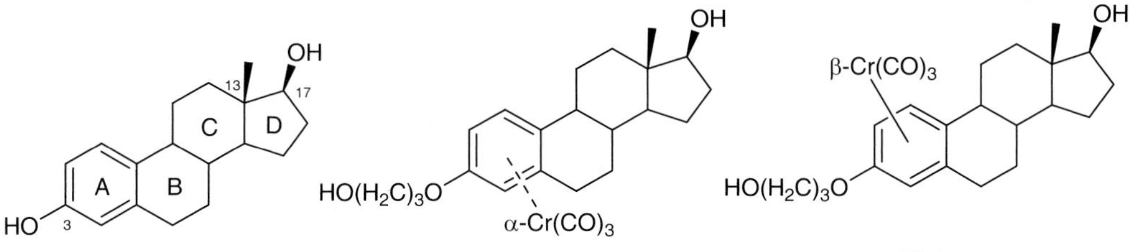

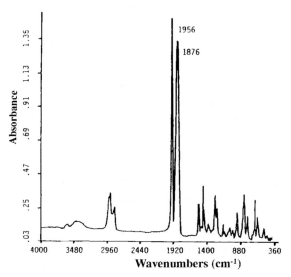

Scheme 1.13 17β-estradiol (E_2) and the α and β benchrotrenyl complexes **26**α and **26**β of an estradiol derivative [110].

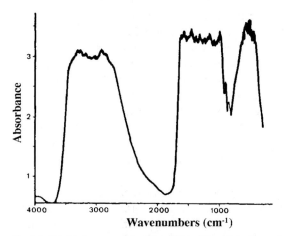

Fig. 1.2 FT–IR spectrum of **26**α in CsI mini pellet [111].

Fig. 1.3 FT–IR spectrum of a mini pellet of lyophilized proteins (from uterine cytosol), off scale, MCT detector [1].

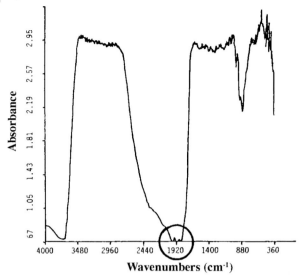

Fig. 1.4 FT–IR spectrum of lamb uterine cytosol following incubation with **26**α and subsequent precipitation from solution with protamine sulfate (off scale, MCT detector) [111].

This unusual property, and in particular its analytical potential, was demonstrated for the first time in 1985 in the case of a hormone bound to its specific receptor [1, 2, 111]. The initial idea was to complex the aromatic A ring of estradiol with a $Cr(CO)_3$ entity (Scheme 1.13). The corresponding complex was successfully prepared, but the presence of a phenol entity makes it unstable and thus unusable in biological media. Moreover, it is well known that this phenol group is essential for the interaction of the hormone with its receptor. Distancing the OH group using a 3-carbon chain produced two stable diastereoisomers, **26**α and β, with the $Cr(CO)_3$ moiety on either one face of the steroid or the other. The β isomer, the one in which the $Cr(CO)_3$ group is on the same side as the 17β-OH and 13β-CH_3 groups, shows only weak recognition for the estrogen receptor (RBA = 1.8%) while recognition of the α complex is much better (RBA = 28%). To date, it is still the only example in the literature that is capable of discrimination, at the level of the A ring, between the α and β faces of estradiol.

Although the amounts of estradiol receptor in target tissues (uterus, certain hormone-dependent breast tumors) are small (in the fmol mg^{-1} protein range), it was possible to detect the characteristic bands in FT–IR of the organometallic hormone **26**α bound to the estrogen receptor and contained in a precipitated cytosol from sheep uterus (Figs. 1.4 and 1.5).

It was difficult to envisage a solid-phase quantitative analysis at this level of concentration. However, for an alkyne cobalt hexacarbonyl complex such as **24** in a chlorinated solvent, use of an ultralow volume gold light-pipe can provide a detection limit as low as 300 fmol (i.e 10 nM), [108], a level fully compatible with the sensitivity range required for the assay of many medications and banned drugs.

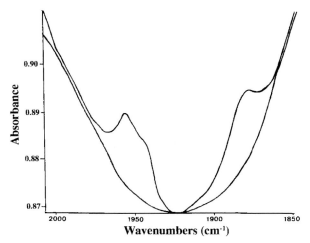

Fig. 1.5 FT–IR spectrum of the metal carbonyl area of lamb uterine cytosol incubated with (a) 10 nM of **26α** and (b) 10 nM of **26α** plus a 500-fold excess of the estrogen diethylstilbestrol (26 000 scans, InSb detector) [110].

This possibility led to the development of the CMIA method to assay antiepileptic medications [101, 112–114] and cortisol [115], a method that offers the additional potential of the simultaneous triple immunoassay, thanks to the distinct and non-overlapping infrared bands of carefully-selected metal carbonyl probes [116]. It is also possible to displace the v_{CO} bands in the infrared by modifying the electron density at the metal, via substitutions of L or Cp ligands [117]. In addition, solvation and environmental phenomena can cause the v_{CO} vibrations to shift. The harnessing of these phenomena is described in Chapter 7.

1.5
Naturally-occurring Organometallics and Synthetic Models

The relationship between organometallics and Nature is very old, and the last few decades have revealed in it new levels of complexity.

It is now accepted that this planet was formed 4.6 billion years ago. The oldest sedimentary rocks containing carbon of biological origin (which can be distinguished by its slightly different isotopic composition) were recently discovered on the island of Akilia, to the south-west of Greenland. These rocks have been shown to be 3.85 billion years old [118].This means that living organisms, presumably very rudimentary and obviously microscopic, must have appeared very quickly on a planet still being bombarded by ultraviolet radiation and with no atmospheric oxygen. The simple molecules present on the earth at that time included H_2, CH_4, H_2O, H_2S, N_2, NH_3, HCN, CO, CO_2. The overall result was a reducing atmosphere very suitable for organometallic chemistry [119]. The question of how the first spark of life arose remains unanswered, but it is conceivable that the bridge between the inorganic and biological worlds, via the effect of metal clusters of the Fe–S type, could have involved the intervention of organometallics [120]. The known capability of FeS and NiS species to catalyze the synthesis of acetyl methyl sulfide from CO and methyl sulfide constituents of hydrothermal fluids, indicates that prebiotic syntheses could have occurred at the interface of these metal–sulfide walled compartments, providing sufficient concentrations of reactants to forge the transition from geochemistry to bio-chemistry [121].

Little by little, research into carbon–metal bonds in natural systems became established [23, 122]. Historical primacy belongs to vitamin B_{12}, its coenzyme and methylcobalamin [16, 123, 124] which were long considered the only naturally-occurring compounds with a metal–carbon bond. The story begins with the observation that raw liver is a cure for an otherwise fatal condition, pernicious anemia. The active component extracted from the liver was first separated and then crystallized in 1948. Dorothy Hodgkin made the first crystallographic structural determination in the series, that of cyanocobalamin, with a tetrapyrrole ligand of the corrin type occupying the horizontal plane. Depending on the example, one of the axial sites may be occupied by a carbon ligand (methyl for methylcobalamin, adenosyl for the coenzyme of vitamin B_{12}) (Scheme 1.14).

The elucidation of the structure of vitamin B_{12} was a tour de force of the early days of X-ray diffraction and was recognized in 1964 by the award of the Nobel Prize to Dorothy Hodgkin. The coenzyme B_{12} acts in concert with a variety of enzymes that catalyze reactions of three main groups. In the first, two substituents on adjacent carbon atoms, –X and –H, are permuted, this is called the isomerase reaction. In the second type, methylcobalamin methylates a substrate, as in the conversion of homocysteine to methionine, for example. Finally, B_{12} is also involved

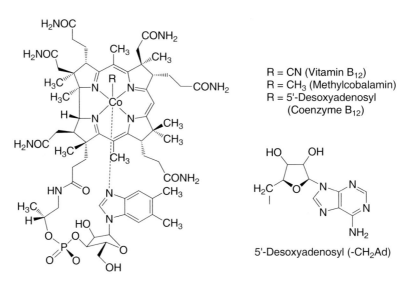

Scheme 1.14 Vitamin B_{12}, methylcobalamin and coenzyme B_{12}.

in the conversion of the ribose ring to the deoxyribose ring related to nucleotides. The coenzyme is required only in small amounts; 2–5 mg is present in the average human, for example, and one of the first signs of deficiency is the failure to form red blood cells.

The role of the coenzyme B_{12} with a reducible cobalt(III)–alkyl bond which can function as nature's Grignard reagent (CR_3^- source), Meerwein's reagent (CR_3^+ source) or as radical source ($CR_3^{\cdot}$) – the latter in the important 1,2-rearrangements e.g. of glutamic acid – is one of the most well studied fields in bioinorganic chemistry (Scheme 1.15). This area therefore will not be covered in this book.

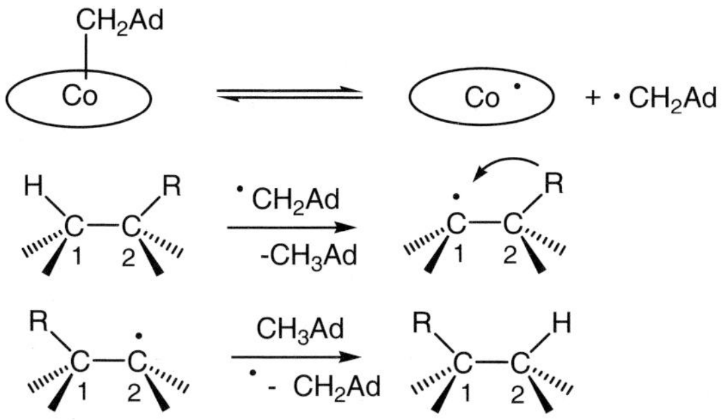

Scheme 1.15 Coenzyme B_{12} catalyzes 1–2 rearrangement.

Many models for vitamin B_{12}, e.g. Schrauzer's cobaloximes [125], have been developed. Methylcobalamin is the agent for the methylation of metals e.g. to $[H_3C\text{-}Hg]^+$ (Scheme 1.16). It is interesting to note that bacteria (mercury lyases) can cleave the mercury carbon bond. This type of cleavage together with the reduction of organometallic compounds by biocatalysts has been reviewed by Ryabov [13].

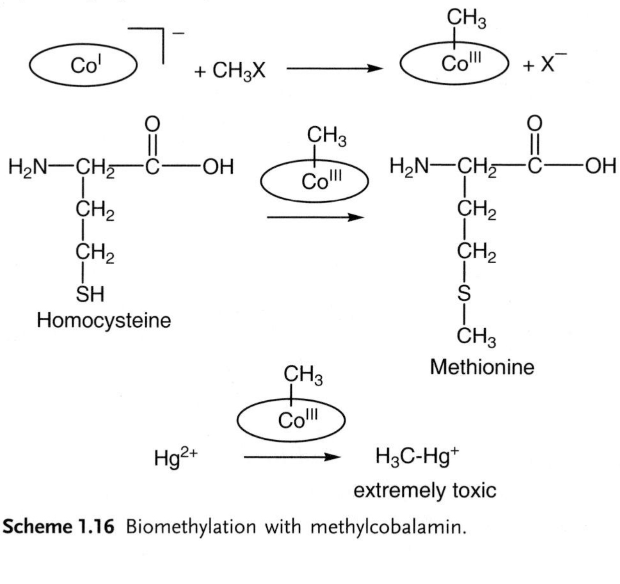

Scheme 1.16 Biomethylation with methylcobalamin.

Although the question of the B_{12} coenzyme remains of current interest [126–128], recent research has focused on other cofactors of metal enzymes, certain of which owe their suprising properties to an organometallic reactivity that is not yet understood. Structural methods have been one of the keys to the recent advances on this front. In particular one may cite the active sites of nitrogenase [129, 130], hydrogenase and CO dehydrogenase/acetyl-coezyme A synthase [131–134]. In this book, J. Fontecilla-Camps, one of the leaders in this area, presents an overview of all the recent mechanistic advances (Chapter 11) while T. Rauchfuss and R. Linck focus on biologically-inspired organometallic models (Chapter 12). These two divisions are currently advancing in parallel, making good progress and feeding off each other.

Nitrogenase, because of its facility for producing NH_3 in mild conditions (Eq. 1.2), was a long-standing challenge for organometallic chemists.

$$N_2 + 8\,H^+ + 8\,e^- + 16\,MgATP \xrightarrow{\text{nitrogenase}} 2\,NH_3 + H_2 + 16\,MgADP + 16\,Pi \quad (1.2)$$

The harsh conditions of the Haber process to produce NH_3 with iron catalysts are well known. Besides this, nitrogen is a poor ligand for coordination of metals. The first recognized dinitrogen complex was $[Ru(NH_3)_5(N_2)]^{2+}$, by Allen and Senoff in Toronto, which dates only from the 1960s [135]. This synthetic challenge to activate molecular nitrogen by organometallic means was undertaken among others by the Russian group of Shilov, and also by Chatt and by Schrock [136–140]. Despite structural determinations to a resolution of 1 Å it is still not completely understood how the hydrogen is reduced by the nitrogenase nor even how it binds initially to the cofactor FeMo, although majority opinion leans towards the belief that the association and the reduction of the nitrogen occur on one of the faces of the Fe_4S_4 cluster.

The hydrogenases are classed into two major families, those containing Ni–Fe and those containing only iron, Fe–Fe. Those in the first category tend to be more implicated in the oxidation of hydrogen and those in the second in the production of H_2. The latter aspect is interesting in the context of new sources of combustibles.

The chemical reaction employed produces the following result (Eq. 1.3).

$$H_2 \rightleftharpoons 2\,H^+ + 2\,e^- \qquad\qquad\qquad\qquad (1.3)$$

This class of enzymes was described for the first time in 1931 by Stephenson and Stickland [141]. A major article published in 1995 reported the X-ray crystallographic structural determination Ni–Fe-H_2ase from *Desulfovibrio gigas* [142, 143]. Among the interesting examples of this type of structure are a heterobimetallic Fe–Ni active site, the presence of CO and CN ligands on iron, and Fe–S clusters on one of the constituents of the dimer. The Fe–only hydrogenase is evolutionarily distinct from Ni–Fe-H_2ase but has some common structural features e.g. CO and CN ligands. The seminal structures of *C. pasteurianum* (Cp) [144] and *Desulfovibrio desulfuricans* [145–147] have increased the focus of work on molecular models. The question of the relevance of these remains open, but the situation is gradually

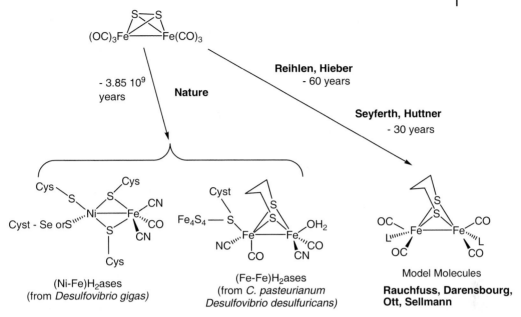

Scheme 1.17 Representation of the evolution of the CO-mobilized iron sulfide to yield the active sites of (Fe–Fe)H$_2$ases and (Ni–Fe)H$_2$ases (adapted from [148]).

improving [4, 148–154]. The parent complexes $(OC)_3Fe(\mu SR)_2Fe(CO)_3$ have a long and ongoing history [155–164] that began 75 years ago (Scheme 1.17).

The evolution of CO-mobilized iron sulfide in nature to yield the [Fe]H$_2$ases can be compared to the biomimetic molecules synthesized in the last 60 years. It is true that the current intense interest in organometallic sulfur chemistry is largely based on a fascinating aspect of biology whose economic potential is just being realized [150]. A recent model molecule of (Fe–Fe)H$_2$ase is a real tour de force with a small Fe–Fe cluster attached to a large Fe$_4$S$_4$ cluster by a bridging cysteinyl unit [165]. We note that the reactions of organometallic enzymes in nature usually occur under anaerobic conditions, exactly as the organometallic chemist is accustomed to carrying out reactions with strict exclusion of oxygen.

The bacterial metalloenzyme acetyl coenzyme A synthase/CO dehydrogenase (ACS/CODH) catalyzes two very important biological processes, namely the reduction of atmospheric CO$_2$ to CO and the synthesis of acetyl coenzyme A from CO, a methyl from a methylated corrinoid iron–sulfur protein, and the thiol coenzyme A [166–168]. This bifunctional enzyme is the key to the Wood-Ljiungahl pathway of anaerobic CO$_2$ fixation (Scheme 1.18) and a major component of the global carbon cycle. Reactions catalyzed by CODH and ACS are shown in Eqs. (1.4) and (1.5) below.

$$CO_2 + 2\ H^+ + 2\ e^- \underset{}{\overset{CODH}{\rightleftharpoons}} CO + H_2O \qquad (1.4)$$

$$CO + CH_3\text{-}CFeSP + CoA \overset{ACS}{\rightleftharpoons} \text{acetyl-CoA} + CFeSP \qquad (1.5)$$

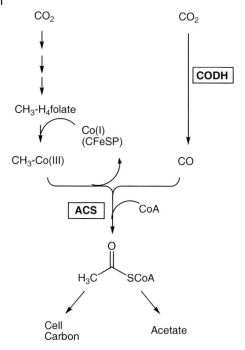

Scheme 1.18 Diagram of the Wood-Ljiungdahl pathway used in one-carbon metabolism of anaerobic organisms [167]. CODH stands for carbon monoxide dehydrogenase and ACS for acetyl-CoA synthase (adapted from [134]).

These reactions exhibit similarities to other well-known organometallic catalytic processes. Thauer, a pioneer in nickel enzymes, emphasized in his lectures the analogy between bioorganometallic chemistry in bacteria and the industrial Reppe and Monsanto processes for the production of acetic acid, both involving metal–carbonyl, metal–methyl, methyl–acetyl bonds [169]. In the Monsanto process carbonylation generates CH_3COI which reacts rapidly with water to give CH_3CO_2H. In the biological reaction, a thioester, also unstable in aqueous media, is easily hydrolyzed.

The nature of the active site of ACS has recently been the subject of much debate [131, 132, 170, 171]. It has now been established that it contains two nickel atoms [131], although the mechanism is still under discussion [172]. For a review of the current situation, see Hegg [134, 173]. Still to be resolved issues are, *inter alia*, the order of introduction of CO and CH_3, the mechanism of the CH_3 migration, the nature of the electron reservoir, the oxidation state of (Ni) (Scheme 1.19), and the question as to the involvement of a mononuclear or binuclear Ni site for the reaction. Fontecilla suggests a Ni(0) species and a mononuclear site [131].

A related mechanism was proposed by Wächtershäuser [174] for the formation of the activated thioester H_3C-$C(O)$-SCH_3 from CO and CH_3SH on coprecipitated

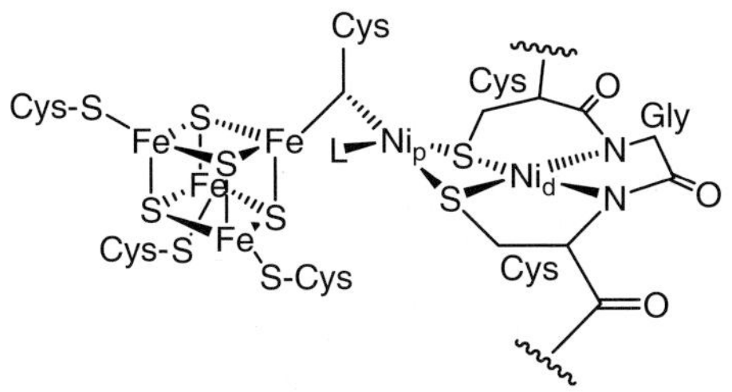

Scheme 1.19 Representation of the ACS active site. The site distal to the (Fe_4S_4) cluster is occupied by a Ni(II) ion in a square-planar environment (Ni_d). (Ni_p) is the catalytically relevant ion in the proximal site but its oxidation state is not clear. The coordination sphere of (Ni_p) is completed by an unknown exogenous ligand L (adapted from [134]).

NiS and FeS. This C–C coupling can be considered as a model reaction from the prebiotic formation of activated acetic acid at the beginning of life. Recently, Weigand, Kreisel et al. [156] observed the formation of ammonia from nitrogen and hydrogen sulfide at the surface of FeS, and at atmospheric nitrogen pressure. This is an important model reaction for the prebiotic production of ammonia.

Another example for the close analogy between biochemistry and organometallic chemistry is found in CO dehydrogenase which has an important ecological role [175]. Soil microbes exploit CO as a carbon and energy source and, using CODH, can remove this toxic gas from the environment at a rate of about 3×10^8 tons per year. According to the Hieber-base reaction, it is suggested that hydroxide (from iron) attacks a Ni coordinated CO ligand to give a metallocarboxylic acid which decomposes to give CO_2 [133]. The original Hieber-base reaction involves the reaction of $Fe(CO)_5$ with hydroxide [176] (Eq. 1.6).

$$Fe(CO)_5 + OH^- \rightarrow [Fe(CO)_4COOH]^- \rightarrow [Fe(CO)_5H]^- + CO_2 \qquad (1.6)$$

This reaction was reported in 1932 and led to the first carbonyl metal hydride, $H_2Fe(CO)_4$. The Hieber-base reaction was first interpreted by Kruck [177], and it is now the accepted mechanism for the metal-catalyzed water gas shift reaction (Eq. 1.7).

$$CO + H_2O \rightleftharpoons CO_2 + H_2 \qquad (1.7)$$

Twenty-five years ago Thauer [178] reported that the growth of methane-producing bacteria (methane from H_2 and CO_2) was influenced by a nickel-containing tubule. This observation finally led to the isolation of Ni complexes from methanogenic bacteria and to their characterization as Ni macrocycles with a reduced porphyrinoid ring (cofactor F_{430}) [179]. The active enzyme contains Ni(I) at which the methyl groups can be oxidatively added. Methane is released by protonation of the Ni-CH_3 species. Many laboratory examples are known whereby methylating agents can be added to metals in low oxidation states; the formation of methane from methyl complexes is also well known [180]. The importance of the methane-producing microorganisms stems from the fact that they release 10^9 tons of methane into the atmosphere every year. However, the anaerobic oxidation of methane has also been reported [179b].

Finding good organometallic models able to take into account the behavior of organometallic enzymes remains a major goal for the chemist. At present the particular relevance of the model chosen often proves to be a key consideration. Novel multidisciplinary approaches are starting to appear, however, and their complexity may provide new sources of inspiration for research. As an example, Böck recently posed the following question: how did nature tame toxic ligands such as CO and CN⁻ so as to fix them on the iron atoms of hydrogenases? [181]. Böck showed that in the biosynthesis of cyanide ligands in Ni–Fe-H$_2$ases the CN⁻ ligand is never free. The key step, after the synthesis of **27**, followed by that of **28** (Eq. 1.8), is a chemical transfer of CN towards the active centre of the protein Ni–Fe—H$_2$ase (with L$_N$ representing the coordination sphere of the active site).

$$2\,e^-$$

$$\boxed{E}\text{–cys-S-CN} \xrightarrow{} \boxed{E}\text{–cys-SH} \qquad (1.8)$$

27 $\qquad$ L$_N$Fe $\quad$ L$_N$Fe-CN $\quad$ **28**

The thiocyanate residue is transported towards the active site, and the transfer of the electrophilic CN entity to the nucleophilic centre of the iron (or vice versa) occurs to form the Fe–CN unit of the enzyme. This seminal work required a combination of radioisotopic labeling, mass spectrometry and reaction modeling studies. Thus, treating phenylthiocyanate with $[(\eta^5\text{-}C_5H_5)Fe(CO)_2Br]$ produces $[(\eta^5\text{-}C_5H_5)Fe(CO)_2CN]$ in good yield and under very mild conditions [15].

Monoxygenases are widely distributed enzymes which catalyze dioxygen activation using two electrons and two protons, with the insertion of one oxygen atom from O$_2$ into a substrate and the formation of water according to Eq. (1.9).

$$RH + O_2 + 2\,e^- + 2\,H^+ \rightarrow ROH + H_2O \qquad (1.9)$$

A great number of these monooxygenases contain a heme protein called cytochrome P-450 which is the site of dioxygen activation. These cytochrome-P-450-dependent monooxygenases are involved in many steps of the biosynthesis and biodegradation of endogeneous compounds such as steroids, fatty acids, prostaglandins and leukotrienes. They also play a key role in the oxidative metabolism of exogenous compounds such as drugs and other enviromental products allowing their elimination from living organisms. Because of their wide distribution in living organisms and their very important role in biochemistry, pharmacology and toxicology, cytochromes P-450 have been the subject of many studies during the last 20 years [182, 183]. For these reasons, the organometallic chemistry of the P-450 cytochromes, widely covered elsewhere, will not be revisited here. This chemistry was started by two groups, that of D. Mansuy on the porphyrin iron carbene complexes [12] and that of R. Guilard on σ-alkyl type complexes [184]. For leading references see [185–191].

The cytochrome P-450 and iron porphyrins exhibit a particular ability to stabilize iron–nitrene and iron–carbene complexes which are formally the nitrogen and carbon equivalents of iron–oxo complexes. Many iron–carbene complexes, Fe=CRR′, have been obtained either by *in situ* reduction of polyhalogenated compounds by iron (II) porphyrins in the presence of an excess of reducing agent, or by direct reaction between iron(II) porphyrins and some carbene precursors like diazo compounds. Among the few metalloporphyrins which have been found able to give stable metal–carbene complexes are Ru and Os porphyrins. On the contrary, Ni(II), Co(II) and Rh(III) porphyrins lead only to bridged carbene complexes upon reaction with carbene donors. These complexes are derived formally from the insertion of the carbene moiety into a metal–nitrogen (pyrrole) bond. Other model reactions indicate the importance of the organometallic chemistry of cytochrome P-450 where iron–carbon σ-bonds can also be formed.

A number of compounds containing carbon–halogen bonds, such as CCl_4, $CF_3CHClBr$ (an anaesthetic agent marketed as halothane), benzyl halides etc. form carbenic or σ-bond complexes, depending on the nature of the reagent. However, other organometallic complexes of cytochrome P-450 may be formed as intermediates during the oxidation of various substrates, such as the epoxidation of propene. Finally, model systems of Fe and Mn are efficient oxidation catalysts, resistant to certain drugs and xenobiotics, and usable in fine chemistry (Scheme 1.20).

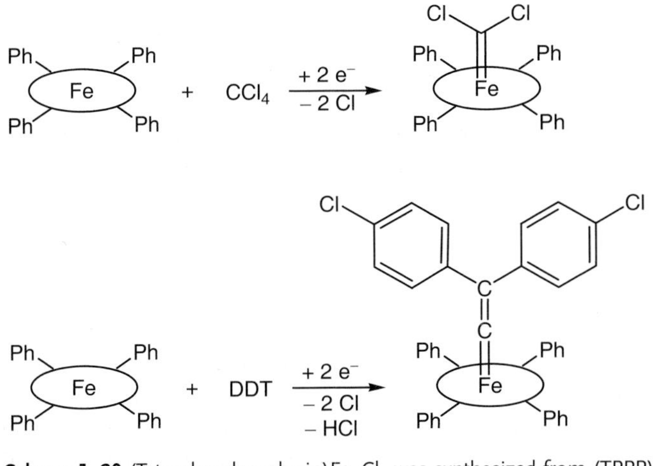

Scheme 1. 20 (Tetraphenylporphyrin)Fe=Cl_2 was synthesized from (TPPP)Fe and CCl_4 in the presence of a reducing agent; an iron carbene complex from DDT has also been reported.

1.6
Organometallic Chemistry and Aqueous Solvents

The compatibility of organometallic species with water, the essential biological solvent, is one of the conditions necessary for the advancement of the discipline. This is shown clearly in the case of anaerobic organometallic enzymes. But for chemists this has required a cultural change, despite the fact that water-soluble "oxo" catalysts of Rh bearing the $P(m\text{-}C_6H_4SO_3Na)_3$ ligand are commercially produced. Aqueous organometallic chemistry [192] is a new and challenging area and ranges from life science to catalysis [198]. The area of organometallic radiopharmaceuticals is particularly representative of this need for compatibility. R. Alberto, one of the principal researchers in this area, deals with this question in Chapter 4.

Here we will recall just two of the advances made in the field of radiopharmaceuticals over the last few years. One shining example of an organometallic pharmaceutical that has been successfully commercialized is technetium-99m sestamibi, a hexakis (alkylisocyanide) complex of $^{99m}Tc(I)$ **29** that has become the most important myocardial imaging agent. The success of this compound, known by its trade name Cardiolite®, has inspired researchers to develop other ^{99m}Tc-based radiopharmaceuticals [193–197].

^{99m}Tc bioorganometallic compounds have to be prepared in water from the readily available precursor, $[^{99m}TcO_4]^-$, and the synthesis should also be rapid because ^{99m}Tc has a half-life of about six hours. In addition, the preparation needs to be quantitative and simple to carry out, with the product solutions ready to inject into the patient.

Alberto has developed a method for directly converting $[^{99m}TcO_4]^-$ into $[^{99m}Tc(OH_2)_3(CO)_3]^+$ **30**, which has attracted much interest as a versatile precursor for ^{99m}Tc radiopharmaceuticals. This water- and air-stable aquo-carbonyl complex **30** contains three labile water ligands that can easily be substituted with a wide

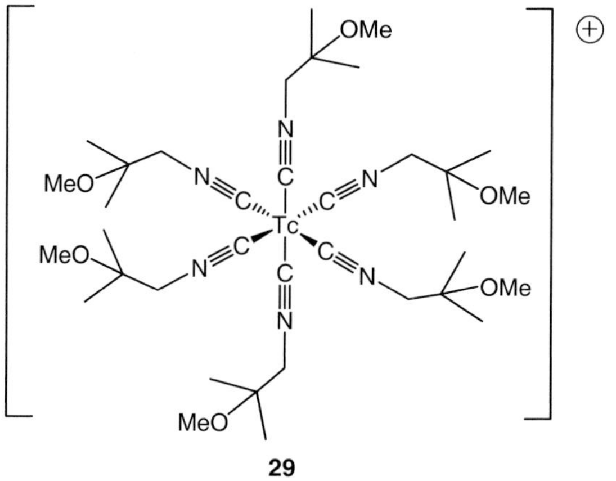

29

Scheme 1.21 Molecular structure of ^{99m}Tc-sestamibi (Cardiolite®) **29**.

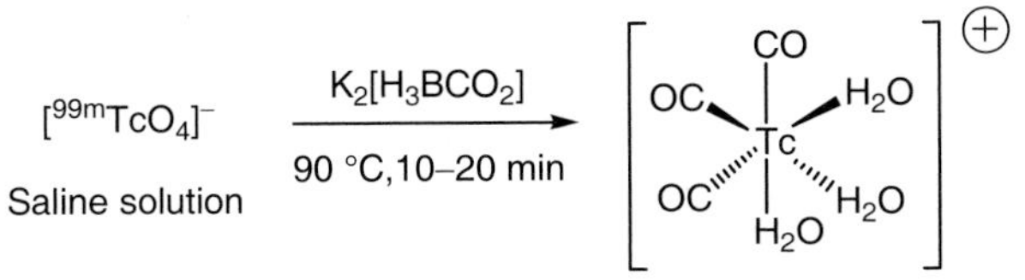

30 (> 98%)

Scheme 1.22 One-pot synthesis of the Alberto's reagent **30**.

variety of organic ligands, thus providing a family of carbonyl complexes that can be used in life sciences and nuclear medicine applications [198–202].

As for organometallic amino acids, the peptides modified with them, and indeed the proteins and carbohydrates, their water solubility is a primary characteristic in terms of their use.

The first organometallic complex with an α-amino acid, $Fe(CO)_2(cysteinate)_2$, was reported in 1929 by Cremer [203], and characterized by Schubert [204]. The first amino acid containing ferrocene was reported as long ago as 1957 by Schlögl [205], and Pauson [206]. For a comprehensive review on ferrocenyl bioorgano-metallics see [207]. The pioneering work of Schlögl was later developed by Sergheraert and Tartar [208] and others [7, 209]. Hieber et al. [210] used the halo-bridged complex $[M(NO)_2Br]_2$ with amino acids for the synthesis of the corresponding N,O-aminocarboxylates. The reaction of halobridged complexes was extremely useful for the preparation of many organometallic complexes of bioligands [7, 209]. Among the first organometallic complexes of α-amino acids were the palladium and platinum complexes (allyl)$Pd(NH_2CHRCO_2)$ [211] and $[Pt(Cl)(NH_2CHRCO_2)(olefin)]$, and the first half sandwich complexes with an α-aminocarboxylate ligand $[(\eta^6\text{-}C_6H_6)(\eta^3\text{-allyl})Mo(NH_2CHRCO_2)]$ [212], $(\eta^6\text{-}C_6H_6)Ru(NH_2CHRCO_2)(Cl)$ [213], $CpMo(CO)_2(NH_2CHRCO_2)$ [214], and the complexes $[Cp_2Mo(NH_2CHRCO_2)]^+$ [215] and $Rh(diene)(NH_2CHRCO_2)_2$ [216]. Sheldrick and Heeb [217], as well as the groups of Carmona, Grotjahn, Werner, and Beck [7] successfully prepared and structurally characterized half-sandwich complexes of α-aminoacidates and small peptides. In these "Brunner type" [218] complexes, the metal is a chiral center and diasteromers arise with optically active α-aminoacids (Scheme 1.23).

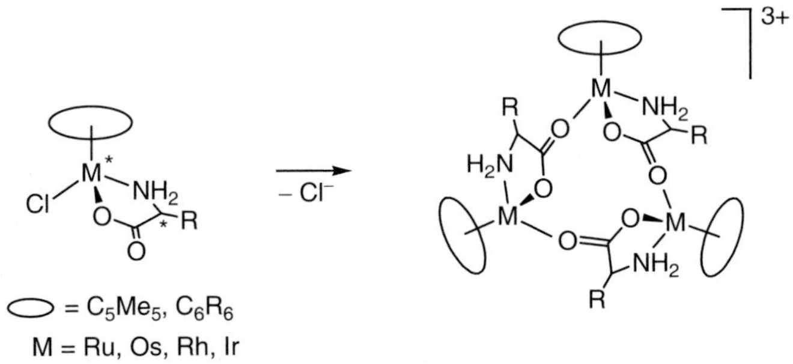

= C_5Me_5, C_6R_6

M = Ru, Os, Rh, Ir

Scheme 1.23 Trimerization of metal complexes of amino acids.

The carbonyl metal complexes which contain ligands of biological origin (nucleobases, amino acids) which were described before 1985 were reviewed by Ioganson [219], and the organometallic chemistry of amino acids and small peptides has been summarized by Beck, Bergs and Severin [7]. The Beck group in Munich has published more than 150 papers in this field and an example is shown on Scheme 1.24. Chapters 5 and 6 give additional examples of organometallic peptides and proteins with chemical, biochemical and biological aims.

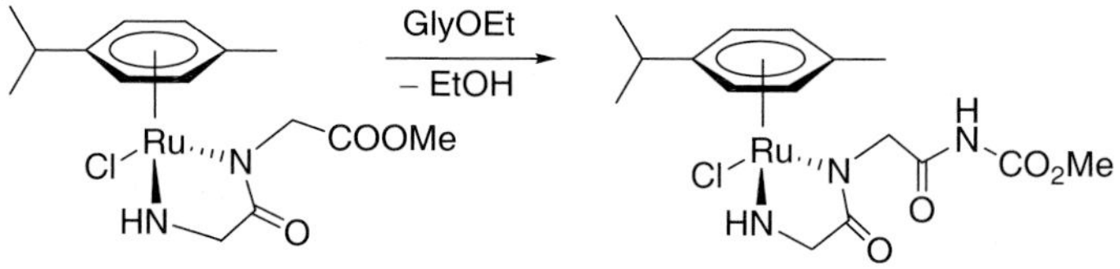

Scheme 1.24 Organometallic complexes of di- and tri-glycine.

Among the first organometallic complexes with nucleobases was a series of metal carbonyl complexes with these bioligands [220–222]. The structures of nucleobase-bridged carbonyl rhodium complexes were determined by Sheldrick [223], who also obtained a series of half-sandwich ruthenium complexes with nucleobases (Scheme 1.25) [224–226].

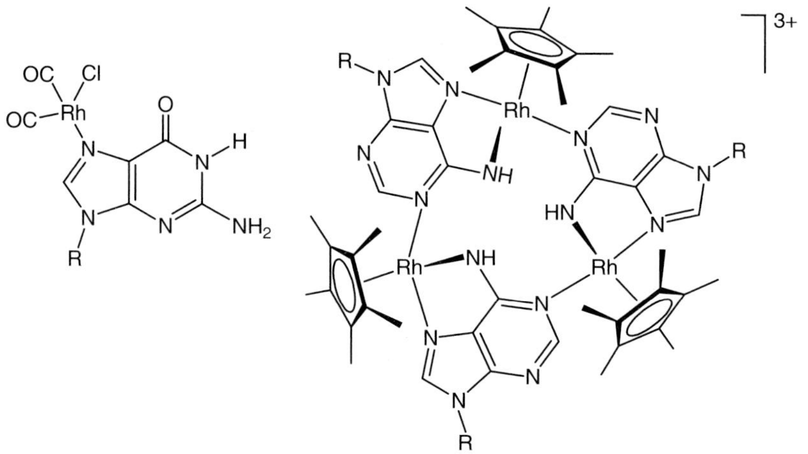

Scheme 1.25 Trimerization of a rhodium nucleobase complex.

Most of these primal studies with bioligands were mainly synthetic in their purpose and devoid of biological applications. However, they paved the way for further investigations. Ionic titanocene amino acid complexes $[Cp_2Ti(NH_2CHRCO_2)_2]^+$ show less pronounced antitumor activity [227, 228] than Cp_2TiCl_2 [227]. Recently, Sadler et al. [229] studied the antitumor activity of the organometallic ruthenium complexes that had first been observed by Sheldrick [230]. Chapter 10 includes other examples of nucleobases complexes displaying unusual supramolecular host–guest recognition properties.

Among the first transition metal complexes (excluding Hg compounds) with metal–carbon σ-bonds to carbohydrates were cobalt carbonyl and $Cp(OC)_2Fe$ complexes [231–234]. These systems were also studied as intermediates in oxo reactions of unsaturated sugars [235–237]. Subsequently, metal carbene complexes of carbohydrates were synthesized [238–242]. In a series of papers, Dötz and coworkers [12, 243] introduced the "Fischer-type carbene functionality" into carbohydrates by using a number of different synthetic methodologies; these complexes can be used for the modification of sugars. Cyclopentadienyl–titanium complexes with chiral σ-coordinated ligands derived from glucose are stereoselective reagents in organic synthesis [244, 245]. A large number of chiral ligands derived from carbohydrates, mainly phosphines, are used in asymmetric catalysis [246–248]. The trimethylplatinum(IV) cation proved to be very well suited for the coordination of unprotected monosacharides (Scheme 1.26) [248].

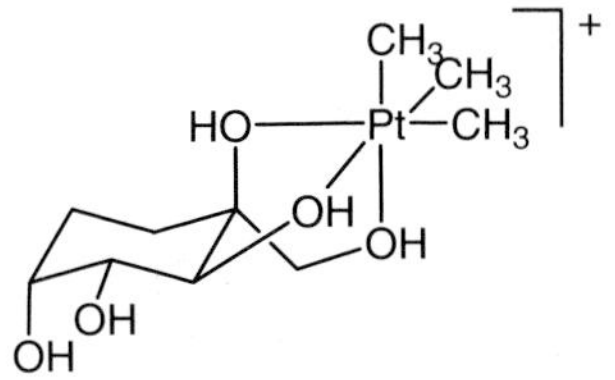

Scheme 1.26 A platinum complex of a carbohydrate.

Another topic related to water and organometallics has been developed by Fish, who is the author of Chapter 10. In particular, Fish has tried to overcome the problem of cofactor regeneration in biocatalysis. At least one-third of all known enzymes require the use of cofactors such as NAD^+ (nicotinamide adenine dinucleotide) and its reduced form, 1,4-NADH, to perform oxidations or reductions. However, these cofactors are expensive and fairly complicated molecules. It has been shown that, in aqueous solution, some organometallic biomimetic models behave chemically like NAD^+ and are able to abstract a hydride ion [249]. A key example is shown below where the hydride **33** reduces **31** into **34** (Scheme 1.27).

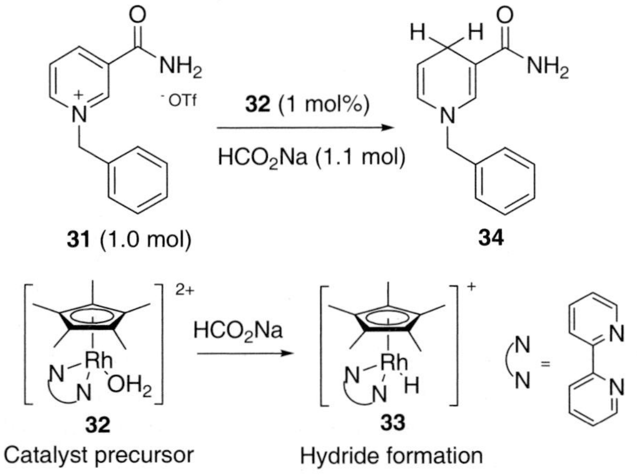

Scheme 1.27 Organometallic cofactor model.

1.7
Conclusions

The panoramic view presented here, which encompasses for the first time the many different aspects of bioorganometallic chemistry, makes it clear that the field now has a solid base from which extensive, multidisciplinary development can reliably be expected over the next few decades. The discipline benefits from a number of positive factors, including its ability to accommodate complexity and diversity, its use of a two-pronged scientific approach, the importance of the problems to be solved, the state of development of contemporary organometallic chemistry, the evolution of chemical thinking towards problems at interfaces and the amount of virgin terrain still to be explored. And perhaps above all, there is the opportunity offered to the imagination of the chemist in this science which, in Berthelot's words, "creates its own purpose".

Although bioorganometallic chemistry has already made great strides, there is much still on the horizon. There is no doubt that this avenue of research is capable of providing a source of inspiration and stimulating the creativity of chemists for the foreseeable future. In fact, some of the apparent obstacles to biological and biomedical applicability (aqueous solvents, time constraints for radiopharmaceuticals, anaerobic conditions, etc.), may be seen less as a brake than as a spur in the search for imaginative solutions.

The social relevance predicted for this discipline is another encouragement to its development. The range of potential commercial applications appears very promising, whether for environmental applications, bioanalysis, enzymology and biomimetic catalysis, in which the models are increasingly efficient, or indeed the development of novel therapies. The creation of new markets to provide solutions to problems yet to be defined will be one of the motors driving growth in this field, together with fundamental research.

The following chapters illustrate these points on several rapidly-evolving fronts which are worthy of attention for this reason alone. It is clear that the choice of topics is somewhat subjective, but this is counterbalanced by the opportunity it affords of presenting ideas within the context of a single body of work, an approach that will become more difficult as the discipline enjoys rapid expansion in the future. This emphasizes the importance of regular international conferences, such as those begun in Paris in 2002 and continued in Zurich in 2004, which can provide a much wider context for a dynamic view of bioorganometallic chemistry.

Acknowledgments

We wish to thank A. Vessières, S. Top and M. Salmain for constructive discussions and their help in editing the manuscript, and B. McGlinchey for her valuable linguistic expertise. G. J. thanks the von Humboldt foundation and the Institut Universitaire de France for beneficial awards.

References

1 G. JAOUEN, A. VESSIÈRES, *Pure Appl. Chem.* **1985**, *57*, 1865–1874.

2 S. TOP, G. JAOUEN, A. VESSIÈRES, J. P. ABJEAN, D. DAVOUST, C. A. RODGER, B. G. SAYER, M. J. McGLINCHEY, *Organometallics* **1985**, *4*, 2143–2150.

3 G. JAOUEN, A. VESSIÈRES, I. S. BUTLER, *Acc. Chem. Res.* **1993**, *26*, 361–369.

4 A. D. RYABOV, *Angew. Chem. Int. Ed. Engl.* **1991**, *30*, 931–941.

5 R. H. FISH, *Coord. Chem. Rev.* **1999**, *185–186*, 569–584.

6 D. MANSUY, *Pure Appl. Chem.* **1980**, *52*, 681–690.

7 K. SEVERIN, R. BERGS, W. BECK, *Angew. Chem. Int. Ed.* **1998**, *37*, 1634–1654.

8 J. C. FONTECILLA-CAMPS, S. W. RAGSDALE, *Adv. Inorg. Chem.* **1999**, *47*, 283–333.

9 G. JAOUEN, S. TOP, A. VESSIÈRES, R. ALBERTO, *J. Organomet. Chem.* **2000**, *600*, 23–36.

10 M. Y. DARENSBOURG, E. J. LYON, J. S. SMEE, *Coord. Chem. Rev.* **2000**, *206*, 533–561.

11 F. GLOAGUEN, J. D. LAWRENCE, T. B. RAUCHFUSS, *J. Am. Chem. Soc.* **2001**, *123*, 9476–9477.

12 K. H. DÖTZ, C. JÄCKEL, W. C. HAASE, *J. Organomet. Chem.* **2001**, *617–618*, 119–132.

13 W. BECK, K. SEVERIN, *Chemie in unserer Zeit* **2002**, *6*, 356–365.

14 R. H. FISH, G. JAOUEN, *Organometallics* **2003**, *22*, 2166–2177.

15 L. D. SLEP, F. NEESE, *Angew. Chem. Int. Ed.* **2003**, *42*, 2942–2945.

16 D. C. HODGKIN, *Proc. R. Soc. London, Ser. A* **1965**, 294.

17 W. KAIM, B. SCHWEDERSKI, in *Bioinorganic Chemistry: Inorganic Elements in the Chemistry of Life*, J. Wiley and Sons, **1994**.

18 P. G. LENHERT, *Proc. R. Soc. London, Ser. A* **1968**, *303*, 45.

19 T. J. KEALY, P. L. PAUSON, *Nature* **1951**, *168*, 1039–1040.

20 G. WILKINSON, M. ROSENBLUM, M. C. WHITING, R. B. WOODWARD, *J. Am. Chem. Soc.* **1952**, *74*, 2125–2126.

21 F. A. COTTON, *J. Organomet. Chem.* **2001**, *637–639*, 18–26.

22 F. A. COTTON, G. WILKINSON, in *Advanced Inorganic Chemistry*, Wiley, **1962**.

23 R. H. CRABTREE, in *The Organometallic Chemistry of the Transition Elements*, Chap. 16, 2nd edition, J. Wiley and Sons, New York, **1994**.

24 C. ELSCHENBROICH, A. SALZER, in *Organometallics: a concise introduction*, VCH, Weinheim, **1989**.

25 In *Acad. Roy. Sci.* "*Mémoires de mathématiques et de physiques*", presented by various scientists, and read in their meetings, **1760**, t.III, pp. 623–637.

26 J. FLAHAUT, in *Charles-Louis Cadet de Gassicourt (1761–1821), royal bastard, pharmacist to the Emperor*, Ed. Historiques Tesseidre, Paris, **2001**.

27 J. S. THAYER, *Adv. Organomet. Chem.* **1975**, *13*, 1–45.

28 L. G. TORAUDE, *Ann. Sci. Chim. Phys.* **1920**, *1*, 1–12.

29 R. W. BUNSEN, *Ann. Pharm.* **1842**, *42*, 15.

30 W. R. CULLEN, *Adv. Organomet. Chem.* **1966**, *4*, 145–242.

31 J. B. DUMAS, *in his presentation of R. W. Bunsen as a foreign member of the Académie des Sciences in Paris* (18 December **1882**).

32 F. HIMMELWEIT, in *The collected papers of Paul Ehrlich*, Vol. 3, Pergamon, London, **1960**.

33 J. MANN, in *The Elusive Magic Bullet: the search for the perfect drug*, Oxford University Press, Oxford, **1999**.

34 J. E. MOORE, in *The Modern Treatment of Syphilis*, 2nd edition, Thomas, pp. 64–133, Baltimore, **1941**.

35 G. W. RAIZISS, J. L. GAVRON, in *Organic Arsenical Compounds*, Chem. Catalog. Co., pp. 492–513, New York, **1923**.

36 D. W. ROBERTSON, *Trans. R. Soc. Trop. Med. Hyg.* **1963**, *57*, 122–133.

37 B. SCHMID, S. NKUNKU, A. MEROLLE, P. VOUATSOU, C. BURRI, *Lancet* **2004**, *364*, 789–790.

38 A. ASTIER, S. GIBAUD, French Patent, 0311919, **2003**.

39 A. KONIG, L. WRAZEL, R. P. WARRELL, R. RIVI, P. P. PANDOLFI, A. JAKUBOWSKI, J. L. GABRILOVE, *Blood* **1997**, *90*, 562–570.

40 Z. G. Wang, R. Rivi, L. Delva,
A. Konig, D. A. Scheinberg,
C. Gambacorti-Passerini,
J. L. Gabrilove, R. P. Warrell Jr.,
P. P. Pandolfi, *Blood* **1998**, *92*,
1497–1504.

41 C. E. Carraher, C. U. Pittman, in
*Macromolecules Containing Metal and
Metal-Like Elements,* Vol. 3, pp. 2–16,
John Wiley & Sons, **2004**.

42 B. Holmstedt, G. Liljestrand, in
Readings in Pharmacology, pp. 22–30,
Raven Press, New York, **1981**.

43 F. C. Whitmore, in *Organic
Compounds of Mercury,* Chem. Catalog.
Co., pp. 368–372, New York, **1921**.

44 J. S. Thayer, in *Organometallic
Compounds and Living Organisms,*
Academic Press, Orlando, **1984**.

45 M. Peyrone, *Ann. Chem. Pharm.* **1844**,
Band LI, 1 ff.

46 B. Rosenberg, L. Vancamp, T. Krigas,
Nature **1965**, *205*, 698–699.

47 B. Rosenberg, L. Vancamp,
J. E. Trosko, V. H. Mansour, *Nature*
1969, *222*, 385–386.

48 B. Lippert, in *Cisplatin: Chemistry and
Biochemistry of a Leading Anticancer
Drug,* John Wiley and Sons, New York,
1999.

49 B. K. Keppler, in *Metal Complexes in
Cancer Chemotherapy,* VCH, Weinheim,
1993.

50 Z. Guo, P. J. Sadler, *Adv. Inorg. Chem.*
2000, *49*, 183–305.

51 M. J. Clarke, F. Zhu, D. R. Frasca,
Chem. Rev. **1999**, *99*, 2511–2533.

52a J. Reedijk, *Chem. Commun.* **1996**,
801–806.

52b D. Wang, S. J. Lippard, *Nat. Rev. Drug
Discov.* **2005**, *4*, 307–320.

53 D. Carmona, M. P. Lamata, L. A. Oro,
Eur. J. Inorg. Chem. **2002**, 2239–2251.

54 A. H. Velders, A. Bergamo,
E. Alessio, E. Zangrando,
J. G. Hasnoot, C. Casarsa,
M. Cocchietto, S. Zorget, G. Sava,
J. Med. Chem. **2004**, *47*, 1110–1121.

55 A. Bergamo, G. Stocco, C. Casarsa,
M. Cocchietto, E. Alessio, B. Serli,
S. Zorget, G. Sava, *Int. J. Oncol.* **2004**,
24, 373–379.

56 M. M. Harding, G. Mokdsi, *Curr. Med.
Chem.* **2000**, *7*, 1289–1303.

57 P. Köpf-Maier, *Eur. J. Clin. Pharmacol.*
1994, *47*, 1–16.

58 P. Köpf-Maier, H. Köpf, *Chem. Rev.*
1987, *87*, 1137–1152.

59 J. B. Waern, M. M. Harding,
J. Organomet. Chem. **2004**, *689*,
4655–4668.

60 S. Top, E. B. Kaloun, A. Vessières,
I. Laïos, G. Leclercq, G. Jaouen,
J. Organomet. Chem. **2002**, *643–644*,
350–356.

61 M. Tacke, L. T. Allen, L. Cuffe,
W. M. Gallagher, Y. Lou,
O. Mendoza, H. Müller-Bunz,
F. J. K. Rehmann, N. Sweeney,
J. Organomet. Chem. **2004**, *689*,
2242–2249.

62 M. Tacke, L. Cuffe, W. M. Gallagher,
Y. Lou, O. Mendoza, H. Müller-Bunz,
F. J. K. Rehmann, N. Sweeney, *J. Inorg.
Biochem.* **2004**, *98*, 1987–1994.

63 C. S. Allardyce, P. J. Dyson,
E. D. J. S. L. Heath, *Chem. Commun.*
2001, 1396–1397.

64 F. Wand, H. M. Chen,
J. A. Parkinson, P. S. Murdoch,
P. J. Sadler, *Inorg. Chem.* **2002**, *41*,
4509–4523.

65 C. S. Allardyce, P. J. Dyson,
D. J. Ellio, P. A. Salter, R. Scopelliti,
J. Organomet. Chem. **2003**, *668*, 35–42.

66 G. Jaouen, S. Top, J. P. Raynaud,
French Patent 2725720–AI, **1994**.

67 S. Top, J. Tang, A. Vessières,
D. Carrez, C. Provot, G. Jaouen,
Chem. Commun. **1996**, 955–956.

68 S. Top, A. Vessières, G. Leclercq,
J. Quivy, J. Tang, J. Vaissermann,
M. Huché, G. Jaouen, *Chem. Eur. J.*
2003, *9*, 5223–5236.

69 G. Jaouen, S. Top, A. Vessières,
G. Leclercq, M. J. McGlinchey,
Curr. Med. Chem. **2004**, *11*, 2505–2517.

70 J. Brocard, J. Lebibi, L. Maciejewski,
French Patent 2733985–A1, **1996**.

71 C. Biot, G. Glorian,
L. A. Maciejewski, J. S. Brocard,
J. Med. Chem. **1997**, *40*, 3715–3718.

72 C. Biot, L. Delhaes, H. Abessolo,
O. Domarle, L. Maciejewski,
M. Mortuaire, P. Delcourt,
P. Deloron, D. Camus, D. Dive,
J. Brocard, *J. Organomet. Chem.* **1999**,
589, 59–65.

73 C. Biot, *Curr. Med. Chem. Anti-Infective Agents* **2004**, *3*, 135–147.

74 P. J. Craig, in *Organometallic Compounds in the Environment*, 2nd ed., Wiley, Chichester, **2003**.

75 T. Tsubaki, K. Irukayama, in *Minamata Disease: Methylmercury Poisoning in Minamata and Niigata Japan*, Kodansha, Tokyo, **1977**.

76 P. A. D'Itric, F. M. D'Itric, in *Mercury Contamination; a Human Tragedy*, Wiley, New York, **1997**.

77 K. E. Jackson, M. A. Jackson, *Chem. Rev.* **1935**, *16*, 439–452.

78 F. W. Oehme, *Clin. Toxicol.* **1972**, *5*, 215–222.

79 R. A. Peters, L. A. Stockton, R. H. S. Thompson, *Nature* **1945**, *156*, 616.

80 C. J. Löwig, *J. P. Chem.* **1853**, *LX*, 304.

81 S. K. Hall, *Environ. Sci. Technol.* **1972**, *6*, 30–35.

82 T. Alojouanine, L. Derbert, S. Thieffrey, *Rev. Neurol.* **1958**, *98*, 85.

83 J. E. Gruner, *Rev. Neurol.* **1958**, *98*, 104.

84 R. J. Fessenden, J. S. Fessenden, *Adv. Organomet. Chem.* **1980**, *18*, 275.

85 A. J. Growe, *Chem. Int. (London)* **1983**, 304–310.

86 M. Tsutsui, N. Kakimoto, D. D. Axtell, H. Oikawa, K. Asai, *J. Am. Chem. Soc.* **1976**, *98*, 8287–8289.

87 J. S. Thayer, *J. Organomet. Chem.* **1974**, *76*, 265–295.

88 F. Challenger, *Chem. Rev.* **1945**, *36*, 315–361.

89 F. Challenger, *Q. Rev. Chem. Soc.* **1955**, *9*, 255.

90 R. S. Yalow, S. A. Berson, *J. Clin. Invest.* **1960**, *39*, 1157–1175.

91 R. S. Yalow, S. A. Berson, *Nature* **1959**, *84*, 1648–1649.

92 R. Edwards, in *Immunoassay*, William Heinemann Medical Books, London, **1985**.

93 T. J. Gill, L. T. Mann, *J. Immunol.* **1965**, *96*, 906–912.

94 J. P. Gosling, *Clin. Chem.* **1990**, *36*, 1408–1427.

95 L. J. Kricka, in *Ligand–Binder Assay*, Marcel Dekker, New York, **1985**.

96 M. Cais, S. Dani, Y. Eden, O. Gandolfi, M. Horn, E. E. Isaacs, Y. Josephy, Y. Saar, E. Slovin, L. Snarsky, *Nature* **1977**, *270*, 534–535.

97 J. F. Hainfield, *Science* **1987**, *236*, 450–453.

98 J. H. W. Leuvering, B. C. Goverde, P. J. Thal, A. H. Schuurs, *J. Immunol. Methods* **1983**, *60*, 9–23.

99 P. Cheret, P. Brossier, *Res. Commun. Chem. Pathol. Pharmacol.* **1986**, *54*, 237–253.

100 K. Di Gleria, H. A. O. Hill, C. J. McNeil, *Anal. Chem.* **1986**, *58*, 1203–1205.

101 M. Salmain, A. Vessières, P. Brossier, I. S. Butler, G. Jaouen, *J. Immunol. Methods* **1992**, *148*, 65–75.

102 M. Cais, N. Tirosh, *Bull. Soc. Chim. Belg.* **1980**, *90*, 27–35.

103 P. Cheret, Ph D thesis, Université de Bourgogne, Dijon, **1987**.

104 B. Limoges, C. Degrand, P. Brossier, R. L. Blankespoor, *Anal. Chem.* **1993**, *65*, 1054–1060.

105 D. H. Farkas, *Clin. Chem.* **2001**, *47*, 1871–1872.

106 S. D. Vernon, D. H. Farkas, E. R. Unger, V. Chan, D. L. Miller, Y. P. Chen, G. F. Blackburn, W. C. Reeves, *BMC Infectious Diseases* **2003**, *3/12*, open access: http://www.biomedcentral.com/ 1471–2334/3/12.

107 C. J. Yu, Y. Wan, H. Yowanto, J. Li, C. Tao, M. D. James, C. L. Tan, G. F. Blackburn, T. J. Meade, *J. Am. Chem. Soc.* **2001**, *123*, 11155–11161.

108 M. Salmain, A. Vessières, G. Jaouen, I. S. Butler, *Anal. Chem.* **1991**, *63*, 2323–2329.

109 I. S. Butler, A. Vessières, M. Salmain, P. Brossier, G. Jaouen, *Appl. Spectrosc.* **1989**, *43*, 1497–1498.

110 A. Vessières, S. Top, A. A. Ismail, I. S. Butler, M. Loüer, G. Jaouen, *Biochemistry* **1988**, *27*, 6659–6666.

111 G. Jaouen, A. Vessières, S. Top, A. A. Ismail, I. S. Butler, *J. Amer. Chem. Soc.* **1985**, *107*, 4778–4780.

112 A. Varenne, A. Vessières, P. Brossier, G. Jaouen, *Res. Commun. Chem. Pathol. Pharmacol.* **1994**, *84*, 81–92.

113 A. Varenne, A. Vessières, M. Salmain, P. Brossier, G. Jaouen, *J. Immunol. Methods* **1995**, *186*, 195–204.

114 A. Vessières, K. Kowalski, J. Zakrzewski, A. Stepien,

M. Grabowski, G. Jaouen, *Bioconjugate Chem.* **1999**, *10*, 379–385.

115 V. Philomin, A. Vessières, G. Jaouen, *J. Immunol. Methods* **1994**, *171*, 201–210.

116 A. Vessières, M. Salmain, P. Brossier, G. Jaouen, *J. Pharm. Biomed. Anal.* **1999**, *21*, 625–633.

117 Z. Wang, B. A. Roe, K. M. Nicholas, R. L. While, *J. Am. Chem. Soc.* **1993**, *115*, 4399–4400.

118 Y. Ueno, H. Yurimoto, H. Yoshioka, T. Komiya, S. Maruyama, *Geochimica Cosmochimica Acta* **2002**, *66*, 1257–1268.

119 R. J. P. Williams, J. J. R. Frausto Da Silva, *J. Theor. Biol.* **2003**, *220*, 323–343.

120 D. C. Rees, J. B. Howard, *Science* **2003**, *300*, 929–931.

121 W. Martin, M. J. Russell, *Phil. Trans. R. Soc. London, Ser. B* **2003**, *358*, 59–85.

122 J. A. Kovacs, S. C. Shoner, J. J. Ellison, *Science* **1995**, *270*, 587–588.

123 D. Dolphin, in *B12*, Vol. 1 and 2, Wiley Interscience, New York, **1982**.

124 S. J. Lippard, J. M. Berg, in *Principles of Bioinorganic Chemistry*, Mill Valley, USA, **1994**.

125 G. N. Schrauzer, *Angew. Chem. Int. Ed. Engl.* **1975**, *15*, 417.

126 S. Gschösser, R. B. Hannak, R. Konrat, K. Gruber, C. Mikl, C. Kratky, B. Kräutler, *Chem. Eur. J.* **2005**, *11*, 81–93.

127 B. Kräutler, D. Arigoni, B. T. Golding, in *Vitamin B12 and B12 proteins*, Wiley-VCH Verlag GmbH, **1998**.

128 L. Randaccio, *Comments Inorg. Chem.* **2000**, *21*, 327–376.

129 O. Einsle, F. A. Tezcan, S. L. Andrade, B. Schmid, M. Yoshida, J. B. Howard, D. C. Rees, *Science* **2002**, *297*, 1696–1700.

130 J. Kim, D. C. Rees, *Science* **1992**, *257*, 1677–1682.

131 C. Darnault, A. Volbeda, E. Kim, P. Legrand, X. Vernede, P. A. Lindahl, J. C. Fontecilla-Camps, *Nat. Struct. Biol.* **2003**, *10*, 446–451.

132 T. I. Doukov, T. Iverson, J. Sevaralli, S. W. Ragsdale, C. L. Drennan, *Science* **2002**, *298*, 567–572.

133 C. L. Drennan, J. W. Peters, *Curr. Opin. Struct. Biol.* **2003**, *13*, 220–226.

134 E. L. Hegg, *Acc. Chem. Res.* **2004**, *37*, 775–783.

135 A. D. Allen, F. Bottomley, *Acc. Chem. Res.* **1968**, *1*, 360–365.

136 J. Chatt, *Chem. Soc. Rev.* **1972**, *1*, 121.

137 J. Chatt, *J. Organomet. Chem.* **1975**, *100*, 17–28.

138 J. Chatt, J. R. Dilworth, R. L. Richards, *Chem. Rev.* **1978**, *78*, 589–625.

139 W. E. Newton, W. H. Orme-Johnson, in *Nitrogen Fixation*, University Park Press, Baltimore, **1980**.

140 R. R. Schrock, *J. Am. Chem. Soc.* **1990**, *112*, 4331–4338.

141 M. Stephenson, L. H. Stickland, *Biochem. J. (London)* **1931**, *25*, 205.

142 M. W. W. Adams, E. I. Stiefel, *Curr. Opin. Chem. Biol.* **2000**, 214–220.

143 A. Volbeda, M. H. Charon, C. Piras, E. C. Hatchikian, M. Frey, J. C. Fontecilla-Camps, *Nature* **1995**, *373*, 580–587.

144 J. W. Peters, W. N. Lanzilotta, B. J. Lemon, L. C. Seefeldt, *Science* **1998**, *282*, 1853–1858.

145 H. Frey, *ChemBioChem* **2002**, *3*, 153–160.

146 Y. Nicolet, C. Piras, P. Legrand, C. E. Hatchikian, J. C. Fontecilla-Camps, *Structure* **1999**, *7*, 13–23.

147 A. Volbeda, J. C. Fontecilla-Camps, *Dalton Trans.* **2003**, 4030–4038.

148 M. Y. Darensbourg, E. J. Lyon, X. Zhao, P. Georgakaki, *Proc. Natl. Acad. Sci. USA* **2003**, *100*, 3683–3688.

149 S. Ott, M. Kritikos, B. Akermark, L. Sum, R. Lomoth, *Angew. Chem. Int. Ed.* **2004**, *43*, 1006–1009.

150 T. B. Rauchfuss, *Inorg. Chem.* **2004**, *43*, 14–26.

151 D. Sellmann, F. Geipel, F. Lauderbach, F. W. Heinemann, *Angew. Chem. Int. Ed.* **2002**, *41*, 632–634.

152 D. Sellmann, F. Lauderbach, F. Geipel, F. W. Heinemann, *Angew. Chem.* **2004**, *116*, 3203–3206.

153 D. Sellmann, R. Prakash, F. W. Heinemann, *Eur. J. Inorg. Chem.* **2004**, 1847–1858.

154 T. D. Weatherill, T. B. Rauchfuss, R. A. Scott, *Inorg. Chem.* **1986**, *25*, 1466–1472.

155 L. F. Dahl, C. H. Wei, *Inorg. Chem.* **1963**, *2*, 334–336.

156 M. Dörr, J. Kässbohrer, R. Grunert, G. Kreisel, W. A. Brand,

R. A. WERNER, H. GEILMANN, C. APFEL, C. ROBL, W. WEIGAND, *Angew. Chem. Int. Ed.* **2003**, *42*, 1540–1543.

157 W. HIEBER, W. BECK, *Z. Anorg. Allg. Chem.* **1960**, *305*, 265.

158 W. HIEBER, J. BRENDEL, J. GRUBER, *Z. Anorg. Allg. Chem.* **1958**, *296*, 68.

159 W. HIEBER, P. SPACU, *Z. Anorg. Allg. Chem.* **1937**, *233*, 353.

160 W. HIEBER, A. ZEIDLER, *Z. Anorg. Allg. Chem.* **1964**, *329*, 92.

161 R. B. KING, *J. Am. Chem. Soc.* **1962**, *84*, 2460.

162 H. REIHLEN, A. GRUHL, G. HESSLING, *Liebigs Ann. Chem.* **1929**, *472*, 268.

163 D. SEYFERTH, R. S. HENDERSON, *J. Am. Chem. Soc.* **1979**, *101*, 508–509.

164 D. SEYFERTH, R. S. HENDERSON, L. C. SONG, *Organometallics* **1982**, *1*, 125–133.

165 C. TARD, X. LIU, S. K. IBRAHIM, M. BRUSHI, L. DE GIOIA, S. C. DAVIES, X. YANG, L. S. WANG, G. SAWERS, C. J. PICKETT, *Nature* **2005**, *433*, 610–613.

166 P. A. LINDAHL, *J. Biol. Inorg. Chem.* **2004**, *9*, 516–524.

167 S. W. RAGSDALE, *Biofactors* **1997**, *6*, 3–9.

168 S. W. RAGSDALE, M. KUMAR, *Chem. Rev.* **1996**, *96*, 2515–2539.

169 R. K. THAUER, *Science* **2001**, *293*, 1264–1265.

170 J. FENG, P. A. LINDAHL, *J. Am. Chem. Soc.* **2004**, *126*, 9094–9100.

171 J. FENG, P. A. LINDAHL, *Biochemistry* **2004**, *43*, 1552–1559.

172 Y. SVETLITCHNYI, H. DOBBEK, W. MEYER-KLAUCKE, T. MEINS, B. THIELE, P. RONER, R. HUBER, O. MEYER, *Proc. Natl. Acad. Sci. USA* **2004**, *101*, 446–451.

173 S. S. MANSY, J. A. CONAN, *Acc. Chem. Res.* **2004**, *37*, 719–725.

174 C. HUBER, G. WÄCHTERSHÄUSER, *Science* **1997**, *276*, 245–247.

175 G. M. KING, *Appl. Environ. Microbiol.* **2003**, *69*, 7266–7272.

176 W. HIEBER, F. LEUTERT, *Z. Anorg. Allg. Chem.* **1932**, *204*, 145.

177 T. KRUCK, M. HÖFLER, M. NOACK, *Chem. Ber.* **1966**, *99*, 1153.

178 P. SCHÖNHEIT, J. MOLL, R. K. THAUER, *Arch. Microbiol.* **1979**, *123*, 105–107.

179a R. K. THAUER, *Microbiology* **1998**, *144*, 2377–2406.

179b M. KRÜGER, A. MEYERDLERKS, R. K. THAUER et al., *Nature* **2003**, *426*, 878–881.

180 W. BECK, K. SÜNKEL, *Chem. Rev.* **1988**, 88, 1405–1421.

181 S. REISSMANN, E. HOCHLEITNER, H. WANG, A. PASCHOS, F. LOTTSPEICH, R. S. GLASS, A. BÖCK, *Science* **2003**, *299*, 1067–1070.

182 D. MANSUY, J. P. MAHY, in F. MONTANARI, L. CASELLA (Eds.), *Metalloporphyrins Catalyzed Oxidations*, p. 175–206, Springer, **1994**.

183 P. R. ORTIZ DE MONTELLANO, in *Cytochrome P450*, Plenum Press, New York, **1995**.

184 P. COCOLIOS, G. LAGRANGE, R. GUILARD, *J. Organomet. Chem.* **1983**, *253*, 65–79.

185 R. GUILARD, K. M. KADISH, *Chem. Rev.* **1988**, *88*, 1121–1146.

186 R. GUILARD, A. TABARD, E. VAN CAEMELBECKE, K. M. KADISH, in K. M. KADISH, K. M. SMITH, R. GUILARD (Eds.), *The Porphyrin Handbook*, Vol. 3, p. 295–345, Academic Press, **2000**.

187 D. MANSUY, *Pure Appl. Chem.* **1987**, *59*, 759–770.

188 D. MANSUY, *Pure Appl. Chem.* **1994**, *66*, 737–744.

189 D. MANSUY, M. LANGE, J. C. CHOTTARD, P. GUÉRIN, P. MORLIÈRE, D. BRAULT, M. ROUGEE, *Chem. Commun.* **1977**, 648–649.

190 D. MANSUY, W. NASTAINCZYK, V. ULLRICH, *Arch. Pharmacol.* **1974**, *285*, 315–324.

191 V. ULLRICH, *Top. Curr. Chem.* **1979**, *83*, 67–104.

192 U. KOELLE, *Coord. Chem. Rev.* **1994**, *135–136*, 623–650.

193 L. I. DELMON-MOINGEON, D. PIWNICA-WORMS, A. D. VAN DEN ABBEELE, B. L. HOLMAN, A. DAVISON, A. G. JONES, *Cancer Res.* **1990**, *50*, 2198–2202.

194 B. L. HOLMAN, V. SPORN, A. G. JONES, S. T. BENJAMIN-SIA, N. PEREZ-BALINO, A. DAVISON, J. LISTER-JAMES, J. F. KRONAUGE, A. E. A. MITTA, L. L. CAMIN, S. CAMBELL, S. J. WILLIAMS, A. T. CARPENTER, *J. Nucl. Med.* **1987**, *28*, 13–18.

195 A. G. Jones, M. J. Abrams, A. Davison, US Patent, N° 4,452,774, Issued 6/5/84, **1984**.

196 A. G. Jones, M. J. Abrams, A. Davison, US Patent, N° 4,707,544, Issued 11/17/87, **1987**.

197 A. G. Jones, M. J. Abrams, A. Davison, US Patent, N° 4,826,961, Issued 5/2/89, **1989**.

198 R. Alberto, *Chimia* **2003**, *57*, 56.

199 R. Alberto, K. Ortner, N. Wheatley, R. Schibli, A. P. Schubiger, *J. Am. Chem. Soc.* **2001**, *123*, 3135–3136.

200 R. Alberto, R. Schibli, R. Waibel, U. Abram, A. P. Schubiger, *Coord. Chem. Rev.* **1999**, *192*, 901–919.

201 J. K. Pak, P. Benny, B. Spingler, K. Ortner, R. Alberto, *Chem. Eur. J.* **2003**, *9*, 2053–2061.

202 B. Salignac, P. V. Grundler, S. Cayemittes, U. Frey, R. Scopelliti, A. E. Merbach, R. Hechinger, K. Hegetschweiler, R. Alberto, U. Prinz, G. Raabe, U. Koelle, S. Hall, *Inorg. Chem.* **2002**, *42*, 3516–3526.

203 W. Cremer, *Biochem. Z.* **1929**, *206*, 208.

204 M. P. Schubert, *J. Am. Chem. Soc.* **1933**, *55*, 4563–4570.

205 K. Schlögl, *Monatsh. Chem.* **1957**, *88*, 601.

206 J. M. Osgerby, P. L. Pauson, *J. Chem. Soc.* **1958**, 656–660.

207 D. R. van Staveren, N. Metzler-Nolte, *Chem. Rev.* **2004**, *104*, 5931–5985.

208 E. Cuingnet, C. Sergheraert, A. Tartar, M. Dautrevaux, *J. Organomet. Chem.* **1960**, *195*, 325–329.

209 K. Severin, *Chem. Eur. J.* **2002**, *8*, 1515–1518.

210 W. Hieber, H. Führling, *Z. Anorg. Allg. Chem.* **1971**, *381*, 235.

211 E. Benetti, G. Maglio, R. Palumbo, C. Pedone, *J. Organomet. Chem.* **1973**, *60*, 189–195.

212 M. L. H. Green, L. C. Mitchard, W. E. Silverthorn, *J. Chem. Soc. Dalton Trans.* **1973**, 1403–1408.

213 D. F. Dersnah, M. C. Baird, *J. Organomet. Chem.* **1977**, *127*, C55–C58.

214 W. Beck, W. Petri, *J. Organomet. Chem.* **1977**, *127*, C40–C44.

215 E. S. Gore, M. L. H. Green, *J. Chem. Soc. A* **1970**, 2315–2319.

216 C. Potvin, L. Davignon, G. Pannetier, *Bull. Soc. Chim. Fr.* **1975**, 507–511.

217 W. S. Sheldrick, S. Heeb, *J. Organomet. Chem.* **1989**, *377*, 357–366.

218 H. Brunner, *Angew. Chem. Int. Ed.* **1999**, *38*, 1194–1208.

219 A. A. Ioganson, *Russ. Chem. Rev.* **1985**, *54*, 277.

220 D. W. Abbott, C. Watts, *Inorg. Chem.* **1983**, *22*, 597–602.

221 W. Beck, K. Kottmair, *Chem. Ber.* **1976**, *109*, 970–993.

222 M. M. Singh, Y. Rosopulos, W. Beck, *Chem. Ber.* **1983**, *116*, 1364.

223 W. S. Sheldrick, B. Guenther, *Inorg. Chim. Acta* **1988**, *151*, 237–241.

224 P. Annen, S. Schildberg, W. S. Sheldrick, *Inorg. Chim. Acta* **2000**, *307*, 115–124.

225 D. Herebian, W. S. Sheldrick, *J. Chem. Soc. Dalton Trans.* **2002**, 966–974.

226 S. Korn, W. S. Sheldrick, *Inorg. Chim. Acta* **1997**, *254*, 85–91.

227 P. Köpf-Maier, H. Köpf, in *Metal Complexes in Cancer Therapy*, Chapman Hall, London, **1994**.

228 P. Köpf-Maier, I. C. Tornieporth-Oetting, *Biometals* **1996**, *9*, 267–271.

229 F. Wang, H. Chen, S. Parsons, I. D. H. Oswald, J. E. Davidson, P. J. Sadler, *Chem. Eur. J.* **2003**, *9*, 5810–5820.

230 W. S. Sheldrick, S. Heeb, *Inorg. Chim. Acta* **1990**, *168*, 93–100.

231 H. H. Baer, H. R. Hanna, *Carbohydr. Res.* **1982**, *102*, 169–183.

232 J. M. Beau, T. Gallagher, *Top. Curr. Chem.* **1997**, *187*, 1–54.

233 A. Rosenthal, H. J. Koch, *Tetrahedron Lett.* **1967**, *8*, 871–874.

234 G. L. Trainor, *J. Organomet. Chem.* **1985**, *282*, C43–C45.

235 I. Arai, G. D. Daves, *J. Am. Chem. Soc.* **1981**, *103*, 7683.

236 L. Jessen, E. T. K. Haupt, J. Heck, *Chem. Eur. J.* **2001**, *7*, 3791–3797.

237 Y. Nagel, W. Beck, *Z. Naturforsch.* **1985**, *40b*, 1181.

238 R. Aumann, *Chem. Ber.* **1992**, *125*, 2773–2778.

239 R. Aumann, *Chem. Ber.* **1994**, *127*, 725–729.

240 H. Fischer, J. Schleu, G. Roth, *Chem. Ber.* **1995**, *128*, 373–378.

241 S. Krawielitzki, W. Beck, *Chem. Ber. Rec.* **1997**, *130*, 1659.

242 T. Pill, K. Polborn, W. Beck, *Chem. Ber.* **1990**, *123*, 11–17.

243 K. H. Dötz, E. Gomes de Silva, *Synthesis* **2003**, 1787–1789.

244 R. O. Duthaler, A. Hafner, P. L. Alsters, P. Rothe-Streit, G. Rihs, *Pure Appl. Chem.* **1992**, *64*, 1897–1910.

245 M. Riediker, R. O. Duthaler, *Angew. Chem. Int. Ed. Engl.* **1989**, *28*, 494–495.

246 M. Diéguez, O. Pamies, C. Clever, *Chem. Rev.* **2004**, *104*, 3189–3216.

247 H. Junicke, D. Steinborn, *Inorg. Chim. Acta* **2003**, *346*, 129–136.

248 D. Steinborn, H. Junicke, *Chem. Rev.* **2000**, *100*, 4283–4318.

249 H. C. Lo, C. Leiva, O. Buriez, J. B. Kerr, M. M. Olmstead, R. H. Fish, *Inorg. Chem.* **2001**, *40*, 6705–6716.

2
Ruthenium Arene Anticancer Complexes

Michael Melchart and Peter J. Sadler

2.1
Introduction

Most new drugs are organic (carbon) compounds but there is an increasing realization that many metal ions are involved in natural biological processes and that there is much scope for the design of therapeutic agents based on both biologically-essential and non-essential metal ions. Metal complexes with their wide spectrum of coordination numbers, coordination geometries, thermodynamic and kinetic preferences (which cover enormous scales of magnitude) for ligand atoms, and in some cases redox activity, offer novel mechanisms of action which are not available to organic compounds. The challenge is to control these properties when metal complexes are introduced into the body, a challenge which is being aided by new insights into the molecular basis for the activity of essential metal ions, for example the mechanism of potassium transport through protein channels in membranes [1], of electron flow through the iron and copper proteins in electron transport chains [2], and the control of transcription by zinc finger proteins [3]. In addition modern genomics should allow the DNA codes which control the selection, uptake and distribution of metals to be deciphered.

The recent clinical success of platinum anticancer drugs has illustrated that metal complexes can be designed to have a specificity of biological activity. For example, not all platinum complexes are active anticancer agents. Some platinum complexes are inert and relatively non-toxic, some attack DNA, some do not. In general, the nature of the metal ion itself, its oxidation state, and the types and number of bound ligands can all play critical roles in the biological activity [4, 5].

Although organomercury complexes were widely used as diuretics [6] until about 40 years ago, only recently have attempts been made to design organometallic complexes (metallocene derivatives) as therapeutic agents [7–9]. Although metal arene complexes offer much potential in drug design, none has yet entered clinical trials. Our work on ruthenium(II) arene complexes is described here and illustrates the rich structural and electronic diversity which can be incorporated into this class of organometallic complexes. The nature of the arene and the other three ligands

Bioorganometallics: Biomolecules, Labeling, Medicine. Edited by Gérard Jaouen
Copyright © 2006 Wiley-VCH Verlag GmbH & Co. KGaA, Weinheim
ISBN: 3-527-30990-X

can all play critical roles in controlling chemical reactions which can have a direct effect on biological activity, redox reactions and the rates and extent of ligand substitution, especially the aquation of chloro complexes, the pK_a values of aqua adducts, the recognition of (poly)nucleotides and amino acids (proteins). Some of these Ru(II) arene complexes have promising *in vitro* and *in vivo* anticancer activity.

2.2
Metal-based Anticancer Complexes

Current interest in the use of metal complexes for the treatment of cancer was triggered by the discovery of the anticancer activity of Pt(II) amine complexes in 1969 by Rosenberg [10]. Now at least four platinum complexes have been approved for clinical use (complexes **1–4**, Fig. 2.1) and platinum complexes are the most widely-used anticancer drugs. Stimulation for the discovery of anticancer complexes of metals other than platinum arises from the cellular resistance to platinum which is sometimes encountered in the clinic, the toxic side-effects of cisplatin (**1**), which can be severe, and the limited spectrum of activity against different types of cancer [11].

The organometallic complex titanocene dichloride, Cp_2TiCl_2 (**5**, where Cp is cyclopentadienyl, Fig. 2.2), was originally investigated because it was believed that the *cis*-$TiCl_2$ motif would react with DNA in a similar manner to cisplatin and lead to the formation of bifunctional cross-links, which might in turn induce apoptosis and cancer cell death. Cp_2TiCl_2 binds only weakly to DNA bases, but more strongly to the phosphate backbone [12]. Cp_2TiCl_2 is difficult to formulate for administration because of its ease of hydrolysis and formation of hydroxy- and oxy-bridged species. Responses to titanocene dichloride in the clinic were not encouraging and the trials have now been abandoned [13, 14].

Other metallocenes are active *in vitro*, e.g. Cp_2VCl_2 and Cp_2NbCl_2 [7], but have not reached clinical trials. Attempts to modify the aqueous solubility and stability of titanocenes are underway in several laboratories [15, 16]. Another Ti(IV) complex, budotitane (**6**, Fig. 2.2), was the first non-platinum complex to be approved for clinical trials, but poor solubility and hydrolysis made formulation difficult even in micelles, and the trials were abandoned [17].

Biological interest in ruthenium has centered on the use of ruthenium red to inhibit Ca^{2+} uptake, on ruthenium complexes as nitric oxide scavengers, as

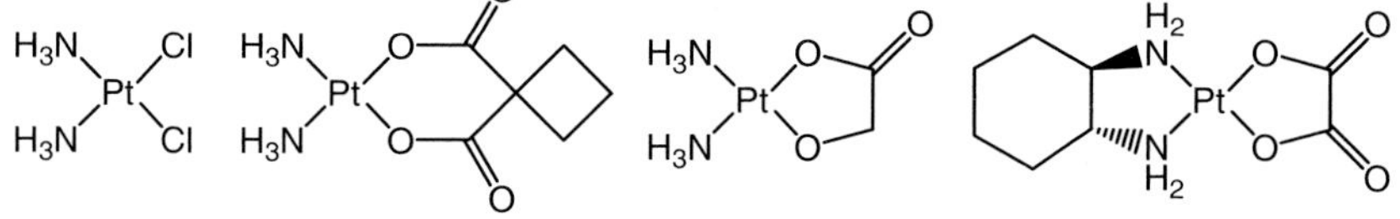

(1) (2) (3) (4)

Fig. 2.1 Platinum(II) complexes approved for clinical use as anticancer drugs: (**1**) cisplatin, (**2**) carboplatin, (**3**) nedaplatin, (**4**) oxaliplatin.

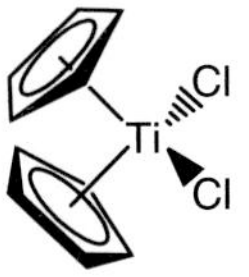

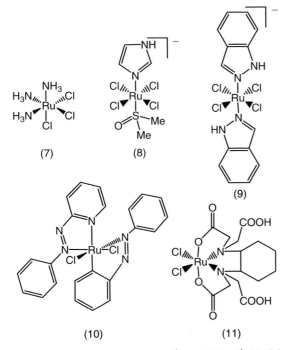

Fig. 2.2 Titanium anticancer drugs which have been on clinical trials:
(5) titanocene dichloride, **(6)** budotitane.

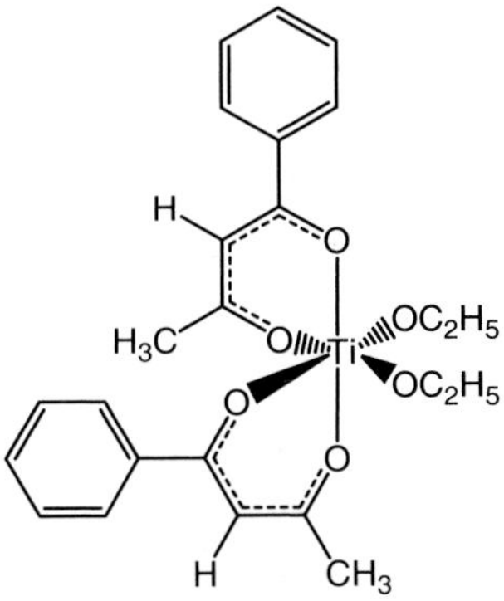

Fig. 2.3 Ruthenium anticancer complexes: **(7)** [RuCl$_3$(NH$_3$)$_3$],
(8) Ru(III) antimetastatic drug NAMI-A, **(9)** Ru(III) KP1019,
(10) Ru(II) azopyridine complex, **(11)** Ru(IV) cdta complex.

immunosuppressants, as radiopharmaceuticals (radioactive ruthenium γ-ray
emitters ^{97}Ru, $t_{1/2}$ 2.9 d, 216 keV, and ^{103}Ru, $t_{1/2}$ 39 d, 497 keV), as well as anticancer agents [18]. Early interest in the anticancer activity of ruthenium complexes
stemmed from the observations of Clarke [19] that Ru(III) ammines are active
anticancer agents (e.g. **7**, Fig. 2.3). However, these were too insoluble for use.

Two other Ru(III) complexes, *trans*-[RuCl$_4$(DMSO)(Im)]ImH (NAMI-A, **8**, where Im is imidazole) [20] and *trans*-[RuCl$_4$(Ind)$_2$]IndH (KP1019, **9**, where Ind is indazole) [21], have now entered trials. KP1019 is cytotoxic to cancer cells, whereas NAMI-A is relatively non-toxic but has antimetastatic activity and prevents the spread of cancer. Several other Ru complexes have shown promise recently [22, 23] as anticancer complexes, e.g. α-[Ru(azpy)$_2$Cl$_2$] (**10**, where azpy is 2-phenylazo-pyridine) and [Ru(H$_2$cdta)Cl$_2$].2H$_2$O (**11**, where H$_2$cdta is 1,2-cyclohexanediamino-tetraacetate, Fig. 2.3).

Clarke has proposed that the activity of Ru(III) complexes, which are usually relatively inert towards ligand substitution, is dependent on *in vivo* reduction to more labile Ru(II) complexes [19, 24]. With this in mind we have explored the activity of Ru(II) complexes. We discovered that Ru(II) aminophosphine complexes were cytotoxic to cancer cells [25], but they had poor aqueous solubility and were difficult to isolate and purify in large quantities. Since arenes are known to stabilize ruthenium in its +2 oxidation state, we have investigated the potential of Ru(II) arene complexes as anticancer agents and their associated aqueous chemistry. We have found that "half-sandwich" Ru(II) mono-arene complexes often possess good aqueous solubility (an advantage for clinical use) and that the Ru(II)-arene bonds are very stable.

2.3
Chemistry of Ru Arenes

2.3.1
Synthesis

The synthesis of Ru(II) arene complexes cannot usually be achieved by simple reaction of Ru(II) with an arene, but requires a redox reaction. The first arene complexes of Ru(II) were synthesized in 1957 by Fischer and Böttcher [26]. They made the sandwich complex [(η^6-mesitylene)$_2$Ru](BPh$_4$)$_2$ (**12**, Fig. 2.4) and the analogous benzene complex by stirring RuCl$_3$ with anhydrous AlCl$_3$, powdered Al, and the respective arene under N$_2$ at 130 °C for 8–10 h. In 1967 Fischer et al. [27] reported the synthesis of [(η^6-naphthalene)$_2$Ru](PF$_6$)$_2$, and Winkhaus and Singer [28] synthesized the polymeric mono-arene complex [(η^6-benzene)RuCl$_2$]$_x$ by reaction of RuCl$_3$ with 1,3-cyclohexadiene in ethanol. Subsequently Bennett et al. [29] reported 'piano-stool' geometries for [(η^6-arene)RuCl$_2$(PMePh$_2$)] with arene = benzene or *p*-cymene (e.g. **13**, Fig. 2.4), as determined by X-ray crystallography, and Zelonka and Baird [30] found that the dimer [(η^6-benzene)RuCl$_2$]$_2$ (**14**, Fig. 2.4) undergoes useful reactions with electrophilic and nucleophilic reagents such as phosphines with formation of monomeric complexes e.g. [(η^6-benzene)-RuCl$_2$(PR$_3$)]. In 'piano-stool' complexes, the arene forms the seat of the 'piano stool' and the other three ligands the legs. Distortion of the bound arene ring was noted for [(η^6-benzene)RuCl$_2$(PMePh$_2$)] (**13**, Fig. 2.4) (four Ru–C bonds of ca. 2.19 Å, two Ru–C bonds of 2.27 Å with the longer bonds *trans* to P). The slight

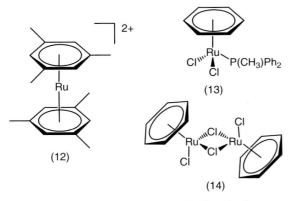

Fig. 2.4 Early examples of sandwich and half-sandwich Ru(II) arene complexes:
(**12**) $[(\eta^6\text{-mesitylene})_2\text{Ru}]^{2+}$, (**13**) $[(\eta^6\text{-benzene})\text{RuCl}_2(\text{PMePh}_2)]$,
(**14**) $[(\eta^6\text{-benzene})\text{RuCl}_2]_2$, a useful synthetic precursor.

bend (dihedral angle 5°) suggested some degree of localization of the ring π-electrons. Later work on other compounds has also suggested that localization of π-electron density can occur, and is accompanied by alternating long and short C–C bonds [31, 32]. The general structural, stereochemical and electronic features of metal–arene complexes have been discussed [33].

Reaction of Ru(III)Cl$_3$ with a precursor 1,4-cyclic diene gives rise to a Ru(II) arene dimer such as **14** (Fig. 2.4). This is often a convenient synthetic route to Ru(II) arenes, and the structures of several such dimers have been determined [34–36]. Recent interest in the design of ruthenium arene complexes as catalyst precursors has led to the exploration of a wide range of synthetic routes to complexes containing various arenes, substituted arenes, together with other ligands such as hydride, phosphines, alkyl and aryl groups [37].

Although synthetic routes using Ru(II) arene dimers can give high yields of relatively pure half-sandwich products, they are limited by the availability of suitable precursor dienes. Alternative routes involve, for example, substitution of weakly bound naphthalene in the Ru(0) complex $[(\eta^6\text{-naphthalene})\text{Ru}(\eta^4\text{-COD})]$ (where COD is cyclo-octa-1,5-diene) by an added arene [38] followed by conversion to the Ru(II) arene dimer on addition of HCl [39]. This pathway can provide access to Ru(II) arene complexes with functional side-chains on the arene [40]. Other routes involve thermal displacement of coordinated *p*-cymene by sterically-demanding arenes such as hexamethylbenzene [41], photoinduced substitution of e.g. benzene in $[(\eta^6\text{-benzene})\text{Ru}(\text{amidinate})(\text{Cl})]$ [42], derivatization of coordinated arenes in $[(\eta^6\text{-arene})\text{Ru}(\text{Cp})]^+$ complexes [43], and stoichiometric cyclotrimerization of alkynes using $[(\eta^6\text{-naphthalene})\text{Ru}(\eta^4\text{-COD})]$ [44, 45]. It is also possible to synthesize mono- and bis-arene Ru(II) complexes starting from $[\text{Ru(II)}(\text{H}_2\text{O})_6]^{2+}$ which can aromatize cyclic olefins when heated in water or other suitable protic solvents [46, 47].

Bis-arene Ru(II) sandwich complexes can be synthesized, for example by removal of the bound chloride ligands from $[(\eta^6\text{-arene})\text{RuCl}_2]_2$ by silver salts in acetone,

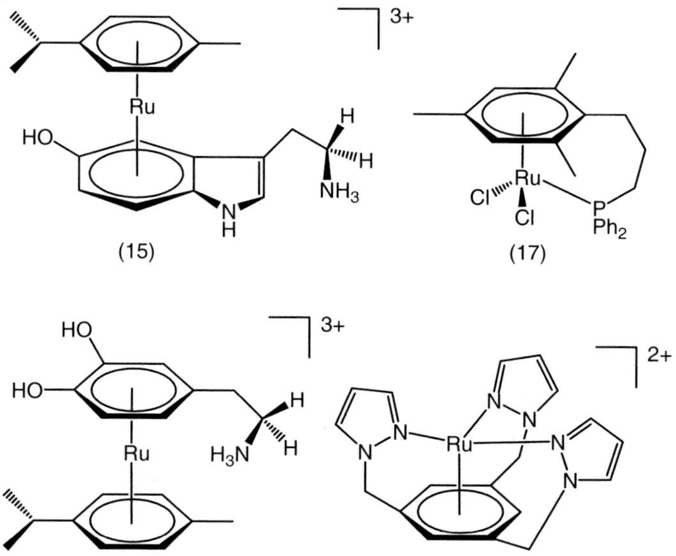

Fig. 2.5 Examples of sandwich (**15**, **16**), tethered (**17**), and encapsulated (**18**) Ru(II) arene complexes.

followed by reaction with an arene in CF_3COOH [48], or by direct reaction of the mono-arene Ru(II) dimer with an arene in CF_3COOH [49]. Indole and catechol sandwich complexes (**15 and 16**, Fig. 2.5) have been made using the former route [50].

It is possible to introduce a bound arene as part of a chelate ring in the form of a tethered complex [51, 52]. Such syntheses often require high pressures, and tethered complexes such as **17** (Fig. 2.5) have been of particular interest in the catalysis field [53, 54]. Strapped sandwich complexes can also be synthesized [55]. Furthermore, it is possible to encapsulate Ru(II) by an η^6-arene possessing a triple strap (**18**, Fig. 2.5) [56, 57].

Interest in the catalytic properties of Ru(II) mono-arene complexes range from alkene- and aromatic-hydrogenation, to Diels–Alder reactions, alkene metathesis, and asymmetric hydrogen transfer reductions of ketones and imines [58]. Catalytic activity usually requires the presence of a labile coordination site on Ru(II) and/ or arene displacement [59, 60]. Such catalysts are likely to be easily poisoned in biological media by the strongly coordinating ligands available, but it is intriguing to speculate that catalytic activity could exist in certain biological compartments to which access of deactivating molecules might be restricted, e.g. membranes.

In this article we will focus mainly on the biological chemistry of ruthenium(II) η^6-mono-arene complexes.

2.3.2
Structure

A typical structure of a half-sandwich 'piano-stool' $[(\eta^6\text{-arene})Ru(X)(Y)(Z)]$ complex is shown in Fig. 2.6.

The X-ray crystal structure [61] of $[(\eta^6\text{-biphenyl})Ru(en)Cl)]^+$ (RM175, where en is ethylenediamine) is shown in Fig. 2.7.

The complex exhibits the normal characteristic features of mono-arene Ru(II) complexes: an alternation of the lengths of the C–C bonds for the coordinated arene and irregular Ru–C bond lengths, with the longest of 2.244 Å (compared with 2.137 Å for the shortest) being for the carbon attached to the phenyl substitutent. Coordinated arenes with extended ring systems can have a flexible structure, as illustrated for dihydroanthracene (DHA) complexes in Table 2.2 (*vide infra*). Such arene flexibility may play a role in interactions of Ru(II) arene complexes with DNA.

Fig. 2.6 Typical structure of a Ru(II) half-sandwich 'piano-stool' complex. Coordination positions X, Y and Z and arene substituent(s) R provide scope for investigation of anticancer structure–activity relationships.

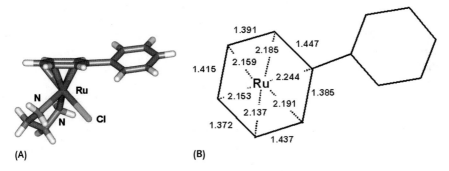

(A)

(B)

Fig. 2.7 (A) X-ray crystal structure of the cation in the anticancer complex $[(\eta^6\text{-biphenyl})Ru(en)Cl)]PF_6$ [61]. The complex has the characteristic 'piano-stool' structure with alternation of C–C bond lengths for the coordinated phenyl ring and irregular Ru–C bond lengths as shown diagrammatically in (B).

2.3.3
Chirality

Ru(II) arene complexes with three non-equivalent ligands bound to $\{(\eta^6\text{-}$ arene)Ru$\}^{2+}$ contain stereogenic centers. In principle, the enantiomers may have different biological properties, and it is therefore important to study the rates of inversion at such centers and to resolve the enantiomers for biological testing if possible. The lability of optically-active Ru(II) arene complexes is dependent on the nature of the bound ligands [62]. Optically-active tethered Ru(II) arene complexes are known, for example 19 (Fig. 2.8), which have very high configurational stability and are difficult to racemize [63].

Davenport et al. [64] have reported that $[(\eta^6\text{-}1,3,5\text{-trimethylbenzene})Ru(pyridyl-oxazoline-}N,N')Cl]^+$ (20, Fig. 2.8) is stable in dichloromethane but epimerizes slowly in methanol or acetone. They proposed that epimerization of the aqua/solvent species is much faster than for the chloro complex. Configurational stability in such complexes appears to depend on the didentate chelating donor in the following order [65]:

neutral N,N' > mono-anionic N,N' > mono-anionic N,O.

The amino acid complexes $[(\eta^6\text{-arene})Ru(aa)Cl]$, where aa = N,O-chelated amino acid, undergo rapid epimerization even at temperatures below 0 °C [66, 67]. Similar lability and epimerization has been found for salicylaldiminato complexes [68]. In contrast, $[(\eta^6\text{-}p\text{-cymene})Ru(imine)Cl]^+$ complexes are configurationally stable unless heated to high temperatures in polar solvents [69]. Configurationally-stable half-sandwich Ru(II) arene complexes are of much interest as potential asymmetric catalysts (e.g. for Diels–Alder and Mukaiyama reactions) [63]. The chloride ligand is usually removed in order to allow the complex to act as a Lewis acid catalyst and exchange of coordinated water on $[(\eta^6\text{-}1,3,5\text{-trimethylbenzene})Ru(pyridyloxazoline-}N,N')(H_2O)]^{2+}$ is slow on the NMR timescale in acetone–water [70]. Chiral chloro arene Ru(II) catalysts for asymmetric transfer hydrogenation can be activated *in situ* by treatment with KOH in 2-propanol [71].

Ligand-based chirality can occur through, for example, a coordinated nitrogen of a secondary amine [72], or can arise from facial chirality induced by unsymmetrical arenes [50].

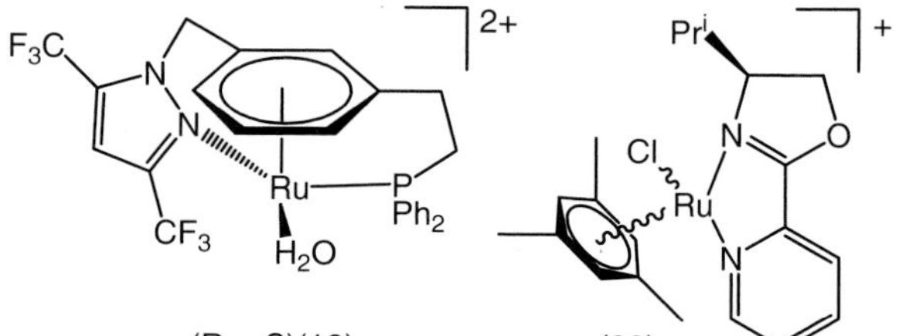

Fig. 2.8 Chiral Ru(II) arene complexes.

2.4
Biological Activity

2.4.1
Antibacterial

There are only a few reports of antimicrobial activity of Ru(II) arene complexes. Tocher et al. [73] found that coordination of metronidazole in [(η^6-benzene)-RuCl$_2$(metronidazole)] (**21**, Fig. 2.9) did not alter the electron affinity of metronidazole, but decreased the lifetime of the one-electron reduction product (the nitro radical anion) and increased the differential cytotoxicity towards *E. coli* grown under hypoxic conditions.

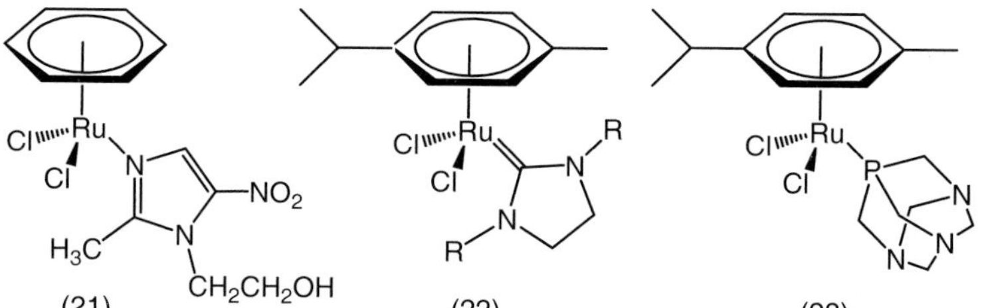

Fig. 2.9 Antimicrobial Ru(II) arene complexes [36, 73, 74].

Çetinkaya et al. [74] have reported antibacterial and antifungal activities of complexes of the type [(η^6-arene)RuCl$_2$(X)], where arene = *p*-cymene or hexamethylbenzene, and X = nitrogen or carbon donors. Compounds containing carbene derivatives were found to be active against Gram-positive bacteria and fungi including *Staphylococcus aureus* over a concentration range of 25–50 µg ml^{-1}. Nitrogen-containing complexes were significantly less active. The conclusion was that compounds with hydrophobic substituents (e.g. **22**, Fig. 2.9) displayed significantly more activity because of an increased ability to cross the cellular membrane.

Allardyce et al. [36] have noted that some phosphine complexes of the type [(η^6-*p*-cymene)RuX$_2$(pta)] (**23**, Fig. 2.9, where X = Cl, NCS, and pta = 1,3,5-triaza-7-phosphatricyclo[3.3.1.1]decane) exhibit antibacterial activity, but no doses were specified. They suggested that proteins rather than DNA may be the target sites. These complexes are inactive against Herpes and Polio viruses but the cluster complex [H$_4$Ru$_4$(η^6-benzene)$_4$]$^{2+}$ was active against Polio virus.

2.4.2
Anticancer

In our initial studies [61], we found that Ru(II) arene complexes with three monodentate ligands [(η^6-arene)Ru(X)(Y)(Z)], where X,Y or Z = halide, acetonitrile or isonicotinamide, were inactive towards A2780 human ovarian cancer cells *in vitro*. These complexes may be too reactive with components of the cell culture

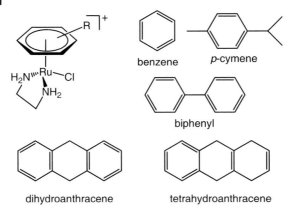

benzene
       p-cymene

biphenyl

dihydroanthracene   tetrahydroanthracene

Fig. 2.10 Five [(η^6-arene)Ru(en)Cl]$^+$ complexes (arenes shown separately) for which activity against A2780 human ovarian cancer cells has been investigated [75].

medium and/or the cells and be deactivated by biomolecules before they reach their target sites.

We obtained [61] reproducible cytotoxicity against A2780 human ovarian cancer cells for chelated diamine complexes of the type [(η^6-arene)Ru(N,N)(X)]$^+$ where N,N is typically ethylenediamine, and X is chloride (Fig. 2.10). Cells were incubated with the drug for 24 h, washed, and then the cell numbers were determined after growth on fresh medium for a further three days.

Activity appeared to increase with the size of the coordinated arene:

benzene < *p*-cymene < biphenyl < dihydroanthracene < tetrahydroanthracene,

such that, in this cell line, the biphenyl complex has similar cytotoxicity to the anticancer drug carboplatin (IC$_{50}$, the dose which inhibits the growth of 50% of the cells, 6 μM) and the tetrahydroanthracene complex is as active as cisplatin (IC$_{50}$ 0.6 μM) (Table 2.1) [75]. Substitution of chloride by other halides such as iodide had only a small effect on cytotoxicity.

Important was the finding [75] that these [(η^6-arene)Ru(N,N)Cl]$^+$ complexes are equally potent towards wild-type and cisplatin-resistant cancer cells in culture. This suggested that the mechanism of action was different from cisplatin.

However, comparisons between wild-type A2780 and adriamycin-resistant A2780^{AD} cells showed varying degrees of cross-resistance between Ru(II) arene complexes and adriamycin. Adriamycin (doxorubicin) is a well-established and widely-used anticancer drug. A2780^{AD} cells over-express the 170 kDa plasma membrane glycoprotein P-gp which is responsible for drug efflux from cells and has a high specificity for hydrophobic positively-charged molecules, features present in [(η^6-arene)Ru(N,N)(Cl/H$_2$O)]$^{+/2+}$ complexes. Knowledge of the active sites of such multi-drug resistance proteins [76] could allow design of Ru(II) arene complexes which circumvent this efflux mechanism. Administration of verapamil, a known inhibitor of P-gp, to cells completely restored sensitivity to Ru(II) arene complexes. P-gp does not transport cisplatin.

Table 2.1 IC_{50} values for Ru(II) arene complexes $[(\eta^6\text{-arene})Ru(X)(Y)(Cl)]A$
$(A = PF_6^-$ for positively-charged complexes) in A2780 human ovarian cancer cells
after 24 h drug exposure, and comparison with carboplatin and cisplatin [75].

Arene/Pt complex	X	Y	IC_{50} (μM)
p-Cymene	CH_3CN	CH_3CN	> 100
p-Cymene	Cl	isonicotinamide	> 100
$C_6H_5CO_2CH_3$		$H_2NCH_2CH_2NH_2$	56
Benzene		$H_2NCH_2CH_2NH_2$	17
p-Cymene		$H_2NCH_2CH_2NH_2$	10
Carboplatin			6
$C_6H_5C_6H_5$		$H_2NCH_2CH_2NH(Et)$	6
$C_6H_5C_6H_5$		$H_2NCH_2CH_2NH_2$	5
Dihydroanthracene		$H_2NCH_2CH_2NH_2$	2
Cisplatin			0.6
Tetrahydroanthracene		$H_2NCH_2CH_2NH_2$	0.5

Ruthenium arene complexes appear to have a wide spectrum of cytotoxicity towards cancer cells. For example, the complexes $[(\eta^6\text{-biphenyl})Ru(en)Cl]PF_6$ and $[(\eta^6\text{-dihydroanthracene})Ru(en)Cl]PF_6$ are active against not only A2780 human ovarian cancer cells, but also HT29 colon, Panc-1 pancreatic and NX02 lung cancer cells with IC_{50} values in the range 1–13 μM [77].

Mice were able to tolerate up to 25 mg kg^{-1} of $[(\eta^6\text{-biphenyl})Ru(en)Cl]PF_6$ (RM175) on days 1 and 5 without significant weight loss, compared with 25 mg kg^{-1} cisplatin as a single intraperitoneal injection. RM175 was active in the A2780 human ovarian cancer xenograph [75] giving a significant growth delay representing a T/C (treated/control) value of 46% on day 16, compared with a T/C value of 23% for cisplatin on day 16. Growth inhibition by this ruthenium complex was maintained in the A2780cis xenograph (T/C 51% on day 13), being significantly more active than cisplatin. However, RM175 was cross-resistant with adriamycin *in vivo* as was found *in vitro*.

2.4.3
Biodistribution and Metabolism

It will be important to gain more knowledge of the *in vivo* metabolism of ruthenium(II) arene complexes of potential interest as drugs. Little is currently known, but some studies on metal cyclopentadienyl (Cp) complexes have been reported. For ferrocene, $[(Cp)_2Fe]$, a coordinated Cp ring is hydroxylated in the liver via the P-450 system (Fig. 2.11) [78].

Ruthenocene complexes have been studied mainly as a way of tagging biomolecules with ruthenium isotopes which have favorable radioactive properties.

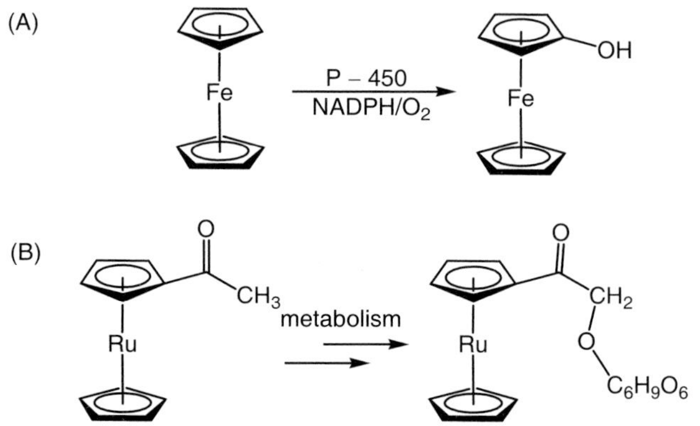

(A)

(B)

Fig. 2.11 Oxidation of (A) ferrocene and (B) acetylruthenocene *in vivo* [78, 79].

The major metabolite present in both bile and urine after oral administration of $[^{103}Ru]$-acetylruthenocene $[(Cp)Ru-(C_5H_4-CO-CH_3)]$ to rats is the glucuronide $[(Cp)Ru-(C_5H_4-CO-CH_2-O-C_6H_9O_6)]$ formed via oxidation of the methyl group to an alcohol followed by conjugation with glucuronic acid (Fig. 2.11) [79]. Carrier-free $[^{103,106}Ru]$-ruthenocenylalanine has been prepared from $^{103,106}RuCl_3$ and evaluated as a pancreatic imaging agent, but no selective uptake in the pancreas was observed [80]. The γ-emitting radionuclide ^{103}Ru has been used to track the activity of bleomycin, an anticancer drug which is thought to be activated *in vivo* by iron [81].

It remains to be seen whether nucleophilic attack on a coordinated arene ring can occur *in vivo*. Under some chemical conditions, $[(\eta^6\text{-}C_6H_6)Ru(N,N)(PR_3)]^{2+}$ complexes, where N,N is 1,10-phenanthroline or 2,2'-bipyridyl, for example, react with H^-, CN^- and OH^- to give η^5-cyclohexadienyl complexes [82].

2.5
Mechanism of Action

DNA and proteins are amongst the possible targets for ruthenium arene complexes in cancer cells. An interesting possibility is that Ru(II) could become involved in blocking Fe(II) pathways but this has yet to be studied.

2.5.1
Nucleobase and DNA Binding

We have investigated [83] reactions of complexes $[(\eta^6\text{-arene})Ru(en)X]^{n+}$, where arene = biphenyl (Bip), tetrahydroanthracene (THA), dihydroanthracene (DHA), *p*-cymene (Cym) and benzene (Ben), X = Cl^- or H_2O, with nucleic acid derivatives using ^1H, ^{31}P and ^{15}N (^{15}N-en) NMR spectroscopy. For mononucleosides, $\{(\eta^6\text{-biphenyl})Ru(en)\}^{2+}$ (Fig. 2.12) binds only to N7 of guanosine (G base), to N7

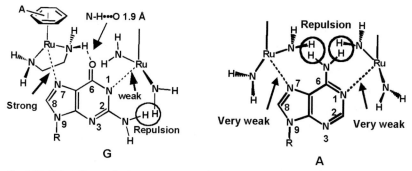

Fig. 2.12 The η^6-biphenyl Ru(II) complex shown (X = H$_2$O) binds to N7 of G, N7 and N1 of I, N3 of T, and very weakly to N3 of C and N7 of A [83]. In competitive studies with G, A, T and C nucleotides (R = sugar phosphate), binding is observed only to G N7.

and N1 of inosine (I), and to N3 of thymidine (T). Binding to N3 of cytidine (C) is weak, and almost no binding to adenosine (A) is observed.

The reactivity of the various binding sites of nucleobases towards Ru(II) at neutral pH decreases in the order:

$$G(N7) > I(N7) > I(N1),\ T(N3) > C(N3) > A(N7),\ A(N1).$$

Therefore, diamino Ru(II) arene complexes can be more highly discriminatory between G and A bases than square-planar Pt(II) complexes. The site-selectivity appears to be controlled by the en-NH$_2$ groups, which H-bond with exocyclic oxygens (e.g. C6O of G, see Fig. 2.15) but are non-bonding and repulsive towards exocyclic amino groups of the nucleobases (e.g. C6NH$_2$ of A, Fig. 2.13). For mononucleotides, the same pattern of site selectivity is observed, and significant amounts of the 5′-phosphate-bound species (40–60%) are present at equilibrium

Fig. 2.13 H-bonding and steric interactions which give rise to strong binding of $\{(\eta^6\text{-arene})\,\text{Ru(en)}\}^{2+}$ to guanine but very weak binding to adenine [83]. For clarity the arene and en ring are omitted in the right-hand (A) structure.

for 5'-TMP, 5'-CMP and 5'-AMP. No binding to the phosphodiester groups of 3',5'-cyclic guanosine monophosphate (cGMP) or cAMP is detected.

Reactions with nucleotides proceed via aquation of $[(\eta^6\text{-arene})Ru(en)Cl]^+$, followed by rapid binding to the 5'-phosphate group, and then rearrangement to give N7, N1 or N3-bound products [83]. In competitive reactions of $[(\eta^6\text{-bi-}$ phenyl)Ru(en)Cl]$^+$ with 5'-GMP, 5'-AMP, 5'-CMP and 5'-TMP the only final adduct is $[(\eta^6\text{-biphenyl})Ru(en)(N7\text{-GMP})]$. The aqua adducts appear to be more reactive than hydroxo adducts. Hence knowledge of the pK_a values of aqua adducts could be useful in drug design.

Binding of $\{(\eta^6\text{-biphenyl})Ru(en)\}^{2+}$ to N7 of 5'-GMP lowers the pK_a of N1H by 1.4 units. Such a lowering is also observed for Pt(II)-G adducts [84, 85]. Metalation of N7 of G, which is accessible from the major groove of B-DNA, can therefore lead to significant electronic perturbations at N1H which is an H-bond donor in G–C base-pairs in a Watson–Crick double helix. This may influence the stability of the duplex.

In kinetic studies (pH 7.0, 298 K, 100 mM NaClO$_4$), the rates of reaction of the diester cGMP with $[(\eta^6\text{-arene})Ru(en)X]^{n+}$ (where X = Cl$^-$ or H$_2$O) decreased in the order (Fig. 2.14):

THA (14.7) > Bip (8.1) > DHA (7.1) $\gg$ Cym (2.5) > Ben (1.1 mM^{-1} s^{-1})

suggesting that N7-binding is promoted by favorable arene–purine hydrophobic interactions in the associative transition state. These experiments showed that the diamine NH$_2$ groups, the hydrophobic arene, and the chloride leaving group can play important roles in reactions of nucleic acids with Ru(II) arene complexes.

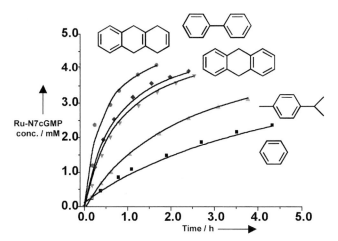

Fig. 2.14 Dependence of the rate of reaction of 3',5'-cyclic guanosine mono-phosphate (cGMP) on the arene in $[(\eta^6\text{-arene})Ru(en)(H_2O)]^{2+}$ complexes, as determined by integration of the nucleotide ^{1}H NMR H8 peak for the product $[(\eta^6\text{-arene})Ru(en)(cGMP-N7)]^+$ [83]. It can be seen that reactions are faster for arenes which can form π-complexes with the purine base. A similar dependence on the arene is observed for reactions of the chloro complex, although the reactions are slower.

We studied [86] the structures of monofunctional adducts of $[(\eta^6\text{-arene})\text{-Ru(en)Cl}][PF_6]$ (where arene = Bip, THA or DHA) with guanine derivatives by X-ray crystallography and in solution using 2D NMR methods. In crystal structures of the 9-ethylguanine (9EtG) complexes $[(\eta^6\text{-DHA})\text{Ru(en)}(9EtG\text{-}N7)][PF_6]_2 \cdot$ (MeOH) (Fig. 2.15A) and $[(\eta^6\text{-THA})\text{Ru(en)}(9EtG\text{-}N7)][PF_6]_2 \cdot 2$ (MeOH), strong π–π arene–nucleobase stacking is present.

The outer ring of the anthracene DHA stacks over the purine base at a distance of 3.45 Å, and for THA at 3.31 Å, with dihedral angles of 3.3° and 3.1°, respectively. In the crystal structure of $[(\eta^6\text{-biphenyl})\text{Ru(en)}(9EtG\text{-}N7)][PF_6]_2 \cdot (\text{MeOH})$, there is intermolecular stacking between the pendent phenyl ring and the purine six-membered ring at a distance of 4.0 Å (dihedral angle 4.5°). This stacking stabilizes a cyclic tetramer structure in the unit cell. The guanosine (Guo) adduct $[(\eta^6\text{-biphenyl})\text{Ru(en)}(Guo\text{-}N7)][PF_6]_2 \cdot 3.75(H_2O)$ exhibits intramolecular stacking of the pendent phenyl ring with the purine five-membered ring (3.8 Å, 23.8°), and intermolecular stacking of the purine six-membered ring with an adjacent pendent phenyl ring (4.2 Å, 23.0°). These occur alternately giving a columnar-type structure. Although the orientation of arene and purine in the crystal structure of $[(\eta^6\text{-biphenyl})\text{Ru(en)}(9EtG\text{-}N7)][PF_6]_2 \cdot (\text{MeOH})$ is *anti*, in solution a *syn* orientation predominates for all the biphenyl adducts as revealed by NMR NOE studies. The predominance of the *syn* orientation can be attributed to hydrophobic interactions between the arene and purine rings.

Significant re-orientations and conformational changes of the arene ligands in $[(\eta^6\text{-arene})\text{Ru(en)}(9EtG\text{-}N7)]^{2+}$ complexes, with respect to those of the parent chloro-complexes $[(\eta^6\text{-arene})\text{Ru(en)}Cl]^+$, are found in the solid state [86], Table 2.2. The arene ligands are flexible through rotation around the arene–Ru π–bonds, and through propeller-twisting for Bip, and hinge-bending for THA and DHA.

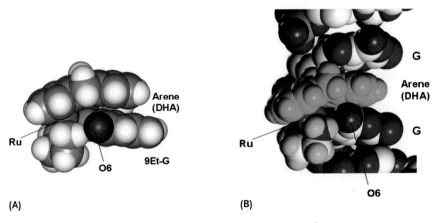

(A) (B)

Fig. 2.15 Arene–purine π–π stacking (A) $[(\eta^6\text{-DHA})\text{Ru(en)}(9EtG\text{-}N7)]^{2+}$ [86]. Ru(II) is coordinated to N7 and O6 of 9EtG and is strongly H-bonded to en-NH. (B) Model of a strand of B-DNA showing how the coordinated arene could intercalate between a pair of G bases (model built by J. A. Parkinson, see [86]).

Table 2.2 Comparison of arene hinge-bending and orientation in $[\eta^6$-arene)Ru(en)X]$^{n+}$ complexes for arene = dihydro- or tetrahydro-anthracene and X = Cl, H_2O or N7-bound 9-ethylguanine [86, 106].

Arene	Orientation angle α (°)			Hinge-bending angle β (°)		
	X =			X =		
	Cl	H_2O	9EtG	Cl	H_2O	9EtG
Dihydroanthracene	64.1	34.0 [a]	20.1	40.6	35.1 [a]	31.9
Tetrahydroanthracene	45.1	63.7	25.6	−7.5	−1.8	27.8

[a] Average for 2 independent cations in unit cell

Propeller-twisting of Bip decreases by ca. 10° so as to maximize intra- or inter-molecular stacking with the purine ring, and stacking of THA and DHA with the purine ring is optimized when their tricyclic ring systems are bent by ca. 30°. This involves increased bending of THA and a flattening of DHA. A dished conformation of coordinated THA has been observed for $[(\eta^6$-THA)RuCl$_2$(DMSO-S)] [87].

Strong stereospecific intramolecular H-bonding between an en NH proton oriented away from the arene and the C6O carbonyl of G is present in the crystal structures of these guanine adducts [86] (Fig. 2.15A; average N$\cdots$O distance 2.8 Å, N–H$\cdots$O angle 163°). In solution, NMR studies provided evidence that en NH protons of the 5′-GMP adduct are involved in strong H-bonding with the 5′-phosphate and C6O of 5′-GMP. Strong H-bonding from G C6O to en NH protons partly accounts for the high preference for binding of $\{(\eta^6$-arene)Ru(en)\}$^{2+}$ to G versus A (adenine). Thus direct coordination to the bases, intercalation, and stereospecific H-bonding are useful features to incorporate into the design of Ru(II) arene complexes to optimize the recognition of DNA (Fig. 2.15B). The strong preference for G bases may allow Ru(II) complexes to bind strongly to G-rich DNA telomeres which play key roles in cell division.

The concept of induced-fit recognition of DNA by organometallic Ru(II) arene complexes containing dynamic stereogenic centers via dynamic epimerization, intercalation and cross-linking may be useful in the design of anticancer drugs [72]. The $R_{Ru}^*R_N^*$ and $S_{Ru}^*R_N^*$ diastereomers of $[(\eta^6$-biphenyl)RuCl(Et-en)]$^+$ (Fig. 2.16) undergo slow interconversion in water ($t_{1/2}$ ca. 2 h, 298 K, pH 6.2) and reactions with 9EtG give rise selectively to $S_{Ru}^*R_N^*$ diastereomers. Dynamic chiral recognition of guanine can lead to high diastereoselectivity of DNA binding. The dinuclear complex $[((\eta^6$-biphenyl)RuCl(en))$_2$-(CH$_2$)$_6]^{2+}$ (Fig. 2.16) has similar dynamic characteristics, induces a large unwinding (31°) of plasmid DNA, and effectively inhibits DNA-directed RNA synthesis *in vitro*.

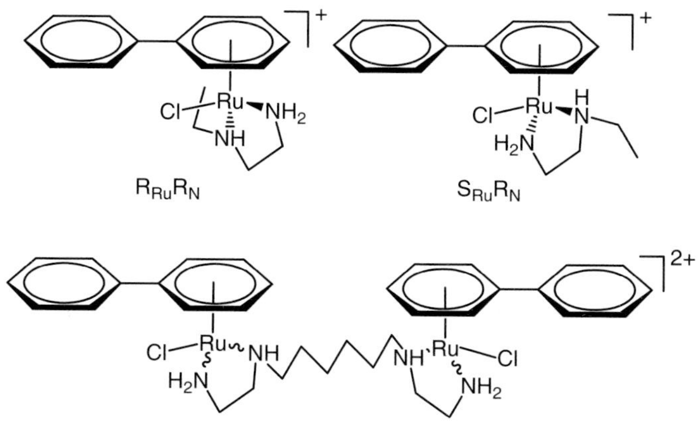

Fig. 2.16 Mono- and di-nuclear complexes containing chiral ruthenium and nitrogen centres.
There are 10 possible configurations for the dinuclear complex: four racemic pairs, $(R^*R^*R^*R^*)$-(±), $(S^*R^*R^*R^*)$-(±), $(R^*S^*R^*R^*)$-(±) and $(S^*R^*R^*S^*)$-(±), and two meso forms, SSRR and SRSR. The synthesis of the mononuclear complex gives two diastereomers in the mol ratio $R_{Ru}*R_N* : S_{Ru}*R_N* = 74 : 26$. Each unit of the dinuclear complex has a similar population of diastereomers [72].

This unwinding angle is more than twice that induced by the mononuclear complex $[(\eta^6\text{-biphenyl})RuCl(Et\text{-en})]^+$ (Table 2.3) and is attributable to cross-linking of the DNA and perturbation of the DNA structure by the pendent phenyl rings. It is notable that Ru(II) complexes containing arenes which cannot intercalate into DNA (e.g. $[(\eta^6\text{-}p\text{-cymene})Ru(en)Cl]^+$) produce only small unwinding angles similar to the monofunctional Pt(II) complex $[Pt(dien)Cl]^+$ (6° [88]). The dinuclear complex also gives rise to interstrand DNA cross-links on a 213-bp plasmid fragment with efficiency similar to bifunctional cisplatin, and to 1,3-GG interstrand as well as to 1,2-GG and 1,3-GTG intrastrand cross-links on site-specifically ruthenated 20-mers. It is also effective in blocking the intercalation of ethidium bromide.

Table 2.3 Unwinding of supercoiled pSP73KB DNA by Ru(II) arene complexes as determined from electrophoresis studies of the comigration of supercoiled and nicked forms [72, 88, 89].

Complex	Unwinding angle (°)
$[(\eta^6\text{-}p\text{-cymene})Ru(en)Cl]^+$	7
$[Pt(dien)Cl]^+$	6
$[(\eta^6\text{-biphenyl})Ru(en)Cl]^+$	14
$[(\eta^6\text{-tetrahydroanthracene})Ru(en)Cl]^+$	14
$[(\eta^6\text{-dihydroanthracene})Ru(en)Cl]^+$	14
$[(\eta^6\text{-biphenyl})Ru(Et\text{-en})Cl]^+$	14
$[((\eta^6\text{-biphenyl})RuCl(en))_2\text{-}(CH_2)_6]^{2+}$	31

Comparative studies [89] have shown that [(η^6-arene)Ru(en)Cl]$^+$ complexes, where arene = Bip, DHA, THA or Cym bind relatively rapidly to calf thymus DNA at 310 K, with 50% binding in 3 h for the *p*-cymene complex and 10–15 min for the others. The major stop sites in assays of *in vitro* RNA synthesis by RNA polymerases on DNA templates modified by these Ru(II) arenes are similar to those exhibited by cisplatin–DNA adducts, although in general reduced relative to cisplatin. The efficiency of the *p*-cymene complex is noticeably lower than that of the other complexes.

Circular dichroism (CD) and differential pulse polarography studies [89] suggest that the biphenyl and anthracene complexes cause non-denaturational changes in the conformation of DNA, in contrast to the changes induced by the benzene and *p*-cymene complexes. For example, binding of the former complexes to DNA gave rise to a CD band at ca. 375 nm associated with Ru(II)-to-arene π^* charge-transfer. Such a band was also seen for binding to poly(dG-dC) and not for poly(dA-dT). Flow linear dichroism (LD) data suggested that these Ru(II) arene complexes cause bending of DNA. The LD data are also consistent with partial intercalation of the biphenyl and anthracene complexes into DNA. In agreement with this, these complexes induced a loss of fluorescence of DNA-ethidium bromide adducts consistent with displacement of intercalated ethidium by the Ru(II) arenes with extended ring systems. Interestingly, the *p*-cymene complex which has methyl and isopropyl substituents on the benzene ring, distorted and thermally destabilized double-helical DNA significantly more that the biphenyl and anthracene complexes. All these data are consistent with a unique profile of DNA binding for ruthenium arene complexes.

Gopal et al. [90] have found that the bifunctional complex [(η^6-benzene)-RuCl$_2$(dmso)] inhibits the DNA relaxation activity of topoisomerase II and forms a drug-induced cleavage complex. This may arise from direct binding of the cleavage complex to Ru and to the enzyme. The [(η^6-benzene)RuCl$_2$(dmso)] complex used in the assay was dissolved in DMSO/water and so it is likely that hydrolyzed species were involved. Its cytotoxicity against Crit-2 (non-Hodgkins human lymphoma) cells was only modest (IC$_{50}$ ca. 150 μM). Morris et al. [61] observed no inhibition of topoisomerase I or II by the monofunctional complexes [(η^6-arene)Ru(en)X]$^+$, where arene = *p*-cymene, X= Cl or I, and arene = Bip or PhCO$_2$Me, X= Cl.

The water-soluble tetranuclear cluster [H$_4$Ru$_4$(η^6-benzene)$_4$][BF$_4$]$_2$ binds to plasmid DNA [91] but the mode of interaction is unknown. The phosphine complex [(η^6-*p*-cymene)RuCl$_2$(pta)] (**23**, Fig. 2.9) binds to DNA at acidic pH values (< 7) but not at basic pH in accordance with the need for protonation of a nitrogen atom of pta (pK_a 6.5) to enhance the interaction with DNA [92]. Such protonation may be enhanced in the vicinity of DNA where the pH may be lowered.

Chelation of {(η^6-arene)Ru}$^{2+}$ to N7 and deprotonated N6 of adenine derivatives to give a 5-membered ring is facile, giving rise to tri- (or tetra-, when N9 available) nuclear adducts (complexes **24** and **25**, Fig. 2.17) [93].

Reaction of [(η^6-benzene)Ru(D$_2$O)$_3$]$^{2+}$ with 5′-AMP gives rise to diastereomeric cyclic trimers [{(η^6-benzene)Ru(5′-AMP)}$_3$] in which 5′-AMP is coordinated by

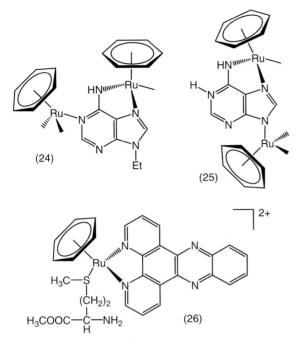

Fig. 2.17 Bridging modes of binding for 9-ethyladenine in cyclic trimer (**24**) and adenine in tetramer (**25**). Complex (**26**) contains a diamine as a DNA intercalator.

N1, N7 and deprotonated N6 of 5′-AMP [94]. Interestingly, reaction of this triaqua Ru(II) arene complex with 5′-ADP leads to initial phosphate/N7 macrochelation followed by cleavage of the β-phosphate and formation of the stable 5′-AMP cyclic trimer. Such cleavage was also observed for the diphosphate 5′-GDP, but for the monophosphate 5′-GMP binding to N7 only occurs giving 1 : 1 and 2 : 1 adducts.

Frodl et al. [95] have incorporated the DNA intercalator dppz (dipyrido[3,2-*a*:2′,3′-*c*]phenazine) as a chelated diamine into [(η^6-arene)Ru(Aa)(dppz)]$^{n+}$ (Aa = L-cysteine or L-methionine) complexes (e.g. **26**, Fig. 2.17). Stable intercalative binding was observed with stability constants which increased from ca. 10^5 to ca. 10^7 as the charge on the complexes increased from +1 to +3.

The strong preference of {(η^6-arene)Ru(II)(en)}$^{2+}$ complexes for guanine N7 binding is attributable to a combination of the high electron density on G N7, the strong H-bonding between en NH and G C6O and possible arene–purine stacking. This led us to replace the ethylenediamine NH H-bond donor by an H-bond acceptor as in the acetylacetonate (acac) ligand with the aim of changing the nucleobase selectivity. This was achieved: the complex [(η^6-*p*-cymene)Ru(acac)Cl] (Fig. 2.18) binds equally well to adenosine as to guanosine [96] and anticancer activity *in vitro* (towards A2780 human ovarian cancer cells) is retained.

2.5.2
Amino Acids and Proteins

The X-ray crystal structures of the amino acid and peptide complexes [(η⁶-benzene)Ru(aa)Cl]ⁿ⁺, where aa = penicillamine, histidine methyl ester and triglycine have interesting features [97]. Penicillamine is chelated via the amino group and thiolate S which bridges between two Ru atoms giving a 4-membered Ru-S-Ru-S ring (**27**, Fig. 2.18). Histidine is chelated via amino and imidazole nitrogens, and in the triglycine complex (prepared in water by reaction with the dimer [(η⁶-benzene)RuCl₂]₂) the tripeptide is chelated via the amino terminus and a deprotonated peptide N.

Sheldrick and Heeb [98] reported the preparation of the amino acid complex [(η⁶-benzene)Ru(L-alanine-*N,O*)Cl]. The crystals contained enantiomers with opposite chirality at the Ru centre. This complex reacted with 9EtG to give [(η⁶-benzene)Ru(L-alanine)(9EtG)]⁺ which contains N7-bound 9EtG stabilized by a C6O··H–N H-bond. Reaction of [(η⁶-benzene)Ru(L-alanine-methyl ester)Cl₂] with 9EtG led to displacement of coordinated monodentate L-alanine methyl ester. These workers observed that aquation of the Ru–Cl bonds occurs readily in water, as noted by Dersnah and Baird [99].

The complex [(η⁶-benzene)Ru(proline)Cl] was reported [98] to have significant antitumor activity towards P388 leukemia *in vivo* but there appears to be no further report on this activity and we have found that similar amino acid complexes are inactive against A2780 human ovarian cancer cells [77].

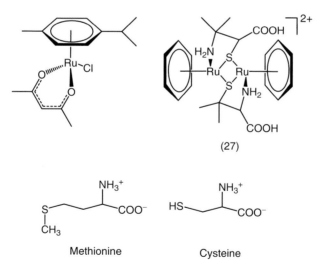

(27)

Methionine   Cysteine

Fig. 2.18 Top: left, acetylacetonate complex which binds to adenine as well as guanine, showing how the chelated ligand can be used to control DNA base selectivity (compared with en complex) [96]; right, dinuclear complex (**27**) with bridging penicillamine thiolate sulfurs. Bottom: structures of the amino acids methionine and cysteine.

Reactions between the sulfur-containing amino acids cysteine and methionine (Fig. 2.18) and ruthenium(II) arene anticancer complexes are of much interest in view of the strong influence of sulfur amino acids on the intracellular chemistry of platinum drugs, their involvement in detoxification and resistance mechanisms [100]. We found [101] that $[(\eta^6\text{-biphenyl})Ru(en)Cl][PF_6]$ reacts slowly with the thiol amino acid L-cysteine in aqueous solution at 310 K, pH 2–5, and only to about 50% completion at a 1 : 2 mM ratio. Reactions appeared to involve aquation as the first step followed by initial formation of 1 : 1 adducts via substitution of water by S-bound or O-bound cysteine. Two dinuclear complexes were also detected as products from the reaction. In these reactions half or all of the chelated ethylenediamine had been displaced and one or two bridging cysteines were present. The unusual cluster species $\{(\text{biphenyl})Ru\}_8$ was also formed especially at higher cysteine concentrations. The reaction was suppressed in 50 mM triethylammonium acetate solution at pH > 5 or in 100 mM NaCl suggesting that thiols may not readily inactivate Ru(II)-en arene complexes in blood plasma or in cells. Similarly, reactions with the thioether sulfur of methionine appeared to be relatively weak. Only 27% of $[(\eta^6\text{-biphenyl})Ru(en)Cl][PF_6]$ reacted with L-methionine (L-MetH) at an initial pH of 5.7 after 48 h at 310 K, and gave rise to only one adduct $[(\eta^6\text{-biphenyl})Ru(en)(\text{L-MetH-}S)]^{2+}$.

Reactions of half-sandwich complexes of Ru (Rh and Ir) with amino acid esters have been used for sequence-specific peptide synthesis. N,N'-Coordinated peptides undergo chain-lengthening at the N terminus by reaction with α-amino acid esters (e.g. in MeOH in the presence of NEt_3). Chain lengths of up to 9 amino acid residues can be synthesized using a $\{(\eta^6\text{-}p\text{-cymene})RuCl\}^+$ fragment [102].

The possibility that Ru(II) can form π-complexes with the aromatic side-chains of proteins is an intriguing one. Such complexes are known for simple amino acids derivatives of e.g. tryptophan [50], but have yet to be characterized for protein adducts of Ru(II) arenes. They are known for cyclopentadiene complexes. Reaction of strapped cyclopentadienyl complexes $[(\eta^5\text{-}C_5,\eta^1\text{-}N\text{-}C_5H_4(CH_2)_nNH_2)\text{-}Ru(CH_3CN)_2]^+$ $(n = 2,3)$ with proteins such as the hormone secretin in water can lead to sandwich complexes by ruthenation of the arene side chain of the phenylalanine residue (Fig. 2.19) [103].

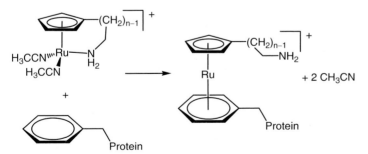

Fig. 2.19 Formation of an η^6-adduct by the phenylalanine side-chain of the protein secretin by reaction with a Ru(II)Cp derivative [103].

2.5.3
Aquation

Since it appears [83, 101] that aquation can be the rate-limiting step in the reactions of Ru(II) arene complexes with both amino acids and nucleotides, we have begun the study in detail the factors which influence the rate and extent of aquation and the pK_a value of the resulting aqua adduct.

Taube et al. [104] carried out early studies on the chemistry of aqua adducts of Ru(II) arene complexes and reported pK_a values of 6.3 for $[(\eta^6\text{-benzene})\text{Ru}(\text{NH}_3)_2(\text{H}_2\text{O})]^{2+}$ and 7.9 for the en analog. They noted the increase in affinity of Ru(II) for chloride when ammonia ligands are replaced by benzene (a comparison of the affinities of $[\text{Ru}(\text{NH}_3)_5\text{H}_2\text{O}]^{2+}$ and $[(\eta^6\text{-benzene})\text{Ru}(\text{H}_2\text{O})_3]^{2+}$ for chloride), attributable to the electron-withdrawing power of the η^6-bonded arene. Dadci et al. [105] reported that introduction of benzene into the Ru(II) hexaaqua complex $[\text{Ru}(\text{H}_2\text{O})_6]^{2+}$ enhances the water exchange rate by a factor of 640, whereas the additional introduction of bipyridine into $[(\eta^6\text{-benzene})\text{Ru}(\text{H}_2\text{O})_3]^{2+}$ to give $[(\eta^6\text{-benzene})\text{Ru}(\text{bpy})(\text{H}_2\text{O})]^{2+}$ reduces the rate by 174 and becomes only ca. four times faster than the hexaaqua ion.

We have reported [106] the X-ray crystal structures of the aqua complexes $[(\eta^6\text{-arene})\text{Ru}(\text{en})\text{Y}][\text{PF}_6]_n$, where arene = biphenyl (Bip), Y = 0.5 H_2O/0.5 OH, $n = 1.5$, arene = tetrahydroanthracene (THA), Y = 0.5 H_2O/0.5 OH, $n = 1.5$, arene = THA, Y = H_2O, $n = 2$, and arene = dihydroanthracene (DHA), Y = H_2O, $n = 2$. In the Bip complex, there is a large propeller twist of 45° of the pendent phenyl ring with respect to the coordinated phenyl ring. The tricyclic THA ligand is relatively flat in these aqua complexes, but the DHA ring system is markedly bent (hinge bend ca. 35°) as it is in the chloro complex (41°) (Table 2.2). The rates of aquation of the chloro complexes $[(\eta^6\text{-arene})\text{Ru}(\text{en})\text{Cl}][\text{PF}_6]$, arene = Bip, DHA and THA ($1.23–2.59 \times 10^{-3}$ s^{-1} at 298 K, I = 0.1 M) were more than 20 times faster than that of cisplatin. The reverse, anation reactions were very rapid on addition of 100 mM NaCl (similar concentration to that in blood plasma). The aquation and anation reactions were ca. two times faster for the DHA and THA complexes compared with the biphenyl complex. The hydrolysis reactions appear to occur *via* an associative pathway. The pK_a values of the aqua adducts ranged from 7.71 to 7.89 and 8.01 for the Bip, DHA and THA aqua complexes, respectively. The rate of hydrolysis of $[(\eta^6\text{-arene})\text{Ru}(\text{en})\text{X}]^+$ complexes can be dramatically slowed down by replacing X = chloride by iodide or azide [107].

At physiologically-relevant concentrations of the ruthenium(II) arene complexes (0.5–5μM), pH 7.4, and body temperature (310 K), the complexes should be present in blood plasma largely as the chloro complex (> 89%), whereas in the cell nucleus ($[\text{Cl}^-]$ = 4 mM) significant amounts (45–65%) of the more reactive aqua adducts would be formed together with smaller amounts of hydroxo complexes (9–25%).

Interestingly, the acetylacetonate (acac) complex $[(\eta^6\text{-}p\text{-cymene})\text{Ru}(\text{acac})\text{Cl}]$ (see Fig. 2.18) hydrolyzes more rapidly than the en analog [96]. The pK_a value of the aqua ligand in the acac complex (9.4) is considerably higher than that for the en analog (8.24 [83]). Therefore the ancillary ligands in Ru(II) arene complexes can be used to fine-tune the steric and electronic properties and thus the hydrolytic behavior of the complexes, features which can be important in drug design.

2.6
Conclusions

Platinum diamine complexes are now widely used for the treatment of cancer, but the development of resistance to platinum, and the lack of activity of platinum compounds against several types of cancer are clinical problems which need to be overcome. The discovery of activity amongst complexes of other metals might lead to further advances in cancer treatment. Organometallic complexes offer much potential for exploration as anticancer agents. Only one organometallic complex, titanocene dichloride, has so far entered clinical trials, but these have not been successful because the complex is unstable in water and is difficult to formulate reliably for administration. Our work has shown that certain ruthenium(II) arene complexes exhibit promising anticancer activity *in vitro* and *in vivo*. The complexes are stable and water-soluble, and their frameworks provide considerable scope for optimizing the design, both in terms of their biological activity and for minimizing side-effects by variations in the arene and the other three coordinated ligands. Initial studies on amino acids and nucleotides suggest that kinetic and thermodynamic control over a wide spectrum of reactions of Ru(II) arene complexes with biomolecules can be achieved. These Ru(II) arene complexes appear to have an altered profile of biological activity in comparison with metal-based anticancer complexes currently in clinical use or on clinical trial.

Acknowledgements

We thank the Edinburgh Technology Fund, EC (COST and Marie Curie programmes), Wellcome Trust and Oncosense Ltd for support. Our work on organometallic anticancer complexes has been made possible by the excellent contributions of our coworkers whose names are listed in our publications, including Dr. Duncan Jodrell and colleagues at the CRUK Oncology Unit at the Western General Hospital in Edinburgh. The ruthenium(II) arene compounds from our laboratory described in this review are the subject of patent applications PCT/GB2003/02879, PCT/GB01/02824 and PCT/GB00/04144.

References

1 R. MacKinnon, *FEBS Lett.* **2003**, *555*, 62–65.

2 For example see articles in *Chem. Rev.* **2004**, *104*, pp. 347–1200.

3 M. Isalan, A. Klug, Y. Choo, *Biochemistry* **1998**, *37*, 12026–12033.

4 Z. Guo, P. J. Sadler, *Angew. Chem. Int. Ed.* **1999**, *38*, 1512–1531.

5 Z. Guo, P. J. Sadler, *Adv. Inorg. Chem.* **2000**, *49*, 183–306.

6 P. Agre, D. Kozono, *FEBS Lett.* **2003**, *555*, 72–78.

7 P. Köpf-Maier, H. Köpf, in S. P. Fricker (Ed.), *Metal Compounds in Cancer Therapy*, Chapman & Hall, London, **1994**, pp. 109–146.

8 G. Jaouen, A. Vessières, I. S. Butler, *Acc. Chem. Res.* **1993**, *26*, 361–369.

9 R. H. Fish, G. Jaouen, *Organometallics* **2003**, *22*, 2166–2177.

10 B. Rosenberg, L. VanCamp, J. E. Trosko, V. H. Mansour, *Nature (London)* **1969**, *222*, 385–386.

11 H. M. Pinedo, J. H. Schornagel (Eds.), *Platinum and Other Metal Coordination Compounds in Cancer Chemotherapy*, Plenum, New York, **1996**.

12 M. Guo, Z. Guo, P. J. Sadler, *J. Biol. Inorg. Chem.* **2001**, *6*, 698–707.

13 K. Mross, P. Robben-Bathe, L. Edler, J. Baumgart, W. E. Berdel, H. Fiebig, C. Unger, *Onkologie* **2000**, *23*, 576–579.

14 N. Kröger, U. R. Kleeberg, K. Mross, L. Edler, G. Sass, D. K. Hossfeld, *Onkologie* **2000**, *23*, 60–62.

15 M. M. Harding, G. Mokdsi, *Curr. Med. Chem.* **2000**, *7*, 1289–1303.

16 O. R. Allen, L. Croll, A. L. Gott, R. J. Knox, P. C. McGowan, *Organometallics* **2004**, *23*, 288–292.

17 T. Pieper, K. Borsky, B. K. Keppler, in M. J. Clarke, P. J. Sadler (Eds.), *Topics in Biological Inorganic Chemistry*, Vol. 1, Springer Verlag, Berlin, **1999**, pp. 171–199.

18 M. J. Clarke, *Coord. Chem. Rev.* **2003**, *236*, 209–233.

19 M. J. Clarke, *Met. Ions Biol. Syst.* **1980**, *11*, 231–283.

20 G. Sava, E. Alessio, A. Bergamo, G. Mestroni, in M. J. Clarke, P. J. Sadler (Eds.), *Topics in Biological*

Inorganic Chemistry, Vol. 1, Springer Verlag, Berlin, **1999**, pp. 143–169.

21 M. Galanski, V. B. Arion, M. A. Jakupec, B. K. Keppler, *Curr. Pharm. Des.* **2003**, *9*, 2078–2089.

22 A. H. Velders, H. Kooijman, A. L. Spek, J. G. Haasnoot, D. de Vos, J. Reedijk, *Inorg. Chem.* **2000**, *39*, 2966–2967.

23 R. A. Vilaplana, F. González-Vílchez, E. Gutierrez-Puebla, C. Ruiz-Valero, *Inorg. Chim. Acta* **1994**, *224*, 15–18.

24 M. J. Clarke, S. Bitler, D. Rennert, M. Buchbinder, A. D. Kelman, *J. Inorg. Biochem.* **1980**, *12*, 79–87.

25 Z. Guo, A. Habtemariam, P. J. Sadler, B. R. James, *Inorg. Chim. Acta* **1998**, *273*, 1–7.

26 E. O. Fischer, R. Böttcher, *Z. Anorg. Allgem. Chem.* **1957**, *291*, 305–309.

27 E. O. Fischer, C. Elschenbroich, C. G. Kreiter, *J. Organomet. Chem.* **1967**, *7*, 481–485.

28 G. Winkhaus, H. Singer, *J. Organomet. Chem.* **1967**, *7*, 487–491.

29 M. A. Bennett, G. B. Robertson, A. K. Smith, *J. Organomet. Chem.* **1972**, *43*, C41–C43.

30 R. A. Zelonka, M. C. Baird, *J. Organomet. Chem.* **1972**, *35*, C43–C46.

31 S. Suravajjala, J. R. Polam, L. C. Porter, *J. Organomet. Chem.* **1993**, *461*, 201–205.

32 S. Bhambri, D. A. Tocher, *Polyhedron* **1996**, *15*, 2763–2770.

33 E. L. Muetterties, J. R. Bleeke, E. J. Wucherer, T. A. Albright, *Chem. Rev.* **1982**, *82*, 499–525.

34 D. Braga, A. Abati, L. Scaccianoce, B. F. G. Johnson, F. Grepioni, *Solid State Sci.* **2001**, *3*, 783–788.

35 R. Baldwin, M. A. Bennett, D. C. R. Hockless, P. Pertici, A. Verrazzani, G. Uccello Barretta, F. Marchetti, P. Salvadori, *J. Chem. Soc., Dalton Trans.* **2002**, 4488–4496.

36 C. S. Allardyce, P. J. Dyson, D. J. Ellis, P. A. Salter, R. Scopelliti, *J. Organomet. Chem.* **2003**, *668*, 35–42.

37 H. Le Bozec, D. Touchard, P. H. Dixneuf, *Adv. Organomet. Chem.* **1989**, *29*, 163–247.

38 G. Vitulli, P. Pertici, P. Salvadori, *J. Chem. Soc., Dalton Trans.* **1984**, 2255–2257.

39 P. Pertici, G. Vitulli, R. Lazzaroni, P. Salvadori, P. L. Barili, *J. Chem. Soc., Dalton Trans.* **1982**, 1019–1022.

40 G. Bodes, F. W. Heinemann, G. Jobi, J. Klodwig, S. Neumann, U. Zenneck, *Eur. J. Inorg. Chem.* **2003**, 281–292.

41 M. A. Bennett, T. N. Huang, T. W. Matheson, A. K. Smith, *Inorg. Synth.* **1982**, *21*, 74–78.

42 T. Hayashida, H. Nagashima, *Organometallics* **2002**, *21*, 3884–3888.

43 J. W. Janetka, D. H. Rich, *J. Am. Chem. Soc.* **1997**, *119*, 6488–6495.

44 P. Pertici, A. Verrazzani, G. Vitulli, R. Baldwin, M. A. Bennett, *J. Organomet. Chem.* **1998**, *551*, 37–47.

45 P. Pertici, A. Verrazzani, E. Pitzalis, A. M. Caporusso, G. Vitulli, *J. Organomet. Chem.* **2001**, *621*, 246–253.

46 M. Stebler-Röthlisberger, A. Ludi, *Polyhedron* **1986**, *5*, 1217–1221.

47 U. Koelle, C. Weissschädel, U. Englert, *J. Organomet. Chem.* **1995**, *490*, 101–109.

48 M. A. Bennett, T. W. Matheson, *J. Organomet. Chem.* **1979**, *175*, 87–93.

49 M. I. Rybinskaya, A. R. Kudinov, V. S. Kaganovich, *J. Organomet. Chem.* **1983**, *246*, 279–285.

50 A. Schlüter, K. Bieber, W. S. Sheldrick, *Inorg. Chim. Acta* **2002**, *340*, 35–43.

51 B. Therrien, T. R. Ward, M. Pilkington, C. Hoffmann, F. Gilardoni, J. Weber, *Organometallics* **1998**, *17*, 330–337.

52 M. A. Bennett, A. J. Edwards, J. R. Harper, T. Khimyak, A. C. Willis, *J. Organomet. Chem.* **2001**, *629*, 7–18.

53 F. Simal, D. Jan, A. Demonceau, A. F. Noels, *Tetrahedron Letters* **1999**, *40*, 1653–1656.

54 Y. Miyaki, T. Onishi, S. Ogoshi, H. Kurosawa, *J. Organomet. Chem.* **2000**, *616*, 135–139.

55 J. R. Miura, J. B. Davidson, G. C. Hincapié, D. J. Burkey, *Organometallics* **2002**, *21*, 584–586.

56 C. M. Hartshorn, P. J. Steel, *Angew. Chem. Int. Ed. Engl.* **1996**, *35*, 2655–2657.

57 W. Y. Sun, J. Xie, T. Okamura, C. K. Huang, N. Ueyama, *Chem. Commun.* **2000**, 1429–1430.

58 J. Soleimannejad, A. Sisson, C. White, *Inorg. Chim. Acta* **2003**, *352*, 121–128.

59 M. Bassetti, F. Centola, D. Sémeril, C. Bruneau, P. H. Dixneuf, *Organometallics* **2003**, *22*, 4459–4466.

60 D. Jan, L. Delaude, F. Simal, A. Demonceau, A. F. Noels, *J. Organomet. Chem.* **2000**, *606*, 55–64.

61 R. E. Morris, R. E. Aird, P. del S. Murdoch, H. Chen, J. Cummings, N. D. Hughes, S. Parsons, A. Parkin, G. Boyd, D. I. Jodrell, P. J. Sadler, *J. Med. Chem.* **2001**, *44*, 3616–3621.

62 H. Brunner, *Eur. J. Inorg. Chem.* **2001**, 905–912.

63 B. Therrien, T. R. Ward, *Angew. Chem. Int. Ed.* **1999**, *38*, 405–408.

64 A. J. Davenport, D. L. Davies, J. Fawcett, S. A. Garratt, D. R. Russell, *Chem. Commun.* **1999**, 2331–2332.

65 D. L. Davies, J. Fawcett, R. Krafczyk, D. R. Russell, *J. Organomet. Chem.* **1997**, 545–546, 581–585.

66 D. Carmona, C. Vega, F. J. Lahoz, R. Atencio, L. A. Oro, M. P. Lamata, F. Viguri, E. San José, *Organometallics* **2000**, *19*, 2273–2280.

67 D. Carmona, A. Mendoza, F. J. Lahoz, L. A. Oro, M. P. Lamata, E. San José, *J. Organomet. Chem.* **1990**, *396*, C17–C21.

68 H. Brunner, R. Oeschey, B. Nuber, *Organometallics* **1996**, *15*, 3616–3624.

69 D. Carmona, C. Vega, F. J. Lahoz, S. Elipe, L. A. Oro, M. P. Lamata, F. Viguri, R. García-Correas, C. Cativiela, M. P. López-Ram de Víu, *Organometallics* **1999**, *18*, 3364–3371.

70 D. L. Davies, J. Fawcett, S. A. Garratt, D. R. Russell, *Chem. Commun.* **1997**, 1351–1352.

71 K. Matsumura, S. Hashiguchi, T. Ikariya, R. Noyori, *J. Am. Chem. Soc.* **1997**, *119*, 8738–8739.

72 H. Chen, J. A. Parkinson, O. Novakova, J. Bella, F. Wang, A. Dawson, R. Gould, S. Parsons, V. Brabec, P. J. Sadler, *Proc. Natl. Acad. Sci. USA* **2003**, *100*, 14623–14628.

73 L. D. Dale, J. H. Tocher, T. M. Dyson, D. I. Edwards, D. A. Tocher, *Anti-Cancer Drug Design* **1992**, *7*, 3–14.

74 B. Çetinkaya, I. Özdemir, B. Binbasioglu, R. Durmaz, S. Günal, *Arzneim.-Forsch./Drug Res.* **1999**, *49*, 538–540.

75 R. E. Aird, J. Cummings, A. A. Ritchie, M. Muir, R. E. Morris, H. Chen, P. J. Sadler, D. I. Jodrell, *Br. J. Cancer* **2002**, *86*, 1652–1657.

76 S. Murakami, R. Nakashima, E. Yamashita, A. Yamaguchi, *Nature* **2002**, *419*, 587–593.

77 A. Habtemariam, R. Fernández, M. Melchart, H. Chen, E. P. L. van der Geer, R. E. Aird, D. I. Jodrell, P. J. Sadler, unpublished results.

78 R. P. Hanzlik, W. H. Soine, *J. Am. Chem. Soc.* **1978**, *100*, 1290–1291.

79 A. J. Taylor, M. Wenzel, *Biochem. J.* **1978**, *172*, 77–82.

80 W. H. Soine, C. E. Guyer, F. F. Knapp Jr., *J. Med. Chem.* **1984**, *27*, 803–806.

81 P. H. Stern, S. E. Halpern, P. L. Hagan, S. B. Howell, J. E. Dabbs, R. M. Gordon, *J. Nat. Cancer Inst.* **1981**, *66*, 807–811.

82 D. R. Robertson, I. W. Robertson, T. A. Stephenson, *J. Organomet. Chem.* **1980**, *202*, 309–318.

83 H. Chen, J. A. Parkinson, R. E. Morris, P. J. Sadler, *J. Am. Chem. Soc.* **2003**, *125*, 173–186.

84 K. Inagaki, Y. Kidani, *J. Inorg. Biochem.* **1979**, *11*, 39–47.

85 G. Schröder, B. Lippert, M. Sabat, C. J. L. Lock, R. Faggiani, B. Song, H. Sigel, *J. Chem. Soc., Dalton Trans.* **1995**, 3767–3776.

86 H. Chen, J. A. Parkinson, S. Parsons, R. A. Coxall, R. O. Gould, P. J. Sadler, *J. Am. Chem. Soc.* **2002**, *124*, 3064–3082.

87 T. J. Beasley, R. D. Brost, C. K. Chu, S. L. Grundy, S. R. Stobart, *Organometallics* **1993**, *12*, 4599–4606.

88 M. V. Keck, S. J. Lippard, *J. Am. Chem. Soc.* **1992**, *114*, 3386–3390.

89 O. Novakova, H. Chen, O. Vrana, A. Rodger, P. J. Sadler, V. Brabec, *Biochemistry* **2003**, *42*, 11544–11554.

90 Y. N. V. Gopal, D. Jayaraju, A. K. Kondapi, *Biochemistry* **1999**, *38*, 4382–4388.

91 C. S. Allardyce, P. J. Dyson, *J. Cluster Sci.* **2001**, *12*, 563–569.

92 C. S. Allardyce, P. J. Dyson, D. J. Ellis, S. L. Heath, *Chem Commun.* **2001**, 1396–1397.

93 S. Korn, W. S. Sheldrick, *Inorg. Chim. Acta* **1997**, *254*, 85–91.

94 S. Korn, W. S. Sheldrick, *J. Chem. Soc., Dalton Trans.* **1997**, 2191–2199.

95 A. Frodl, D. Herebian, W. S. Sheldrick, *J. Chem. Soc., Dalton Trans.* **2002**, 3664–3673.

96 R. Fernández, M. Melchart, I. Hanson, S. Parsons, R. E. Aird, J. Cummings, D. I. Jodrell, P. J. Sadler, ISBOMC'02 (First International Symposium on Bioorganometallic Chemistry), Paris, 18–20 July **2002**, Abs. PB6, p. 122.

97 W. S. Sheldrick, S. Heeb, *J. Organomet. Chem.* **1989**, *377*, 357–366.

98 W. S. Sheldrick, S. Heeb, *Inorg. Chim. Acta* **1990**, *168*, 93–100.

99 D. F. Dersnah, M. C. Baird, *J. Organomet. Chem.* **1977**, *127*, C55–C58.

100 J. Reedijk, *Proc. Natl. Acad. Sci. USA* **2003**, *100*, 3611–3616.

101 F. Wang, H. Chen, J. A. Parkinson, P. del S. Murdoch, P. J. Sadler, *Inorg. Chem.* **2002**, *41*, 4509–4523.

102 K. Severin, R. Bergs, W. Beck, *Angew. Chem. Int. Ed.* **1998**, *37*, 1634–1654.

103 D. B. Grotjahn, *Coord. Chem. Rev.* **1999**, *190–192*, 1125–1141.

104 Y. Hung, W. J. Kung, H. Taube, *Inorg. Chem.* **1981**, *20*, 457–463.

105 L. Dadci, H. Elias, U. Frey, A. Hörnig, U. Koelle, A. E. Merbach, H. Paulus, J. S. Schneider, *Inorg. Chem.* **1995**, *34*, 306–315.

106 F. Wang, H. Chen, S. Parsons, I. D. H. Oswald, J. E. Davidson, P. J. Sadler, *Chem. Eur. J.* **2003**, *9*, 5810–5820.

107 R. Fernández, M. Melchart, E. P. L. van der Geer, F. Wang, A. Habtemariam, P. J. Sadler, unpublished results.

3
Organometallics Targeted to Specific Biological Sites: the Development of New Therapies

Gérard Jaouen, Siden Top, and Anne Vessières

3.1
Introduction

Medicinal inorganic chemistry's current position has been the subject of a number of recent reviews and books [1–10]. Until now, this field has generally focused on coordination complexes.

The first successful application of organometallic compounds (i.e. compounds with a direct M–C bond) in therapy could be considered both a social and a scientific breakthrough. Paul Ehrlich (1854–1915), in the course of the work that would lead in 1910 to the anti-syphilis drug Salvarsan® and related compounds, introduced the concepts of chemotherapy, "magic bullets", and receptors, *corpora non agunt nisi fixata* [11]. He also demonstrated the great flexibility of this type of chemistry, which can take advantage of an almost unlimited set of organic ligands in order to modulate activity, solubility and toxicity. During the 1940s, however, this initial breakthrough was overwhelmed by the advent of penicillin.

In the second half of the twentieth century, two scientific advances led to a renaissance in the field of biomedical organometallic chemistry. These were the discovery by Rosenberg of the antitumoral properties of an inorganic complex, cisplatin 1 (Scheme 3.1) [12, 13], together with the development of a theoretical understanding of new types of organometallic bonds. The latter provided access to a range of new complexes and better control of the kinetic stability of the species in water, the essential biological solvent [14–16].

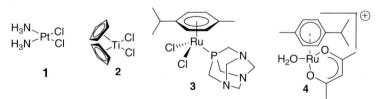

Scheme 3.1 Comparing structures **1–4**, shows the evolution of the organometallic molecules tested as pharmaceuticals.

Bioorganometallics: Biomolecules, Labeling, Medicine. Edited by Gérard Jaouen
Copyright © 2006 Wiley-VCH Verlag GmbH & Co. KGaA, Weinheim
ISBN: 3-527-30990-X

Although cisplatin itself **1** is a purely inorganic compound, with an unfortunately high level of general toxicity, its structural analogy to metallocene-type complexes of other metals, Ti, Fe, V, Mo, Re, Nb, Hf [17–28] soon became apparent. Dichlorotitanocene **2** has been the subject of a number of studies but clinical testing was recently discontinued due principally to difficulties of formulation [29, 30].

Other types of organometallic complexes, including several ruthenium complexes such as **3**, **4** have recently been reported [31–35]. Complex **3** shows antibacterial and antiviral properties while several complexes of the type $[(\eta^6\text{-arene})\text{Ru(chelate)Cl}]^{n+}$ (e.g. **4**) exhibit anticancer activity both *in vitro* and *in vivo* [31], including activity against cisplatin-resistant cancer cells. Modification of the ligands, for example by chelates, permits fine control of the properties of these molecules by modulating their interaction with DNA. This approach appears very promising for the development of clinically active molecules that benefit from a reduced level of general toxicity. A detailed analysis of these ideas can be found in Chapter 2 of this book and a review outlining the therapeutic use of organometallics has been published recently [36].

This chapter will outline another approach to the treatment of specific diseases, whether infectious, cancerous or parasitic, using organometallic moieties that are accurately targeted in order to reduce the problems of general toxicity. The approach is to use a known vector as a Trojan horse to transport the organometallic function into the target cell.

3.2
Overview of Previous Developments

The first attempts to use targeted organometallics for medical purposes were made 30 years ago [37]. The metallocenes were already established at that time as the ideal organometallic entities for use in this aim. The metallocenes, of which ferrocene is the archetype, are small, rigid, lipophilic molecules which can easily penetrate cellular membranes even when substituted. In addition, metallocenes have an external surface that resembles that of an aromatic nucleus, although they are thicker due to their sandwich structure. From this sprang the idea of substituting an aromatic component of a bioactive molecule by a metallocene, in order to modify its effects. The relative robustness of metallocenes in biological media is one of their advantages, whether used as tracers or as vectored species.

One of the first applications of this approach was an attempt to resolve, at least temporarily, the problem of the acquired resistance to antibiotics of certain bacteria. These bacteria produce β-lactamase which degrades antibiotics of the penicillin or cephalosporin type. Edwards et al. [38–41] developed a series of ferrocenyl-penicillins **5** and ferrocenyl-cephalosporins **6** as β-lactamase inhibitors, in which the initial aromatic group was replaced by a ferrocene entity. Some representative examples are shown below (Scheme 3.2).

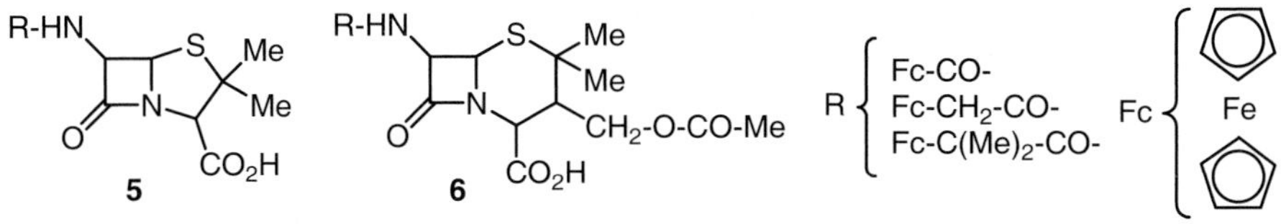

Scheme 3.2 Ferrocenyl-penicillins **5** and ferrocenyl-cephalosporins **6**.

When the ferrocenyl group is situated too close sterically to the β-lactam nucleus, where R = FcC(O)-, no antibiotic activity is observed for the modified penicillins and cephalosporins. However, the addition of a methylene spacer to distance the ferrocenyl moiety from the β-lactam nucleus restores the antibiotic activity to the same high level as that of the controls (Penicillin G and methicillin). Replacement of the methylene by CMe_2 increases β-lactamase inhibition and decreases the antibiotic activity. For a recent and exhaustive review of metallic antibiotics in general, see Ming [10].

To the best of our knowledge, no derivative of the type **5** or **6** has been commercially developed. The number of ferrocene compounds actually used in medicine up to the present day remains very limited, although there have been a few instances. Ferrocerone **7** (sodium salt of *o*-carboxybenzoyl ferrocene) (Scheme 3.3) developed in USSR to treat iron-deficiency anemia is one of the rare cases to come to fruition [42]. This product is well tolerated for oral administration, and can also be prescribed in the treatment of gum diseases.

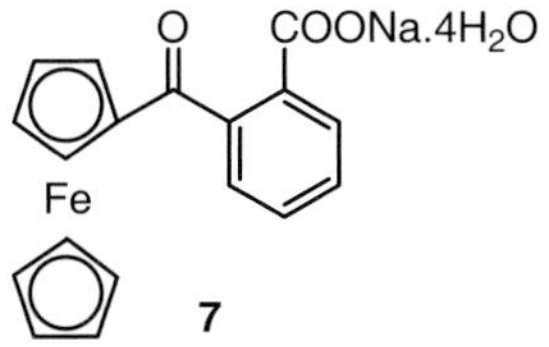

Scheme 3.3 Ferrocerone **7,** a ferrocene-base drug used for treatment of iron-deficiency anemia.

Because of its lipophilicity, ferrocene was considered as an interesting alternative to the use of iron salts for administration of Fe(II) to patients. For this reason, several *in vivo* toxicology studies have been carried out on ferrocene derivatives [43–45]. Most of these studies are in agreement in finding only a fairly low level of toxicity [46] despite acknowledged liver-related problems [47]. The result obtained in the case of ferrocerone **7** above is a good omen for the future development of drugs incorporating the ferrocenyl group.

The pioneering work of Edwards et al. was followed by research studies on ferrocene modifications to antibiotics such as penicillins, cephalosporins and rifamycins [48–50]. Certain of these products have proved to have good activity against gram-positive bacteria, while activity against gram-negative bacteria is negligible. Nevertheless, the last decade has seen interest focused chiefly on other biological systems, in particular hormone receptor-mediated organometallic antitumoral agents, and organometallic antiparasitic agents.

3.3
Metal Complex SERMs (Selective Estrogen Receptor Modulators)

In western countries, breast cancer affects one in eight women, and despite increasingly early detection it remains the most lethal of female cancers in many countries, including France. In two-thirds of these tumors, which contain the alpha form of the estrogen receptor (ERα), growth is stimulated by estrogens. The mechanism of action of the estrogens in these tumors is due to the specific interaction of the hormone with this receptor, which belongs the family of nuclear receptors. The binding of the hormone to its receptor is followed by a sequence of events: dimerization, attachment of specific co-activators, interaction with the Estrogen Response Element (ERE), a specific region of DNA, which causes the expression of a large number of specific genes responsible for a number of processes including cellular growth [51, 52]. Adjuvant treatment of these hormone-dependent cancers includes administration of antiestrogens also known as SERMs or more recently of antiaromatases [53]. The standard antiestrogen for the past 20 years has been hydroxytamoxifen (OH-Tam, 8) which, for reasons of bioavailability, is administered in its non-hydroxylated form, tamoxifen 9 (Scheme 3.4).

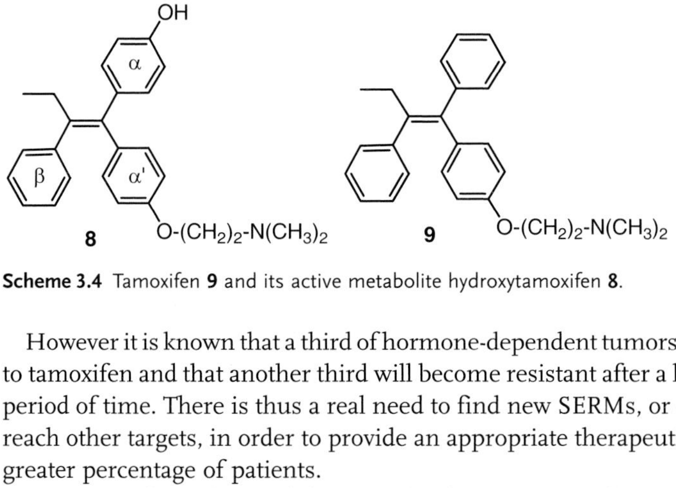

Scheme 3.4 Tamoxifen **9** and its active metabolite hydroxytamoxifen **8**.

However it is known that a third of hormone-dependent tumors will not respond to tamoxifen and that another third will become resistant after a longer or shorter period of time. There is thus a real need to find new SERMs, or compounds that reach other targets, in order to provide an appropriate therapeutic response for a greater percentage of patients.

It is generally understood that in order for a compound to act via the estrogen receptor, it must be recognized by this receptor. Recognition is measured by the relative binding affinity (RBA) value of the compound for the ER, the reference value being that of estradiol (RBA = 100% by definition). It should also be noted that a second form of the receptor, ERβ, was recently discovered in various tissues and has been found in breast cancers classified as hormone-independent. However, the exact role of this receptor in the cell is not yet well understood and remains a subject of debate [54, 55].

3.3.1
Inorganic Complexes of Platinum

The idea of using molecules with affinity for the estrogen receptors as vectors for cytotoxic agents arose in the 1970s. It was applied in the first instance to nitrogen mustards (*N*-mustard) [56], and then to cisplatin. Several inorganic complexes derived from estradiol, estrone or the synthetic estrogen hexestrol were prepared in this way (Scheme 3.5) [57–63].

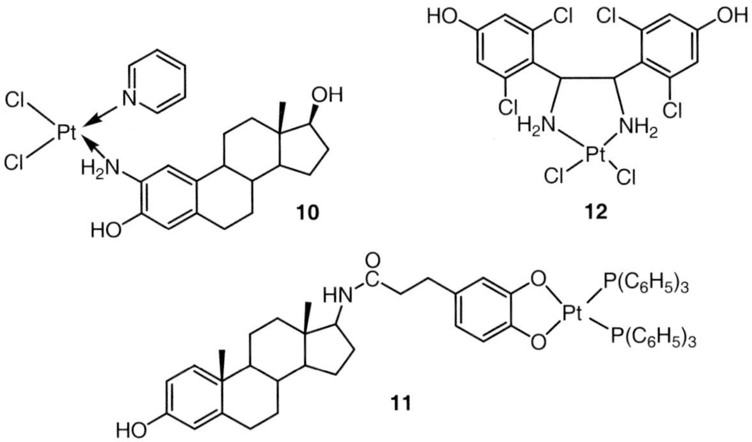

Scheme 3.5 Selected platinum derivatives of estradiol, **10** and **11** and of hexestrol **12**.

Some of these complexes at high molarities (around 5 μM) have antiproliferative effects on various breast cancer cell lines, MCF7 for **10** and **11**, EVSA-T for **10**, DMBA-induced rat mammary carcinoma for **12**. However these effects are very similar to those observed in the presence of the corresponding platinum complexes alone, and thus seem to be fundamentally linked to the cytotoxicity of these complexes. None of these inorganic complexes of platinum, to the best of our knowledge, has been the subject of further development.

Recently complexes such as **13** and **14** have been prepared. They are based on the coupling of the DACH-Pt [(*R*,*R*)-trans-1,2-diaminocyclohexane-platinum(II)] fragment, the active part of oxaliplatin which is used in the treatment of colorectal cancer [64, 65], to tamoxifen (Scheme 3.6) [66].

Complex **14** recognizes the estradiol receptor reasonably well (RBA = 6.4%) but the antiproliferative effect observed *in vitro* on MCF7 cells, derived from hormone-dependent breast cancers, this time appears to be essentially anti-hormonal, with (DACH)PtCl$_2$ itself showing only slight cytoxic activity on these cells.

In the case of breast cancers, therefore, the inorganic complexes of platinum still do not appear to be cytotoxic agents capable of being successfully coupled to a hormonal delivery system.

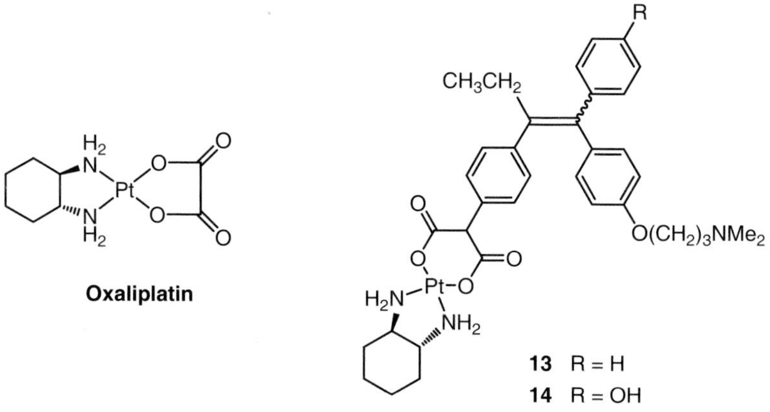

Oxaliplatin

13 R = H
14 R = OH

Scheme 3.6 Chemical structures of tamoxifen derivatives **12**, bearing a DAC-Pt fragment.

3.3.2
Carborane Derivatives with Estrogenic Properties

The incorporation of a large number of boron atoms into a tumor for boron neutron capture therapy (BNCT) has been accorded considerable interest over the last few years [67]. It is known that if boron-containing cells are exposed to a neutron flux, they are destroyed without causing irreparable harm to adjacent healthy tissue. In the context of SERMs, for example, Endo et al. have prepared complex **18** [1-(hydroxymethyl)-12-(4-hydroxyphenyl)-1,12-dicarba-closo-dodeca-borane] in which the hydrophobic carborane cage mimics the C and D rings of estradiol (Scheme 3.7) [68].

This complex was prepared from 1,12-dicarba-closo-dodecaborane, **15**. The coupling of the C-copper derivative of **15** with 4-methoxyiodobenzene gave the mono-C-arylated compound **16**, which was then converted to the C-methoxy-carbonyl derivative **17** by reaction of the lithiate of **16** with methyl chloroformate.

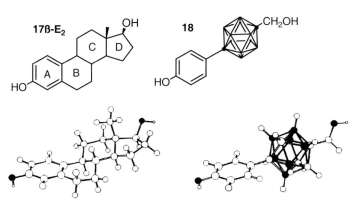

Scheme 3.7 Comparison of the chemical and X-ray structures of 17β-estradiol (17β-E$_2$) and borane complex **18** [69].

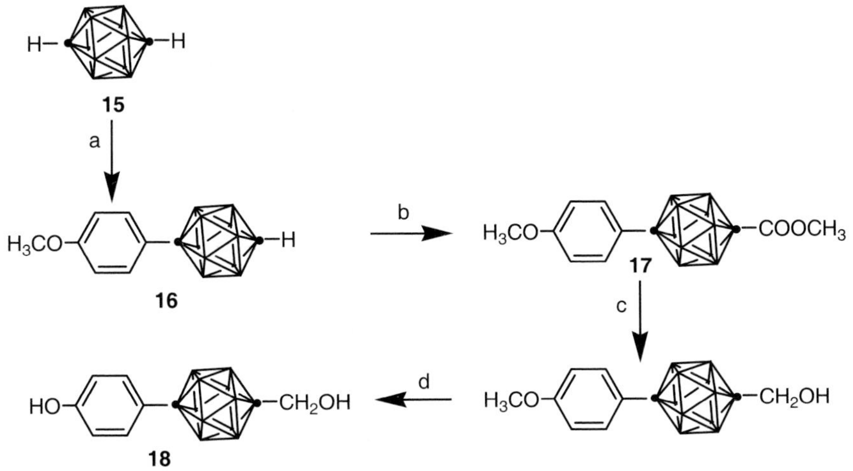

Scheme 3.8 Preparation of the carborane derivative **18** [68].
(a) 1. *n*-BuLi, CuCl/DME, 2. *p*-iodonanisole/pyridine reflux;
(b) 1. *n*-BuLi/benzene-Et$_2$O, 2. ClCOOCH$_3$; (c) LiAlH$_4$; (d) BBr$_3$/CH$_2$Cl$_2$.

After reduction of **17** with LiAlH$_4$, demethylation of the methoxy group gave **18** (Scheme 3.8).

The X-ray crystal structure of this complex is highly analogous to that of estradiol, particularly in terms of the respective positions of the two OH groups, which are known to play an essential role in the interaction of the hormone with its specific receptor [69]. *In vitro*, this complex behaves as an estrogen more powerful than estradiol itself. The estrogenic effect is also found *in vivo*, with ovariectomized mice injected with 10 ng day^{-1} of product **18** showing an increase in both uterus weight and bone mineral density. In addition to the appropriate positioning of the two OH groups, which allows the molecule to sit comfortably inside the active receptor site, the effect can also be explained by the existence, supported by docking calculations, of hydrophobic interactions, stronger than those observed for estradiol, between the carborane cage and the interior of the active site.

This carborane cage has also been used to prepare complex **19** which, with its dimethylamino chain, identical to that of hydroxytamoxifen, acts as an efficient antiestrogen [68], while complex **20**, based on the structure of (*Z*)-tamoxifen [70] and thus without an OH group, has no affinity whatsoever for the estradiol receptor [71].

The *in vivo* results show that these estrogenic carboranes are capable of specific interaction with the estrogen receptors. However, the reason their therapeutic use is limited seems to be the low level of estradiol receptors present in the cells of target tissues. In fact, taking into account the receptor/cell density, which is of the order of 10^4–10^5, the maximum amount of ^{10}B atoms that could be introduced into a given cell is in the order of 10^5–10^6, at least 3 orders of magnitude lower than the amount required to produce a lethal effect on the tumor cell [67].

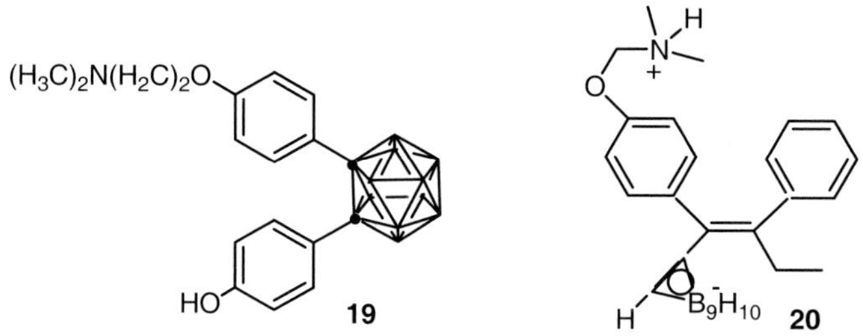

Scheme 3.9 Structures of carborane analogs of hydroxytamoxifen and tamoxifen.

3.3.3
Titanocene Dichloride Derivative of Tamoxifen

The initial idea in the preparation of complex **25** was to use tamoxifen as a vector for the potentially cytotoxic Cp_2TiCl_2 **2** (section 2, this book). Preparation of complex **25**, in which the aromatic β ring of tamoxifen is substituted by a Cp_2TiCl_2 entity, required development of a novel synthesis (Scheme 3.10) [72]. This synthesis starts with a McMurry coupling reaction between propionyl cyclopentadienyl manganese tricarbonyl **21** and the ketone **22**, permitting preparation of the complex diphenyl-ethylene **23**. This is converted to a cyclopentadienyl **24** via photochemical decomplexation, and immediately recomplexed in the presence of $CpTiCl_3$ to give **25**, which is isolated as the chloride.

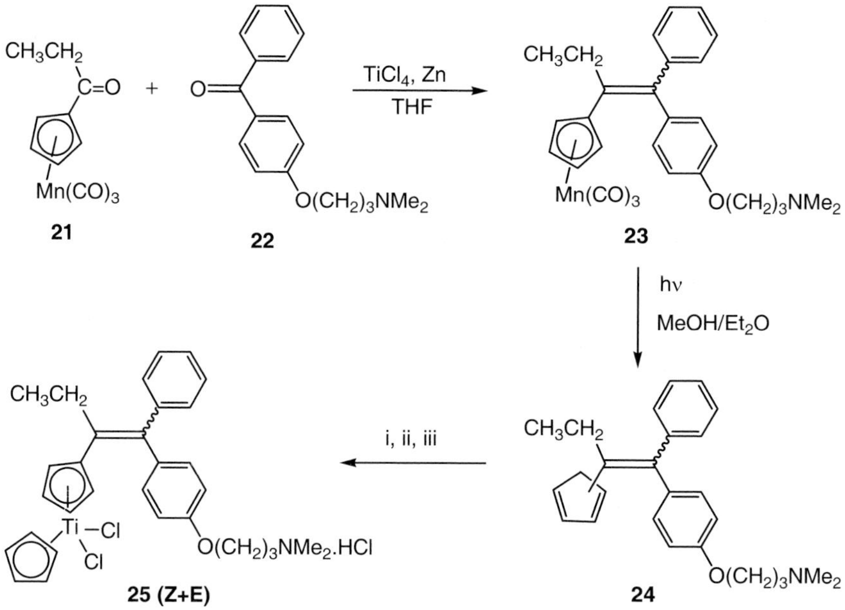

Scheme 3.10 Synthesis of the Cp_2TiCl_2 derivative of tamoxifen **25** [72].
i. *n*-BuLi, THF, −70 °C; ii. $CpTiCl_3$; iii. HCl.

This complex still has a marked affinity for the estrogen receptor (RBA = 8.5%) but more importantly, it reveals unexpected estrogenic effects on hormone-dependent breast cancer cells (MCF7) *in vitro*, effects as strong as those of estradiol itself (Fig. 3.1). This estrogenic activity is observed at extremely low molarities (10 nM and 0.1 μM) and is also found for the Cp_2TiCl_2 entity alone, suggesting that it is effectively linked to that molecule. In fact, the hydrolysis of Cp_2TiCl_2 is a complex phenomenon that has been the subject of many studies [73, 74]. This hydrolysis is strongly dependent on the experimental conditions, but in biological media a recent study by Sadler showed that it leads to the generation of Ti^{IV}, a hard metal similar to Fe^{III}, which is quickly captured and stabilized by transferrin, one of the iron transporters in the blood [75].

Although titanium has not yet been studied in this context, the role of certain metal ions has already been the subject of occasional analyses in relation to estrogen receptors. For example, treatment of the receptor with Cd^{2+} at very low (1 nM) concentration results in coordination of this metal at the level of the ER ligand binding domain [76–78]. The complex formed with this exogeneous metal activates ER alpha just as binding to estradiol would. Actually, cadmium is able to coordinate with cysteins 381 and 447, glutamic acid 523, histidine 524 and aspartic acid 538 of the receptor ligand binding domain. The idea is that this type of coordination may cause the receptor to act in the same way as the natural ligand does, by snapping shut the trap between the two helices H4 and H12 and causing activation of the transcriptional machinery and proliferation of the hormone-dependent breast cancer cells. This hypothesis could also apply to the case of Ti^{IV} and would provide an explanation for the surprising estrogenic effect observed for Cp_2TiCl_2. Overall, these results show the impossibility of using these complexes

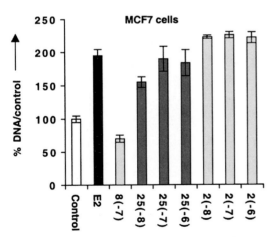

Fig. 3.1 Effect of E_2 (17β-estradiol), **8** (OH-tamoxifen), **25** and **2** (Cp_2TiCl_2) on the proliferation of MCF7 cells (estrogen receptor-positive cells). The results are expressed as the percentage of DNA in the sample versus the value of the control. The values in brackets correspond to the Log of molarity of incubation [72]. A value higher than 100% is indicative of an estrogenic effect.

of Ti^{IV} in the treatment of breast cancer and pose the problem of the possible role of titanium salts as endocrine disruptors.

3.3.4
Cyclopentadienyl Rhenium Tricarbonyl Derivatives of Tamoxifen Derivatives

In this series the stable $CpRe(CO)_3$ moiety has been incorporated into the tamoxifen framework [79, 80]. Substitution of a phenol by a sterically bulkier $CpRe(CO)_3$ moiety is likely to cause changes, inside the active receptor site, to the interactions of the dimethyl amino side chain, known to be responsible for the antiestrogenic effect of tamoxifen. For this reason complexes bearing side chains of varying lengths ($n = 2$–5 and 8) were prepared. These complexes can be obtained via a McMurry coupling reaction as shown in Scheme 3.11.

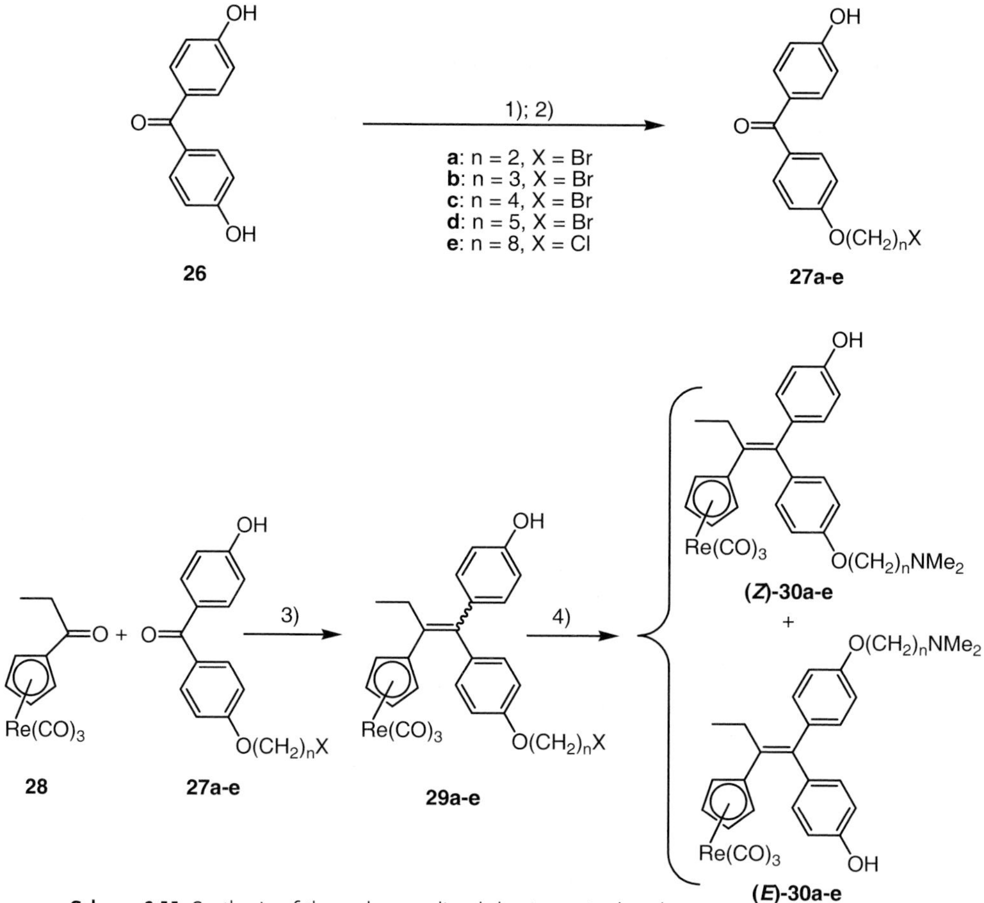

Scheme 3.11 Synthesis of the cyclopentadienyl rhenium tricarbonyl derivatives of OH-tamoxifen by McMurry cross-coupling.
1. KH, THF; 2. $X(CH_2)_nX$; 3. $TiCl_4/Zn$, reflux; 4. $HNMe_2/MeOH$, 60 °C autoclave.
The Z and E isomers are separated by semi-preparative HPLC [79].

In the first step 4,4′-dihydroxybenzophenone **26** was monoalkylated with the selected halogenoalkyl chain by using the corresponding dihalide and the monopotassium salt of **26**. McMurry coupling of the corresponding ketone **27a–e** with ketone **28** gave the alkenes **29a–e** in good yield (>60%). These alkenes were converted to amines **30a–e** in an autoclave by treatment with dimethylamine at 60 °C. The McMurry coupling reaction gives a mixture of Z and E isomers with slight excess of the latter. These isomers can be separated by semi-preparative HPLC. They do not isomerize in solution, which has made it possible to study the biological properties of the Z and E isomers separately [79]. It was found that addition of a CpRe(CO)$_3$ entity increases the lipophilicity of the compound, an important consideration in terms of its ability to penetrate the lipid membrane of the cell. Recognition of the complexes by one of the forms of the estrogen receptor is measured by its RBA value. The addition of the CpRe(CO)$_3$ entity produces a net decrease in RBA value (38.5 and 6.4 respectively for (Z+E)-OH-tamoxifen and (Z+E)-**30a** (n = 2)). The value decreases steadily as the side chain is lengthened, so that for n = 5, for example, the RBA value is 2%. However the RBA values do not seem to be a determining factor in the antiproliferative effect of these complexes. On MCF7 breast cancer cells the observed antiproliferative effect is in fact very similar for any length of chain (n = 3 or 5) or either isomer (Z or E) (Fig. 3.2A). The effect is slightly higher than that observed with OH-tamoxifen, the standard antiestrogen. On hormone-independent breast cancer cells (MDA-MB231), i.e. those without the alpha form of the estrogen receptor, estradiol and OH-tamoxifen have no effect and rhenium complexes have only a marginal effect (Fig. 3.2B).

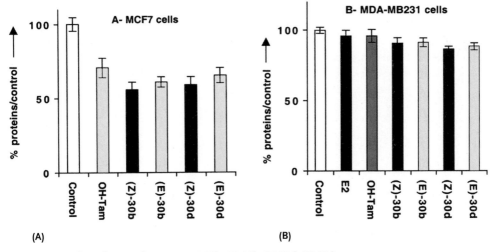

Fig. 3.2 Effect of 1 μM of **OH-Tam**, **(Z)-30b**, **(E)-30b**, **(Z)-30d**, **(E)-30d** on (A) MCF7 cells (estrogen receptor-positive cells) and (B) MDA-MB231 (estrogen receptor-negative cells). The results are expressed as the percentage of protein in the sample versus value of the control [79].

Substitution of a phenyl by the $CpRe(CO)_3$ moiety, bulkier and not easily oxidizable but more lipophilic (log $P_{O/W}$: 4.3 for (Z)-**30d**; 3.2; (Z)-OH-tamoxifen) thus produces complexes with an antiestrogenic effect very close to that of OH-tamoxifen itself, an effect that is not influenced by the length of the side chain or by the isomer used. Radioactive forms of these complexes could be envisaged for use either as [188]Re β emitters in radiotherapy, or, since the chemical behavior of technetium is known to be similar to that of rhenium, as technetium γ emitters for use in radioimaging. Owing to the very short half-life required in radioisotopes for medical applications, new synthetic routes will have to be found that allow the radioactive entity to be incorporated easily and in good yield at the final step of the synthesis.

3.3.5
Ferrocene Tamoxifen Derivatives (Ferrocifens)

In this series, the β-phenyl ring of hydroxytamoxifen was substituted by a ferrocenyl to give complexes **31a-e**, also called ferrocifens. The synthesis of these complexes employs McMurry coupling reactions [81–85] identical to those described for rhenium. The complexes are obtained in the form of a mixture of the isomers (Z+E), which can be separated by HPLC. However, since these complexes rapidly isomerize in solution, most of the biological studies were performed on mixtures of the two isomers.

The complexes with the shortest dimethyl amino side chains (n = 2, 3, 4) retain the best affinity for the two forms of the estrogen receptor (RBA for ERα respectively 14.6, 11.5, 12). Substitution of a phenyl by a more lipophilic ferrocenyl, followed by an increase in the number of CH_2 links in the side chain, produces as expected a corresponding progressive increase in the lipophilic value of the complexes compared to hydroxytamoxifen (log $P_{O/W}$ = 3.8 for (Z)-**31a** and 6.0 for (Z)-**31e**). On MCF7 hormone-dependent breast cancer cells, the antiproliferative effect of these complexes at molarities of 1 μM is a little stronger than that of OH-Tam for complexes where n = 3 and 5 (**31b** and **31d**) (Fig. 3.3) and this effect is not entirely suppressed by addition of estradiol. This seems to indicate that the

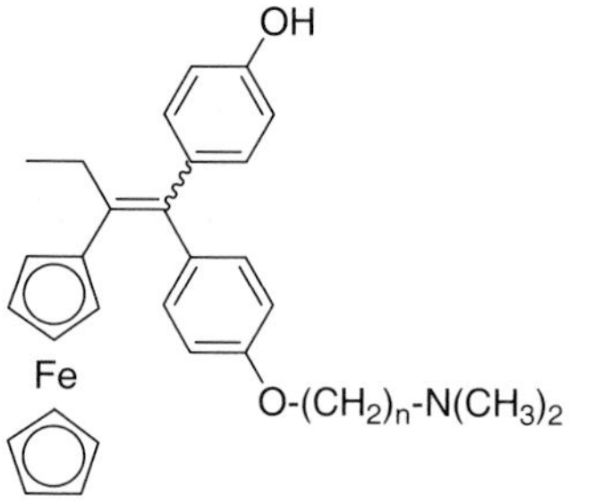

(Z+E)-31a-e (n = 2–5, 8)

Scheme 3.12 General formulae of hydroxyferrocifens.

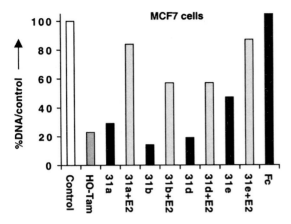

Fig. 3.3 Antiproliferative effect of 1 μM of OH-tamoxifen (HO-Tam), ferrocene (Fc), ferrocifens **31a** ($n = 2$), **31b** ($n = 3$), **31d** ($n = 5$) and **31e** ($n = 8$) in the absence or presence of 10 nM of 17β-estradiol (E$_2$) on hormone dependent breast cancer cells (MCF7 cells with a high level of ERα) after five days of culture (adapted from [85]).

observed antiproliferative effect is the combination of an antihormonal effect linked to the tamoxifen skeleton and mediated by the estradiol receptor, plus a cytotoxic effect induced by the presence of the ferrocenyl substituent. In fact, the cytotoxic properties of the ferricinium cation (Fc$^{+\cdot}$), which is easily formed by oxidation in biological media, are well known [18, 25, 86, 87]. It should however be noted that ferrocene alone has no effect, although its lipophilic value (log P$_{0/W}$ = 3.3) allows it to enter the cell without difficulty.

The novel feature of ferrocifens lies in the strong cytotoxic effect observed on MDA-MB231 cells, which are hormone-independent breast cancer cells (Fig. 3.4). In these cells without ERα, OH-tamoxifen has no effect, while ferrocifens **31b** and **31d** have a remarkable antiproliferative effect (IC$_{50}$ = 0.5 μM).

Here again, ferrocene alone is not toxic. In addition, it has been shown that derivatives of estradiol bearing a ferrocenyl substituent in position 17α, **32** and **33** have an estrogenic effect *in vitro*, and are devoid of any cytotoxic effect either on hormone-dependent or hormone-independent cells [71]. Simply delivering an estrogenic molecule bearing a ferrocenyl substituent payload into the interior of a target cell is not sufficient to obtain a cytotoxic effect. The production of this effect seems to be linked to a specific structure that allows the ferrocenyl – double bond – phenol pattern to come into play.

The mechanism of action of ferrocifens in the cell has not yet been elucidated. However, molecular modeling studies have revealed that ferrocifens such as (Z)-**31b** may be inserted into the antagonist configuration of the active site of the estrogen receptor (Fig. 3.5 left). The interior of the active site is of sufficient size to accommodate the ferrocenyl group, which is bulkier than a phenyl (respective volumes of (Z)-**31b** and OH-Tam are 572 Å and 413 Å), and interactions between

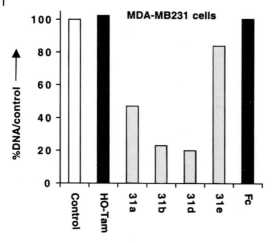

Fig. 3.4 Effect of 1 μM of OH-tamoxifen (HO-Tam), **31a** (*n* = 2), **31b** (*n* = 3), **31d** (*n* = 5) and **31e** (*n* = 8) and ferrocene (Fc) on hormone independent breast cancer cells (MDA-MB231 cells ERα-negative) [85].

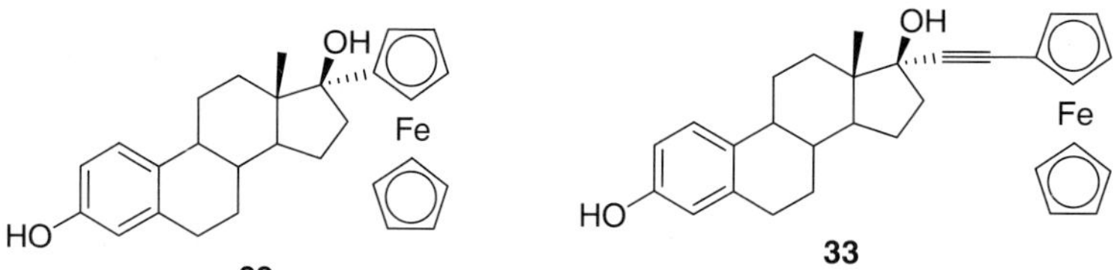

Scheme 3.13 Ferrocenyl estrogen derivatives.

His524 and the ferrocenyl group, and between Asp351 and the nitrogen of the basic chain, provide correct positioning of the organometallic hormone and particularly of the dimethyl amino side chain which is known to be the source of the antihormonal effect. This explains why the antiestrogenic effect observed is comparable to that of OH-Tam. As far as the cytotoxic effect of these molecules is concerned, this is clearly linked to the presence of a ferrocenyl substituent which may be reversibly oxidized under certain conditions. The cascade of events that begins with the oxidation of the ferrocene and leads finally to apoptosis or senescence of the cells has not yet been elucidated, in particular in terms of identification of the targets. Is it perhaps linked to the presence of ERβ, the second form of the receptor which seems to play a role in oxidation phenomena in the cell [88, 89], or does it involve the regulation of particular enzymes or other proteins implicated in the cellular cycle? Studies currently in progress should provide an answer to this question. The most promising ferrocifens, **31b** and **31c** (*n* = 3 and 4) are in preclinical studies.

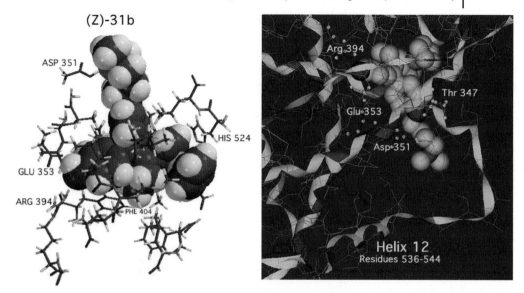

Fig. 3.5 (Z)-**31b** (left) and (Z)-**31c** (right) in the binding site of the antagonist form of the ERα ligand binding domain. The basis of the calculation is the x-ray structure described by Shiau et al. [90]. The modelling studies are performed with the Mac Spartan Pro Software [91] for (Z)-**31b** and with the Molview software [92] for (Z)-**31c** (data adapted from [84, 85]).

3.3.6
Ruthenocene Tamoxifen Derivatives

In order to examine the hypothesis of a mechanism of action for the ferrocifens based on oxidation of the iron, their equivalents in the ruthenium series, the ruthenocifens **34a–d** were prepared (Scheme 3.14) [93]. In fact, ruthenium's position in group 8 just below iron would suggest that the corresponding complexes would be more easily oxidized and thus would have an increased antiproliferative effect relative to the ferrocifens. These complexes were prepared, like those of iron and rhenium, via McMurry coupling.

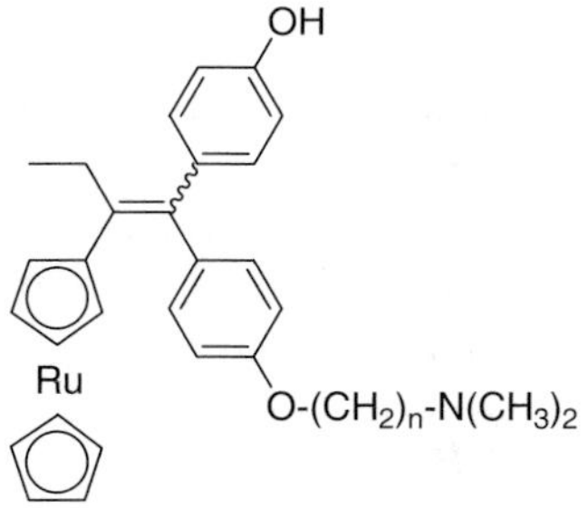

(Z+E)-34a-d (n = 2–5)

Scheme 3.14 General formulae of ruthenocifens [93].

They are obtained as a mixture of the two isomers (Z+E) which cannot be separated by HPLC, presumably because interconversion is too rapid. The complexes for which $n = 2$ and 3 have very high RBA values for ERα (85% and 53% respectively), significantly higher than that of OH-Tam (RBA = 38.5). *In vitro*, complexes **34b** and **34c** have an antiproliferative effect slightly higher than that of OH-Tam on hormone-dependent MCF7 cells and no effect on hormone-independent MDA-MB231 cells. Unlike ferrocifens, these complexes thus behave essentially as antiestrogens. The electrochemical study of the complexes provides a possible explanation for this difference, since it shows that the ferrocenyl oxidizes reversibly while oxidation of the ruthenocenyl is irreversible and leads to rapid decomposition of the organometallic entity [94].

3.3.7
Conclusion for SERMs

The various organometallic SERMs derived from tamoxifen have led to the discovery of products with novel, varied and sometimes unexpected biological properties [95]. One such is the surprising estrogenic effect of titanium complexes, which appears to be linked only to the Cp_2TiCl_2 moiety and draws attention to the potential risks of these compounds to act as endocrine disruptors. In neither the rhenium nor the ruthenium complexes does the organometallic entity confer any benefit in terms of antiproliferative effect, but the reasons for this are different: the stable entity, $CpRe(CO)_3$, is merely a spectator, while CpRuCp seems the victim of irreversible oxidative decomplexation. The use of rhenium complexes as radiopharmaceuticals, however, is conceivable. There is finally the case of the ferrocifens where the organometallic entity behaves as a genuine cytotoxic agent, in addition to the antiestrogenic properties of the vector. This type of molecule brings an enhancement to the SERM which may prove useful in the treatment of tamoxifen-resistant cancers. These products are worthy of thorough study for the prospects thus offered.

3.4
The Alkyne Cobalt Carbonyl Complexes

Biomolecules labeled with (alkyne)$[Co_2(CO)_6]$ entities are among the most widely studied molecules in bioorganometallic chemistry, thanks to their ease of preparation and reasonable stability [96]. It is thus no surprise that the literature provides examples of peptides [97, 98], proteins [99], therapeutic drugs [100], abortifacient medications such as RU486, **35** [101], corticoids and androgens [102], methotrexate **37** [103], the mycotoxin zearalenone **36** [104, 105], estrogens [106–110] and pesticides [111] modified by this entity (Scheme 3.15).

A large amount of the work on these complexes is inspired by the interest of this type of probe in assays and biological characterizations, and in particular on their application in the carbonyl metallo immunoassay (CMIA), based on the

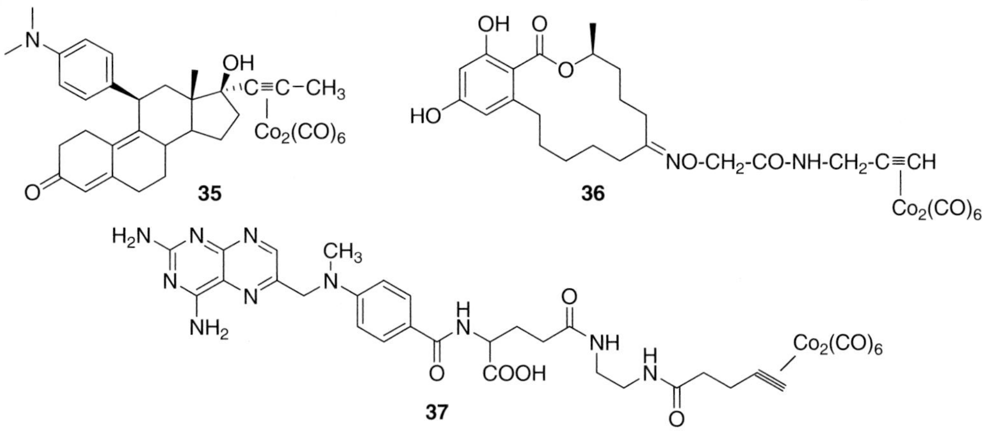

Scheme 3.15 Alkyne cobalt carbonyl complexes of the steroid RU 486 **35**, zearalenone **36**, methotrexate **37**.

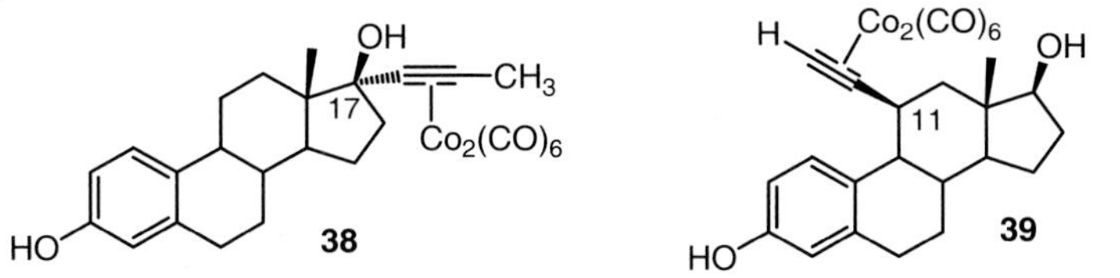

Scheme 3.16 Alkyne cobalt carbonyl complexes of 17α-propynyl ethynylestradiol **38**.

concept of strong absorption of M–CO vibrations in the 2000 cm^{-1} region in infrared where proteins do not absorb [112] (see Chapter 8).

In the steroid series, the alkyne cobalt carbonyl estrogen derivatives **38** and **39**, substituted in position 17α and 11β, have proved to be the most interesting (Scheme 3.16).

These complexes are both well recognized by estrogen receptor alpha (ERα) (RBA = 12% and 18% respectively) and are strongly estrogenic at molarities between 0.1 nM and 1 μM. But most notable is an anomaly found in their interaction with ERα [106, 109, 113]. Compound **38** bearing an alkyne cobalt carbonyl group in the 17α position displays an abnormally long residence time on the active site of the receptor, resembling the behavior of an affinity marker. A suggested explanation for this is the possible generation *in situ* of a carbenium ion stabilized by the adjacent organometallic entity. The carbenium ion, easily generated by Zn^{++} present in the receptor, could alkylate the cysteine 530 which is known to be the alkylation site for affinity markers bearing an aziridine function [114]. One can also consider the $C_2/Co_2(CO)_6$ cluster as a nido trigonal bipyramid with 12 electrons delocalized over the skeleton. This leaves a coordination site vacant, making it possible, depending on the local environment, to increase the residence time of the hormone on the receptor, for example by interaction with methionines 343 and 421 which are able to interfere with the cluster at the 17alpha position [115]. This type of mechanism based on particular interactions related to

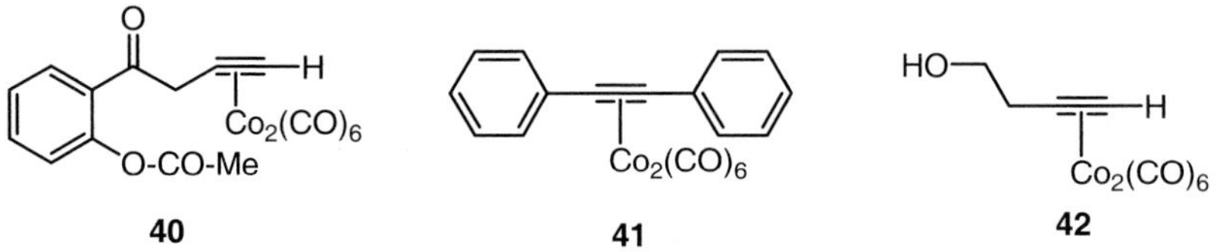

Scheme 3.17 Alkyne cobalt carbonyl complexes of aspirin **40**, diphenyl acetylene **41** and 2-propyn-1-ol **42**.

the nido nature of the cluster could also explain the unusual behavior of **39** which, at 25 °C, has a RBA value of 65%, much higher than the 18% observed at 0 °C, and a long residence time on the receptor [106]. From this it is clear that the interaction between the $Co_2(CO)_6$-labeled hormone and the receptor protein can produce surprising results and, for this reason alone, it merits particular attention.

A series of alkyne cobalt hexacarbonyl complexes was synthesized by Jung et al. [116] and Gust et al. [117–119]. These include, for example, $Co_2(CO)_6$ complexes derived from aspirin **40**, from diphenyl acetylene **41** and from 2-propyn-1-ol **42**.

The antiproliferative effect of these compounds has been studied on various cell lines, including melanomas and lung cancers [116], leukemias and lymphomas [118], and also recently on hormone-dependent (MCF7) and hormone-independent (MDA-MB231) breast cancer cell lines [119]. The most active of these compounds is still the aspirin derivative **40** with IC_{50} values between 1.4 and 10 μM. Its activity however remains lower than that of cisplatin, except in the case of breast cancer cells, which in any case are known not to respond optimally to cisplatin. The cytotoxic effect of these complexes does not appear to be linked to the accumulation of cobalt in the cells, nor to its bond to DNA. On the other hand, **40**, the most effective compound, is a powerful cyclooxygenase (COX-1 and COX-2) inhibitor, which could explain its mode of action, although this has not yet been confirmed [119].

Doubtless these preliminary results will encourage complementary studies with this type of cobalt complex or with others in order to better elucidate and understand their behavior.

3.5
Ferroquine, a New Weapon in the Fight Against Malaria: the Archetypical Bioorganometallic Approach

3.5.1
The Problem of Malaria

The search for new antimalarials is of intense current concern, and the contribution of an organometallic approach to the problem is well documented [120].

The disease remains the most widespread parasitic pathological condition in the world, with an estimated 1 million deaths per year, the majority children under 5 years of age, 90% inhabitants of sub-Saharan Africa. More than 300 million

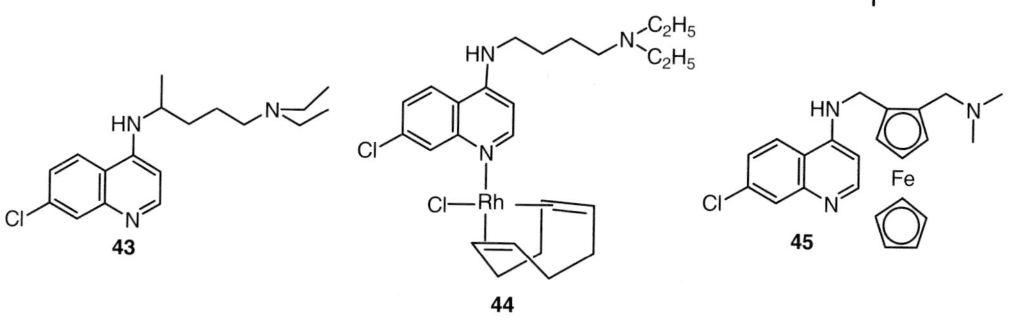

Scheme 3.18 Chloroquine **43**, its rhodium and ferrocenyl derivatives **44** and **45**.

clinical cases of malaria are reported each year, five times more than tuberculosis, AIDS, measles and leprosy combined. The effects of tourism and global warming may also lead to an increased incidence of malaria in western countries.

The parasite *Plasmodium falciparum*, which is responsible for the disease, has proved to be highly adaptable in developing resistance to all of the current range of drugs, such as chloroquine **43** and this resistance is spreading worldwide [121–124].

It is clear that these problems of (multiple) resistance are a serious concern. Although the discovery of a new antimalarial that would not eventually succumb to resistance would seem an impossible goal, new strategies based on the chemistry of bioorganometallics have been developed, which may help to hold the line until a vaccine becomes available [125].

Many ferrocene complexes derived from artemisinine **46** [126], quinine **47** and mefloquine **48** [127], chalcones **49** [128], benzylimidazolium **50** [129] and sugars **51** [130] have recently been prepared with an eye to their antimalarial properties (Scheme 3.19).

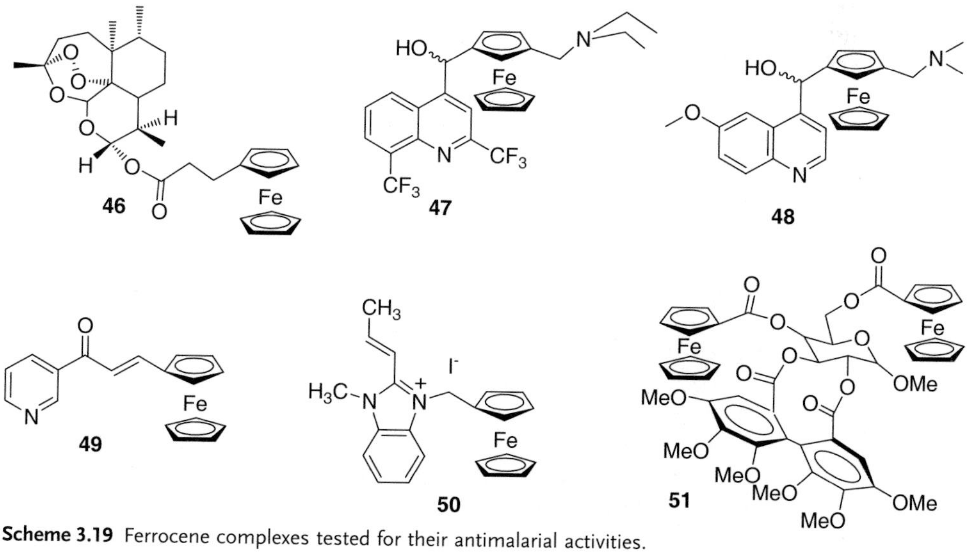

Scheme 3.19 Ferrocene complexes tested for their antimalarial activities.

This is in the chloroquine series that the advantages of the organometallic labeling have been by far the most fully explored [120, 131]. Sanchez-Delgado et al., for example, have obtained the rhodium complex **44** by direct complexation of the chloroquine (Scheme 3.19) [131]. This complex has an efficacy *in vitro* comparable to that of chloroquine (IC_{50} = 72 and 73 nM respectively) but it is chiefly in the ferrocene series that the results are most spectacular.

3.5.2
Ferroquine: a Bioorganometallic Approach

Earlier results obtained with the ferrocifens [81] led Brocard et al. to apply the same concept to the chloroquine series, to produce an analog, ferroquine **45**, a ferrocenic structure close to that of chloroquine **43** [132, 133].

The synthesis of ferroquine is cheap and simple, an issue of major importance if it is to be considered for development as a treatment for use in the impoverished countries of the third world. The synthesis is shown in Scheme 3.20.

Ferrocenylmethylamine **52**, which is commercially available, gives 2-dimethyl-aminomethyl formyl ferrocene by metallation and formylation. This aldehyde **53** is converted to an oxime, then reduced to a primary amine **54**. An S_NAr reaction between this amine and 4,7-dichloroquinoline yields the expected ferroquine **45** in the form of a yellow solid. Finally, the free base **45** is converted by L(+) tartaric acid or HCl to the salt **55** which is water-soluble and thus more convenient for use.

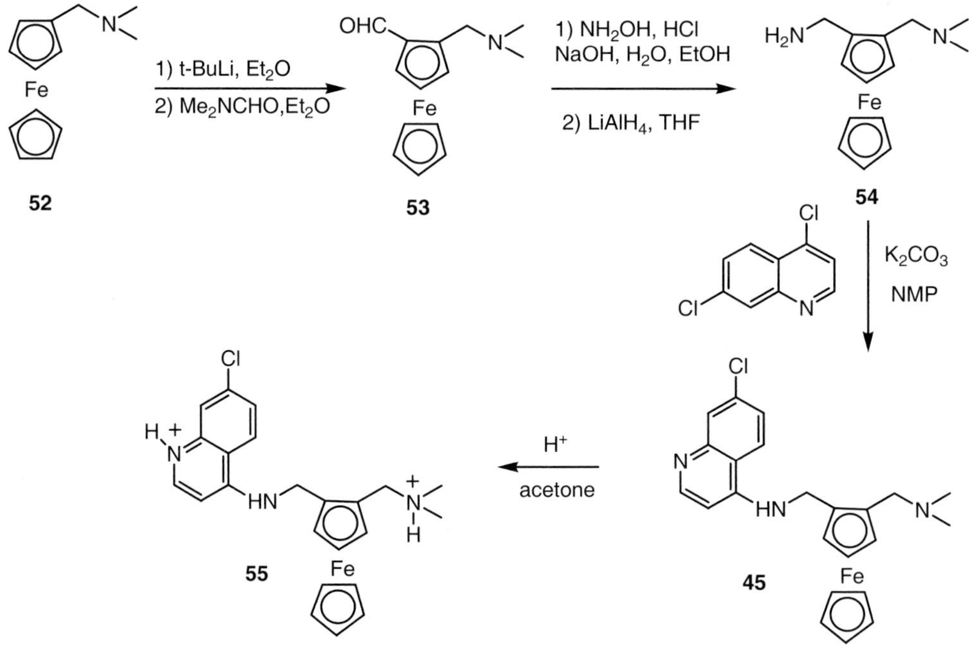

Scheme 3.20 Synthesis of ferroquine **45** and of its salt **55**.

The antimalarial activity of ferroquine, as compared to that of chloroquine, has been studied *in vitro* on the chloroquine-sensitive *P. falciparum* strain, HB3 5CQS, and in this case ferroquine and chloroquine were shown to have a comparable level of activity. However, a different situation is observed when the activity of the two compounds is compared on chloroquine-resistant strains such as Dd2. In this case IC_{50} values of 22 nM for ferroquine as against 130 nM for chloroquine are obtained [134]. This interesting effect of the ferrocenes is also observed on other isolates of *P. falciparum*, originating in Gabon [135, 136] and Senegal [137]. *In vivo*, ferroquine administered to mice for 4 days at a dose of 8.4 mg Kg^{-1} protects them from fatal infection [138].

It is worth noting that ferroquine possesses planar chirality due to the non-symmetrical 1,2 substitution of the ferrocene entity. The pure enantiomers (+)45 and (−)45 were obtained by enzymatic resolution using the lipase from *Candida rugosa* as a biocatalyst, in a key step [134] (Scheme 3.21). The enantiomeric purity levels obtained were greater than 98%.

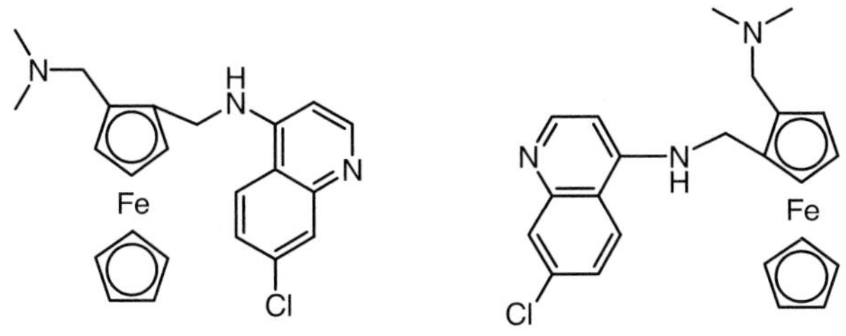

Scheme 3.21 the two enantiomers of ferroquine **45.**

The two optical isomers display identical activity *in vitro* at the nanomolecular level. *In vivo*, however, either of the enantiomers alone is less active than the racemic mixture against both chloroquine-sensitive and chloroquine-resistant strains. In addition, (+)45 displays better curative effects than (−)45, suggesting different pharmacokinetic properties.

The reasons for the enhanced behaviour of racemic ferroquine have not yet been elucidated. It is still not clear whether perhaps compound 45 is oxidized by the parasite to give the ferricinium ion, thus initiating Fenton-type reactivity. There is a clear need for further information on the redox properties of 45 in the food vacuole of the parasite.

In order to study the distribution of ferroquine, ruthenoquine 56, an analog of ferroquine was prepared, in which the iron is replaced by ruthenium, a well-known contrasting agent in electron microscopy (Scheme 3.22) [139]. Chemically, ruthenoquine is similar to ferroquine and possesses similar antimalarial activity. In infected mice treated with ruthenoquine, Ru atoms have been detected in the food vacuole, not only close to the malarial pigment but also in the membrane of the parasite. This clear difference from the case of chloroquine, which is never found in the membrane, could account for the difference in activity.

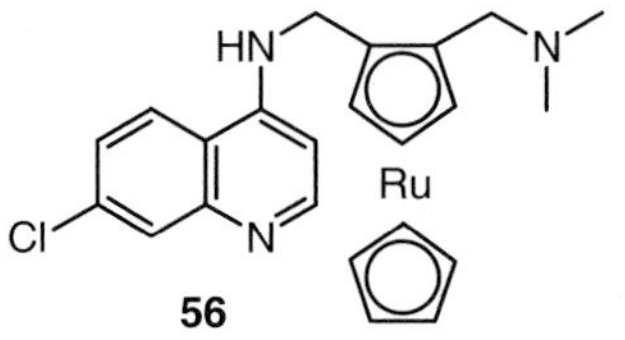

Scheme 3.22 Ruthenocenyl derivative of chloroquine **56**.

3.5.3
Conclusion for Ferroquine

The specific properties of ferroquine, certain of which encompass the structure–function relationship proposed for chloroquine [140] are the following. (1) Maintenance of a 4-aminoquinoline entity for vacuolar accumulation by pH trapping and haematin association; (2) A side chain comprising a basic amino group, to provide strong antiplasmodial activity; (3) A ferrocenyl group for its lipophilic (log Po/w = 3.3) and redox (ferrocene/ferricinium E_0 = +460 mV) properties (4) A rapid and cheap synthetic method, involving few steps.

Based on these principles, several analogs of ferroquine have been developed [120, 141, 142], one of which **57**, is shown in Scheme 3.23. This is a ferrocene triazacyclononane conjugate active against the chloroquine-resistant *P. falciparum* strain Dd2 [143], which gives a good general indication of the approach.

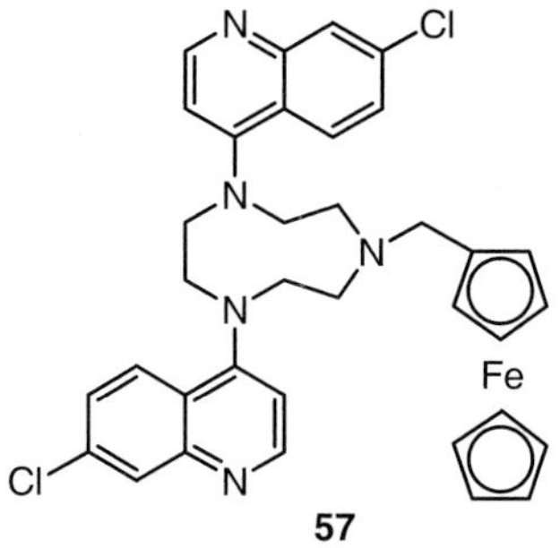

Scheme 3.23 Ferrocenyl derivative of triazacyclononane **57**.

3.6
Other Examples of Organometallics Complexes Tested for their Biological Activities

The ferrocifens and the ferroquines are currently the most advanced examples of targeted organometallics for therapeutic applications. Ferroquine is already in phase I clinical development by Sanofi-Aventis, and the ferrocifens are in preclinical studies. These successes represent a solid scientific base that gives credence to a number of recent studies on the modification by organometallics of skeletons with known biological properties. To name only those organometallic

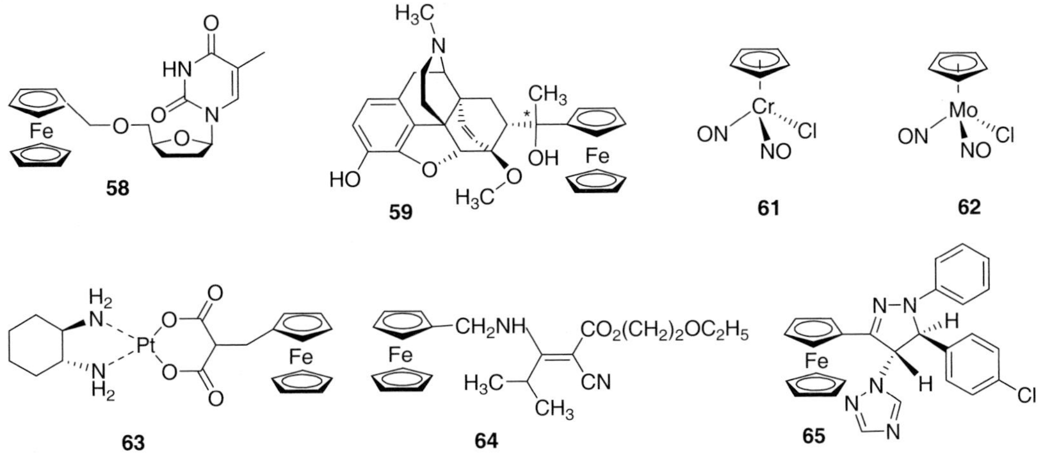

Scheme 3.24 Examples of organometallic complexes with established biological activities.

complexes whose biological properties have been studied, there are, in addition to many examples from the antimalarial family already noted (Scheme 3.19), organometallic complexes derived from AZT **58** [144], from alkaloids **59** [145], from vasodilatators **61** and **62** [146], from the anticancer drug oxaliplatin **63** [147], from dihydropyrazol-derived antibacterial agents **65** [148], and from cyanoacrylate-derived herbicides **64** [149] (Scheme 3. 24).

The biological effects observed for these complexes are at best equivalent to the organic starting molecule. From this it is apparent that merely delivering an organometallic entity to a biological target is not sufficient in itself to obtain a beneficial effect or an improvement over the initial product. It is important to use a mechanistic approach in examining the behavior of the effective entities.

3.7
Conclusions

The following example illustrates the importance of the positioning of the organometallic group in order to obtain antiproliferative effects.

The organic diphenols **66** (diethylstilbestrol) and **67** (Scheme 3.25) are estrogens that act via the specific estrogen receptor (ER). In an attempt to endow them with a cytoxic effect the ferrocene diphenols **68** and **69** were prepared [150].

Surprisingly, these complexes, which differ only in the relative position of the two phenol groups, have totally different behavior *in vitro* on cell lines derived from breast cancers. While complex **68** shows a very marked antiproliferative effect on both hormone-dependent MCF7 ($IC_{50} = 0.7\ \mu M$) and hormone-independent MDA-MB231 cells ($IC_{50} = 0.6\ \mu M$), complex **69** demonstrates an estrogenic effect on MCF7 cells and a weak antiproliferative effect on MDA-MB231 cells (Fig. 3.6 A and B).

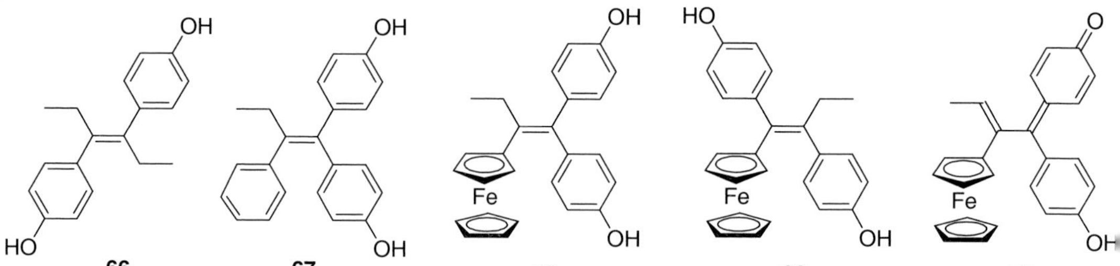

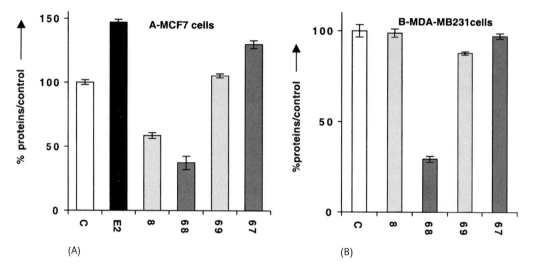

Scheme 3.25 Some organic diphenols and their corresponding ferrocenyl complexes and the ferrocenyl quinone methide **70**.

Fig. 3.6 Effect of 1 μM, OH-tamoxifen (**8**), the ferrocenyl diphenols **68** and **69**, the organic diphenol **67** and of 10 nM of 17β-estradiol (E$_2$) compared to the control (C) on (A) MCF7 cells (B) and MDA-MB231 after 5 days of culture [150].

The structural design of **68** seems to be the key to the problem. In fact, an electrochemical study has shown that the biological effect seems triggered by the reversible oxidation of the ferrocene entity, which could then be followed by a premature transformation of the phenol via generation of an intermediate carbenium ion, leading to a fairly stable quinone methide **70** [94]. It is well documented that electrophilic species such as quinone methides have the potential to alkylate cellular macromolecules to produce a cytotoxic effect [151–153]. Complex **69**, however, which is isolated in the form of the stable *trans* isomer, cannot produce a stabilized quinone methide. It is thus clear that for targeted organometallic products such as **68** and **69**, structural considerations come to the fore.

Indeed, other aspects of organometallic targeting for therapeutic application are rapidly emerging. A particularly promising example is the role of certain metal

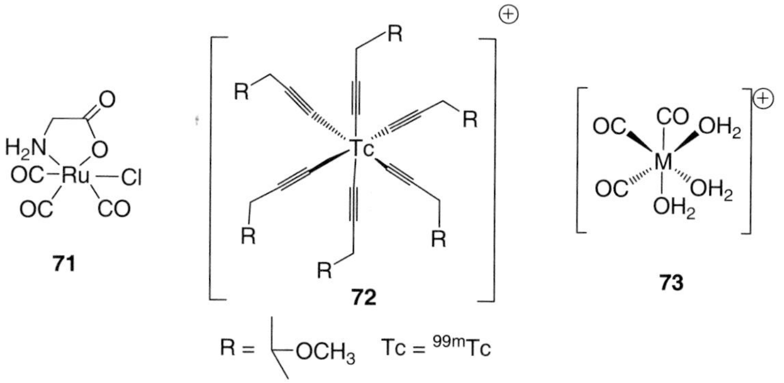

Scheme 3.26 Ru(CO)$_3$Cl(glycinate) **71**, ^{99m}Tc sestamibi (cardiolite®) **72**, Alberto's reagent **73**.

carbonyls in the targeted biological release of CO [154]. It is known that carbon monoxide, which is generated in mammals during the degradation of heme by the enzyme heme oxygenase, is an important signaling mediator, to a certain extent similar to the messenger molecule NO, despite its high toxicity [155, 156]. The use of Fe(CO)$_5$, Mn$_2$(CO)$_{10}$, and [RuCl$_2$(CO)$_3$]$_2$ to act as CO releasing molecules has been reported [151, 152]. In the case of Fe(CO)$_5$ and Mn$_2$(CO)$_{10}$, photolysis was necessary to release CO, but in the case of [RuCl$_2$(CO)$_3$]$_2$, CO was released directly to myoglobin (Mb) to form CO-Mb. Among the Ru complexes examined so far, Ru (CO)$_3$Cl(glycinate) **71** (Scheme 3.26) was shown to suppress organ graft rejection and protect tissues from ischemic injury and apoptosis [154]. This approach, paired with targeting by M-NO complexes [157, 158], is certainly destined for development over the next few years.

In the area of organometallic radiopharmaceuticals, the idea of selective targeting is again a source of new advances. The importance of cardiolite® **72** is well known in imaging of the cardiac muscle [159] and a recent study suggests that this could be extended to molecular breast imaging [160]. Alberto's reagent [(H$_2$O)$_3$M(CO)$_3$]$^+$ **73** which, at low concentrations and in water, allows preparation of biomolecules selectively labeled with organometallics of ^{99m}Tc et ^{188}Re tricarbonyl, is another major breakthrough in this area [161]. This work is reviewed in Chapter 4 of the present work. The number of targets examined is rapidly increasing [162–164]. One example is the folate receptor (FR), a high-affinity membrane protein, overexpressed in many tumor cells but underexpressed in healthy tissue, that has recently benefited from a radiopharmaceutical approach termed a Trojan horse [162]. The near future should see a multiplicity of such examples. The future appears bright and depends essentially on the imagination of the chemist.

The constraints and requirements (aqueous solvents, relative stability, rapid reaction time for radionucleides, selectivity, biological recognition, etc.) inherent in the synthesis of organometallic molecules designed for biological targeting should not be viewed as an inhibitory factor in the development of the discipline, but instead as a spur to creativity. The chemist must often be prepared to leave the beaten track and explore new synthetic routes to the desired destination [161,

165]. The example of the derivatives of tamoxifen complexed by titanocene dichloride [72] or $[(H_2O)_3{}^{99m}Tc(CO)_3]^+$ is particularly enlightening in this regard. This type of research thus also contributes to the sum of synthetic possibilities for organometallic chemistry as a whole.

Acknowledgments

We wish to thank Barbara McGlinchey for her linguistic talent and our co-workers named in the cited papers. Michael J. McGlinchey (University College of Dublin, Ireland), Domenico Osella (Universita del Piemonte Orientale, Alessandria, Italy), Guy Leclercq (Institut Jules Bordet, Brussels, Belgium), Roger Alberto (University of Zurich, Switzerland) and Patricia de Cremoux (Institut Curie, Paris) are acknowledged for fruitful discussions and collaborations. G. J. wants to thank the "Institut Universitaire de France", Bertrand Chastenet (4M), Ipsen and the COST D20 (Metal compounds in the treatment of cancer and viral diseases)for financial support.

References

1 M. J. CLARKE, F. ZHU, D. R. FRASCA, *Chem. Rev.* **1999**, *99*, 2511–2533.

2a M. GUO, P. J. SADLER, *Angew. Chem. Int. Ed. Engl.* **1999**, *38*, 1513–1531.

2b Z. GUO, P. J. SADLER, *Adv. Inorg. Chem.* **2000**, *49*, 183–305.

3 H. T. CHIFOTIDES, K. R. DUNBAR, *Acc. Chem. Res.* **2005**, *38*, 146–156.

4 R. H. HOLM, E. I. SOLOMON, *Chem. Rev.* **1996**, *96* (7), Issue dedicated to Bioinorganic Enzymology.

5 H. E. HOWARD-LOCK, C. J. L. LOCK, in G. WILKINSON, R. GILLARD, J. MCCLEVERTY (Eds.), *Comprehensive Coordination Chemistry*, 755–778, Pergamon, Oxford, **1987**.

6 B. K. KEPPLER, in *Metal Complexes in Cancer Chemotherapy*, VCH, Weinheim, **1993**.

7 B. LIPPERT, in *Cisplatin: Chemistry and Biochemistry of a Leading Anticancer Drug*, John Wiley and Sons, New York, **1999**.

8 C. ORVIG, M. J. ABRAMS, *Chem. Rev.* **1999**, *99* (5), Issue dedicated to Medicinal Inorganic Chemistry.

9 P. J. SADLER, *Adv. Inorg. Chem.* **1991**, *36*, 1–48.

10 L. J. MING, *Med. Res. Rev.* **2003**, *23*, 697–762 (for a recent review on inorganic metalloantibiotics).

11 J. MANN, in *The Elusive Magic Bullet: the search for the perfect drug*, Oxford University Press, Oxford, **1999**.

12 B. ROSENBERG, L. VANCAMP, *Cancer Res.* **1970**, *30*, 1799–1802.

13 B. ROSENBERG, L. VANCAMP, T. KRIGAS, *Nature* **1965**, *205*, 698–699.

14 W. BECK, K. SEVERIN, *Chemie in unserer Zeit* **2002**, *6*, 356–365.

15 C. ELSCHENBROICH, A. SALZER, in *Organometallics: a concise introduction*, VCH, Weinheim, **1989**.

16 R. H. FISH, G. JAOUEN, *Organometallics* **2003**, *22*, 2166–2177.

17 J. AUBRECHT, R. K. NARLA, P. GHOSH, J. STANEK, F. M. UCKUN, *Organometallics* **1999**, *154*, 228–235.

18 G. CALDWELL, M. G. MEIRIM, E. W. NEUSE, C. E. J. VAN RENSBURG, *Appl. Organomet. Chem.* **1998**, *12*, 793–799.

19 C. V. CHRISTODOULOU, A. G. ELIOPOULOS, L. S. YOUNG, L. HODGKINS, D. R. FERRY, D. J. KERR, *Br. J. Cancer* **1998**, *77*, 2088–2097.

20 P. GHOSH, O. J. D'CRUZ, R. K. NASLA, F. M. UCKUN, *Clin. Cancer Res.* **2000**, *6*, 1536–1545.

21 M. M. HARDING, M. PRODIGALIDAD, M. J. LYNCH, *J. Med. Chem.* **1996**, *39*, 5012–5016.

22 M. M. HARDING, G. MOKDSI, *Curr. Med. Chem.* **2000**, *7*, 1289–1303.

23 P. KÖPF-MAIER, *Eur. J. Clin. Pharmacol.* **1994**, *47*, 1–16.

24 P. KÖPF-MAIER, T. KLAPOTKE, *Cancer Chemother. Pharmacol.* **1992**, *29*, 361–366.

25 P. KÖPF-MAIER, H. KÖPF, E. W. NEUSE, *Angew. Chem. Int. Ed. Engl.* **1984**, *23*, 456–457.

26 E. W. NEUSE, F. KANZAWA, *Appl. Organomet. Chem.* **1990**, *4*, 19–26.

27 J. H. TONEY, M. S. MURTHY, T. J. MARKS, *Chem. Biol. Interact.* **1985**, *56*, 45–54.

28 J. B. WAERN, M. M. HARDING, *J. Organomet. Chem.* **2004**, *689*, 4655–4668.

29 N. KRÖGER, U. R. KLEEBERG, K. MROSS, L. EDLER, D. K. HOSSFELD, *Onkologie* **2000**, *23*, 60–62.

30 K. MROSS, P. ROBBEN-BATHE, L. EDLER, J. BAUMGART, W. E. BERDEL, H. FIEBIG, C. UNGER, *Onkologie* **2000**, *23*, 576–579.

31 R. E. AIRD, J. CUMMINGS, A. A. RITCHIE, M. MUIR, R. E. MORRIS, H. CHEN, P. J. SADLER, D. I. JODRELL, *Br. J. Cancer* **2002**, *86*, 1652–1657.

32 D. N. AKBAYEVA, L. GONSALVI, W. OBERHAUSER, M. PERUZZINI, F. VIZZA, P. BRUEGGELLER, A. ROMEROSA, G. SAVA, A. BERGAMO, *Chem. Commun.* **2003**, 264–265.

33 C. S. ALLARDYCE, P. J. DYSON, D. J. ELLIO, P. A. SALTER, R. SCOPELLITI, *J. Organomet. Chem.* **2003**, *668*, 35–42.

34 C. S. ALLARDYCE, P. J. DYSON, E. D. J. S. L. HEATH, *Chem. Commun.* **2001**, 1396–1397.

35 L. D. DALE, J. H. TOCHER, T. M. DYSON, D. I. EDWARDS, D. A. TOCHER, *Anti-Cancer Drug Des.* **1992**, *7*, 3–14.

36 C. S. ALLARDYCE, A. DORCIER, C. SCOLARO, P. J. DYSON, *Appl. Organomet. Chem.* **2005**, *19*, 1–10.

37 K. E. DOMBROWSKI, W. BALDWIN, J. E. SHEATS, *J. Organomet. Chem.* **1986**, *302*, 281–306 (for a review covering all the initial period).

38 E. I. EDWARDS, R. EPTON, G. MARR, *J. Organomet. Chem.* **1979**, *168*, 259–272.

39 E. I. EDWARDS, R. EPTON, G. MARR, *J. Organomet. Chem.* **1976**, *122*, C49–C53.

40 E. I. EDWARDS, R. EPTON, G. MARR, *J. Organomet. Chem.* **1975**, *85*, C23–C25.

41 E. I. EDWARDS, R. EPTON, G. MARR, *J. Organomet. Chem.* **1976**, *107*, 351–357.

42 A. N. NESMEYANOV, L. G. BOGOMOLOVA, V. VILTCHEVSKAYA, N. PALITSYNE, I. ANDRIANOVA, O. BELOZEROVA, US Patent, 119 356, **1971**.

43 L. L. GERSHBEIN, *Res. Commun. Chem. Pathol. Pharmacol.* **1980**, *27*, 139–145.

44 R. W. MASON, K. MCGROUTHER, P. R. R. RANATUNGE-BANDARAGE, B. H. ROBINSON, J. SIMPSON, *Appl. Organomet. Chem.* **1999**, *13*, 163–173.

45 R. A. YEARY, *Toxicol. Appl. Pharmacol.* **1969**, *15*, 666–676.

46 J. T. YARRINGTON, K. W. HUFFMAN, G. A. LEESON, D. J. SPRINKLE, D. E. LOUDY, C. HAMPTON, G. J. WRIGHT, J. P. GIBSON, *Fundam. Appl. Toxicol.* **1983**, *3*, 86–94.

47 N. P. BUU-HOI, D. P. HIEN, H. T. HIEU, *C. R. Acad. Sci. Paris, ser. D* **1970**, *270*, 217–219.

48 D. SCUTARU, I. MAZILU, L. TATARU, M. VATA, T. LIXANDRU, *J. Organomet. Chem.* **1991**, *406*, 183–187.

49 D. SCUTARU, I. MAZILU, M. VATA, L. TATARU, A. VLASE, T. LIXANDRU, C. SIMIONESCU, *J. Organomet. Chem.* **1991**, *401*, 87–90.

50 D. SCUTARU, L. TATARU, I. MAZILU, E. DIACONU, T. LIXANDRU, C. SIMIONESCU, *J. Organomet. Chem.* **1991**, *401*, 81–85.

51 A. H. CHARPENTIER, A. K. BEDNAREK, R. L. DANIEL, K. A. HAWKINS, L. J. S. GADDIS, M. C. MACLEOD, C. M. ALDAZ, *Cancer Res.* **2000**, *60*, 5977–5983.

52 C. K. OSBORNE, H. ZHAO, S. A. W. FUQUA, *J. Clin. Oncol.* **2000**, *18*, 3172–3186.

53 V. C. JORDAN, *J. Med. Chem.* **2003**, *46*, 883–908.

54 E. A. VLADUSIC, A. E. HORNBY, F. K. GUERRA-VLADUSIC, R. LUPU, *Cancer Res.* **1998**, *58*, 210–214.

55 C. PALMIERI, G. J. CHENG, S. SAJI, M. ZELADA-HEDMAN, A. WÄRRI, Z. WEIHUA, S. VAN NOORDEN, T. WAHLSTROM, R. C. COOMBES, M. WARNER, J. A. GUSTAFSSON, *Endocr. Relat. Cancer* **2002**, *9*, 1–3.

56 J. Raus, H. Martens, G. Leclercq, in *Cytotoxic estrogens in hormone receptive tumors*, Academic Press, London, **1980**.

57 C. Chesne, G. Leclercq, P. Pointeau, H. Patin, *Eur. J. Med. Chem.* **1986**, *21*, 321–327.

58 R. Gust, H. Schönenberger, U. Klement, K. J. Range, *Arch. Pharm. (Weinheim)* **1993**, *326*, 967–976.

59 J. Karl, R. Gust, T. Spruss, M. R. Schneider, H. Schönenberger, J. Engel, K. H. Wrobel, F. Lux, S. Trebert Haeberlin, *J. Med. Chem.* **1988**, *31*, 72–83.

60 O. Gandolfi, J. Blum, F. Mandelbaum-Shavit, *Inorg. Chim. Acta* **1984**, *91*, 257–261.

61 E. von Angerer, Platinum complexes with specific activity against hormone dependent tumors, in B. K. Keppler (Ed.), *Metal Complexes in Cancer Chemotherapy*, 73–83, VCH, Weinheim (Germany), **1993**.

62 A. Jackson, J. Davis, R. J. Pither, A. Rodger, M. J. Hannon, *Inorg. Chem.* **2001**, *40*, 3964–3973.

63 G. Grenier, G. Bérubé, C. Gicquaud, *Chem. Pharm. Bull.* **1998**, *46*, 1480–1483.

64 S. G. Chaney, *J. Oncology* **1995**, *6*, 1291–1305.

65 Y. Kidani, K. Inagaki, M. Igo, A. Hashi, K. Kuretani, *J. Med. Chem.* **1978**, *21*, 1315–1318.

66 S. Top, E. B. Kaloun, A. Vessières, G. Leclercq, I. Laïos, M. Ourevitch, C. Deuschel, M. J. McGlinchey, G. Jaouen, *ChemBioChem* **2003**, *4*, 754–761.

67 A. H. Soloway, W. Tjarks, B. A. F. G. Rong, R. F. Barth, I. M. Codogni, J. G. Wilson, *Chem. Rev.* **1998**, *98*, 1515–1562.

68 Y. Endo, T. Iijima, Y. Yamakoshi, M. Yamaguchi, H. Fukasawa, K. Shudo, *J. Med. Chem.* **1999**, *42*, 1501–1504.

69 Y. Endo, T. Iijima, Y. Yamakoshi, H. Fukasawa, C. Miyaura, M. Inada, A. Kubo, A. Itai, *Chem. Biol.* **2001**, *8*, 341–355.

70 J. F. Valliant, P. Schaffer, K. A. Stephenson, J. F. Britten, *J. Org. Chem.* **2002**, *67*, 383–387.

71 A. Vessières, unpublished results.

72 S. Top, E. B. Kaloun, A. Vessières, I. Laïos, G. Leclercq, G. Jaouen, *J. Organomet. Chem.* **2002**, 643–644, 350–356.

73 G. Mokdsi, M. M. Harding, *J. Organomet. Chem.* **1998**, *565*, 29–35.

74 J. H. Toney, T. J. Marks, *J. Am. Chem. Soc.* **1985**, *107*, 947–953.

75 M. Guo, H. Sun, H. J. McArdle, L. K. Gambling, P. J. Sadler, *Biochemistry* **2000**, *39*, 10023–10033.

76 A. Stoica, B. S. Katzenellenbogen, M. B. Martin, *Mol. Endo.* **2000**, *14*, 545–553.

77 M. D. Johnson, N. Kenney, A. Stoica, L. Hilakivi-Clarke, B. Singh, G. Chepko, R. Clarke, P. F. Sholler, A. A. Lirio, C. Foss, R. Reiter, B. Trock, S. Paik, M. B. Martin, *Nat. Med.* **2003**, *9*, 1081–1084.

78 M. B. Martin, R. Reiter, T. Pham, Y. R. Avellanet, J. Camara, M. Lahm, E. Pentecost, K. Pratap, B. A. Gilmore, S. Divekar, R. S. Dagata, J. L. Bull, A. Stoica, *Endocrinology* **2003**, *144*, 2425–2436.

79 S. Top, A. Vessières, P. Pigeon, M. N. Rager, M. Huché, E. Salomon, C. Cabestaing, J. Vaissermann, G. Jaouen, *ChemBioChem* **2004**, *5*, 1104–1113.

80 G. Jaouen, S. Top, A. Vessières, P. Pigeon, G. Leclercq, I. Laïos, *Chem. Commun.* **2001**, 383–384.

81 S. Top, J. Tang, A. Vessières, D. Carrez, C. Provot, G. Jaouen, *Chem. Commun.* **1996**, 955–956.

82 S. Top, B. Dauer, J. Vaissermann, G. Jaouen, *J. Organomet. Chem.* **1997**, *541*, 355–361.

83 G. Jaouen, S. Top, A. Vessières, G. Leclercq, J. Quivy, L. Jin, A. Croisy, *C. R. Acad. Sci. Paris*, **2000**, Série IIc, 89–93.

84 S. Top, A. Vessières, C. Cabestaing, I. Laios, G. Leclercq, C. Provot, G. Jaouen, *J. Organomet. Chem.* **2001**, *637*, 500–506.

85 S. Top, A. Vessières, G. Leclercq, J. Quivy, J. Tang, J. Vaissermann, M. Huché, G. Jaouen, *Chem. Eur. J.* **2003**, *9*, 5223–5236.

86 D. Osella, M. Ferrali, P. Zanello, F. Laschi, M. Fontani, C. Nervi, G. Cavigiolio, *Inorg. Chim. Acta* **2000**, *306*, 42–48.

87 G. Tabbi, C. Cassino, G. Cavigiolio, D. Colangelo, A. Ghiglia, I. Viano, D. Osella, *J. Med. Chem.* **2002**, *45*, 5786–5796.

88 M. M. Montano, A. K. Jaiswal, B. S. Katzenellenbogen, *J. Biol. Chem.* **1998**, *273*, 25443–25449.

89 S. H. Yang, R. Liu, E. J. Perez, Y. Wen, S. M. Stevens, T. Valencia, A. M. Brun-Zinkernagel, L. Prokai, Y. Will, J. Dykens, P. Koulen, J. W. Simpkins, *Proc. Natl. Acad. Sci. USA* **2004**, *101*, 4130–4135.

90 A. K. Shiau, D. Barstad, P. M. Loria, L. Cheng, P. J. Kushner, D. A. Agard, G. L. Greene, *Cell* **1998**, *95*, 927–937.

91 Mac Spartam Pro, Wavefunction Co., Irvine CA 92612, USA.

92 J. M. Cense, *Phys. Theor. Chem.* **1990**, *71*, 763–766.

93 P. Pigeon, S. Top, A. Vessières, M. Huché, E. A. Hillard, E. Salomon, G. Jaouen, *J. Med. Chem.* **2005**, *48*, 2814–2821.

94 C. Amatore, E. A. Hillard, G. Jaouen, L. Thouin, A. Vessières, Manuscript in preparation.

95 G. Jaouen, S. Top, A. Vessières, G. Leclercq, M. J. McGlinchey, *Curr. Med. Chem.* **2004**, *11*, 2505–2517.

96 H. Greenfield, H. W. Sternberg, R. A. Friedel, J. H. Wotiz, R. Markby, I. Wender, *J. Am. Chem. Soc.* **1956**, *78*, 120–124.

97 N. A. Sasaki, P. Potier, M. Savignac, G. Jaouen, *Tetrahedron Lett.* **1988**, *29*, 5759–5762.

98 M. Savignac, N. A. Sasaki, P. Potier, G. Jaouen, *J. Chem. Soc. Chem. Commun.* **1991**, 615–617.

99 A. Varenne, M. Salmain, C. Brisson, G. Jaouen, *Bioconjugate Chem.* **1992**, *3*, 471–476.

100 A. Vessières, M. Salmain, P. Brossier, G. Jaouen, *J. Pharm. Biomed. Anal.* **1999**, *21*, 625–633.

101 A. Vessières, S. Tondu, G. Jaouen, S. Top, A. A. Ismail, G. Teutsch, M. Moguilewsky, *Inorg. Chem.* **1988**, *27*, 1850–1852.

102 V. Philomin, A. Vessières, M. Gruselle, G. Jaouen, *Bioconjugate Chem.* **1993**, *4*, 419–424.

103 M. Salmain, G. Jaouen, *J. Organomet. Chem.* **1992**, *445*, 237–243.

104 M. Gruselle, J. L. Rossignol, A. Vessières, G. Jaouen, *J. Organomet. Chem.* **1987**, *328*, C12–C15.

105 M. Gruselle, P. Deprez, A. Vessières, S. Greenfield, G. Jaouen, J. P. Larue, D. Thouvenot, *J. Organomet. Chem.* **1989**, *359*, C53–C56.

106 D. Osella, G. Cavigiolio, M. Vincenti, A. Vessières, I. Laïos, G. Leclercq, E. Napolitano, R. Fiaschi, G. Jaouen, *J. Organomet. Chem.* **2000**, *596*, 242–247.

107 G. Palyi, G. Varadi, A. Viziorosz, L. Marko, *J. Organomet. Chem.* **1975**, *90*, 85–91k.

108 A. Vessières, G. Jaouen, M. Gruselle, J. L. Rossignol, M. Savignac, S. Top, S. Greenfield, *J. Steroid Biochem.* **1988**, *30*, 301–306.

109 A. Vessières, S. Top, C. Vaillant, D. Osella, J. P. Mornon, G. Jaouen, *Angew. Chem. Int. Ed. Engl.* **1992**, *31*, 753–755.

110 A. Vessières, C. Vaillant, M. Salmain, G. Jaouen, *J. Steroid Biochem.* **1989**, *34*, 301–305.

111 N. Fischer-Durand, A. Vessières, J. M. Heldt, F. Le Bideau, G. Jaouen, *J. Organomet. Chem.* **2003**, *668*, 59–66.

112 M. Salmain, A. Vessières, G. Jaouen, I. S. Butler, *Anal. Chem.* **1991**, *63*, 2323–2329.

113 A. Vessières, C. Vaillant, M. Gruselle, D. Vichard, G. Jaouen, *J. Chem. Soc. Chem. Commun.* **1990**, 837–839.

114 K. W. Harlow, D. N. Smith, J. A. Katzenellenbogen, G. L. Greene, B. S. Katzenellenbogen, *J. Biol. Chem.* **1989**, *264*, 17476–17485.

115 S. Top, H. El Hafa, A. Vessières, M. Huché, J. Vaissermann, G. Jaouen, *Chem. Eur. J.* **2002**, *8*, 5241–5249.

116 M. Jung, D. E. Kerr, P. D. Senter, *Arch. Pharm. Pharm. Med. Chem.* **1997**, *330*, 173–176.

117 R. Gust, I. Ott, P. D. K. Sommer, *J. Med. Chem.* **2004**, *47*, 5837–5846.

118 I. Ott, B. Kircher, R. Gust, *J. Inorg. Biochem.* **2004**, *98*, 485–489.

119 I. OTT, K. SCHMIDT, B. KIRCHER,
P. SCHUMACHER, T. WIGLENDA,
R. GUST, *J. Med. Chem.* **2005**, *48*,
622–629.

120 C. BIOT, *Curr. Med. Chem. Anti-Infective Agents* **2004**, *3*, 135–147.

121 R. G. RIDLEY, *Curr. Biol.* **1998**, *8*,
R346–349.

122 C. P. SANCHEZ, P. HORROCKS,
M. LANZER, *Cell* **1998**, *92*, 601–602.

123 D. J. SULLIVAN JR., H. MATILE,
R. G. RIDLEY, D. E. GOLDBERG,
J. Biol. Chem. **1998**, *273*, 31103–31107.

124 P. A. WINSTANLEY, *Parasitol. Today*
2000, *16*, 146–153.

125 T. L. RICHIE, A. SAUL, *Nature* **2002**, *415*,
694–701.

126 L. DELHAES, C. BIOT, L. BERRY,
L. MACIEJEWSKI, D. CAMUS, J. BROCARD,
D. DIVE, *Bioorg. Med. Chem.* **2000**, *8*,
2739–2745.

127 C. BIOT, L. DELHAES,
L. A. MACIEJEWSKI, M. MORTUAIRE,
D. CAMUS, D. DIVE, J. S. BROCARD,
EUR. *J. Med. Chem.* **2000**, *35*, 707–714.

128 X. WU, P. WILAIRAT, M. L. GO, *Bioorg.
Med. Chem. Lett.* **2002**, *12*, 2299–2302.

129 J. HOWARTH, K. HANLON, *Tetrahedron
Lett.* **2001**, *42*, 751–754.

130 T. ITOH, S. SHIRAKAMI, N. ISHIDA,
Y. YAMASHITA, T. YOSHIDA, H. S. KIM,
Y. WATAYA, *Bioorg. Med. Chem. Lett.*
2000, *10*, 1657–1659.

131 R. A. SANCHEZ-DELGADO, M. NAVARRO,
H. PÉREZ, J. A. URBINA, *J. Med. Chem.*
1996, *39*, 1095–1099.

132 C. BIOT, G. GLORIAN,
L. A. MACIEJEWSKI, J. S. BROCARD,
J. Med. Chem. **1997**, *40*, 3715–3718.

133 O. DOMARLE, G. BLAMPAIN,
H. AGNANIET, T. NZADIYABI, J. LEBIBI,
J. BROCARD, L. MACIEJEWSKI, C. BIOT,
A. J. GEORGES, P. MILLET, *Antimicrob
Agents Chemother.* **1998**, *42*, 540–544.

134 L. DELHAES, C. BIOT, L. BERRY,
P. DELCOURT, L. A. MACIEJEWSKI,
D. CAMUS, J. S. BROCARD, D. DIVE,
ChemBioChem **2002**, *3*, 418–423.

135 C. ATTEKE, J. M. NDONG, A. AUBOUY,
L. MACIEJEWSKI, J. BROCARD, J. LEBIBI,
P. DELORON, *J. Antimicrob. Chemother.*
2003, *51*, 1021–1024. Epub.

136 B. PRADINES, T. FUSAI, W. DARIES,
V. LALOGE, C. ROGIER, P. MILLET,
E. PANCONI, M. KOMBILA, D. PARZY,
J. Antimicrob. Chemother. **2001**, *48*,
179–184.

137 B. PRADINES, A. TALL, C. ROGIER,
A. SPIEGEL, J. MOSNIER, L. MARRAMA,
T. FUSAI, P. MILLET, E. PANCONI,
J. F. TRAPE, D. PARZY, *Trop. Med. Int.
Health* **2002**, *7*, 265–270.

138 L. DELHAES, H. ABESSOLO, C. BIOT,
L. BERRY, P. DELCOURT,
L. MACIEJEWSKI, J. BROCARD, D. CAMUS,
D. DIVE, *Parasitol. Res.* **2001**, *87*,
239–244.

139 H. ABESSOLO, PhD thesis, University of
Lille1, Villeneuve-d'Ascq, France, **2000**.

140 T. J. EGAN, R. HUNTER, C. H. KASCHULA,
H. M. MARQUES, A. MISPLON,
J. WALDEN, *J. Med. Chem.* **2000**, *43*,
283–291.

141 M. BLACKIE, P. BEAGLEY, K. CHIBALE,
C. CLARKSON, J. R. MOSS, P. J. SMITH,
J. Organomet. Chem. **2003**, *688*,
144–152.

142 K. CHIBALE, J. R. MOSS, M. BLACKIE, V.
S. D. P. J. SMITH, *Tetrahedron Lett.* **2000**,
41, 6231–6235.

143 C. BIOT, J. DESSOLIN, I. RICARD,
D. DIVE, *J. Organomet. Chem.* **2004**, *689*,
4678–4682.

144 M. DE CHAMPDORÉ, G. DI FABIO,
A. MESSERE, D. MONTESARCHIO,
G. PICCIALLI, R. LODDO, M. LA COLLA,
P. LA COLLA, *Tetrahedron* **2004**, *60*,
6555–6563.

145 G. LAUS, J. SCHUTZ,
H. SHOTTENBERGER, M. ANDRE,
K. WORST, M. SPETEA, K. H. ONGANIA,
A. G. MULLER, H. SCHMIDHAMMER,
Helv. Chim. Acta **2003**, *86*, 3274–3280.

146 Y. X. WANG, P. LEGZDINS, J. S. POON,
C. C. PANG, *J. Cardiovasc. Pharmacol.*
2000, *35*, 73–77.

147 A. ROSENFELD, J. BLUM, D. GIBSON,
A. RAMU, *Inorg. Chim. Acta* **1992**, *201*,
219–221.

148 J. FANG, Z. JIN, Z. LI, W. LIU,
J. Organomet. Chem. **2003**, *674*, 1–9.

149 S. HUIKAI, W. QINGMIN, H. RUNQIU,
L. HENG, L. YONGHONG, *J. Organomet.
Chem.* **2002**, *655*, 182–185.

150 A. VESSIÈRES, S. TOP, P. PIGEON,
E. A. HILLARD, L. BOUBEKER, D. SPERA,
G. JAOUEN, *J. Med. Chem.* **2005**, *48*,
3937–3940.

151 P. W. Fan, F. Zhang, J. L. Bolton, *Chem. Res. Toxicol.* **2000**, *13*, 45–52.

152 L. Yu, H. Liu, W. Li, F. Zhang, C. Luckie, R. B. van Breemen, G. R. Thatcher, J. L. Bolton, *Chem. Res. Toxicol.* **2004**, *17*, 879–888.

153 F. Zhang, P. W. Fan, X. Liu, L. Shen, R. B. van Breeman, J. L. Bolton, *Chem. Res. Toxicol.* **2000**, *13*, 53–62.

154 T. R. Johnson, B. E. Mann, J. E. Clark, R. Foresti, C. J. Green, R. Motterlini, *Angew. Chem. Int. Ed.* **2003**, *42*, 3722–3729.

155 R. Motterlini, R. Foresti, C. J. Green, in R. E. Wang (Ed.), *Carbon monoxide and cardiovascular functions*, CRC, Boca Raton, FL, **2002**.

156 L. E. Otterbein, *Antioxid. Redox Signaling* **2002**, *4*, 309–319.

157 D. Rattat, A. P. Schubiger, H. Berke, H. Shmalle, R. Alberto, *Cancer Biother. Radiopharm.* **2001**, *16*, 339–343.

158 D. Rattat, A. Verbruggen, H. Schmalle, H. Berke, R. Alberto, *Tetrahedron Lett.* **2004**, *45*, 4089–4092.

159 L. I. Delmon-Moingeon, D. Piwnica-Worms, A. D. van den Abbeele, B. L. Holman, A. Davison, A. G. Jones, *Cancer Res.* **1990**, *50*, 2198–2202.

160 D. J. Rhodes, M. K. O'Connor, S. W. Phillips, R. L. Smith, D. A. Collins, *Mayo Clin. Proc.* **2005**, *80*, 24–30.

161 G. Jaouen, S. Top, A. Vessières, R. Alberto, *J. Organomet. Chem.* **2000**, *600*, 23–36.

162 C. Müller, C. Dumas, U. Hoffmann, P. A. Schubiger, R. Schibli, *J. Organomet. Chem.* **2004**, *689*, 4712–4721.

163 S. Mundwiler, M. Kundig, K. Ortner, R. Alberto, *Dalton Trans.* **2004**, 1320–1328.

164 D. R. van Staveren, S. Mundwiler, U. Hoffmanns, J. K. Pak, B. Spingler, N. Metzler-Nolte, R. Alberto, *Org. Biomol. Chem.* **2004**, *2*, 2593–2603.

165 S. Top, E. B. Kaloun, G. Jaouen, *J. Am. Chem. Soc.* **2000**, *122*, 736–737.

4
Radiopharmaceuticals

Roger Alberto

4.1
What are Radiopharmaceuticals?

In modern medicine, early diagnosis of socially relevant diseases plays a major role in improving life quality and reducing health care costs to society. Just as important is the detection of curative progress, made after therapeutic treatment of a patient's disease, e.g. the regression of cancer after chemotherapy or the influence of pharmaceuticals on some neurodegenerative diseases such as Alzheimer's or related diseases [1, 2]. On the molecular level, biochemical analytic methods are well established. Since the discovery of X-rays, visualizing methods became more and more important and routine X-ray diagnosis is nowadays complemented with magnetic resonance imaging (MRI), ultrasound and radio-isotope-based scintigraphy. While X-ray and MRI are basically structural methods, positron emission tomography (PET) or single photon emission computed tomography (SPECT) which permit the visualization of organ functions and are, hence, functional diagnostic methods. Although the resolution of MRI has improved strongly during recent years, the contrast agents actually available do not yet allow a relatively good functional diagnosis. Both methods, PET and SPECT, are based on radionuclides and are included in nuclear medicine and/or radio-pharmacy. The compounds used in these disciplines are radioactive and a radio-pharmaceutical is nothing else but the combination of a radioisotope of an element with a new molecule (complexes, perfusion agents) or with a molecule of well known biological behavior (targeting). If the radionuclide is, for example, ^{11}C the structure of the original pharmaceutical remains unchanged if ^{11}C replaces a ^{12}C in the original (lead) structure. For practical reasons it is often not possible to perform this 'substitution' on a reasonable time scale and the ^{11}C is introduced as, for example, a $-CH_3$ group which will (or will not) change the biological behavior of the lead structure to a certain extent. Reviews on this topic have appeared recently and will not be discussed here [3].

As outlined elsewhere, ^{99m}Tc is the radionuclide of choice for SPECT from a economic point of view, but other market factors, such as availability and waste treatment, are also relevant. Furthermore, its decay characteristics are almost ideal for imaging and its other physico-chemical properties, such as half-life time of decay, are very favorable. Actually, most of the applications in nuclear medicine are performed with this radionuclide, despite growing competition from other methods. ^{99m}Tc-based radiopharmaceuticals can be divided roughly into two classes. The first class of compounds which are well established on the market are the so-called perfusion agents: metal complexes that for some reason follow a particular biological pathway and may accumulate in distinct organs, where their function can be visualized. Complexes of this kind are found more or less by chance and are no longer the focus of research interest. A number of such complexes is shown in Scheme 4.1.

The second class of ^{99m}Tc-based radiopharmaceuticals are labeled targeting agents. The central part consists of an organic molecule of natural or artificial origin which binds with high affinity to a specific receptor in an organism [4–6]. These molecules are also called vectors or targeting agents since it is believed that they accumulate at specific receptors on cancer cells. They might be used in the normal therapy of diseases and relevant examples are antibodies or peptides, and also pharmaceuticals. Apoptosis markers, central nervous system (CNS) receptor-binding molecules represent further attractive vectors for diagnostic nuclear medicine. These targeting molecules are the 'active' part of the second generation of radiopharmaceuticals and they are combined with a ^{99m}Tc-containing metal complex by means of a bifunctional chelator; bifunctional because one function is covalent attachment to the vector, while the second function is coordination to ^{99m}Tc. As expected, the introduction of the chelator (and hence the complex)

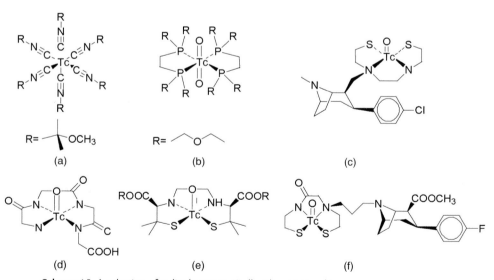

Scheme 4.1 A selection of radiopharmaceutically relevant complexes,
a) Cardiolite$^{®}$ (DuPont, R=CH$_2$C(CH$_3$)$_2$OCH$_3$), b) Myoview$^{®}$ (Nycomed-Amersham, R=CH$_2$CH$_2$OC$_2$H$_5$), c) TRODAT-1, d) TechneScan$^{®}$ MAG3 (Mallinckrodt), e) Tc-ECD Neurolite, f) Variation of TRODAT-1.

significantly changes the behavior of the vectors (mostly in an unfavorable way) and this represents the starting point and the challenge for radiopharmaceutical chemistry in general, and the importance of bioorganometallic chemistry in particular.

Hardly any ^{99m}Tc-based targeting radiopharmaceuticals have been introduced to the market [7]. This is surprising since not only in academic institutions but also in companies, much effort is invested in these developments. Obviously, as in the normal pharmaceutical chemistry, the physiological behavior of compounds has to be optimized, a procedure that is well-known under the term "drug finding and/or development".

In this article, the chemistry of organometallic $^{99(m)}$Tc complexes as relevant for radiopharmaceutical chemistry will be reviewed and discussed. Beside the many possibilities of synthesizing new complexes as such or attached to targeting molecules, the potential for drug development will be the focus of the different topics discussed in more detail.

4.1.1
Radiopharmaceutical Drug Finding and Drug Development

In "normal" drug development, a known lead structure is derivatized with libraries of functional groups to enable a systematic and rational structure–activity relationship. Target interaction (e.g. inhibition of an enzyme) is optimized and the fate of the molecule in the rest of the body is of lower importance, unless side effects occur limiting the therapeutic index of the drug. The situation with radiopharmaceutical chemistry is quite different and makes drug development incomparably more difficult, since biodistribution in the system as an entity is limiting from a biological point of view. In contrast to a normal pharmaceutical, which is ideally biologically inactive at non-target sites, a radiopharmaceutical is active everywhere, as is visible by its radiation. Thus, beside the requirement of high target localization it should be excreted as fast as possible from the rest of the body in order to allow imaging with a high target/non-target ratio and, thus, an accurate diagnosis for the good of the patient. High target accumulation and fast excretion are conflicting factors and a compromise needs to be found.

Limiting organs for high resolution images are the blood pool, kidneys and liver. To optimize the biological behavior of a radiopharmaceutical, the portion essentially represented by the metal complex must be varied. The leading structure, the vector, is in general already optimized but there are also examples where the biomolecule needs to be optimized while the label is kept constant.

Systematic variation of a metal complex is not routine. A particular metal center prefers a particular set of donors in order to be stable under physiological conditions. The label must be stable and biologically inactive or be cleaved from the vector in a predictable pharmacokinetic way, and then rapidly excreted for the sake of good imaging. These conditions point towards organometallic complexes or moieties for radiopharmaceutical application. Radiopharmaceutical drug development is only possible if the ligand shape can be systematically altered while keeping the set of donors constant. The larger the set of potential donors yielding stable complexes, the larger the respective donor combinations and, hence, of ligands which shape can be varied. Again, this points to organometallic complexes since in general they have closed shells and the ligands are therefore difficult to substitute, i.e. they are expected to be stable (or robust). Systematic variation of the ligand sphere with a broad variety of (easily available) ligands would clearly enable the elucidation of structure–activity relationships, an essential basis for any drug development. The proposed procedure of radiopharmaceutical drug development based on organometallic complexes implies a precursor complex that contains a stable set of ligands, preferably those with M–C bonds, and an additional set of ligands, preferably water molecules, that can be exchanged by the bifunctional chelator. Later, we will discuss the complex $[^{99(m)}Tc(OH_2)_3(CO)_3]^+$ which represents a prototype of such a basic structure, but the experienced or intuitive chemist will readily recognize that CO is not the only ligand that could be part of such a basic framework.

A further and probably more relevant issue in radiopharmaceutical drug finding and drug development is the fact that the (radio)pharmaceutical must be synthesized in the clinic immediately prior to its use. This means that the organometallic precursor must be synthesized from saline in high purity and under feasible conditions. Chemists familiar with organometallic compounds are well aware that this task is very difficult since the reaction conditions usually result in different end products and a wild mixture of potentially coordinating

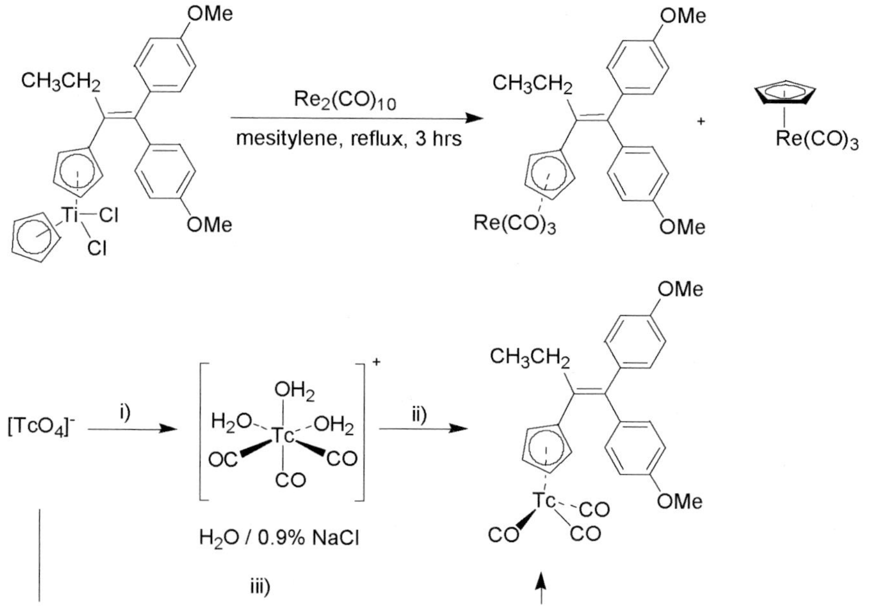

Scheme 4.2 Preparation of a bioorganometallic radiopharmaceutical (above) and a radiopharmaceutical (below). The steps (1) and (2) are in research while step (3) is ideally performed in the routine application.

molecules and water. In a second step (or ideally in one overall reaction) the targeting molecule must then be labeled at the coordinating function of the bifunctional chelator to produce the radiopharmaceutical, again in quantitative yield and of high stability. Of course the procedure must be highly reproducible otherwise the compound will fail to satisfy any drug administration authority. In case of ^{99m}Tc the whole procedure must start from $[^{99m}TcO_4]^-$, the only chemical form routinely available in a hospital.

These requirements make radiopharmaceutical bioorganometallic chemistry different from most of the other topics in this field. Generally, some organometallic molecules such as ferrocene Fc or $[(C_5H_5)Re(CO)_3]$ are attached to a vector for special purposes and there is no need to perform the synthesis in water from inconvenient starting materials. These organometallic complexes obviously must obey the same rules as radiopharmaceuticals but their synthesis is not a limiting and essential point for application in bioorganometallic chemistry. Bioorgano-metallic chemistry with "cold" compounds is still an important factor in radio-pharmaceutical drug development since the biological behavior such as receptor binding reveals leading structures for radiopharmaceuticals. The observation that the cyclopentadienyl ring in $[(C_5H_5)Re(CO)_3]$ mimics a phenyl substituent in a leading structure initiated the research for preparing this class of compounds with ^{99m}Tc [8]. The different issues and requirements for radiopharmaceutical drug development are summarized in Scheme 4.2.

4.1.2
Organometallic Complexes in Radiopharmaceutical Routines

So far the only organometallic ^{99m}Tc complex that has been introduced to the market is $[Tc(CN-R)_6]^+$ (Sestamibi or Cardiolite®, Du Pont) with R = $-CH_2-C(CH_3)_2(OCH_3)$. The ligand is known under the name MIBI or methoxy-isobutyl-methoxy-isonitrile. The complex contains six M–C bonds and the metal center is Tc(I). The pioneering work performed by Davison and coworkers in the early 1980s proved how important it is to overcome chemical paradigms, since he synthesized a low-valent organometallic complex directly from water, in one step and in quantitative yield along a convenient route [9]. Despite the convenience of the approach which was readily transferred to routine production in clinics, the final radiopharmaceutical was only introduced to the market in December 1990, underlining the difficulty of drug development, kit production and registration procedures even with "simple" complexes. The synthesis requires boiling water making clear how stable this complex is. It also shows the importance of kinetic stability. The energy of the Tc–C bond hardly exceeds about 150 kJ mol^{-1}. Normal coordination compounds without a chelate effect would readily decompose since their thermodynamic stability would not be high enough. Drug development consisted here in the variation of the organic residues attached to the isocyanide group in order to optimize uptake in the target organ. The complex belongs to the class of perfusion agents and is nowadays the most widely applied radio-pharmaceutical in myocardial imaging with a market potential of several hundred millions of dollars. A great deal of basic chemistry could also be performed with $[Tc(CN–R)_6]^+$ although the complex is very stable and the ligands diffcult to replace unless oxidation to higher states is included in the synthesis.

Unfortunately, the robust nature of the complex makes its use as a precursor for the labeling of vectors impossible but it exemplifies the importance of the special nature of organometallic complexes for the good of biological applications. Still nowadays, sestamibi remains the only inorganic radiopharmaceutical that has a fixed position in the market and it remains a prototype of how bioorgano-metallic chemistry must advance and be developed.

Although they have not found their way into the market, two additional compounds merit brief mention here, since both are further examples of how drug finding and drug development works and are prototypes for unusual but chemically successful approaches. The first is of the general class of bis-arene-Tc(I) cation $[Tc(C_6H_6)_2]^+$ whose synthesis was attempted to produce a competitor to sestamibi, due to its lipophilic character and its monocationic charge [10]. Reasonably, Wester and coworkers synthesized these complexes with different arenes under conditions not compatible with any future demands for routine application, just in order to test their biological properties. $[^{99m}TcO_4]^-$ was reacted with ultrasound in neat ligand and in the presence of Al° as a reducing agent. This unusual reaction indeed gave the sandwiches in very good yield. Since the compounds did not show the expected *in vivo* behavior (accumulation in the heart) further development stopped. One can learn two important things for bioorganometallic radiopharmaceutical chemistry: firstly, a compound should be synthesized under any conditions and be tested before adapting the reaction conditions to user requirements (e.g. the synthesis of $[Tc(C_6H_6)_2]^+$ from water); and secondly, that the organometallic compound is stable under physiological conditions and therefore represents a potential precursor for further (labeling) chemistry. Despite the water stability of $[Tc(C_6H_6)_2]^+$ no efforts were made to further explore its chemistry. Presumably, one of the arene ligands could be substituted and the hypothetical organometallic aqua-ion $[Tc(C_6H_6)(OH_2)_3]^+$ would represent an extremely attractive target not only for radiopharmacy but for technetium (and rhenium) chemistry as well.

The second complex is a seven-coordinate Tc(III) complex which contains one CO ligand that has been generated *in situ*. Baldas and coworkers in the early 1980s, reacted $[^{99m}TcO_4]^-$ with dithiocarbamates in the presence of formamidine-sulfinic acid. They produced the complex $[Tc(dtc)_3(CO)]$ in which one CO, released from the decomposition of the reducing agent was coordinated [11]. Although this reaction was not further explored either, it exemplifies the possibility of producing ligands leading to organometallic complexes from the decomposition of an organic precursor. As will be shown later in the article, such a possibility is of enormous importance. Most of the ligands used in organometallic chemistry need to be produced in such a way, since most of them are not stable enough to persist under the required reaction conditions. In Davison's preparation of $[Tc(CN-R)_6]^+$ the isocyanide ligand is present in the form of a Cu(I) complex, also a kind of protecting group and *in situ* source for this water-sensitive ligand. Important findings and procedure are summarized in Scheme 4.3.

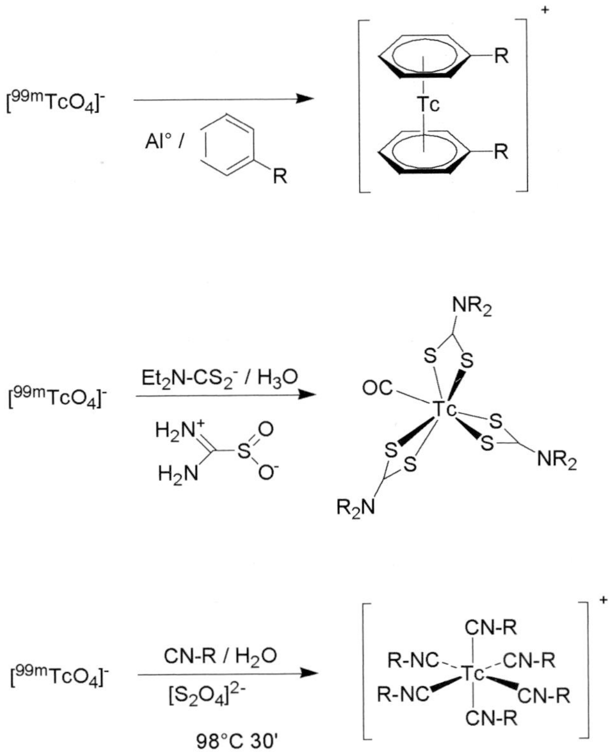

$[^{99m}TcO_4]^-$ $\xrightarrow{\quad Al^\circ\ /\quad}$

$[^{99m}TcO_4]^-$ $\xrightarrow{\quad Et_2N\text{-}CS_2^-\ /\ H_3O\quad}$

$[^{99m}TcO_4]^-$ $\xrightarrow[\quad [S_2O_4]^{2-}\quad]{\quad CN\text{-}R\ /\ H_2O\quad}$

98°C 30'

Scheme 4.3 Relevant organometallic complexes or precursors in organometallic radiopharmacy [9–11].

4.2
Organometallic Aquo-ions

What has been shown so far comprises the synthesis and some properties of bioorganometallic complexes. As described in the introductory section, radio-pharmaceutical chemistry requires precursor complexes that contain ligands which can be replaced by donors from chelators attached to a targeting molecule. The corresponding chemistry takes place in biological media, therefore such re-placeable ligands are preferably water itself. Besides the water-stable organometallic compounds that are described in related chapters in this book, we are now heading towards a rare class of complexes, the mixed organometallic aquo-ions. Whereas water is omitted in general organometallic chemistry, it is a necessity in bioorgano-metallic or radiopharmaceutical chemistry. Water can easily be tolerated by most of the classical carbon ligands provided that they are not prone to protonation and subsequent cleavage. On the other hand, water as a proton source might oxidize and subsequently decompose low-valent complexes with concomitant H_2 formation. Accordingly, typical ligands for which stability in the presence of water can be expected are CO, $(\eta^5\text{-}C_5H_5)$, $\eta^3\text{-}C_3H_5$, C_6H_6 and some others, such as isocyanides but alkyls (despite methyl-cobalamin and coenzyme B12) are less likely to exist. Carbenes and carbynes are in principal possible as well but only one example has been described to date [12, 13]. The metal center with such an arrangement of C-bound ligands and H_2O is preferentially not in a very low oxidation state and carries a positive charge which makes it more difficult to be oxidized. In fact, organometallic aquo-ions have almost exclusively d^6 or d^3 electron configuration and are found with the above mentioned ligands. Chemistry that has been performed with these precursors, has been reviewed elsewhere [14, 15]. Aiming at the development of novel precursors, these characteristics should be considered and the inspired chemist can in fact predict many so far unknown complexes that could be stable and from which very interesting chemistry is likely to emerge. Some of the most important organometallic aquo-ions are depicted in Scheme 4.4. Recently a comprehensive review has been published covering bioorganometallic aspects of $[Cp*Rh(OH_2)_3]^{2+}$ [16].

For all these examples CO ligands are specially favorable, since their ability to stabilize low-valent metal centers by π-backbonding means that the metal less easily oxidized than would be expected for the corresponding valence. An excellent example is $[^{99}Tc(OH_2)_3(CO)_3]^+$ for which oxidation occurs at a potential > 1.5 V *vs.* SCE in water. This electron depletion leads consequently to increased Lewis acidity and the complexes depicted in Scheme 4.4 are indeed medium-to-strong Broenstedt acids. This would not be expected from their charge or their position in the periodic table. All these organometallic aquo-ions represent a very versatile class of complexes for the development of radiopharmaceuticals if the corresponding elements posses a radioisotope with the corresponding characteristics. ^{105}Rh and ^{97}Ru belong to this class but they have been explored to only a small extent in research and even in the context of bioorganometallic chemistry.

The features of organometallic aquo-ions are two-fold: first they display a hemisphere (usually one face of the coordination octahedron) that is shielded against substitution, thus, invariant; and second they possess a number of water ligands that can be substituted. The positions occupied by water represent the

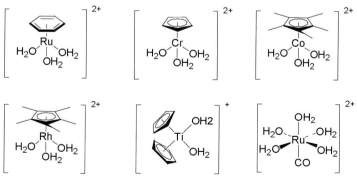

Scheme 4.4 Important organometallic aquo-ions, $[(C_6H_6)Ru(OH_2)_3]^{2+}$ [17, 18], $[CpCr(OH_2)_3]^{2+}$ [19], $[Cp*Co(OH_2)_3]^{2+}$ [20], $[Cp*Rh(OH_2)_3]^{2+}$ [21, 22], $[(Cp*)_2Ti(OH_2)_2]^{2+}$ [23] and $[Ru(CO)(OH_2)_5]^{2+}$ [24].

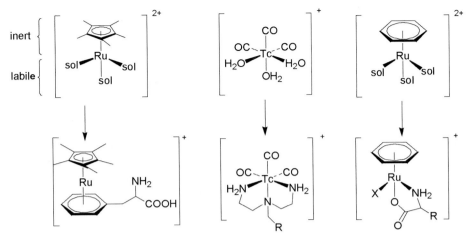

Scheme 4.5 General reactivity patterns of organometallic aquo- (or solvent) ions leading to novel bioorganometallic compounds and/or (radio)pharmaceuticals [15, 25].

variable portion in the complex. Exchanging these for incoming chelators permits labels to be designed that match the properties of the vector they are attached to. Since the resulting complexes are 18 e$^-$ and closed shell they will be inert, regardless of the nature of ligands that have replaced the coordinated H_2O molecules. The mechanism of substituting a once-coordinated ligand is in general dissociative D or interchange dissociative I_d with d^6 compounds. Accordingly, the activation barriers are high and the exchange reactions slow-to-very-slow, a feature that is experimentally confirmed by the known behavior of some of the aquo-ions in Scheme 4.4 in general, and of $[^{99}Tc(OH_2)_3(CO)_3]^+$ in particular. Furthermore, water has a low tendency to coordinate to electron-rich metal centers, which also supports the stability against the first step in a presumed decomposition mechanism, the substitution of one ligand by water. The features are summarized in Scheme 4.5.

Coming back to the arguments underlying the advantages of bioorganometallic complexes mentioned at the beginning of this article, it should now be emphasized that the properties of organometallic aquo-ions enable more systematic radio-pharmaceutical drug development than that with the classical precursors. Assuming kinetic stability as the basis for application, the metal center can now adopt a wide variety of different ligand types, from classical Werner to non-classical organic ligands with M–C bonds. The ligands can be mono- or multidentate, can stem from a bifunctional chelator or be part of the vector itself. They can be chosen to mimic part of a structure of the vector or be completely different, and finally physico-chemical properties such as charge can be varied. Obviously, this possibility is based on the organometallic nature of the precursor since similar behavior with classical complex fragments is rarely encountered.

Although we may become excited about the properties of organometallic compounds, we should not ignore the limitations imposed by the invariant portion of the complex. Desirably, the facially coordinating ligands should be a C_6H_6 or a $[C_5H_5]^-$ ligand since neither is sterically demanding and they are of low molecular weight. The three CO ligands are small but they stick out, away from the metal center and make the whole complex sterically relatively demanding. Although not yet realized in radiopharmaceutical bioorganometallic chemistry, this consideration gives an incentive to the development of novel precursors.

4.3
The Prototype [^{99}Tc(OH$_2$)$_3$(CO)$_3$]$^+$, Synthesis and Properties

Complexes containing CO ligands are in general prepared from the binary carbonyl complexes synthesized under high pressure and high temperature conditions. Single CO ligands can also be introduced at ambient conditions in electron-deficient coordination compounds. Binary carbonyls are rarely prepared at ambient conditions. The binary carbonyls are the thermodynamic end product from which a number of CO groups can be replaced to get other complexes for particular purposes. For most of the chemistry developed with low-valent transition elements, mixed halide–CO complexes are sufficiently versatile since the halides can also be replaced by incoming ligands. Organometallic Tc chemistry is relatively underdeveloped, since it is nowadays almost impossible to synthesize [Tc$_2$(CO)$_{10}$] for radiation safety reasons. Still, even 20 years ago Mazzi and coworkers described the preparation of fundamental mixed phosphine–halide–CO complexes of Tc(I) and Tc(III) [26, 27]. Also [ReCl(CO)$_5$] and probably its Tc-congener as well were prepared from concentrated HX and formic acid as a CO source. None of these complexes were investigated for their behavior with water since there was no reason to do so. The exploration of Tc(I) chemistry demands a more convenient route to a stable but reactive precursor. Some basic structures of low-valent Tc complexes prepared at ambient pressure are shown in Scheme 4.6.

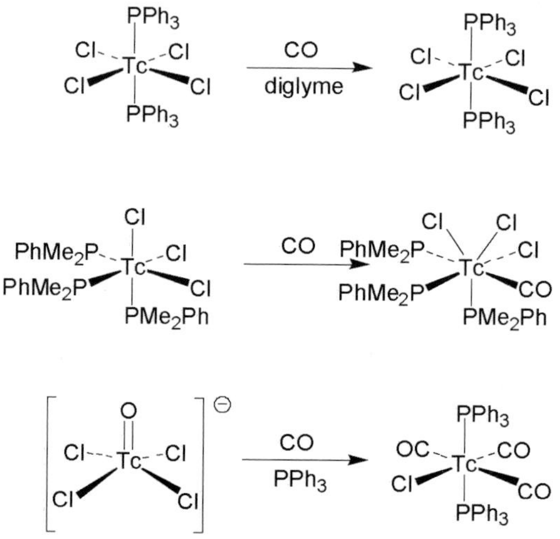

Scheme 4.6 Preparation of some fundamental CO complexes at 1 atm [26–28].

Clearly, a more convenient way is required. Beside the usual reducing agents, boranes have infrequently been used in this respect with transition elements although they are easy to handle and very oxophilic. We reacted $[^{99}TcO_4]^-$ with $BH_3 \cdot THF$ under 1 atm of CO and in the presence of Cl$^-$ as a potential counterion. At room temperature and after a relatively short time period, the complex $[^{99}TcCl_3(CO)_3]^{2-}$ could be isolated in good yield [29, 30]. The reaction pathway is unclear since no intermediates could be authenticated. With the exception of minor amounts of $[TcH(CO)_4]_3$ no higher carbonyl complexes were found or isolated [30]. The *fac*-$[Tc(CO)_3]^+$ moiety is thermodynamically the most stable unit, since any additional CO trans to another CO is labile and readily cleaved. The reaction mechanism at this point still remains unclear since the CO concentration in solution requires delicate control and other complexes might be present as well to explain the reaction pathway.

Surprisingly, when $[^{99}TcCl_3(CO)_3]^{2-}$ is dissolved in water, the three Cl$^-$ ligands are replaced by H_2O to form the basic organometallic aquo-ion $[^{99}Tc(OH_2)_3(CO)_3]^+$. This is the point when bioorganometallic chemistry came in. The step of exposing an organometallic moiety to biological conditions (water and air) was a crucial decision which would probably be worth testing with other organometallic complexes as well, since novel or biologically relevant organometallic aquo-ions could emerge. The complex $[^{99}Tc(OH_2)_3(CO)_3]^+$ proved to be exceptionally stable and did not decompose in air under very alkaline or very acidic conditions [31]. This nature is the essential base for the introduction of $[^{99}Tc(OH_2)_3(CO)_3]^+$ in bioorganometallic chemistry. The different points outlined earlier are perfectly fulfilled to enable radiopharmaceutical drug finding based on the *fac*-$[^{99}Tc(CO)_3]^+$ moiety. Three of the coordination sites are "blocked" by the CO ligands and three coordination sites are available for coordination to a wide variety of ligands and chelators.

In that respect, $[^{99}Tc(OH_2)_3(CO)_3]^+$ resembles a true aquo-ion and compares with e.g. the lanthanides where H_2O ligands also have to be substituted to form the radiopharmaceutical.

In our experience it was not possible to substitute one single CO ligand. A recent paper describes the cleavage of one CO at high temperature and in the presence of the trans-effect ligand CN–R which leads to compounds of general composition $[^{99m}Tc(CO)_2(CN-R)_4]^+$ [32]. The CO ligands on the other hand are exchangeable for ^{13}CO on a reasonable time scale and all the intermediates $[^{99}Tc(OH_2)_3(CO)_n(^{13}CO)_{(3-n)}]^+$ ($n = 0-3$) could be detected by NMR spectroscopy [33]. The exchange experiment even gave access to the binary Tc(I) cation $[Tc(CO)_6]^+$ which was isolated after a longer time period of reaction, implying that the complexes $[^{99}Tc(OH_2)_2(CO)_4]^+$ and $[^{99}Tc(OH_2)(CO)_5]^+$ are in fact intermediates and not transitions states. Beside the synthesis, theoretical calculation recently confirmed the existence of these intermediates [34].

Despite many efforts it has not so far been possible to crystallize $[^{99}Tc(OH_2)_3(CO)_3]^+$ but spectroscopic evidence is strong enough to confirm the proposed composition together with many X-ray structures that have subsequently been elucidated. Some of them are described later in this text.

This basic synthesis of $[^{99}TcCl_3(CO)_3]^{2-}$ encouraged the development of further basic chemistry of Tc(I) but it does not enable the preparation of a radiopharmaceutically useful precursor. The synthesis still demands THF as a solvent, which is not possible in routine applications. Therefore a fully aqueous synthesis was developed for ^{99m}Tc. Replacing $BH_3 \cdot THF$ by $Na[BH_4]$ and THF by saline (0.9% NaCl in water) gave $[^{99m}Tc(OH_2)_3(CO)_3]^+$ in yields better than 98% under an atmosphere of CO and 30 min at 75 °C [35]. The mechanism is not known and requires a 6 e$^-$ reduction and the coordination of 3 CO groups at low concentration. The clean procedure probably relies on the very high dilution of ^{99m}Tc (usually $< 10^{-6}$ M) which prevents the formation of polymers such as TcO_2, the thermodynamic trap during reduction processes in water. The importance of dilution is also mirrored by the observation that this reaction with higher concentration of ^{99}Tc gives $< 20\%$ of $[^{99}Tc(OH_2)_3(CO)_3]^+$, the rest being polymeric $TcO_2 \cdot nH_2O$ as a black precipitate. Still, Tc–CO complexes can be synthesized and it would be worth investigating whether the yield can be increased, and if this approach is also possible with other transition elements to produce novel mixed CO–H_2O complexes.

The conditions outlined here are still not sufficient for routine synthesis since the use of free CO in a kit poses problems. We have developed a new compound which was described in the earlier days of boron chemistry but which has not been further explored [36–39]. Boranocarbonate $[H_3BCO_2H]^-$ (BC) is an intriguing compound since it includes the $[BH_4]^-$ function for reduction and *in situ* CO. The carboxylate portion, when protonated, releases CO, a reaction the compares well with metalla-carboxylates that do the same. BC when reacted with $[^{99m}TcO_4]^-$ at 95 °C for 20 min gives $[^{99m}Tc(OH_2)_3(CO)_3]^+$ in quantitative yield. [40]. The convenience of this approach enabled the production of a kit for the routine synthesis of $[^{99m}Tc(OH_2)_3(CO)_3]^+$ in clinics. The kit's name is Isolink® (Mallinckrodt Med. BV, Petten NL). A summary of the different reactions is given in Scheme 4.7.

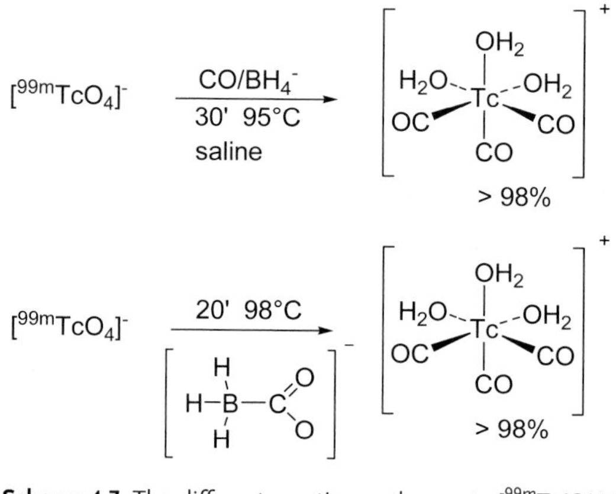

Scheme 4.7 The different reaction pathways to $[^{99m}Tc(OH_2)_3(CO)_3]^+$.

The complex [^{99m}Tc(OH$_2$)$_3$(CO)$_3$]$^+$ is very hydrophilic and elutes from HPLC at almost the same retention time as [^{99m}TcO$_4$]$^-$. If not exposed for too long time it is air stable in alkaline and acidic conditions. After preparation, the solvent has a pH of about 10 and must be neutralized with phosphoric acid to get physiological pH and a buffered environment. At the same time, excess BC is also destroyed. Both of these aspects are still disadvantages and merit attention in the further development of the kit.

4.3.1
Coordination Chemistry with [^{99}Tc(OH$_2$)$_3$(CO)$_3$]$^+$

With the background of developing novel radiopharmaceuticals it is necessary that a complex core binds to a variety of ligands to form stable complexes with different size, charge and hydrophilicity. This will enable investigations about structure–activity relationships in which biodistribution can be studied as a function of the physico-chemical properties of the metal-complexes. At this point the important question arises, if structural units are accessible which can mimic biological structures as e.g. proposed with Tc(V) compounds in the context of steroid hormones [41–43].

To forestall a general observation, it can be stated that the "*fac*-[$^{99(m)}$Tc(CO)$_3$]$^+$" core coordinates to almost any potential donor site from a chelator and many examples of such complexes have been described [44–48]. Regardless of denticity, the complexes are kinetically stable and do not decompose on a time scale required for the process of imaging a patient. The ligands, however, differ significantly in their efficiency for H$_2$O substitution in [^{99m}Tc(OH$_2$)$_3$(CO)$_3$]$^+$. For the labeling of vectors or for the synthesis of model complexes, the rate of substitution as well as the lowest possible concentration of chelator (and hence of targeting agent) to be employed for quantitative complex formation are crucial factors. If the labeling is slow or requires high ligand concentrations, the radiopharmaceutical can only be applied after purification by e.g. HPLC, a procedure that cannot usually be performed in a hospital. A reasonable upper concentration limit for receptor targeting biomolecules is 10^{-5} M. The concentration of ^{99m}Tc as eluted from a generator and used for the preparation of a radiopharmaceutical is in general about 10^{-6}–10^{-7} M. These numbers mean that the biomolecule is still in 100-fold excess. A typical amount of radiopharmaceutical administered to a patient still does not lead to receptor saturation with cold material since more receptors are theoretically available to prevent competition of cold and hot receptor ligand. At these low concentrations, heating is almost always required in order to achieve a reasonable rate of labeling.

We found that many tridentate ligands are very efficient and afford quantitative complex formation at 10^{-6} M and for some examples even down to concentrations of 10^{-7} M, i.e. at stoichiometric ratios [49]. The drawback of tridentate chelators is the need to introduce the second function for covalent coupling to the biomolecule. Sometimes this functional group might already be present in the native ligand, e.g. as in nitrilo-triacetic acid (NTA), but tailor-made chelators in general demand

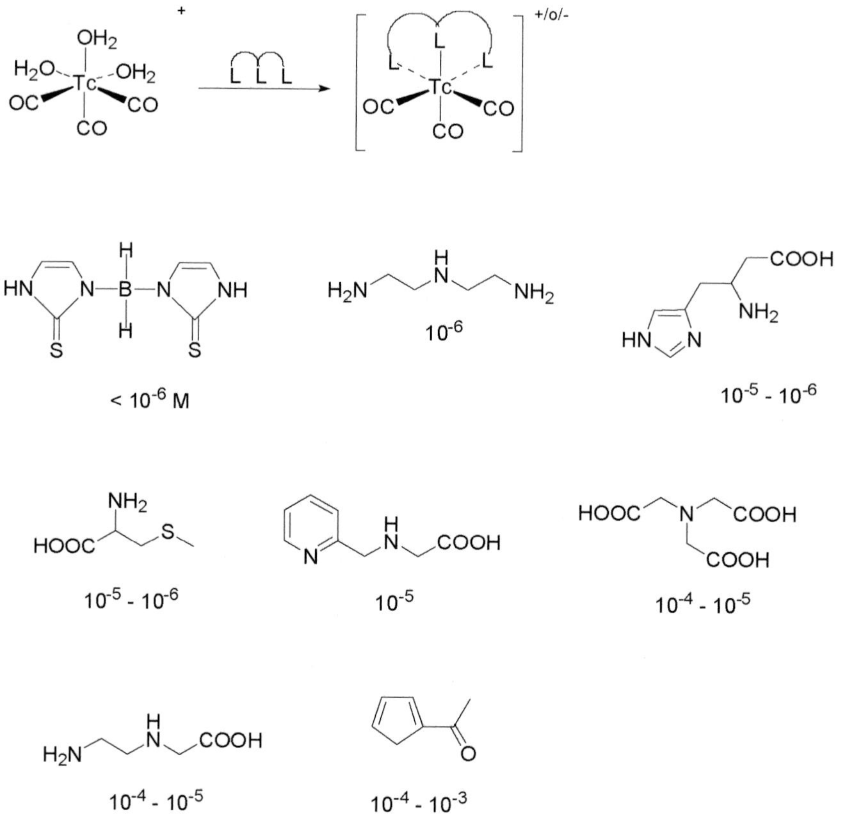

Scheme 4.8 Typical tridentate chelators with the lowest possible concentration to afford quantitative complex formation after 30 min and 98 °C.

separate introduction. To achieve selectivity in the coupling process, protecting group chemistry is required which sometimes makes the preparation of the final ligand a multistep synthesis. Examples for tridentate ligands together with the lowest concentration limits for labeling reaction at 98 °C and 30 min reaction time are given in Scheme 4.8.

The high temperatures are a consequence of the high dilution of the reaction components but the strongest ligands also react quantitatively at 10^{-4} M at room temperature. If receptor binding is not a limiting factor then these reaction conditions can conveniently be employed. The most efficient ligands are histidine, pure aliphatic or mixed aromatic–aliphatic triamines and methionine. Mixed imidazole-thiones in combination with borohydride have been described as even more efficient chelators [50]. The introduction of a linking functionality in histidine and its conjugation to a biomolecule is given in Scheme 4.9 [51].

Obviously, a multistep synthesis is required but for histidine the efficiency and versatility of this chelator merits the preparation of the derivatized precursor. More conveniently but unusual, histidine can be protected by the organometallic core

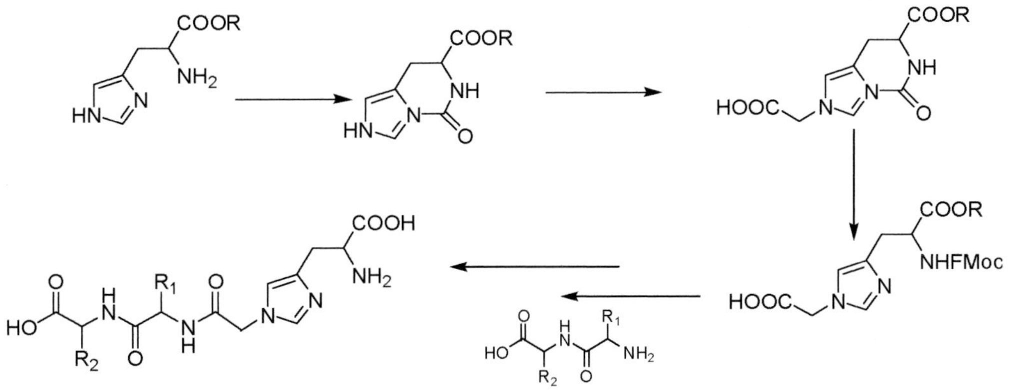

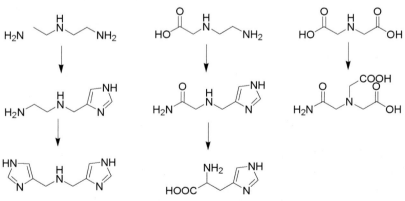

Scheme 4.9 N^ε-derivatization of histidine and conjugation to a biomolecule.

[Re(CO)₃]⁺ itself. The use of organometallic moieties for protecting functionalities in organic molecules are known in literature but only a few, very special examples have been described. After the one-step synthesis of [Re(his)(CO)₃], alkylation at the N^ε in histidine occurs easily and e.g. a carboxylate function can be introduced for further biomolecule conjugation. The complex is then coupled to the vector, for instance a peptide, and the [Re(CO)₃]⁺ protecting group is then released at low pH and in the presence of an oxidizing agent such as H₂O₂. Of course, the biomolecule must persist under these conditions but peptides are not very susceptible to acids and/or H₂O₂ unless they include methionine or other sulfur-containing groups.

As can be seen from Scheme 4.8, the complexes attached to the biomolecules may be anionic, neutral or cationic and it is possible to systematically vary the nature of the donors without significantly varying the topology of the complex. A sequence of such a possible variation is given in Scheme 4.10.

Scheme 4.10 Examples of systematically developed ligands without varying the nature of the donor set.

Although bioorganometallic, the chemistry with the type of donors outlined is normal coordination chemistry and does not include the formation of new Tc–C bonds in water. The high tendency to form coordination compounds does however not exclude the possibility that Tc–C bonds are formed in this way as well and the corresponding reactions will be described later in this text.

To bypass relatively complicated derivatizations, one of the functionalities in tridentate chelators can be directly selected for conjugation. In general, this then leads to bidentate chelators. Alternatively, bidentate chelators can be derivatized and coupled to biomolecules. Some examples of bidentate chelators are given in Scheme 4.11.

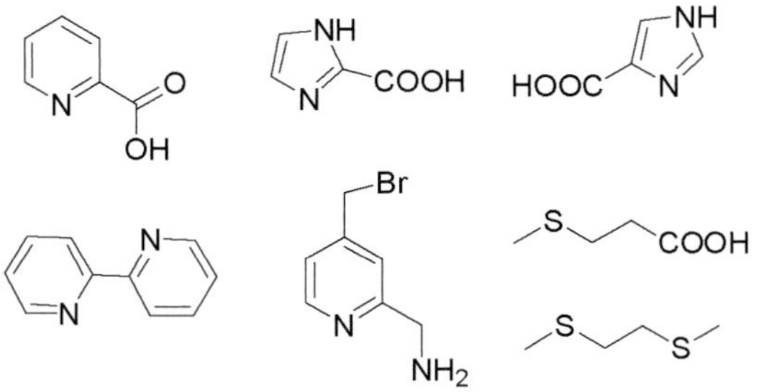

Scheme 4.11 Highly efficient bidentate chelators introduced in biomolecules.

Natural amino acids with two coordinating functionalities (one in the side-chain) can be coupled to the N terminus of peptides. They are then potent bidentate chelators that can be labeled at low concentrations. They can also be coupled at the C terminus via the N^{α}-amino group but the combination of functional side-chain and carboxylate does not yield very efficient chelators since the ring size is larger than six. Preferably, histidine and methionine are coupled in the former way and give bidentate N,N and N,S ligands. If the bidentate ligands are neutral, the last site on the metal center is occupied by a Cl^- for electrostatic reasons to yield an overall zero charge of the complex. Depending on the bidentate ligand, these complexes are often rather lipophilic. In contrast to the ligand, the Cl^- tends to be slowly exchanged and potential competing coordinating sites from biomolecules (e.g. serum proteins) might bind instead. This might result in unfavorably long residence time in the blood pool and to a high background in imaging. The lability of the Cl^- depends on the bidentate chelator but systematic studies of its influence have not been performed so far [52]. Typical chelators are given in Scheme 4.4.

In contrast to the halides, bidentate chelators are not exchanged at all under physiological conditions demonstrating once more the importance of kinetic stability. Many of these ligand can be labeled at 10^{-5} M concentration but they are generally less efficient than tridentate chelators.

Monodentate ligands such as imidazole, thioethers, pyridine etc. form complexes of all stoichiometric compositions 1 : 1, 1 : 2 and 1 : 3. In the presence of monodentate ligands the substitution reaction stops at complexes of composition

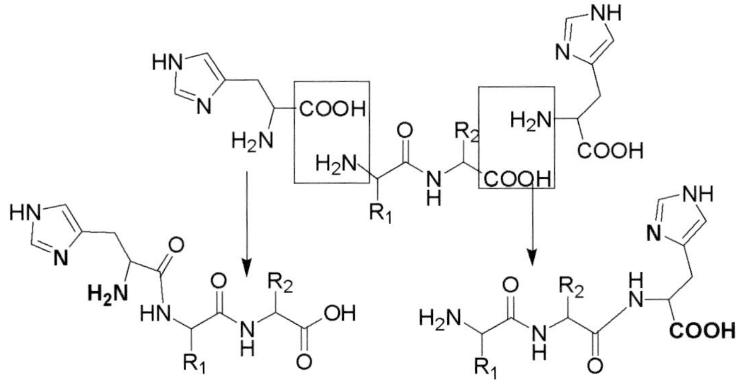

Scheme 4.12 Derivatization of a peptide at the N and C terminus respectively with histidine to yield a bidentate N,O and N,N ligand respectively.

[^{99m}TcCl(L)$_2$(CO)$_3$] but extended reaction time or higher concentrations also afford [^{99m}Tc(L)$_3$(CO)$_3$]$^+$. It would be very convenient to derivatize biomolecules with just a monodentate ligand, however, because double coordination occurs, the reactions often yield cross-linking by coordination to two different biomolecules. This is an unwanted aspect of monodentate ligands, hence, they are not considered as very useful in the context of radiopharmacy.

An interesting feature is the interaction of naturally occurring monodentate ligands with [^{99m}Tc(OH$_2$)$_3$(CO)$_3$]$^+$. The imidazole or thioether side-chain in histidine or methionine respectively offer monodentate coordination sites. If several of these functionalities are available in close vicinity, these monodentate ligands can be used for multi-coordination without providing a chelate effect, but occupying more than one site with groups from the same biomolecule, thus preventing cross-links as described before. In such a situation, one can envisage direct labeling of the biomolecule if the corresponding coordinating sites can bind to one metal center. This is the case in histidine-tagged recombinant antibodies which bear at the C terminus a chain of up to five histidines. These histidines can provide imidazoles for coordination in close vicinity and these proteins can indeed be labeled directly. Although the exact site of coordination and the ligand sphere are not known (and are probably not of one single composition) the fact that no cross-links are found supports the hypothesis of multiple coordination from the same biomolecule without providing chelate effect [53].

A further example is the interaction of *fac*-[M(CO)$_3$]$^+$ with the purine base guanosin. On a macroscopic level *fac*-[M(CO)$_3$]$^+$ binds to G through N7 affording [M(OH$_2$)(9Me-G)$_2$(CO)$_3$]$^+$ with head-to-head or head-to-tail configuration of the two Gs [54]. This implies the preparation of directly labeled G with the corresponding ^{99m}Tc precursor. N7 is however not a very good donor and reaction on the microscopic level affords only mono-coordinated complex [^{99m}Tc(OH$_2$)$_2$(G)(CO)$_3$]$^+$. It was found that plasmid DNA scrambled when exposed to small concentrations of [^{99m}Tc(OH$_2$)$_3$(CO)$_3$]$^+$ which confirms an affinity to G also incorporated in DNA.

Cisplatinum also caused scrambling but already at smaller concentration than with $[^{99}Tc(OH_2)_3(CO)_3]^+$. It can be concluded at this stage that $[^{99}Tc(OH_2)_3(CO)_3]^+$ can form inter- or intra-strand cross-links in DNA although the true form of interactions still has to be authenticated.

Although mono- and bidentate ligands have some disadvantages in the preparation of novel radiopharmaceuticals, the combination of both gave rise to a new mixed-ligand concept [55]. The reaction of $[^{99}Tc(OH_2)_3(CO)_3]^+$ with an anionic bidentate ligand gives neutral complexes of general composition $[^{99}Tc(OH_2)(\eta^2\text{-}L^2)(CO)_3]$. One water ligand remains bound and coordination of a second bidentate ligand was never found, probably for steric reasons. The H_2O ligand can now be exchanged by a monodentate ligand giving neutral complexes of composition $[^{99}Tc(L^1)(L^2)(CO)_3]$. Ligand L^1 can be e.g. an imidazole, a pyridine, thioethers isocyanide and many others giving complexes of a similar structural type but with different substituents L^1. Assuming that L^2 is a bidentate chelator attached to a targeting molecule, the variation of L^1 affords complexes of different characteristics and systematic biodistribution studies can be performed. This approach is called $[2_B+1]$ mixed ligand concept and is shown in Scheme 4.13. The combination of a mono- and a bidentate ligand is a useful concept and can be extended to another form. Assuming that a monodentate chelator is attached to the biomolecule, the variable portion is then represented by the bidentate chelator. This is the so-called $[2+1_B]$ concept and in particular useful since the direct labeling precursor $[^{99}Tc(OH_2)(\eta^2\text{-}L^2)(CO)_3]$ can be prepared in one step directly in the kit. The bidentate ligand is added to the carbonylation solution and the complex $[^{99m}Tc(OH_2)(\eta^2\text{-}L^2)(CO)_3]$ forms in one step. Subsequently, this new precursor is reacted with the biomolecule bearing the monodentate ligand to produce the radiopharmaceutical. The $[2+1_B]$ approach includes the potential of drug development as well. While the biomolecule with the monodentate ligand is kept constant, the bidentate ligand can be varied and in fact many versatile bidentate ligands are commercially available. As mentioned at the beginning, monodentate ligands are not so efficient in terms of required concentration and rate of complex formation. Therefore this approach needs relatively high (biomolecule) concentrations for the reaction to go to completion, unless the monodentate ligand is an isocyanide. To our experience, isocyanides are very efficient and form complexes according to the $[2+1_B]$ concept at a concentration as low as 10^{-5} M. Isocyanides can include a second functionality that allow their coupling to biomolecules. Not all types of monodentate ligands have been explored and there remains much work to do in order to identify the best kind of monodentate ligand.

Although not yet investigated in detail, the presented mixed ligand concept would also allow the combination of two different types of targeting molecules with different kind of biological function. The two biological principles can be coupled together by the metal core in the center, bridging the two. One of the functions can be e.g. targeting and the second one e.g. intracellular trapping. Of course, such conjugates are also possible by simple covalent coupling of the corresponding two organic molecules, however, the separation of the two will affect their original function less.

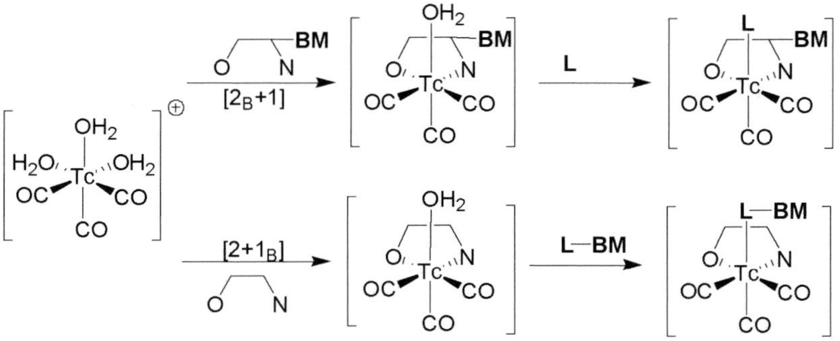

Scheme 4.13 The mixed ligand $[2_B+1]$ and $[2+1_B]$ concept based on the fac-$[^{99(m)}Tc(CO)_3]^+$ moiety.

4.3.2
Organometallic Chemistry in Water with $[^{99}Tc(OH_2)_3(CO)_3]^+$

From the previous sections it is obvious that "natural" ligands such as amino acids can be employed for coordination to the fac-$[^{99}Tc(CO)_3]^+$ moiety. Although coordinated to a metal, these ligands are "known" in biological systems and behave as a shield to the metal. Still, the ligands are relatively large and it is desirable to design smaller ones, probably also mimicking structural features that are known to nature. One of the smallest one can think of is the cyclopentadienyl Cp⁻ ring. Beside its size, the topology of a (coordinated) Cp⁻ resembles a benzene ring and corresponding piano-stool Tc complexes can potentially replace benzene in biological molecules such as phenylalanine provided that the surface of the molecule is more important than the 3D-shape. This potential replacement can be shown in different examples in bioorganometallic chemistry and is described elsewhere in this book. Reduced size is desirable for the design of labeled CNS-receptor ligands and mimicking of benzene rings in the labeling of e.g. steroid hormones or cytotoxic agents. Since these receptor-binding molecules are relevant in modern diagnosis or therapy, the preparation of corresponding radiopharmaceutical represented a strong incentive for the finding of synthetic strategies to provide Cp⁻ type complexes. From classical bioorganometallic chemistry one knows a lot about structure–activity relationships and it could be shown that [CpRe(CO)₃] derivatized estradiol lost some RBA in comparison to native estradiol but was still efficient enough to merit investigations of the corresponding ^{99m}Tc analog [56, 57].

Many attempts have been made to introduce Cp⁻ into radiopharmaceutical chemistry. The limiting factor to overcome is surely represented by the need to synthesize the corresponding complexes from water and $[^{99m}TcO_4]^-$. Cp⁻, the classical organometallic ligands and water, seem to exclude each other since Cp⁻ is readily protonated and decomposes rapidly by di- or polymerization. The prohibitively high pK_a of about 14 efficiently prevents the presence of the deprotonated form, Cp⁻, which is more stable than protonated HCp. Many attempts from our research to react $[^{99m}Tc(OH_2)_3(CO)_3]^+$ in water with HCp or HCp* failed and, if at all, only minor amounts of $[(Cp)^{99m}Tc(CO)_3]$ could be detected. Attempts from the early 1990s performed by Wenzel and coworkers in the double ligand transfer (DLT) reaction proved the possibility of synthesizing $[(Cp)^{99m}Tc(CO)_3]$ and also showed its radiopharmaceutical versatility in some preliminary biological experiments [58, 59]. The DLT reaction of Wenzel was a reaction in molten solid and was further developed by Katzenellenbogen and coworkers [60, 61]. His approach was based on the DLT approach of Wenzel and could now be performed in CH_3OH or other organic solvents, but still had the drawback of rather drastic conditions. Piano-stool complexes prepared in these works and attached to biomolecules again supported the versatility of $[(Cp)^{99m}Tc(CO)_3]$ for radiopharmaceutical purposes [62, 63].

Protonated HCp is a weak ligand since its η^5-coordination to a metal center requires initial alkene binding but alkenes show difficulty in competing with large amounts of solvent such as water for all known ^{99m}Tc precursors. A negative charge would electrostatically support coordination or a functional group attached to the Cp-ring would initialize coordination and induce ring slippage to η^5-coordination. Based on this concept, Cp^- was derivatized with an acetyl group to yield Acetyl-Cp (ACp) which is well-known. Surprisingly the pK_a value of ACp did reduce by six orders of magnitude and has a value of about 8.4 [64]. In its protonated or deprotonated form it is reasonably stable in aqueous solution and can be considered as a kind of Werner-type ligand. ACp now directly reacts with $[^{99m}Tc(OH_2)_3(CO)_3]^+$ in water to yield the corresponding piano-stool complexes $[ACp^{99(m)}Tc(CO)_3]$. The different approaches for the synthesis of Cp complexes are shown in Scheme 4.14.

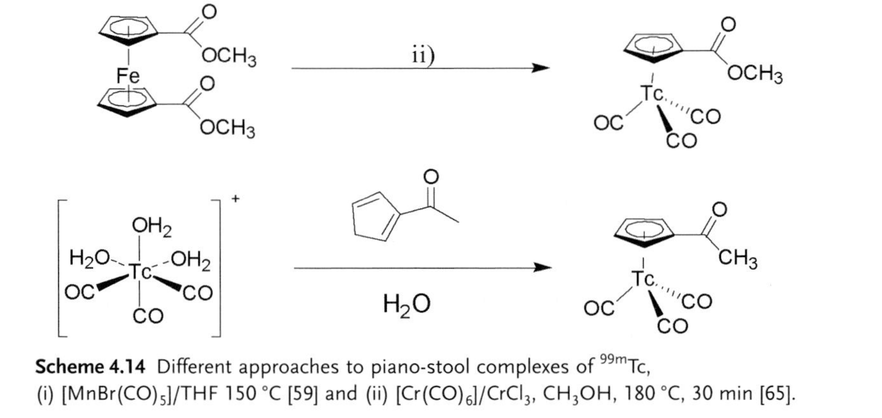

Scheme 4.14 Different approaches to piano-stool complexes of ^{99m}Tc, (i) [MnBr(CO)$_5$]/THF 150 °C [59] and (ii) [Cr(CO)$_6$]/CrCl$_3$, CH$_3$OH, 180 °C, 30 min [65].

The yield with macroscopic amounts of $[^{99}Tc(OH_2)_3(CO)_3]^+$ is in the range of 20–40%. The reason for not being quantitative is not related to instability of the ACp ligand but because there is a competing side reaction of hydrolytic cluster formation. The rest of the starting material is present as $[^{99}Tc(\mu^3\text{-}OH)(CO)_3]_4$. Since the corresponding reaction with ^{99m}Tc occurs at high dilution, the formation of tetranuclear cluster is highly unlikely and indeed not observed. Hence, the corresponding reaction with ^{99m}Tc is quantitative after 30 min at 90 °C and at a ligand concentration between 1 mM and 0.1 mM, depending on the type of Cp-derivative [66].

The general synthetic pathway to ACp offers a convenient possibility of attaching biomolecules by the same route. The reaction between NaCp and an ester function affords a keto group attached to Cp and a link to the biomolecule. In this way we have introduced CNS-receptor ligands such as the serotonergic receptor ligand WAY. The Cp derivatives could be labeled to yield the piano-stool derivatized biomolecule. Depending on the spacer length, receptor affinity was fully retained and the radiopharmaceuticals are actually under biological testing. Some labeled CNS-receptor ligands are depicted in Scheme 4.15. Beside the investigations with piano-stool complexes of ^{99m}Tc, the synthesis of the corresponding ^{188}Re complexes according to the DLT method have been described [67].

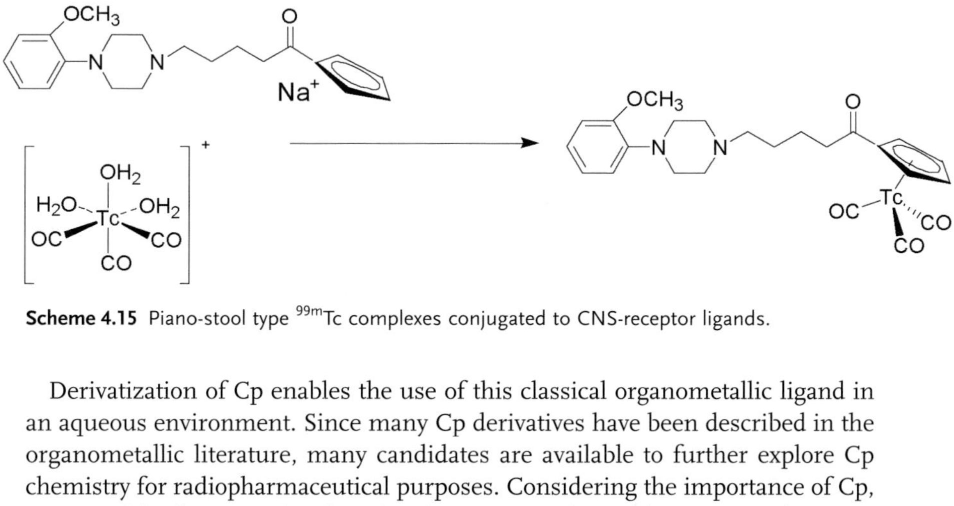

Scheme 4.15 Piano-stool type ^{99m}Tc complexes conjugated to CNS-receptor ligands.

Derivatization of Cp enables the use of this classical organometallic ligand in an aqueous environment. Since many Cp derivatives have been described in the organometallic literature, many candidates are available to further explore Cp chemistry for radiopharmaceutical purposes. Considering the importance of Cp, supported by the examples that already exist, reveals its advantages and it must be emphasized that Cp merits broader and more thorough investigation than is actually the case.

An alternative approach to the introduction of piano-stool ^{99m}Tc complexes into biomolecules has recently been described. Besides free Cp attached to a bio-molecule one can also choose ferrocene as a Cp source as already shown by Wenzel or Katzenellenbogen and coworkers. The difficulty consists in the cleavage of the very stable "$[CpFe]^+$" moiety from the original ferrocene. Still, it could be shown that $[Re(CO)_6]^+$ in hot dmso (probably present as $[Re(CO)_3(dmso)_3]^+$ can cleave ferrocene and produce [(R-Cp)Re(CO)$_3$] (where R is e.g. an estradiol derivative) to form the corresponding rhenium compound. It was also found that the same procedure in a dmso/water mixture at 98 °C produces the analog ^{99m}Tc radiopharmaceuticals [68]. Clearly, dmso is not "biocompatible" but is required to dissolve the highly lipophilic estradiol derivatives rather than playing an active role in the reaction mechanism. For less lipophilic vectors this approach might well be a valuable alternative to the first strategy (Scheme 4.16).

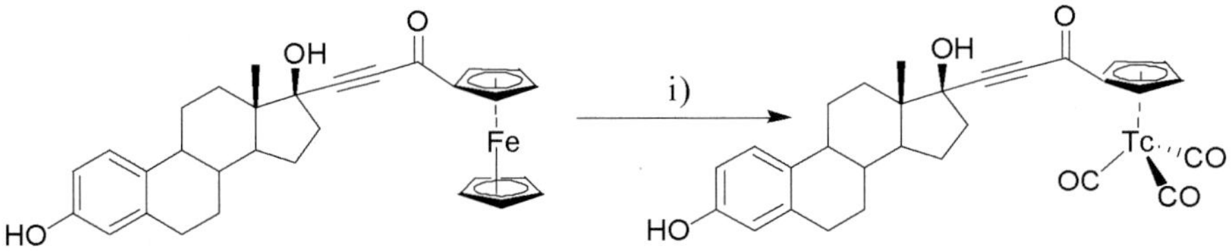

Scheme 4.16 Synthesis of a piano-stool type ^{99m}Tc complex starting from ferrocene conjugates [68].

Isocyanides are further organometallic ligands that react with [$^{99(m)}$Tc(OH$_2$)$_3$(CO)$_3$]$^+$ in aqueous solution. Isocyanides are surprisingly efficient ligands and can form mono-, bis- and tri-substituted complexes [29]. They have already been touched in the discussion of the mixed ligand concept. Although potentially useful in the context of bioorganometallic chemistry, alkenes, dienes, arenes or other π-ligands have not been described yet. The reason for this lack might be unreactivity towards the [^{99}Tc(OH$_2$)$_3$(CO)$_3$]$^+$ or the simple statement that no investigations have been performed so far. Another type of extremely interesting organometallic ligand is the anionic nido-carborane ligands of the type [C$_2$B$_9$H$_{12}$]$^-$ and derivatives thereof [69]. Valliant and coworkers recently showed that these Cp-analogs react in water to form the corresponding complexes [(C$_2$B$_9$H$_{11}$)Tc(CO)$_3$]$^-$ [70, 71]. They also introduced functionalities in the carborane framework which conveniently allows conjugating to a biomolecule [72]. The complexes are perfectly stable under physiological conditions. It can be predicted that this class of potential radiopharmaceutical will become very important in future, not only in the context of bioorganometallic radiopharmaceuticals but also in the context of boron neutron capture therapy (BNCT).

4.4
Combining [^{99}Tc(OH$_2$)$_3$(CO)$_3$]$^+$ with Targeting Vectors

Despite the many advantages outlined for [$^{99(m)}$Tc(OH$_2$)$_3$(CO)$_3$]$^+$ for the development of radiopharmaceuticals, complexes with targeting vectors are still at an early stage but their utility is recognized [73]. Potential reasons for this have recently been discussed extensively [7, 74]. One of the main reasons is the newness of the approach. Reliable kits have only been available for about one year. A new method takes time for adoption and optimization. It should be emphasized at this point that it is not anticipated that development of radiopharmaceuticals with the *fac*-[^{99m}Tc(CO)$_3$]$^+$ moiety will be the unique solution, but will complement the possibilities in this field by an additional versatile approach. In the following Section, some recent and selected studies for the labeling of targeting vectors with *fac*-[^{99m}Tc(CO)$_3$]$^+$ will be described without claiming to be comprehensive.

Research with a number of peptides has been actively pursued, in particular research with neurotensin, bombesin [75, 76] and octreotide is worthwhile. Neurotensin has been derivatized according to strategies described earlier. Original work with bidentate ligand (histidine and the N terminus) gave very good labeling yield and stability but suffered from high accumulation in non-targeted organs, probably due to exchange of the chloride by competing groups in proteins or by the relatively high lipophilicity. *In vivo* stability (of the peptide) also proved to be a problem but this characteristic could be improved by chemically stabilizing the peptide. Later work in which a tridentate histidine ligand is coupled through alkylation of the α-amino group in histidine improved the features of labeling, of stability and of tumor/non-tumor ratio. Currently, some neurotensin compounds are being introduced in preclinical trials.

Similarly, bombesin [77–79], octreotide and derivatives [80] and surfactant proteins [81] have been investigated with the *fac*-[^{99m}Tc(CO)$_3$]$^+$ moiety. Beside the fact that suboptimal chelators have been chosen, stability and binding studies elucidated that the behavior of a ligand attached to a targeting vector and the free ligand respectively are comparable. This confirms that studies with model ligands are useful and relevant for extrapolation of complexation conditions to vectors.

Further *in vitro* and *in vivo* studies have been performed with derivatized glucose [82], estradiol [83–85], folic acid [86], fatty acids [87], recombinant single chain fragments scFv [88] and many other biomolecules. [89, 90]. Since the corresponding labeling chemistry is adopted from model chemistry, biological results will not be discussed here and the reader is encouraged to have a look at the corresponding literature.

From our own projects we would like to present recent studies performed with vitamin B12. Vitamin B12 is a very attractive target since any living human cell in general and fast proliferating cancer cells in particular are strongly dependent on it. Several papers about ^{99m}Tc labeling have appeared in recent literature, and labeled B12 has been the subject of many patents. We studied B12 with the carbonyl approach by derivatizing B12 at the 5′-OH group of the ribose moiety. The synthetic pathway to introduce a histidine at the corresponding functionality is outlined in Scheme 4.17.

The tridentate histidine ligand seemed convenient since it is a powerful ligand requiring only low concentrations. Vitamin B12 is sensitive to high temperatures, hence, the chelator should trap the *fac*-[^{99m}Tc(CO)$_3$]$^+$ moiety at temperatures below 50 °C. The derivatives presented in Scheme 4.18 labeled quantitatively at 50 °C and a concentration between 10^{-4} M and 10^{-5} M within 30 min. No unspecific labeling was observed and the radiopharmaceutical is stable in serum. Both compounds are enzymatically active and bind to intrinsic factor IF. Preliminary *in vivo* mice studies showed good tumor uptake but also a high liver and/or kidney

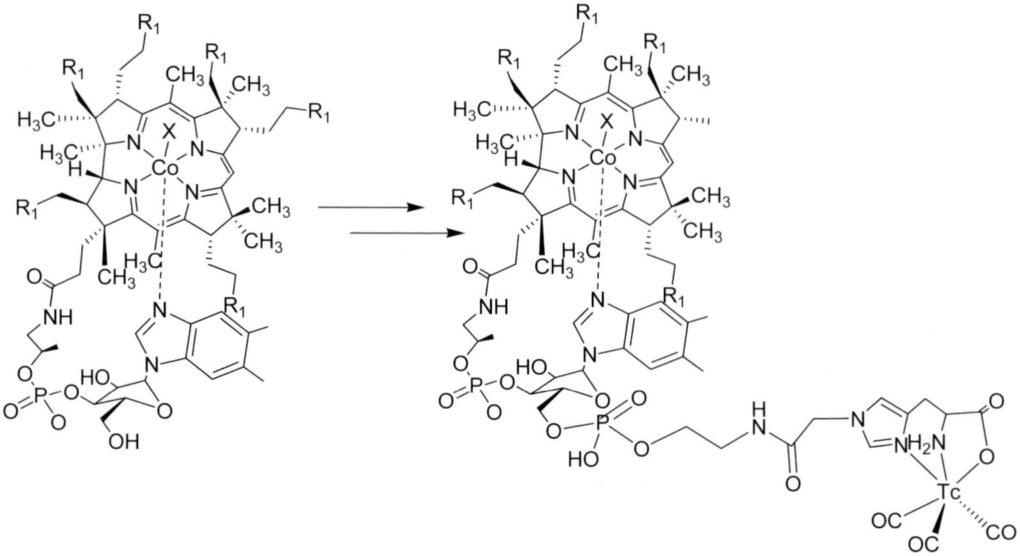

Scheme 4.17 Derivatization of vitamin B12 with tridentate chelators at the 5′-OH group.

uptake, a characteristic that has been observed with other B12 radiopharmaceuticals. Still, this is a good starting point for investigating structure–activity relationships, good receptor affinity but unsatisfying biodistribution. It will be interesting to learn how a chelator slightly or significantly different in charge or donors will influence the biodistribution.

So far, only the chemistry of the fac-$[^{99m}Tc(CO)_3]^+$ moiety has been discussed. As widely accepted in radiopharmacy, it is considered as a paradigm that the ^{188}Re analogs of any ^{99m}Tc radiopharmaceutical could potentially be used for radiotherapeutic purposes. However, despite belonging to the same group in the periodic table, the two elements show distinctly different properties concerning in particular their reactivity towards redox processes. Rhenium is much more difficult to reduce and reacts in general between 10 and 100 times slower. Correspondingly, the synthesis of $[^{188}Re(OH_2)_3(CO)_3]^+$ does not work in the same kit as its ^{99m}Tc homolog. Completely new conditions had to be found and have been described [91]. The yields are not as good and reproducible as in the ^{99m}Tc case but $[^{188}Re(OH_2)_3(CO)_3]^+$ is at least available. Thus, systematic studies about the reactivity and labeling chemistry are not as frequent as for ^{99m}Tc to confirm that stronger reaction conditions are required to achieve quantitative labeling of a biomolecule or a model ligand. Clearly, a challenge remains for the future development of radiotherapeutic agents, to find a synthesis of $[^{188}Re(OH_2)_3(CO)_3]^+$ that can be performed on a routine basis. It is anticipated that such a successful synthesis must proceed by a completely different reaction mechanism than that worked out for ^{99m}Tc, an incentive for innovative inorganic chemists.

4.5
Perspectives

Bioorganometallic chemistry for radiopharmaceutical purposes is at an early stage. Organometallic complexes however have a strong potential for application in this field of life science. The related difficulties for the preparation and optimization of novel precursors or complexes are compensated by the rich fundamental chemistry revealed by this search. Bioorganometallic chemistry should not be a dead-end user chemistry but stimulate more basically oriented researchers to continue the work or to apply the strategies and compounds for other investigations. Reactivity as found here can have many objectives and does not need to be limited to a particular purpose. Of course the question arises: what other fragments could be the goal of such a search? One of the closest potential fragments is already under investigation. Replacing CO with the isoelectronic NO^+ will change the electronic properties of the metal center completely and lead to a novel chemistry in water, interesting for radiopharmacy but also for other fields. The complex $[^{99(m)}Tc(OH_2)_3(CO)_2(NO)]^{2+}$ has been synthesized and some basic properties been described [92]. It will be very interesting to study the use of this complex for the labeling of biomolecules. Based on the structures of some other organometallic aquo-ions (Scheme 4.4) one could anticipate a hypothetical complex $[(C_6H_6)^{99(m)}Tc(OH_2)_3]^+$ as a precursor,or $[(C_6H_6)^{99(m)}Tc(CO)_3]^+$ as a complex in which a biomolecule is attached to the arene. The second compound has been described by Fischer et al. to be stable *in* water [93]; the challenge remains as to how this complex can be synthesized *from* water. Allyl complexes of molybdenum are widely used in bioorganometallic chemistry, would it also be possible to use the allyl group for radiopharmaceutical purposes? Beside all these nice ideas, the enthusiastic chemist should not forget that the ultimate goal is to bring a promising compound finally to the market for the benefit of health care. There is a long way to go and intimate interaction between scientists from many disciplines is required. Bioorganometallic research is in particular difficulty since the language of organometallic chemists is different from that of the biologist or physician, but it is also very attractive for young students to be active among many scientists and to learn from their experience to achieve a common goal.

Acknowledgment

The author wishes to express his gratitude to his former and present coworkers who enthusiastically did most of the experimental work presented in this chapter, namely Sylwia Wyss, Pascal Häfliger, JaeKyong Pak, Susanne Kunze, Paul Benny, Fabio Zobi, Dave van Staveren, Stefan Mundwiler, Bernhard Spingler, Nikos Agorastos, Kirstin Ortner, Jon Bernard, Joachim Wald and Roger Schibli. Mallinckrodt Med. BV, Petten NL, Solidago AG, Switzerland and the University of Zürich supported the work financially.

References

1 R. E. Weiner, M. L. Thakur, *Appl. Radiat. Isot.* **2002**, *57*, 749–763.

2 H. F. Kung, M. P. Kung, S. R. Choi, *Semin. Nucl. Med.* **2003**, *33*, 2–13.

3 A. Alavi, J. W. Kung, H. M. Zhuang, *Semin. Nucl. Med.* **2004**, *34*, 56–69.

4 S. S. Jurisson, J. D. Lydon, *Chem. Rev.* **1999**, *99*, 2205–2218.

5 J. R. Dilworth, S. J. Parrott, *Chem. Soc. Rev.* **1998**, *27*, 43–55.

6 R. C. Mease, C. Lambert, *Semin. Nucl. Med.* **2001**, *31*, 278–285.

7 M. Welch, *Eur. J. Nucl. Med. Mol. I.* **2003**, *30*, 1302–1304.

8 G. Jaouen, S. Top, A. Vessières, R. Alberto, *J. Organomet. Chem.* **2000**, *600*, 23–36.

9 M. J. Abrams, A. Davison, A. G. Jones, C. E. Costello, H. Pang, *Inorg. Chem.* **1983**, *22*, 2798–2800.

10 D. W. Wester, J. R. Coveney, D. L. Nosco, M. S. Robbins, R. T. Dean, *J. Med. Chem.* **1991**, *34*, 3284–3290.

11 J. Baldas, J. Bonnyman, P. M. Pojer, G. A. Williams, M. F. Mackay, *J. Chem. Soc. Dalton* **1982**, 451–455.

12 H. Braband, U. Abram, *Chem. Commun.* **2003**, 2436–2437.

13 H. Braband, T. I. Zahn, U. Abram, *Inorg. Chem.* **2003**, *42*, 6160–6162.

14 U. Koelle, *Coordin. Chem. Rev.* **1994**, *135*, 623–650.

15 K. Severin, R. Bergs, W. Beck, *Angew. Chem. Int. Ed.* **1998**, *37*, 1637–1654.

16 R. H. Fish, G. Jaouen, *Organometallics* **2003**, *22*, 2166–2177.

17 Y. Hung, W. J. Kung, H. Taube, *Inorg. Chem.* **1981**, *20*, 457–463.

18 M. Stebler-Rothlisberger, W. Hummel, P. A. Pittet, H. B. Burgi, A. Ludi, A. E. Merbach, *Inorg. Chem.* **1988**, *27*, 1358–1363.

19 L. O. Spreer, I. Shah, *Inorg. Chem.* **1981**, *20*, 4025–4027.

20 U. Koelle, B. Fuss, *Chem. Ber.-Recl.* **1984**, *117*, 753–762.

21 A. Nutton, P. M. Bailey, P. M. Maitlis, *J. Chem. Soc. Dalton* **1981**, 1997–2002.

22 U. Koelle, W. Kläui, *Z. Naturforsch. B* **1991**, *46*, 75–83.

23 U. Thewalt, H. P. Klein, *J. Organomet. Chem.* **1980**, *194*, 297–307.

24 U. Koelle, G. Flunkert, R. Gorissen, M. U. Schmidt, U. Englert, *Angew. Chem. Int. Edit.* **1992**, *31*, 440–442.

25 A. J. Gleichmann, J. M. Wolff, W. S. Sheldrick, *Journal of the Chemical Society* **1995**, *Dalton Trans.*, 1549–1554.

26 U. Mazzi, A. Bismondo, N. Kotsev, D. A. Clemente, *J. Organomet. Chem.* **1977**, *135*, 177–182.

27 G. Bandoli, D. A. Clemente, U. Mazzi, *J. Chem. Soc.* **1978**, *Dalton Transactions*, 373–380.

28 R. Alberto, W. A. Herrmann, P. Kiprof, F. Baumgärtner, *Inorg. Chem.* **1992**, *31*, 895–899.

29 R. Alberto, R. Schibli, P. A. Schubiger, *Polyhedron* **1996**, *15*, 1079–1089.

30 R. Alberto, R. Schibli, U. Abram, R. Hubener, H. Berke, T. A. Kaden, P. A. Schubiger, *J. Chem. Soc. Chem. Commun.* **1996**, 1291–1292.

31 A. Egli, K. Hegetschweiler, R. Alberto, U. Abram, R. Schibli, R. Hedinger, V. Gramlich, R. Kissner, P. A. Schubiger, *Organometallics* **1997**, *16*, 1833–1840.

32 M. Dyszlewski, H. M. Blake, J. L. Dahlheimer, C. M. Pica, D. Piwnica-Worms, *Molecular Imaging* **2002**, *1*, 24–25.

33 N. Aebischer, R. Schibli, R. Alberto, A. E. Merbach, *Angew. Chem.* **2000**, *39*, 254–256.

34 X. Y. Wang, Y. Wang, X. Q. Liu, T. W. Chu, S. W. Hu, X. H. Wei, B. L. Liu, *Phys. Chem. Chem. Phys.* **2003**, *5*, 456–460.

35 R. Alberto, R. Schibli, A. Egli, U. Abram, T. A. Kaden, P. A. Schubiger, *J. Am. Chem. Soc.* **1998**, *120*, 7987–7988.

36 L. J. Malone, M. R. Manley, *Inorg. Chem.* **1967**, *6*, 2260–2262.

37 L. J. Malone, R. W. Parry, *Inorg. Chem.* **1967**, *6*, 817–822.

38 J. C. Carter, R. W. Parry, *J. Am. Chem. Soc.* **1965**, *87*, 2354–2358.

39 E. Mayer, *Monatsh. Chem.* **1971**, *102*, 940–945.

40 R. Alberto, K. Ortner, N. Wheatley, R. Schibli, A. P. Schubiger, *J. Am. Chem. Soc.* **2001**, *123*, 3135–3136.

41 D. Y. Chi, J. A. Katzenellenbogen, *J. Am. Chem. Soc.* **1993**, *115*, 7045–7046.

42 D. Y. Chi, J. P. Oneil, C. J. Anderson, M. J. Welch, J. A. Katzenellenbogen, *J. Med. Chem.* **1994**, *37*, 928–937.

43 R. K. Hom, J. A. Katzenellenbogen, *J. Org. Chem.* **1997**, *62*, 6290–6297.

44 R. Alberto, R. Schibli, R. Waibel, U. Abram, A. P. Schubiger, *Coordin. Chem. Rev.* **1999**, *192*, 901–919.

45 R. Schibli, R. La Bella, R. Alberto, E. Garcia-Garayoa, K. Ortner, U. Abram, P. A. Schubiger, *Bioconjugate Chem.* **2000**, *11*, 345–351.

46 H. J. Pietzsch, A. Gupta, M. Reisgys, R. Alberto, B. Johannsen, *Bioconjugate Chem.* **2000**, *3*, 414–424.

47 S. R. Banerjee, M. K. Levadala, N. Lazarova, L. H. Wei, J. F. Valliant, K. A. Stephenson, J. W. Babich, K. P. Maresca, J. Zubieta, *Inorg. Chem.* **2002**, *41*, 6417–6425.

48 L. H. Wei, S. R. Banerjee, M. K. Levadala, J. Babich, J. Zubieta, *Inorg. Chem. Commun.* **2003**, *6*, 1099–1103.

49 R. Alberto, R. Schibli, A. P. Schubiger, U. Abram, H. J. Pietzsch, B. Johannsen, *J. Am. Chem. Soc.* **1999**, *121*, 6076–6077.

50 R. Garcia, A. Paulo, A. Domingos, I. Santos, K. Ortner, R. Alberto, *J. Am. Chem. Soc.* **2000**, *122*, 11240–11241.

51 J. K. Pak, P. Benny, B. Spingler, K. Ortner, R. Alberto, *Chem-Eur. J.* **2003**, *9*, 2053–2061.

52 S. Seifert, J.-U. Künstler, A. Gupta, H. Funke, T. Reich, C. Hennig, A. Rossberg, H. J. Pietzsch, R. Alberto, *Radiochim. Acta* **2000**, *3–4*, 239–245.

53 R. Waibel, R. Stichelberger, R. Alberto, P. A. Schubiger, K. A. Chester, R. H. J. Begent, *Eur. J. Nucl. Med.* **2000**, *27*, S86.

54 F. Zobi, B. Spingler, T. Fox, R. Alberto, *Inorg. Chem.* **2003**, *42*, 2818–2820.

55 S. Mundwiler, M. Kündig, K. Ortner, R. Alberto, *Bioconjugate Chem.* **2004**, *15*, 195–202.

56 S. Top, H. E. Hafa, A. Vessières, J. Quivy, J. Vaissermann, D. D. Hughes, M. J. McGlinchey, J.-P. Mornon, E. Thoreau, G. Jaouen, *J. Am. Chem. Soc.* **1995**, *117*, 8372–8380.

57 G. Jaouen, S. Top, A. Vessières, P. Pigeon, G. Leclercq, I. Laios, *Chem. Comm.* **2001**, 383–384.

58 M. Wenzel, *J. Label. Compd. Radiopharm.* **1992**, *31*, 641–650.

59 M. Wenzel, C. Klinge, *J. Label. Compd. Radiopharm.* **1994**, *34*, 981–987.

60 R. R. Cesati, G. Tamagnan, R. M. Baldwin, S. S. Zoghbi, R. B. Innis, J. A. Katzenellenbogen, *J. Labelled Compd. Radiopharm.* **1999**, *42*, S150–S152.

61 R. R. Cesati, G. Tamagnan, R. M. Baldwin, S. S. Zoghbi, R. B. Innis, N. S. Kula, R. J. Baldessarini, J. A. Katzenellenbogen, *Bioconjugate Chem.* **2002**, *13*, 29–39.

62 T. W. Spradau, W. B. Edwards, C. J. Anderson, M. J. Welch, J. A. Katzenellenbogen, *Nucl. Med. Biol.* **1999**, *26*, 1–7.

63 E. S. Mull, V. J. Sattigeri, A. L. Rodriguez, J. A. Katzenellenbogen, *Bioorgan. Med. Chem.* **2002**, *10*, 1381–1398.

64 J. Wald, R. Alberto, K. Ortner, L. Candreia, *Angew. Chem. Int. Edit.* **2001**, *40*, 3062–3066.

65 T. W. Spradau, J. A. Katzenellenbogen, *Organometallics* **1998**, *17*, 2009–2017.

66 J. Bernard, K. Ortner, B. Spingler, H. J. Pietzsch, R. Alberto, *Inorg. Chem.* **2003**, *42*, 1014–1022.

67 T. Uehara, M. Koike, H. Nakata, S. Miyamoto, S. Motoishi, K. Hashimoto, N. Oku, M. Nakayama, Y. Arano, *Nucl. Med. Biol.* **2003**, *30*, 327–334.

68 S. Masi, S. Top, L. Boubekeur, G. Jaouen, S. Mundwiler, B. Spingler, R. Alberto, *Eur. J. Inorg. Chem.* **2004**, in press.

69 P. Morel, P. Schaffer, J. F. Valliant, *J. Organomet. Chem.* **2003**, *668*, 25–30.

70 J. F. Valliant, P. Morel, P. Schaffer, J. H. Kaldis, *Inorg. Chem.* **2002**, *41*, 628–630.

71 P. SCHAFFER, J. F. BRITTEN,
A. DAVISON, A. G. JONES, J. F. VALLIANT,
J. Organomet. Chem. **2003**, *680*, 323–328.

72 J. F. VALLIANT, O. O. SOGBEIN,
P. MOREL, P. SCHAFFER,
K. J. GUENTHER, A. D. BAIN,
Inorg. Chem. **2002**, *41*, 2731–2737.

73 A. P. SATTELBERGER, R. W. ATCHER,
Nat. Biotechnol. **1999**, *17*, 849–850.

74 R. ALBERTO, *Eur. J. Nucl. Med. Mol. I.*
2003, *30*, 1299–1302.

75 R. LA BELLA, E. GARCIA-GARAYOA,
M. BAHLER, P. BLAUENSTEIN,
R. SCHIBLI, P. CONRATH, D. TOURWE,
P. A. SCHUBIGER, *Bioconjugate Chem.*
2002, *13*, 599–604.

76 S. R. KARRA, R. SCHIBLI, H. GALI,
K. V. KATTI, T. J. HOFFMAN,
C. HIGGINBOTHAM, G. L. SIECKMAN,
W. A. VOLKERT, *Bioconjugate Chem.*
1999, *10*, 254–260.

77 J. SMITH, S. PEREZ, P. A. RUSHING,
G. P. SMITH, J. GIBBS, *Peptides* **1997**, *18*,
1465–1467.

78 C. J. SMITH, G. L. SIECKMAN,
N. K. OWEN, D. L. HAYES, D. MAZURU,
W. A. VOLKERT, T. J. HOFFMAN,
Anticancer Res. **2003**, *23*, 63–70.

79 R. LA BELLA, E. GARCIA-GARAYOA,
M. LANGER, P. BLAUENSTEIN,
A. G. BECK-SICKINGER,
P. A. SCHUBIGER, *Nucl. Med. Biol.* **2002**,
29, 553–560.

80 J. DU, J. HILTUNEN, M. MARQUEZ,
S. NILSSON, A. R. HOLMBERG, *Appl.
Radiat. Isotopes* **2001**, *55*, 181–187.

81 A. AMANN, C. DECRISTOFORO, I. OTT,
M. WENGER, D. BADER, R. ALBERTO,
G. PUTZ, *Nucl. Med. Biol.* **2001**, *28*,
243–250.

82 J. PETRIG, R. SCHIBLI, C. DUMAS,
R. ALBERTO, P. A. SCHUBIGER,
Chem.-Eur. J. **2001**, *7*, 1868–1873.

83 M. REISGYS, F. WUST, R. ALBERTO,
R. SCHIBLI, P. A. SCHUBIGER,
H. J. PIETZSCH, H. SPIES,
B. JOHANNSEN, *Bioorg. Med. Chem. Lett.*
1997, *7*, 2243–2246.

84 L. G. LUYT, H. M. BIGOTT, M. J. WELCH,
J. A. KATZENELLENBOGEN, *Bioorgan.
Med. Chem.* **2003**, *11*, 4977–4989.

85 J. B. ARTERBURN, C. CORONA,
K. V. RAO, K. E. CARLSON,
J. A. KATZENELLENBOGEN, *J. Org. Chem.*
2003, *68*, 7063–7070.

86 D. P. TRUMP, C. J. MATHIAS, Z. F. YANG,
P. S. W. LOW, M. MARMION,
M. A. GREEN, *Nucl. Med. Biol.* **2002**, *29*,
569–573.

87 C. M. JUNG, W. KRAUS, P. LEIBNITZ,
H. J. PIETZSCH, J. KROPP, H. SPIES,
Eur. J. Inorg. Chem. **2002**, 1219–1225.

88 J. KRIANGKUM, B. W. XU, L. P. NAGATA,
R. E. FULTON, M. R. SURESH, *Biomol.
Eng.* **2001**, *18*, 31–40.

89 J. F. TAIT, D. S. BROWN, D. F. GIBSON,
F. G. BLANKENBERG, H. W. STRAUSS,
Bioconjugate Chem. **2000**, *11*, 918–925.

90 S. HAMILTON, J. ODILI, O. GUNDOGDU,
G. D. WILSON, J. M. KUPSCH,
Hybridoma Hybridom. **2001**, *20*, 351–360.

91 R. SCHIBLI, R. SCHWARZBACH,
R. ALBERTO, K. ORTNER, H. SCHMALLE,
C. DUMAS, A. EGLI, P. A. SCHUBIGER,
Bioconjugate Chem. **2002**, *13*, 750–756.

92 D. RATTAT, P. A. SCHUBIGER,
H. G. BERKE, H. SCHMALLE, R. ALBERTO,
Cancer Biother. Radio. **2001**, *16*, 339–343.

93 E. O. FISCHER, M. W. SCHMIDT, *Chem.
Ber.* **1969**, *102*, 1954–1960.

5

Conjugates of Peptides and PNA with Organometallic Complexes: Syntheses and Applications

Nils Metzler-Nolte

5.1
Introduction

This chapter describes the labeling of peptides and peptide nucleic acids with transition metal organometallic compounds. The first organometallic bioconjugate of ferrocene was reported as early as 1957 [1], only six years after the first report on ferrocene. In this paper, Schlögl reported the synthesis and characterization of various ferrocene amino acids, including ferrocenylalanine and *p*-ferrocenyl-phenylalanine. Not only the reaction of ferrocene carboxylic acid with amino acid esters was described in this early paper, but also the first *N*-ferrocenylmethyl (Fem) amino acid derivative was reported. Several dozen papers were later published on these three groups of compounds, some of which are reviewed here. It is quite remarkable that this very first paper already covered the three most important groups of ferrocene amino acid and peptide derivatives. Since then, a large body of work has developed on conjugates of organometallic compounds with amino acids. These compounds will not be covered in this chapter, which is instead devoted to peptides. In addition, the conjugates of organometallics with di- and tripeptides will be excluded, since these are frequently just an extension of the simple amino acids. This work has been reviewed by Beck and coworkers [2].

In order for organometallic compounds to be suitable for the labeling of peptides, some general requirements must be fulfilled [3]. First and most importantly, the compounds must be stable in air and water in order to be of (potential) use under physiological conditions. From a synthetic point of view, the metal fragments should in most cases be amenable to standard peptide chemistry, that is coupling and deprotection reactions. Furthermore, at least some of the peptide conjugates were prepared via solid phase synthesis methods. In this case, the organometallic compound must also withstand the cleavage procedure. In addition to common means of purification in organometallic chemistry like sublimation and recrystallization, HPLC is an additional powerful means of purification for organometallic bioconjugates. Section 5.5 shows some examples of how organometallic groups

Bioorganometallics: Biomolecules, Labeling, Medicine. Edited by Gérard Jaouen
Copyright © 2006 Wiley-VCH Verlag GmbH & Co. KGaA, Weinheim
ISBN: 3-527-30990-X

may actually help in the preparation of peptides, as well as facilitate the purification by HPLC.

Conceptually, there are several ways in which an organometallic group can be attached to amino acids or peptides. The metal may be directly bound to the peptide backbone, i.e. the N-terminal amino group, nitrogen or oxygen atoms of the amide group(s), the C-terminal carboxylate or even the α-carbon atom. A metal fragment may also coordinate to functional groups in the side chain of amino acids, for example a cysteine thiol group, a lysine amino group or even an aromatic ring as a π donor. Finally, there are many examples in which a functionalized metal fragment (e. g. the cyclopentadienyl ring in ferrocene) is attached to the peptide via a suitable organic linker either to the α-carbon, one of the termini (most frequently the N terminus) or a functional side chain like a lysine amino group.

Organometallic compounds are of course known for a very long time, starting with Zeise's salt, Frankland's discovery of diethyl zinc, and the large family of metal carbonyl complexes. The real beginning of organometallic chemistry is, however, arguably dated with the discovery of ferrocene [4, 5] and the elucidation of its structure by Wilkinson [6] and Fischer [7]. Ferrocene has a remarkable chemical and physical stability [8]. Its chemistry is well established, and a number of derivatives such as ferrocene carboxylic acid or ferrocene carbaldehyde are commercially available. In that sense, it is almost ideally suited for bioconjugates along the criteria defined above. Consequently, more than half of the conjugates described in this chapter contain ferrocene as the organometallic moiety. A current review covers the bioorganometallic chemistry of ferrocene in almost all its aspects [9].

The organization of this chapter is as follows: in Section 5.2, conjugates of organometallics with small peptides without any (known) physiological function will be discussed; Section 5.3 summarizes work on organometallic bioconjugates with bioactive peptides, such as peptide hormones and neuropeptides, and deals with special aspects of solid phase synthesis with organometallics; Section 5.4 covers organometallic conjugates with DNA analogs, peptide nucleic acids (PNA). PNA is a topological analog of DNA. PNA oligomers interact with complementary DNA and RNA via Watson–Crick and Hoogsteen base pairing just as the homologous DNA does. Organometallic DNA derivatives are covered elsewhere in this book. Organometallic PNA derivatives are nonetheless included in this chapter because the chemistry of PNA is basically peptide chemistry, as PNA has an amino and a carboxy terminus. The last section (Section 5.5) covers applications of organometallics in peptide chemistry. Naturally, there is some overlap between this last section and the previous ones.

5.2
Conjugates of Organometallics with Small Peptides

Coupling of an organometallic carboxylic acid to amino acids and peptides via amide formation is probably the most extensively studied conjugation reaction. The first amino acid and dipeptide ferrocenoyl derivatives of this kind, Fc-CO-Aaa-OEt (Fc = $(C_5H_5)FeC_5H_4$, Aaa = Gly, Leu, Phe, Met), Fc-CO-Gly-OH and Fc-CO-Gly-Leu-OEt, were reported by Schlögl as early as 1957 [1]. Since then, well over two dozen similar ferrocenoyl amides have been prepared from ferrocene carboxylic acid 3 and amino acids, di-, tri- and tetrapeptides. These compounds, as exemplified for the glycine derivative Fc-CO-Gly-OH (4, Scheme 5.1), have been studied extensively with respect to their electrochemical properties and their structure in the solid state as well as in solution. A comprehensive discussion is found in a recent review from my group [9]. Kraatz has also recently summarized his work in the field [10].

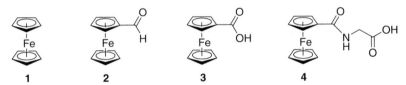

Scheme 5.1 Molecular structure of ferrocene (**1**), ferrocene carbaldehyde (**2**), ferrocene carboxylic acid (**3**) and ferrocenoyl glycine methyl ester (**4**, Fc-CO-OMe).

A 21-mer helical peptide with six Lys residues has also been studied [11, 12]. The Lys-N^{ε}-amino groups were derivatized with ferrocenoyl groups in order to investigate whether the overall conformation of the peptide would change upon derivatization with these large lipophilic groups. Spectroscopic data showed the conjugate to adapt a helical conformation similar to that of the unmodified peptide. Related results were observed upon derivatization of this 21-mer peptide with ferrocene carbaldehyde **2** to afford Schiff bases of the amino groups of lysine (Scheme 5.1). Another paper reported the reaction of a L-lysine polymer of MW 126.2 kDa with **3** [13]. In this case, about 6% of the lysine residues were carrying the ferrocenoyl tag. In the same work, a 52.1 kDa lysine copolymer, having the amino acid molar ratio Lys : Ala : Glu : Tyr of 34 : 45 : 14 : 7, was reacted with **3**. This resulted in derivatization of 14% of the lysine amino groups. These two water-soluble ferrocenoyl polymers were used for the construction of an amperometric glucose-sensing electrode by immobilizing them together with the enzyme Glucose Oxidase (GOD) on the surface of a glassy carbon electrode.

A number of bimetallic amino acid derivatives have been reported, mainly to explore the related chemistry. Our group has prepared di- and tripeptides **5** and **6** with organometallic labeling groups on both the N and the C terminus [14, 15]. Scheme 5.2 shows these compounds, which were prepared by coupling (activated) benzoic acid chromium tricarbonyl **7** or ferrocenylmethyl phenylalanine **8**, respectively, to the amino terminus of a ferrocenylmethyl amido dipeptide derivative.

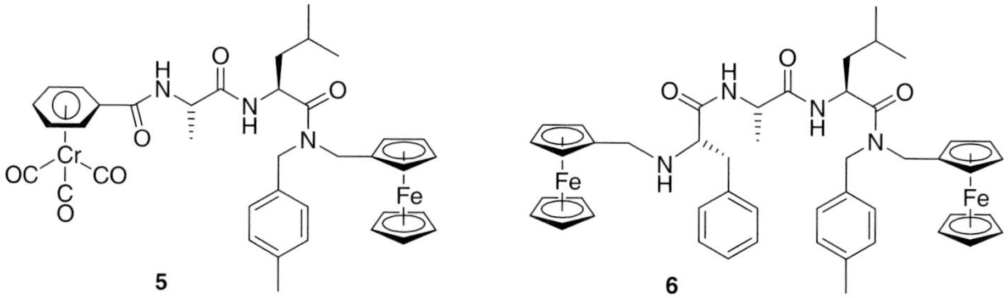

5   **6**

Scheme 5.2 Two bimetallic conjugates of the dipeptide Ala- Leu and the tripeptide Phe-Ala-Leu.

5.2.1
Organometallics as Templates for the Induction of Secondary Structural Elements in Peptides

5.2.1.1 Derivatives of 1,1′-Ferrocene Dicarboxylic Acid

In addition to monosubstituted ferrocene amino acid derivatives discussed above, a large number of 1,1′-disubstituted amino acid derivatives have been reported. Again, not all compounds will be presented herein. I will only concentrate on derivatives with longer peptide chains and their peculiar properties. In 1996, Herrick et al. pointed out that compounds of the general formula $Fe(C_5H_4$-CO-Aaa-OMe$)_2$, with amino acids (Aaa) others than proline, have an ordered structure in CH_2Cl_2 and $CHCl_3$ (Scheme 5.3) [16]. This ordered structure comprises two equivalent hydrogen bonds between the amide NH and the methyl ester carbonyl moiety of another strand which we will denote a "Herrick-type" conformation. Geometrically, such an arrangement resembles turn structures in peptides, which are not only ubiquitious structural motifs with importance for the biological function but are also a lead structure for drugs [17]. This observation has stimulated research efforts in order to elucidate the requirements for the use of organometallic compounds as turn mimetics in peptides. An overview of compounds in which each Cp ring is at least substituted by a dipeptide is given in Table 5.1.

Hirao and coworkers were the first to report and investigate compounds of the general composition $Fe(C_5H_4$-CO-Ala-Pro-OR$)_2$, with (R = Me, Et, n-Pr, Bn) [18, 19].

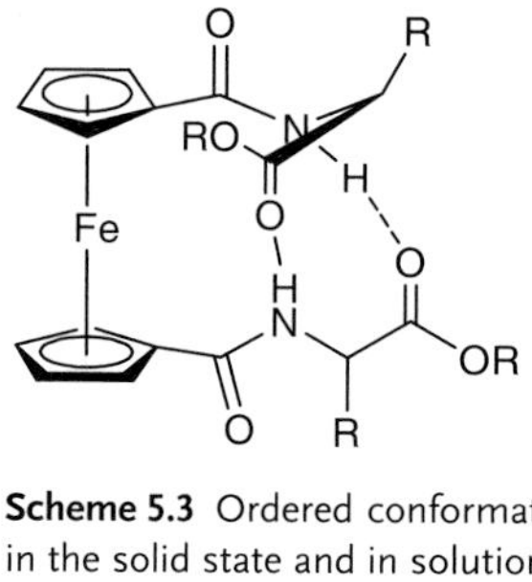

Scheme 5.3 Ordered conformation observed for $Fe(C_5H_4$-CO-AA-OMe$)_2$ in the solid state and in solution in non-polar solvents.

Table 5.1 Dipeptide derivatives of 1,1'-metallocene dicarboxylic acid M (C_5H_4-CO-substituent1)(C_5H_4-CO-substituent2) characterized by X-ray crystallography.

Metal M	Substituent 1	Substituent 2	Chirality of amino acids	Chirality at metallo-cene	Ref.
Fe	Ala-Pro-OMe	Ala-Pro-OMe	all *L*	*P*	[19]
Fe	Ala-Pro-OEt	Ala-Pro-OEt	all *L*	*P*	[18–20]
Fe	*D*-Ala-*D*-Pro-OEt	*D*-Ala-*D*-Pro-OEt	all *D*	*M*	[20]
Fe	Ala-Pro-OPr	Ala-Pro-OPr	all *L*	*P*	[19]
Fe	Ala-Pro-OBz	Ala-Pro-OBz	all *L*	*P*	[19]
Fe	Ala-Pro-NHpy	Ala-Pro-NHpy	all *L*	*P*	[32]
Fe	Ala-Pro-NHpy($PdCl_2$)	Ala-Pro-NHpy	all *L*	*P*	[32]
Fe	Ala-Pro-NHpyMe	Ala-Pro-NHpyMe	all *L*	*P*	[34]
Fe	Gly-Pro-OEt	Gly-Pro-OEt	all *L*	*P*	[22]
Fe	Gly-Phe-OEt	Gly-Phe-OEt	all *L*	*P*	[20]
Fe	Gly-Leu-OEt	Gly-Leu-OEt	all *L*	*P*	[20]
Fe	Ala-Phe-OMe	Ala-Phe-OMe	all *L*	*P*	[23]
Co^+	Ala-Phe-OMe	Ala-Phe-OMe	all *L*	*P*	[23]

As a representative example, a plot of the X-ray crystal structure of Fe(C_5H_4-CO-Ala-Pro-OEt)$_2$ **9** is depicted in Fig. 5.1 [18, 20]. The structure reveals an ordered conformation related to the one shown in Scheme 5.3, having intramolecular hydrogen bonds between the Ala-NH and the Ala-CO of the other strand, with NH$\cdots$OC contacts of 2.06 Å ("Herrick-type" conformation).

All these derivatives with varying R groups display a similar type of ordered structure in the solid state as well as in solution in aprotic solvents such as CDCl$_3$ or CH$_3$CN. This was derived by combining the results from NMR, IR and CD spectroscopic investigations [19, 20]. Ferrocene itself is an achiral compound. 1,1'-disubstituted derivatives, however, it may exist in two enantiomeric forms as shown in Scheme 5.4. In ferrocene, the barrier of rotation about the metal-Cp

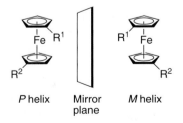

P helix   Mirror plane   *M* helix

Scheme 5.4 Helical chirality of 1,1'-disubstituted ferrocenes.

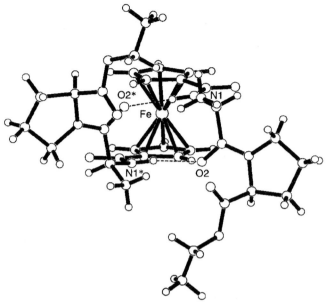

Fig. 5.1 Solid state structure of Fe(C$_5$H$_4$-CO-Ala-Pro-OEt)$_2$.
(Reprinted with modifications from [20], with permission from The American Chemical Society).

axis is only a few kcal mol^{-1} and therefore, rotation about this axis is usually fast, making the two enantiomeric forms interconvert rapidly [21]. This situation is changed by hydrogen bonds between the two peptide strands which will help to fix the two Cp rings in one position. CD spectroscopy showed that indeed such an ordered conformation exists in solution as was shown by the induction of chiral bands around 420 nm, corresponding to the absorption of ferrocene [19, 20]. In this picture, the orientation of the two substituents on ferrocene (*M* or *P* helix) is induced by the chirality of the amino acids. It is interesting to note that all but one compound in Table 5.1 are prepared from the naturally occuring L-amino acids. Consequently, all these derivatives display *P*-helical conformation at the ferrocene. Hirao and coworkers prepared and structurally characterized the D-amino acid containing compound Fe(C$_5$H$_4$-CO-D-Ala-D-Pro-OEt)$_2$ [20]. This compound is in an enantiomeric relationship to Fe(C$_5$H$_4$-CO-L-Ala-L-Pro-OEt)$_2$ (**9**) mentioned above, which is nicely illustrated by CD spectroscopy: Cotton effects of identical intensity, but opposite sign, were observed (Fig. 5.2).

The compounds Fe(C$_5$H$_4$-CO-Ala-Pro-OR)$_2$ (R = Me, Et, Pr and Bn) and Fe(C$_5$H$_4$-CO-Gly-Pro-OEt)$_2$ [22] possess only one NH group per peptide strand because proline is the only naturally occuring amino acid with a secondary amino group. In addition, peptide derivatives with two NH amide moieties have been prepared. These include Fe(C$_5$H$_4$-CO-Gly-Leu-OEt)$_2$, Fe(C$_5$H$_4$-CO-Gly-Phe-OEt)$_2$ [20] and Fe(C$_5$H$_4$-CO-Ala-Phe-OMe)$_2$ [23]. The amide NH moieties of the second amino acid are not involved in additional hydrogen bonding because they are orientated in such a way that the peptide strands are pointing away from each other in the

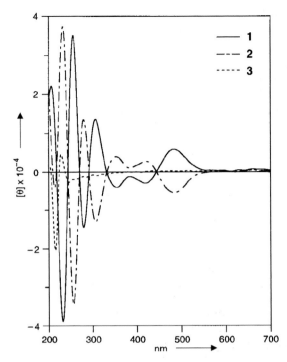

Fig. 5.2 CD spectra of Fe(C$_5$H$_4$-CO-L-Ala-L-Pro-OMe)$_2$ (solid line, **1**) Fe(C$_5$H$_4$-CO-D-Ala-D-Pro-OMe)$_2$ (hatched line, **2**) and Fe(C$_5$H$_4$-CO-C$_3$H$_6$-L-Ala-L-Pro-OMe)$_2$ (dotted line, **3**) in MeCN (0.1 mM). The propyl chain in **3** prevents formation of strong intramolecular hydrogen bonds and thus the last compound does not have a fixed chiral conformation at the ferrocene. (Reprinted from [20], with permission from The American Chemical Society).

solid state. These NH groups are involved in intermolecular hydrogen bond interactions, resulting in supramolecular assemblies.

We have investigated the strength of the hydrogen bonds in Fe(C$_5$H$_4$-CO-Ala-Phe-OMe)$_2$ (**10**) in comparison to that of its positively charged cobaltocenium derivative [23]. It appears that the hydrogen bonding interactions for the cobalt-ocenium derivative are slightly stronger than those for the ferrocene compound **10**. However, the differences are small and sensitive to experimental conditions. This finding correlates to the observation that positively charged amino acids (such as Lys or Arg) do not occur more frequently in turn structures in peptides [24].

To some extent, the structures discussed above resemble turn structures in peptides. A typical peptide turn structure is depicted in Scheme 5.5. Ring constraints by hydrogen bonds define γ-turns (7-membered ring) or β-turns (10-membered ring), whereas 13-membered rings are present in α-helices. Before and after the turn, the peptide extends as anti-parallel strands, often in β-sheets

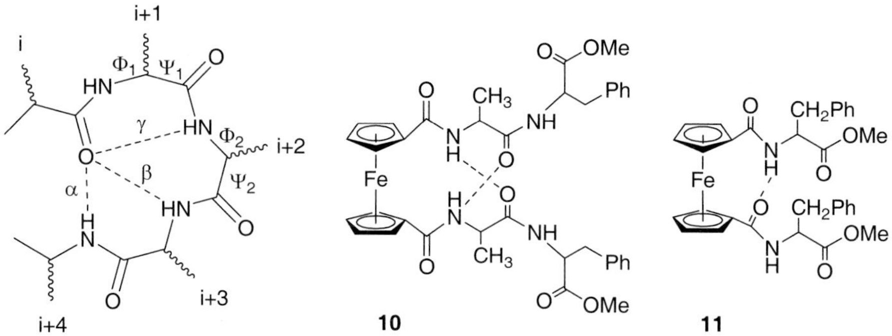

Scheme 5.5 Naturally occurring peptide turn structures which are stabilized by hydrogen bonds in comparison to intramolecular hydrogen bonds in **10** and **11**.

[25]. In all peptide derivatives of ferrocene-1,1'-dicarboxylic acid a parallel orientation of the peptide strands is enforced. Although these compounds may not be regarded as true β-turn mimetics [16], geometrical similarities remain. Based on this observation, we have suggested a nomenclature for such parallel turn mimetics [23], and this nomenclature is exemplified by the solid state structures of **11** and **10** (Scheme 5.5). For **11**, an unusual conformation is seen which we will denote "van Staveren-type" conformation. An 8-membered ring is formed across the hydrogen bond (H-N-$C_{C=O}$–$C_{Cp\text{-}ipso}$-Fe-$C_{Cp\text{-}ipso}$-C_α-O). Only one hydrogen bond forms between the *i* (C=O) and *i+2* (NH) amino acid. This type of turn is denoted as pseudo-γ_p (p for parallel). If a similar formalism is applied, **10** forms 11-membered rings across the intramolecular hydrogen bond between amino acids *i* and *i+3* and we denote this a pseudo-β_p turn. Because of the approximate C_2 symmetry of this molecule, the same rules apply to both Cp rings and their substituents.

Recently, Kraatz and coworkers prepared oligo-proline derivatives of the general formula $Fe(C_5H_4\text{-}CO\text{-}Pro_n\text{-}OBn)_2$, with $n = 1$–4 [26]. As observed for the mono substituted compounds $Fc\text{-}CO\text{-}Pro_n\text{-}OBn$ [27], the proline chains of the 1,1'-di-substituted compounds display the left-handed helical arrangement characteristic for polyproline II. For the compound $Fe(C_5H_4\text{-}CO\text{-}(Pro)_2\text{-}OBn)_2$, the ordered conformation adapted by the other 1,1'-disubstituted derivatives discussed above cannot form because the Pro-amide linkages lack NH moieties. The redox potential for the $Fe(C_5H_4\text{-}R)_2/Fe(C_5H_4\text{-}R)_2^+$ couple was found to shift to a small extent to less positive values in going from $Fe(C_5H_4\text{-}CO\text{-}Pro\text{-}OBn)_2$ to $Fe(C_5H_4\text{-}CO\text{-}Pro_4\text{-}OBn)_2$. A similar trend was observed for the monosubstituted derivatives $Fc\text{-}CO\text{-}(Pro)_n\text{-}OBn$ [27].

Interestingly, it was reported that the synthesis of $Fe(C_5H_4\text{-}CO\text{-}Pro_n\text{-}OBn)_2$ with $n = 2$, 3 or 4, yielded a second class of compounds, i.e. the "incomplete" substituted derivatives $Fe(C_5H_4\text{-}CO\text{-}Pro_n\text{-}OBn)(C_5H_4\text{-}CO\text{-}OBt)$ [26]. In a subsequent paper, Kraatz and coworkers reacted the "incompletely substituted" derivatives with $n = 3$ and 4 with a pepstatin analog to obtain the conjugates shown in Scheme 5.6 [28].

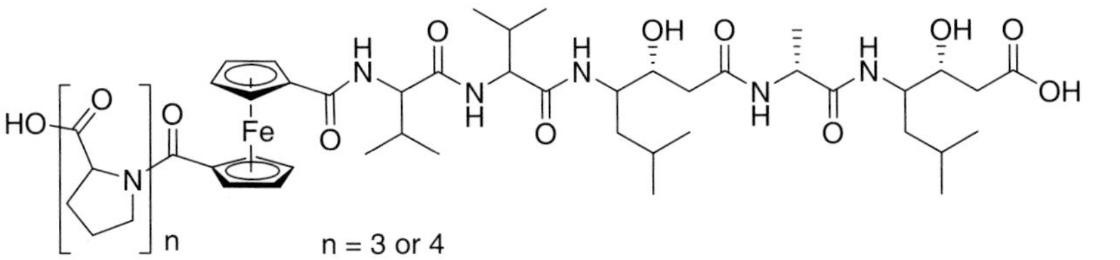

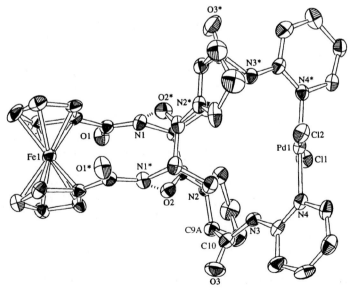

Scheme 5.6 Structure of a 1-oligoprolyl-1'-pepstatin ferrocene derivative.

Pepstatin is a potent naturally occurring inhibitor of aspartic proteases, including HIV-1 protease and pepsin [29–31]. It will be interesting to investigate the physiological applications of these ferrocene peptide conjugates (see also Section 5.3).

By incorporation of a suitable coordinating group, peptides may serve as (large) ligands for transition metals. The incorporation of phosphine ligands into helical peptides for organometallic compounds is discussed later in this chapter. Along the same idea, Hirao and coworkers prepared the compound Fe(C$_5$H$_4$-CO-Ala-Pro-NHPy)$_2$ (**12**, NHPy is the amide of 2-amino-pyridine) and subsequently transformed it into the *trans*-PdCl$_2$ complex **13**, with the pyridine nitrogen atoms coordinated to the Pd atom [32–34]. X-ray crystal structures for both compounds were reported; the one for **13** is reproduced in Fig. 5.3. Both compounds **12** and **13** adapt the usual "Herrick-type" conformation in the solid state as well as in solution, with the Ala-NH group being involved in intramolecular hydrogen interactions with the Ala-CO moiety of the other strand. In fact, coordination of the pyridine rings to the Pd atom stabilizes the intramolecular hydrogen bonds in solution as well as in the solid state.

Fig. 5.3 X-ray single crystal structure for the PdCl$_2$ complex **13**.
(Reprinted from [32], with permission from The American Chemical Society).

5.2.1.2 **Other Derivatives**

As discussed above, ferrocene-1,1′-dicarboxylic acid will almost inevitably induce parallel peptide strands. Anti-parallel peptide strands, which are indeed more frequently observed in nature, can be induced by 1′-aminoferrocene-1-carboxylic acid (Fca, **14**) (Scheme 5.7). This interesting compound, which represents one of the simplest ferrocene amino acids one can conceive, has been reported in 1998 independently by two groups, but could not be obtained in pure form [35] or only in low yield [36]. Subsequently, Rapic and coworkers were able to obtain **14**, along with differently protected derivatives from ferrocene-1,1′-dicarboxylic acid [37]. Very recently, Heinze and Schlenker published an improved synthesis of **14** [38]. In collaboration with Rapic' group, we have investigated a number of peptide derivatives, in which **14** serves as a pseudo-amino acid [39]. These peptides show interesting intramolecular hydrogen bonds in the solid state as well as in solution. An example is shown in Fig. 5.4. This compound, which is formally a (pseudo) tetrapeptide with the sequence Boc-Ala-Fca-Ala-Ala-OMe, was prepared by segment coupling reactions in solution and could be crystallized after purification by preparative thin layer chromatography. As shown in Fig. 5.4, two intramolecular hydrogen bonds form, which serve to stabilize this conformation also in solution. The compound has a *P*-helical conformation at the metallocene. In contrast to the derivatives of 1,1′-ferrocene dicarboxylic acid discussed above, the peptides extending from the C and N terminus of **14** form antiparallel strands. As such, **14** is not just an organometallic amino acid but also the first "real" organometallic turn mimetic.

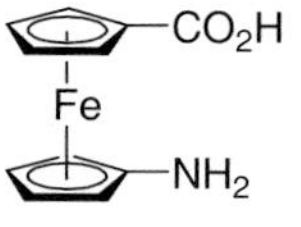

14

Scheme 5.7 The organometallic amino acid 1′-aminoferrocene-1-carboxylic acid **14**.

Another interesting example in which a C_3-symmetric peptide bundle is capped by an organometallic fragment of the same symmetry has been reported by Beck, Steglich and coworkers. The helix bundle is built up on a tri(benzylglycine) template with a central nitrogen atom (BzGly*). It is then capped by reaction with Cp^*TiCl_3 to yield compounds **15** (Scheme 5.8) [40]. Unpolar amino acids such as Leu, Phe and Gly were chosen for the peptide bundle which consisted of either a pseudo-nonapeptide ($n = 2$) or even a pseudo pentadecapeptide ($n = 4$). Depending on the length of the peptides ($n = 2$ or 4 for **15**), very large macro-bicyclic systems with high symmetry are formed [41]. The stereochemistry of the peptide bundles depends on the template and can be elucidated by CD spectroscopy in solution. Because these compounds are not only built on a rigid template but also capped by the half-sandwich Cp^*Ti fragment, very strong hydrogen interactions between individual peptide strands are observed.

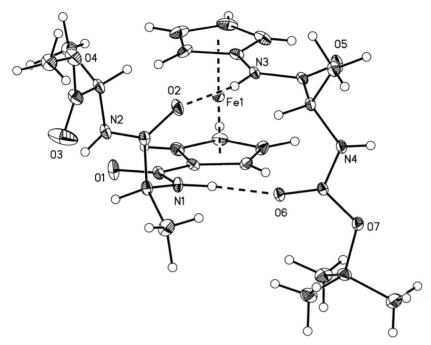

Fig. 5.4 X-ray single crystal structure of the tetrapeptide Boc-Ala-Fca-Ala-Ala-OMe containing the unnatural 1′-aminoferrocene-1-carboxylic acid (Fca).
Note the *P*-helical conformation at the metallocene and the intramolecular hydrogen bond.
(Reprinted from [39], with permission from The Royal Society of Chemistry).

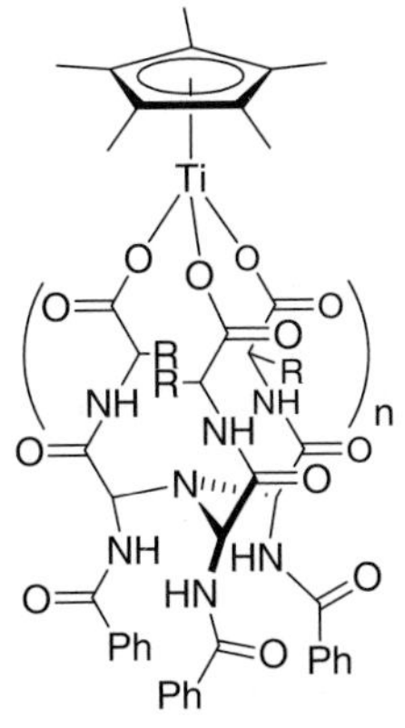

Scheme 5.8 Synthesis of a C_3 helical peptide bundle capped by Cp*Ti (n = 2 or 4, see text).

5.2.2
Peptides as Ligands for Organometallics

Amino acids in proteins serve as natural ligands for metal complexes. It is therefore surprising that chemists have only recently started to design larger peptides as ligands for metals. Very few attempts have been made to design natural peptides such that they may serve as ligands for *organometallic* metal fragments. The labeling of a single chain antibody fragment with $Tc(CO)_3$, which binds to a $(His)_6$ tag of the antibody fragment, would be one rare example (see Section 5.3.2 below). Examples where natural proteins do indeed serve as ligands for organometallic groups, such as in the hydrogenase enzyme family, are covered elsewhere in this book.

Gilbertson and coworkers have published a series of papers which comes very close to using peptides as ligands for organometallic metal fragments [42–49]. They have introduced phosphine-containing amino acids during the solid phase synthesis of peptides. After cleavage and deprotection, these phosphines may serve as ligands for late transition metal fragments such as Rh and Pd. Originally, peptide sequences were chosen such that a high helical content of the resulting peptide is expected [42–46]. Phosphino amino acids, for which several synthetic routes have been proposed [42, 44, 50, 46, 51, 48], were introduced into positions *i* and *i*+4. This ensures that the two phosphine ligands are close to each other in an α-helical structure and may readily bind a metal (Scheme 5.9). Reaction with $[Rh(NBD)Cl]ClO_4$ yields the metal complex as clearly shown by ^{31}P NMR spectroscopy [42–44]. Although inter-strand bridging of the Rh atom cannot be excluded, dilution and variable temperature NMR experiments and MS data suggest a chelating coordination mode as shown schematically in Scheme 5.9. More recently, the X-ray single crystal structure of a P-protected phosphorus-containing dodeca-peptide Ac-Ala-Aib-Ala-Pps(S)-Ala-Ala-Aib-Pps(S)-Ala-Ala-Aib-Ala-OH (**16**) could be obtained [45]. This structure shows that the first few residues were in a 3_{10}-helical conformation. The residues in the middle of the sequence are α-helical, while the last two C-terminal residues are in an extended conformation. As expected, the two phosphine groups are placed above each other along the helical backbone, thus well suited for metal chelation. Analysis of the solution structure by NMR revealed a similar conformation [45]. However, the conformation seems to change

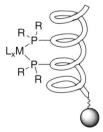

Scheme 5.9 Schematic drawing of a helical peptide on a solid support with two unnatural phosphino amino acids conveniently placed for metal coordination (ML_x).

somewhat upon coordination of Rh to the phosphine ligands. Unfortunately, it has not been possible so far to obtain either single crystals suitable for an X-ray structure determination or unambiguous NMR data for a solution structure of such a metalated phosphino peptide.

The Rh(diphosphino)peptide complexes with ligands similar to deprotected **16** were tested for their activity in the hydrogenation of the prochiral methyl 2-acetamidoacrylate, from which N-acetyl-alanine methyl ester was obtained in 100% yield. Initially, the enantiomeric excess (ee) was below 10%. It is interesting to note that the reaction could be performed with similar results either in solution or with the Rh(diphosphino)peptide complex still attached to the peptide synthesis resin [43]. Subsequently, the peptide ligands were optimized in a parallel combinatorial approach and ee values up to almost 40% for the asymmetric hydrogenation of enamides were obtained [46]. Again, similar selectivity was observed with purified peptide ligands in solution and with the peptide bound on the solid support. Because solid phase peptide synthesis protocols are optimized to give high purity products in good yield, the Gilbertson approach allows for the rapid development of selective catalysts for specific reactions or individual substrates by parallel synthesis. This is exemplified by the sequence optimization of peptide ligands with β-turn structures for the asymmetric allylic addition of malonate to acetoxycyclopentene, which was optimized to 75% ee [47]. Further optimization, including the phosphine ligands, yielded 95% ee under optimized conditions in solution for ligand **17** (Scheme 5.10) [48]. Peptide ligands were also optimized for the Pd-catalyzed desymmetrization of 2,4-cyclopentenediol [49]. In this case, β-turn structures yielded only moderate selectivity up to 37%. Far better selectivity was achieved with a Pps-Pro-Pps motif. A stepwise screening through 75 ligands identified peptide **18** as the best ligand with 76% ee on the solid support. Additional optimization of the reaction conditions in solution is expected to improve this number further. It is worth noting that these peptide ligands were specifically optimized for one single reaction. Even small changes in the substrate,

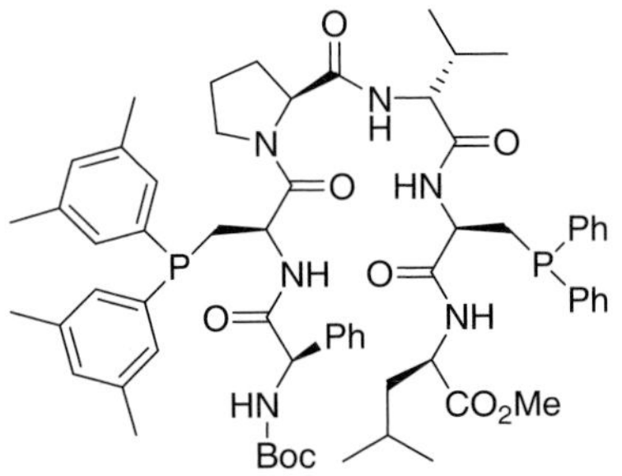

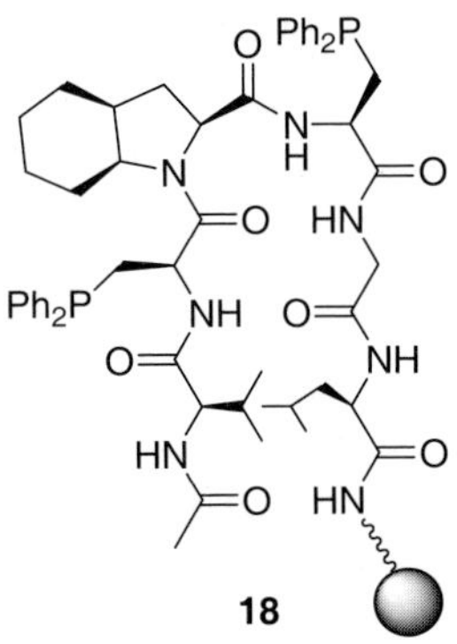

Scheme 5.10 Optimized peptide ligands for a Pd-catalyzed allylic addition and a desymmetrization reaction.
Left: Boc-D-Phg-Xps-Pro-D-Val-Pps-D-Leu-OMe (**17**).
Right: Ac-D-Val-Pps-Oic-Pps-Gly-D-Leu-NH-resin (**18**).

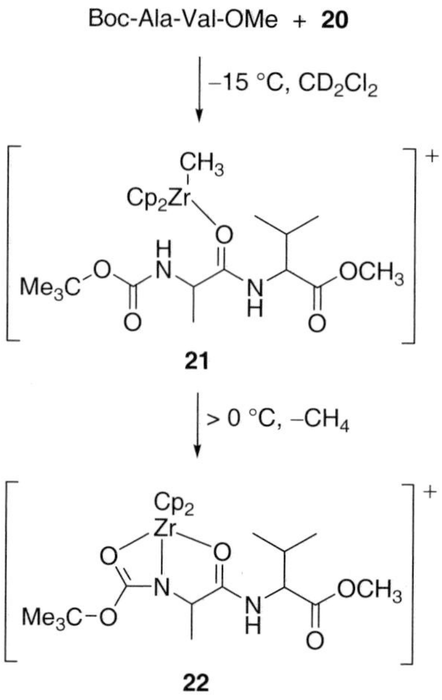

Boc-Ala-Val-OMe + **20**

21

22

Scheme 5.11 Optimized peptide ligands with for a Pd-catalyzed allylic
addition and a desymmetrization reaction.
Left: Boc-D-Phg-Xps-Pro-D-Val-Pps-D-Leu-OMe (**17**).
Right: Ac-D-Val-Pps-Oic-Pps-Gly-D-Leu-NH-resin (**18**).

for example from acetoxycyclopentene to acetoxycyclohexene in the reaction with
ligand **17**, causes a drop in selectivity from 95% ee to only 81% ee in solution
under otherwise identical conditions [48].

A different approach was chosen by Erker and coworkers. They have studied
the reaction of the organometallic cations $[Cp_2MCH_3(THF)]^+$ (M = Ti, **19**, or Zr,
20, "Jordan's cation") with protected di- and tripeptides and their derivatives. In
the case of Zr, a kinetic adduct **21** forms at low temperatures by coordination to
one of the carbonyl oxygen atoms [52]. Upon warming above 0 °C, this primary
adduct **21** eliminates CH_4 and the Cp_2Zr fragment coordinates in a chelating
mode to yield **22** as shown in Scheme 5.11 for the dipeptide Boc-Ala-Val-OMe.
Coordination to the N terminus yields the thermodynamically most stable adduct
as depicted for **22** in Scheme 5.11, although other coordination modes were
observed by NMR as transient species. Reaction of $[Cp_2ZrCH_3(THF)]^+$ **20** with
oligopeptide derived isocyanates yields the *N*-acetyl oligopeptides by methyl group
transfer to the isocyanate carbon atom [53]. Again, a similar N-terminal coordination mode as for compound **22** is observed. Figure 5.5 shows the result of an
X-ray single crystal structure analysis of the product from the reaction of **20** with
the isocyanate of Val-Val-OMe. The reaction of **19** with a range of 16 tripeptides
has been studied in detail [54]. In principle, similar products as for **20** were
obtained. Compared with the Zr compound **20**, however, the Ti analog **19** is far
less reactive. Consequently, more intermediates could be detected, and compounds

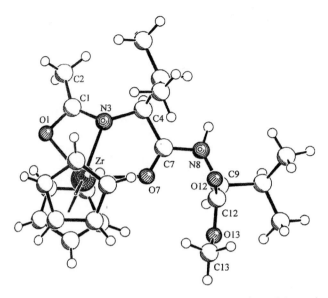

Fig. 5.5 X-ray single crystal structure of the Cp_2Zr-coordinated dipeptide Ac-Val-Val-OMe. (Reprinted from [53]).

with the N-terminal coordination mode were not in all cases found as the thermodynamic product.

5.3
Conjugates of Organometallics with Natural Peptides

There has been some interest in the labeling of natural peptides with organometallic compounds. Towards this aim, two main strategies have been followed. First, one or more amino acids in the peptide are replaced by organometallic derivatives. Scheme 5.12 shows organometallic amino acids which have been used to prepare analogs of natural peptides. For instance, ferrocenylalanine **23** may replace phenylalanine. Beck and coworkers published the synthesis of several other organometallic amino acids from reaction of bromoglycine with metalate anions [55]. Alternatively, the intact peptide has been labeled with a suitable organometallic derivative at the N terminus, the C terminus, or a side chain. Of these later possibilities, the first (derivatization of the N terminus) is by far the most frequently used, whereas only scare literature for the second possibility (derivatization of the C terminus) is available so far. This section deals mostly with peptides that have some biological relevance. In addition to compounds discussed in Section 5.2 above, ferrocene derivatives with peptides lacking a biological function have also been reported, including cyclo(D-ferrocenylalanyl-L-prolyl), cyclo(L-ferrocenylalanyl-L-prolyl) [56] and polypeptides containing repeating L-Fer-[Glu(OBzl)]$_4$ or L-Fer$_2$-[Glu(OBzl)]$_4$ units [57].

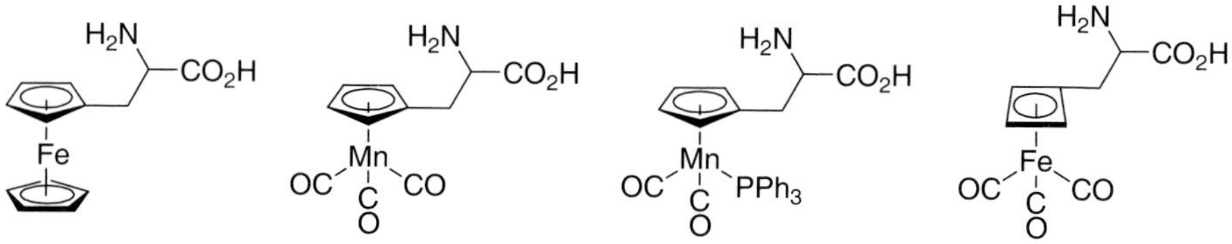

Scheme 5.12 Organometallic amino acids: Ferrocenylalanine (Fer, **19**), cymantrenylalanine (Cym, **20**), triphenylphosphine cymantrenylalanine (Cym(PPh$_3$), **21**), cyclobutadiene iron tricarbonyl alanine (**22**).

5.3.1
Organometallic Derivatives of Enkephalins

The first biogenic peptide that was modified with an organometallic compound is the pentapeptide [Leu^5]-Enkephalin. It has since been the most popular target peptide for derivatization with organometallics. [Leu^5]-Enkephalin has the primary structure H-Tyr-Gly-Gly-Phe-Leu-OH. [Leu^5]-Enkephalin was first isolated as a mixture with its position 5 methionine analog ([Met^5]-Enkephalin) by Hughes et al. from pig brain in 1975 [58]. These pentapeptides were detected in various human tissues but mainly in the brain and the central nervous system (CNS) [59]. The enkephalins are neuropeptides with an action similar to morphine, exhibiting high affinity for the opioid receptor [60]. In contrast to morphine, which has the highest activity at the μ and κ receptor subtypes, enkephalins bind preferentially to the δ receptor subtype. Differences in subtype specificity have been associated with different physiological functions.

Our group has prepared a number of [Leu^5]-Enkephalin derivatives with organometallic compounds attached to the N terminus via amide bond formation. In addition to the ferrocenoyl derivative **27** and the related cobaltocenium derivative **28** [61], we have also published the synthesis and characterization of two molybdenum derivatives **29** and **30** (see Schemes 5.13 and 5.14) [62]. In all cases, the enkephalin was prepared by standard solid phase peptide synthesis (SPPS) techniques the by Fmoc strategy. For compounds **27–29**, the metal complex was attached to the deprotected N terminus of the peptide as the final step on the solid support. Subsequent cleavage from the resin was achieved by conc. NH$_3$ solution in MeOH (**27** and **29**) or conc. TFA (**28**), respectively, in accordance with

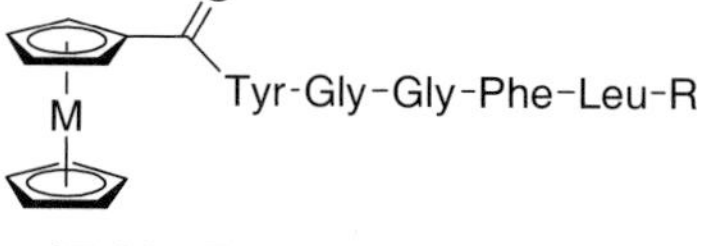

27: M = Fe, R = NH$_2$
28: M = Co$^+$, R = OH

Scheme 5.13 N-terminal metallocenoyl enkephalin derivatives.

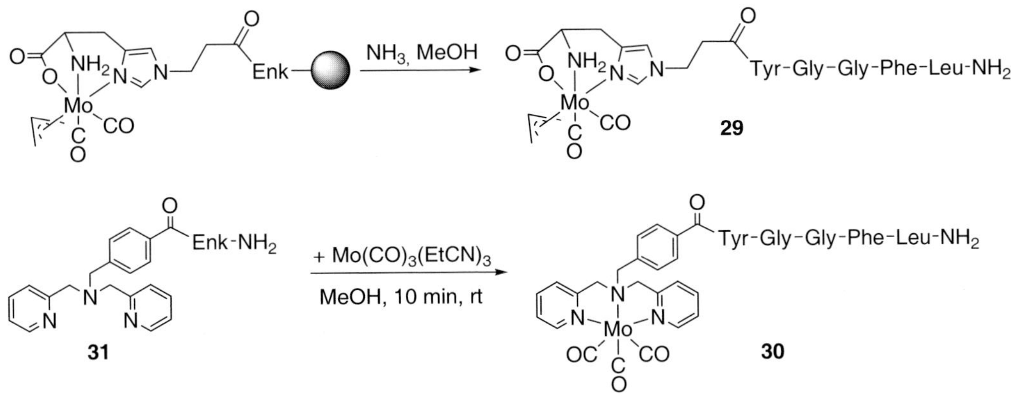

Scheme 5.14 Two different enkephalin derivatives with organometallic Mo carbonyl labeling groups (Enk = Tyr-Gly-Gly-Phe-Leu).

the chemical stability of the metal complexes. Naturally, resins and linkers were chosen accordingly. In contrast, compound **30** was prepared on the resin with only the bis(picolyl)amine (bpa) ligand attached to the N terminus of the peptide. The bpa-Enk derivative **31** was cleaved from the resin and purified by preparative HPLC. It was then reacted in solution with $Mo(CO)_3(CH_3CH_2CN)_3$ as the transfer reagent for the $Mo(CO)_3$ moiety [63]. The metal conjugate **30** precipitated from solution and was > 95% pure by HPLC. We have thus demonstrated two different strategies for the labeling of enkephalin, which we have called "on-resin labeling" (for **27–29**) and "post-SPPS labeling" (for **26**). The former extends the advantages of solid phase synthesis to the metal complex, that is high coupling yield and purity of the conjugate. The latter "post-SPPS" approach is more suitable for very labile metal complexes or in cases where a labeling is desirable immediately before use of the conjugate, e. g. for radioimaging purposes (see also Section 5.3.2 for further examples). The Mo histidinate complex that was used for the synthesis of **29** has very interesting spectroscopic properties [64, 65]. We were able to distinguish at least two independent fluxional processes, that is an endo–exo flip of the allyl ligand and a trigonal twist of the allyl-dicarbonyl face vs. the histidinate ligand, in the neutral as well as the one-electron oxidized form of this complex. Finally, we have prepared enantiomerically pure isomers with L- and D-His and used those for the labeling of enkephalin [66].

For all enkephalin conjugates, we have determined the water/octanol partition coefficient (log P values) [61]. These values are a relative measure of lipophilicity of a compound. It must be noted that enkephalin as such is not able to cross the blood–brain barrier (BBB), very likely because of its fairly polar nature. It is only released in the CNS in proximity to the receptors from a larger peptide called pre-enkephalin. In comparison, the conjugates **27–29** are much more lipophilic, with log P values over 3.5 for **27** and 3.0 for the Mo complex **29**. No difference in log P was found between the enkephalin diastereomers with Mo complexes derived from L-His or D-His.

The other class of enkephalin derivatives comprises compounds in which the phenylalanine in position 4 has been substituted by ferrocenylalanine (Fer, **23**) or cymantrenylalanine (Cym, **24**, see Scheme 5.15). As indicated before, the former was prepared by Schlögl as early as 1957 [1]. Later, improved syntheses for the racemic compound as well as for enantiomerically pure Fer were published [56, 57, 67, 68]. Compared with Phe, the cylindrically shaped Fer is much more bulky and lipophilic. The manganese tricarbonyl derivative Cym is also considered to be much more lipophilic than Phe [69]. Owing to these differences, results from binding assays for the modified peptide in comparison to those for the natural peptide could reveal important information about substrate–receptor interactions.

Three reports on the synthesis of [Fer^4, Leu^5]-Enkephalin appeared briefly one after another in the literature [70–72]. The compounds were synthesized by SPPS on a solid support using the classical Merrifield technique, that is Boc protection/TFA deprotection and cleavage from the resin with anhydrous HF. It was noted that ferrocene is oxidized to the green ferrocenium cation under the harsh conditions of cleavage, but this oxidation could be reversed by addition of ascorbic acid and sufficient material could be obtained for physiological studies [72]. Oxidative decomposition did not pose a problem for [Cym^4, Leu^5]-Enkephalin, which must however be protected from direct sunlight. Diastereomerically pure [D-Fer^4] and [L-Fer^4] enkephalins were obtained either by HPLC purification of the racemic [DL-Fer^4, Leu^5]-Enkephalin mixture as for the Cym derivatives [72], or by performing the solid phase peptide synthesis with enantiomerically pure L-Fer and D-Fer [71]. All metal enkephalins displayed a much reduced affinity for enkephalin receptors compared with natural [Leu^5]-Enkephalin, although the [D-Fer^4]-diastereomer was significantly more potent than the L-Fer analog [70, 72]. Of the two [Cym^4, Leu^5]-Enkephalin diastereomers, one isomer (presumably D-Cym) did indeed bind about one order of magnitude better than the Fer derivatives, while the other (presumably L-Cym) was comparable to D-Fer [72].

Sergheraert's group also attempted the synthesis of an enkephalin derivative with the organometallic amino acid cyclobutadiene iron tricarbonyl alanine **26** [73]. This compound, however, did not withstand the conditions of solid phase peptide synthesis and could not be isolated in pure form.

5.3.2
Organometallic Derivatives of Peptide Hormones

5.3.2.1 Substance P and Neurokinin A
Substance P (SP) is an undecapeptide, belonging to the class of tachykinins, with two consecutive Phe residues in positions 7 and 8 (Scheme 5.15). The peptide was first discovered by von Euler and Gaddum in 1931 [74], but it took another four decades before its amino acid sequence was determined by Chang and coworkers [75, 76]. SP is involved in several important physiological processes, such as contraction of the smooth muscles, pain transmission and activation of the immune system [77]. Tartar and coworkers synthesized the following three Fer-modified SP-derivatives by SPPS in which one or both Phe residues were

H-Arg-Pro-Lys-Pro-Gln-Gln-Xxx-Yyy-Gly-Leu-Met-NH$_2$

Xxx = Phe, Yyy = Phe : SP

Xxx = Fer, Yyy = Phe : [Fer^7]-SP

Xxx = Phe, Yyy = Fer : [Fer^8]-SP

Xxx = Fer, Yyy = Ile   : [Fer^7, Ile^8]-SP

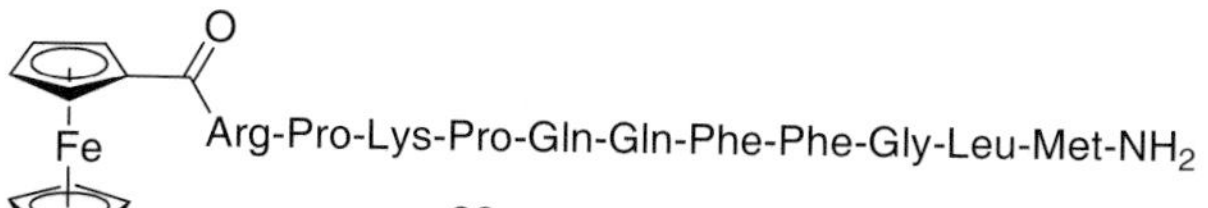

Scheme 5.15 Organometallic analogs of substance P (SP).

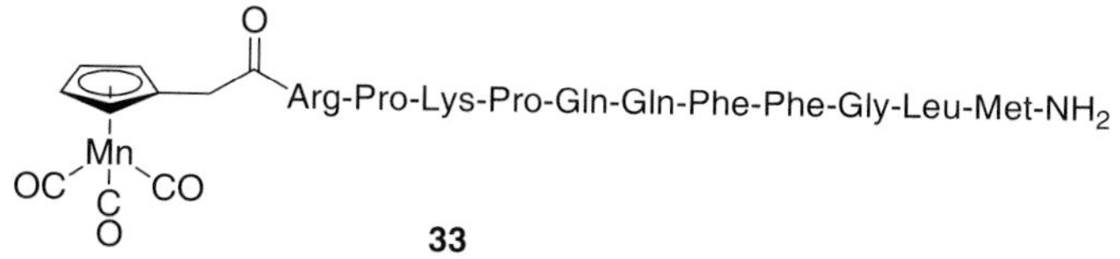

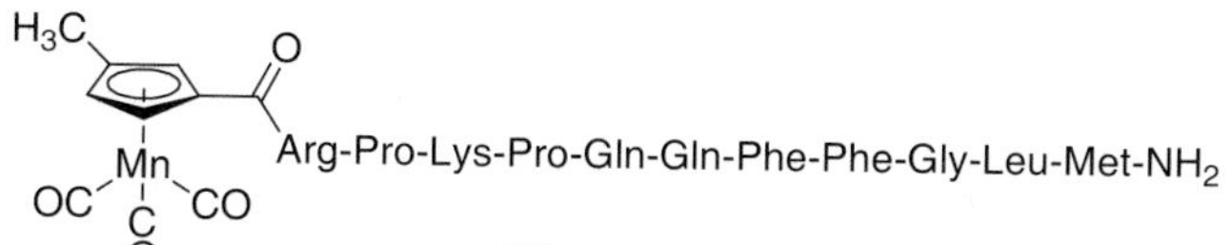

replaced by Fer: [DL-Fer^7]-SP, [DL-Fer^8]-SP and [DL-Fer^7, Ile^8]-SP (Scheme 5.15) [73]. Racemic Fer was used for the syntheses, resulting in diastereomeric mixtures, which could not be resolved by preparative HPLC. Results from binding assays show a decreased activity of the ferrocenylalanine substituted peptides compared with native SP in the following order: [DL-Fer^8]SP (2 · 10^{-2} relative to SP) > [DL-Fer^7]SP (3 · 10^{-3} relative to SP) > [DL-Fer^7, Ile^8]. (6 · 10^{-4} relative to SP).

In addition to the Fer derivatives discussed above, three organometallic derivatives of SP have been prepared via classical Merrifield SPPS methods in which the N-terminal amino group is acylated by metallocene carboxylic acid derivatives (Scheme 5.15) [78]. In contrast to the ferrocenylalanine derivatives presented above, no serious oxidation or decomposition was observed upon cleavage of the ferrocenoyl conjugate **32** from the resin with HF. The higher robustness of **32** has been attributed to the increase in oxidation potential of the ferrocene moiety, on account of the electron density withdrawing effect of the amide group. Structure activity studies on SP revealed that only the seven C-terminal amino acids are crucial for biological activity. Unlike enkephalin, in which the N-terminal amino group is crucial for receptor binding, a free amino group is not necessary for the biological activity of SP. Consequently, the spasmogenic potency of all three metallocenoyl derivatives was similar to that of native SP (0.93 ± 0.10 for **32** (Fc-CO), 0.51 ± 0.10 for **33** (Cym-CH$_2$-CO), and even 1.32 ± 0.10 for **34** (3-Me-Cym-CO) relative to native SP), determined by a guinea pig ileum assay [78].

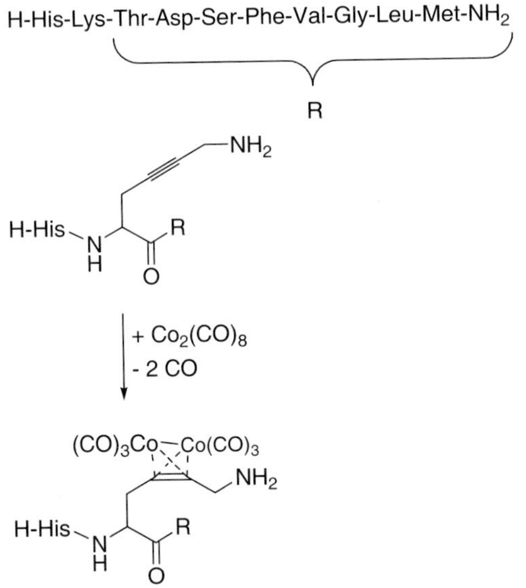

H-His-Lys-Thr-Asp-Ser-Phe-Val-Gly-Leu-Met-NH$_2$

R

—NH$_2$

H-His—N—H   R
         O

+ Co$_2$(CO)$_8$
- 2 CO

(CO)$_3$Co—Co(CO)$_3$
                —NH$_2$

H-His—N—H   R
         O

Scheme 5.16 Primary sequence of Neurokinin A (NKA) and labeling of an alkyne NKA derivative with dicobalt-hexacarbonyl.

Dicobalt-hexacarbonyl-alkyne complexes are another class of organometallic compounds with good stability under physiological conditions. Complexation of the alkyne proceeds smoothly under mild conditions by reaction with Co$_2$(CO)$_8$ under loss of two molecules of CO [79]. The applicability of this reaction to peptides was shown by Jaouen and coworkers by the reaction of Co$_2$(CO)$_8$ with protected 2-amino-4-hexynoic acid (Aha) and dipeptides thereof (Boc-Phe-Aha-OMe and Ac-Aha-Phe-OMe) [80]. Similarly, Cp$_2$Mo$_2$(CO)$_4$ complexes of these alkynes were obtained. It has been shown that the C-terminal Met[11] in SP can be replaced by isostere analogs without appreciable loss of physiological activity. The same is true for the C-terminal Met[10] in neurokinin A (NKA), another tachykinin peptide hormone (Scheme 5.16). Alkyne analogs of SP and NKA were obtained by replacement of these methionines with norleucine acetylene residues. Alternatively, Lys[2] in NKA may be replaced by an alkyne derivative which can also be complexed to Co$_2$(CO)$_8$ as shown in Scheme 5.16. Complexation with Co$_2$(CO)$_8$ proceeds smoothly in about 50% yield for all derivatives [81]. After HPLC purification, these cobalt alkyne peptides were comprehensively characterized spectroscopically. Most notably, they exhibit typical IR absorptions for the metal carbonyl moieties between 2000–2100 cm^{-1} [3]. Recently, there is renewed interest in Co$_2$(CO)$_6$(alkyne) complexes because of their cytotoxicity [82–84].

5.3.2.2 Angiotensin

The family of Angiotensin II (AT II) related compounds was also investigated by Tartar and coworkers [85, 86]. The amino acid sequence of AT II is H-Asp-Arg-Val-Tyr-Ile-His-Pro-Phe-OH with a phenylalanine in position 8. The octapeptide

Table 5.2 Relative biological activity of organometallic Angiotension II derivatives, expressed in % of the ID_{50} value vs. Saralasin® (ID_{50} = 2.57 nmol) as the standard. Values are taken from [86].

Angiotensin derivative	Relative affinity (L)	Relative affinity (D)
[Sar¹, Val⁵, Ala⁸]AT II (Saralasin®)	100	–
[Sar¹, Phe⁸]AT II	agonist	38
[Sar¹, Fer⁸]AT II	1.3	1.3
[Sar¹, Cym⁸]AT II	8.5	5.6
[Sar¹, Cym(PPh₃)⁸]AT II	12	7.8
[Sar¹, Car⁸]AT II	0.15	–

strongly increases blood pressure in mammals [87]. Several Angiotensin II derivatives were optimized for medicinal applications, such as the agonist [Asn¹, Val⁵]-AT II (Hypertensin®), which has been used to normalize blood pressure as quickly as possible after a shock or collapse. The antagonist [Sar¹, Val⁵, Ala⁸]-AT II (Saralasin®), on the other hand, has been used for the diagnosis of AT II-dependent forms of hypertonia. The following AT II derivatives were prepared and compared in their activity (see Table 5.2): Ferrocenylalanine [Sar¹, Fer⁸]-AT II, Cymantrenylalanine [Sar¹, Cym⁸]-AT II and [Sar¹, Cym(PPh₃)⁸]-AT II, where one carbonyl ligand has been photochemically replaced by a triphenylphosphine group [86]. The synthesis of all metallogenic AT II analogs was performed with racemic mixtures of the metallocene amino acids and the diastereomers were separated by HPLC for physiological testing in assays with rabbit aorta strips [86]. As shown in Table 5.2, both diastereomers of the Fer analogs show only about 1% affinity compared with Saralasin. The diastereomeric mixture, on the other hand, showed an activity of 6.1% relative to [Sar¹]-AT II in a binding assay to purified bovine adrenocortical membranes [85]. Unfortunately, it is not clear whether an un-expected additive effect prevails in this case as no data for the separated diastereo-mers is available. Even so, this value is far less than the [Sar¹, D-Phe⁸]-AT II derivative. Interestingly, the [Sar¹, L-Phe⁸]-AT II diastereomer is an agonist. This behavior is not reproduced by any of the metallocenic L analogs. The [Sar¹, Cym⁸]-AT II derivative had a slightly higher affinity, and some difference between the diastereomers is observed. It seems that the sterically more demanding Fer derivative decreases binding affinity and the $Mn(CO)_3$ moiety is better accommo-dated in the binding pocket. This view is substantiated by the very low binding affinity of the carboranylalanine derivative [Sar¹, Car⁸]-AT II, which is even bulkier than Fer. On the other hand, photochemical substitution of one CO ligand in Cym with a PPh_3 group, although dramatically increasing hydrophobicity and bulkiness, yields a derivative with even higher affinity. These results are in qualitative agreement with structure activity studies which demonstrate that positions 1–7 define specificity, duration and intensity of the biological effect, whereas residue 8 does not influence these parameters but establishes the hormone derivative as an agonist or antagonist. These results also resemble the findings for enkephalin discussed above, in that replacement of Phe by Cym is much better tolerated than Fer substitution.

5.3.2.3 Bradykinin

The same group also prepared ferrocenylalanine Bradykinin analogs as shown in Scheme 5.17 [73]. The nonapeptide Bradykinin (BK) is a tissue hormone involved in the blood clotting process. When the hormone is released from the precursor kininogen, it induces a lowering of the blood pressure via dilation of the blood vessels [88, 89]. Either of the Phe residues at positions 5 or 9 has been substituted by DL-Fer, resulting in the conjugates shown in Scheme 5.17. Although separation of the diastereomers was accomplished by HPLC, it was not possible to assign their absolute configuration. Again, binding studies revealed a significant decrease of activity upon substitution of Phe by Fer. The two diastereomers of $[Fer^5]$-BK and $[Fer^8]$-BK showed activities relative to native BK of $7 \cdot 10^{-2}$ and $3.5 \cdot 10^{-2}$, and $1.8 \cdot 10^{-2}$ and $1.7 \cdot 10^{-2}$, respectively. These results show the same trend as those obtained by Couture and coworkers, who found that introduction of the bulky amino acid carboranylalanine in position 9 had a larger impact on the activity than carboranylalanine substitution of the position 5 Phe residue [90, 91]. Considering the low activities of the diastereomeric [Fer]-BK analogs, it appears that the ferrocenyl moiety poorly interacts with the receptor.

H-Arg-Pro-Pro-Gly-Xxx-Ser-Pro-Yyy-Arg-OH

Xxx = Phe, Yyy = Phe : BK

Xxx = Fer, Yyy = Phe : $[Fer^5]$-BK

Xxx = Phe, Yyy = Fer : $[Fer^8]$-BK

Scheme 5.17 Organometallic analogs of bradykinin (BK).

5.3.2.4 Gonadotropin-releasing Hormone

Cymantrenylalanine has also been incorporated in the gonadotropin-releasing hormone GnRH, also known as luteinizing hormone-releasing hormone LHRH (Glp-His-Trp-Ser-Tyr-Gly-Leu-Arg-Pro-Gly-NH$_2$) [69]. It has been shown that replacement of Gly in position 6 by bulky hydrophobic amino acids increases the biological activity. In this case, stereochemistry plays a major role as for instance $[D\text{-Phe}^6]$GnRH is 10 times more active than GnRH [92]. The diastereomers of $[Cym^6]$GnRH and $[Cym(PPh_3)^6]$GnRH could be separated by reverse-phase HPLC. For both compounds, the slower eluting isomer displayed higher biological activity. This seems to imply that it is the D-isomer, in line with the observation that the D-diastereomer of metallocenic peptides generally elutes slower than the L-diastereomer. On the other hand, the potency of these organometal derivatives is only 147% ($[Cym^6]$GnRH) and 193% ($[Cym(PPh_3)^6]$GnRH) of GnRH, respectively. While the more hydrophobic triphenylphosphine derivative is indeed more active as expected, steric hindrance is also a relevant factor as shown by the much lower potency compared with $[D\text{-Phe}^6]$GnRH. On the synthetic side it is interesting to note that the PPh$_3$-substitution of Cym could not be achieved on the peptide in good yield. The amino acid was rather prepared by irradiation of *N*-Boc-Cym-OMe in the presence of excess PPh$_3$ in a benzene/cyclohexane mixture, followed by ester hydrolysis. It was, however, possible to incorporate Cym(PPh$_3$) into a peptide by solid phase peptide synthesis techniques. Protection from sunlight was the only necessary precaution for this amino acid.

5.3.2.5 Secretin

The [CpRu]$^+$ fragment binds strongly to arene rings such as benzene. This high arenophilicity has been exploited for the labeling of aromatic amino acids by [CpRu]$^+$ as outlined in Scheme 5.18 [93–96]. This reaction is also highly selective in the presence of other metal binding peptide side chains such as His, Ser, Arg or Glu, as was demonstrated recently by Grotjahn and coworkers for the labeling of secretin [97, 98]. Secretin is a gastrointestinal peptide hormone with 27 amino acid residues and a molecular weight of about 3 kDa. Secretin has a unique phenylalanine residue at position 6 along with an *N*-terminal His, four Arg, one Asp, and two Glu residues. In the labeling experiment, the complex **35** was added to an aqueous solution of secretin in an NMR tube at room temperature (Scheme 5.19). After standing for 8 h at room temperature, a relatively clean NMR spectrum was obtained. Most significantly, the phenylalanine signals in the aromatic region are quantitatively shifted to higher field (ca. 6.2 ppm) which is diagnostic for coordination of arenes to the [CpRu]$^+$ moiety. The conjugate could be purified by HPLC and characterized by ESI-MS. Edman degradation provides additional proof that the site of labeling is indeed the phenyl ring of Phe, since the peak for the phenylthiohydantoin of Phe is clearly missing in the HPLC of partially degraded Ru-secretin. The site-selectivity of this reaction in the presence of over 25 concurring amide bonds, carboxylate groups, and even the imidazole ring of His is astonishing. Grotjahn and coworkers observed a "water-effect" in

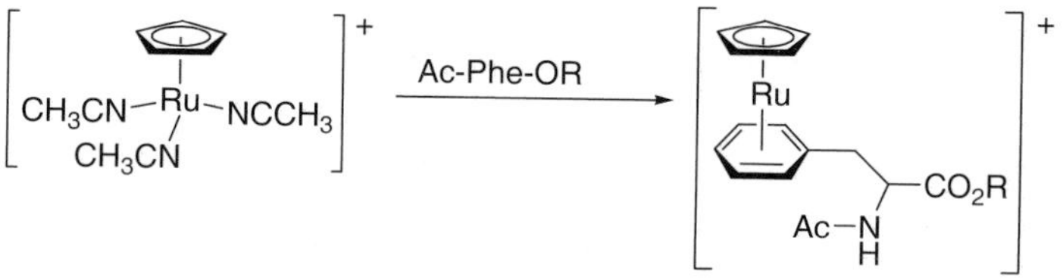

Scheme 5.18 Coordination of the [CpRu]$^+$ fragment to aromatic amino acids, exemplified by reaction of [CpRu(CH$_3$CN)$_3$]$^+$ with phenylalanine.

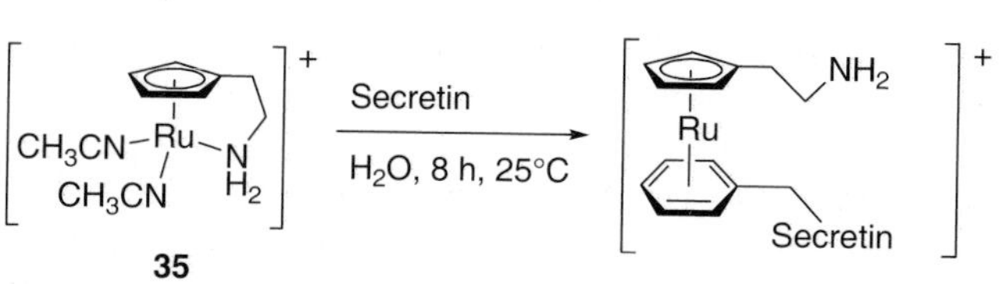

Scheme 5.19 Labeling of the Phe side chain of secretin with [CpRu]$^+$.

that the migration of the [CpRu]$^+$ fragment from the primary products with N, S, or O coordination to the thermodynamically favorable π complex is considerably faster for secretin in water than for model peptides such as Boc-Phe-His-OMe in water/methanol mixtures. Another interesting application of Ru-arene complexes in the synthesis of (pseudo-)peptides is discussed below in Section 5.2.3.

5.3.3
Organometallic Derivatives of Other Peptides

5.3.3.1 Nuclear Localization Signal

Noticeably, all metallocene derivatives treated so far in this section were of peptide hormones. We have recently initiated a programme to use cellular signalling peptides for the directed delivery of organometallic moieties inside a cell. In this project, we have prepared metallocenoyl derivatives of a nuclear localization signal (NLS, Scheme 5.20) [99]. NLS conjugates with an N-terminal ferrocenoyl (**36**) or cobaltocenium carbonyl group (**37**) were prepared by appropriate SPPS techniques on the resin and comprehensively characterized.

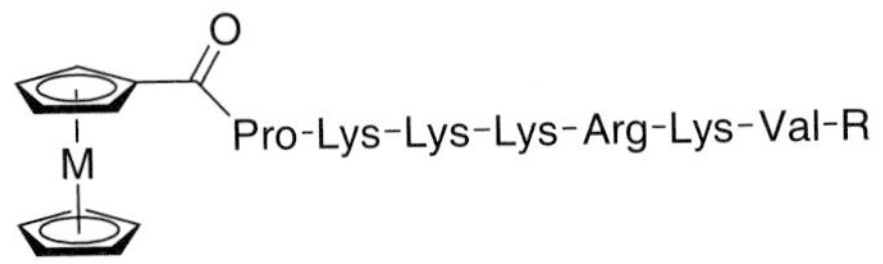

36: M = Fe, R = NH$_2$
37: M = Co$^+$, R = OH

Scheme 5.20 Two metallocene derivatives of the intracellular signaling sequence NLS.

The prudent choice of appropriate side chain protecting groups for Lys and Arg did pose a particular problem in combination with the chemical stability of the metallocenes and available resins and linkers. In order to study cellular uptake and localization in the nucleus of living cells, derivatives of **36** and **37** with an additional fluorescent dye attached to the peptide were also prepared. Cellular uptake and localization of these conjugates was then followed in different cell lines by fluorescence microscopy. This is the first study of sub-cellular localization of organometallic peptide derivatives with non-radioactive metals (see also Section 5.5.3.2 below for an example of a dual radiometal- and fluorescent-labeled peptide).

5.3.3.2 Glutathione

Thiol derivatization of cysteine residues in Glutathione (γ-glutamylcysteinylglycine, GSH) has been achieved with CpFe(CO)$_2$(η^1-*N*-succinimidate) **38** (Scheme 5.21) [100, 101]. The sulfhydryl group adds to the succinimidate double bond of **38** as shown by reaction with GSH. Similar to results for the Fed protecting group described below (Section 5.1.1), the reaction is not entirely selective for cysteine residues. At higher temperature and metal-to-protein ratios, lysine groups also

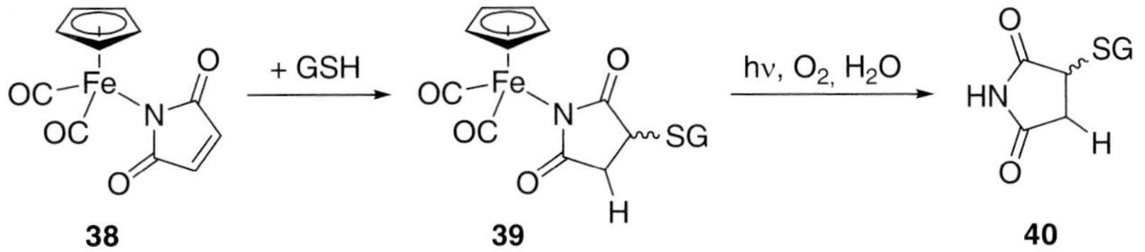

Scheme 5.21 Labeling of glutathione with $CpFe(CO)_2$(imidate) and photochemical decomplexation.

reacted with complex **38** as shown in a detailed investigation of the reactivity of **38** with BSA [100]. Interestingly, the Glutathione conjugate **39** (**38**-GS) was found to be a competitive inhibitor of the enzyme glutathione-(*S*)-transferase (GST). This enzyme catalyzes the nucleophilic addition of glutathione to electrophiles. For the model electrophile 1-chloro-2,4-dinitrobenzene, the organometallic conjugate **39** was a competitive inhibitor with a K_i of about 30 μM (equine liver GST) [101]. In contrast, the strong competitive inhibtor (*S*)-hexylglutathione had a $K_i = 1.1$ μM. After irradiation with visible light from a tungsten lamp, complex **39** is decomposed. The remaining fragment **40** is *not* an effective inhibitor of GST anymore. This is the first demonstration that an enzyme inhibiting effect in a peptide structure is due to the presence of an organometallic moiety and disappears after photochemical removal of this moiety [101].

5.3.3.3 Papain Inhibitors

Papain is a 212 amino acid proteolytic enzyme from the papaya fruit. The X-ray crystal structure of several papain-inhibitor complexes have been reported [102–104]. Even earlier, Kaiser et al. showed that papain can be inhibited by the tetrapeptide H-Gly-Gly-Tyr-Arg-OH [105]. The optimum pH for inhibition is 6.2. Furthermore, the inhibitor could even be immobilized on a polymer resin without loosing its binding affinity to papain. In order to develop an electrochemical sensor, Kraatz and coworkers prepared the *N*-ferrocenoyl peptide, Fc-CO-Gly-Gly-Tyr-Arg-OH (**41**) by fragment condensation of Boc-Gly-Gly-OH and H-Tyr(Bzl)-Arg(NO_2)-OMe, followed by Boc deprotection, coupling to Fc-CO_2H, methyl ester cleavage and reductive removal of the protecting groups by H_2 with Pd/C in MeOH/H_2O (3 : 1) [106]. The crude product was purified by column chromatography and comprehensively characterized. Compound **41** is indeed a competitive inhibitor of papain with an inhibition constant $K_i = 9$ μM at pH 6.2. Compared with Kaiser's results, the presence of the N-terminal ferrocenoyl group increases the affinity to papain. Binding of **41** to papain could also be demonstrated by significant changes of the CV of **41** in the presence of papain. First, a shift of the cathodic wave of about 30 mV is observed. Second, the peak current of both waves decreases as a consequence of the smaller diffusion coefficient of **41** bound to the larger papain molecule. Finally, an increase in the peak separation is attributed to a slowing of the electron transfer process from the **41**-papain complex to the electrode. The same group has published electron transfer studies on helical ferrocenoyl-oligoproline-cystamine conjugates which were bound to a gold surface [107].

It will be interesting to follow the development of surface-bound electrochemical protein sensors based on metallocene-peptide conjugates.

5.3.3.4 **Alamethicin**

Another interesting application of the electrochemical properties of ferrocene is related to the formation of ion channels. Eisenthal and coworkers prepared two ferrocenoyl derivatives of the channel forming peptide alamethicin (ALM) [108]. Alamethicin is produced by the fungus *Trichoderma viride*, and belongs to the class of peptaibols. Alamethicin self-associates in lipid bilayers, in this way forming voltage-dependent ion channels [109–111]. One of the derivatives was prepared from ferrocene carboxylic acid, whereas the other was synthesized from 1,1′-ferro-cene-dicarboxylic acid. Therefore, the compounds differ in the unconjugated cyclopentadienyl ring; one of the compounds has a Cp ring, whereas the other has a η-C_5H_4-COOH moiety (Scheme 5.22). In both cases, the ferrocenoyl groups were linked via ester formation to the C terminus of the 20-mer peptide in solution. This is indeed one of the very rare cases of C-terminal derivatization.

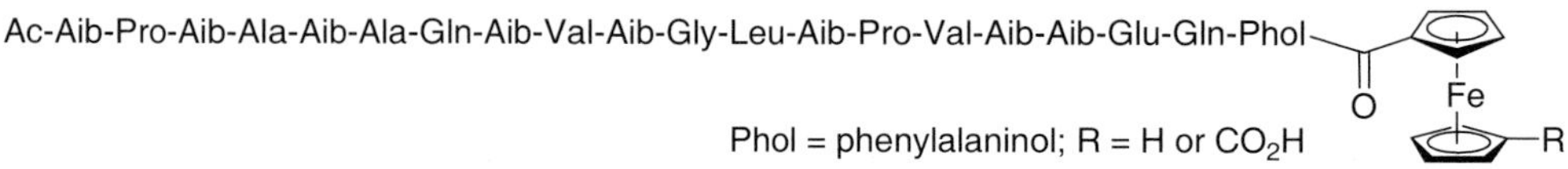

Ac-Aib-Pro-Aib-Ala-Aib-Ala-Gln-Aib-Val-Aib-Gly-Leu-Aib-Pro-Val-Aib-Aib-Glu-Gln-Phol

Phol = phenylalaninol; R = H or CO_2H

Scheme 5.22 Alamethicin ferrocene derivatives. Note that the ferrocenoyl group is bound to the C terminus via ester formation.

In the reduced (neutral) form, both ferrocenoyl alamethicin derivatives are shown to form voltage-dependent channels in planar lipid bilayers at positive potentials, with conductance properties similar to unmodified alamethicin. When they were incorporated in the lipid bilayer, oxidation of the ferrocenoyl peptide ALM-CO-CpFe(Cp-COOH) results in a shorter lived channel whereas the oxidation of ALM-CO-Fc causes a time-dependent elimination of channel-openings, which can be restored by increasing the trans-bilayer potential. Pretreatment of the ferrocenoyl peptides with oxidizing agents alters their single-channel properties in a qualitatively similar manner.

5.3.3.5 **Others**

In addition to ferrocenylalanine, the difunctionalized derivative 1,1′-ferrocenyl-*bis*-alanine **42** (Scheme 5.23) has been reported [68, 112–117]. Via a sophisticated synthetic route, Frejd and coworkers synthesized an enantiomerically pure 1,1′-ferrocenyl-*bis*-alanine derivative **43** which contains four orthogonal protecting groups (Scheme 5.24) [115–118]. By catalytic hydrogenation, optically pure **43** was conveniently transformed into *S, S*-**44**. This compound is basically the methyl ester of *N*-Boc 1,1′-ferrocenyl-*bis*-alanine with a free amino group and a free carboxylic acid group. When **44** was reacted with the peptide coupling reagent PyAOP, a mixture of the 1,1′-ferrocenophane lactam **45** and two macrocyclic peptides, namely a dimer **46** and trimer **47**, was obtained [118].

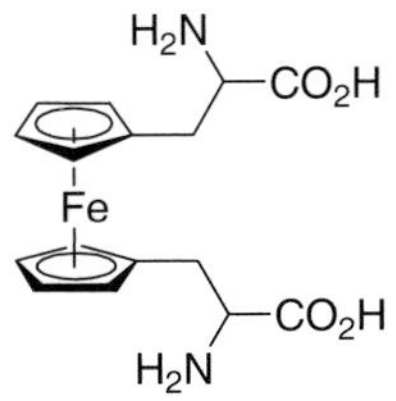

Scheme 5.23 Molecular structure of 1,1′-ferrocenyl-*bis*-alanine **42**.

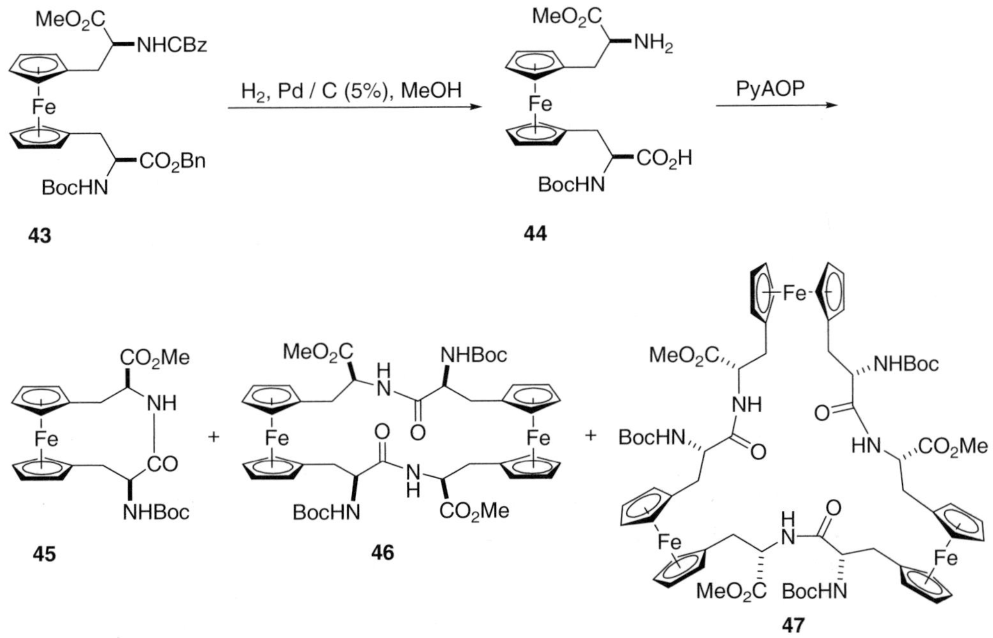

Scheme 5.24 Synthesis of the methyl ester of *N*-Boc-1,1′-ferrocenyl-*bis*-alanine (**44**) and formation of cyclic peptides.

Frejd and coworkers then used lactam **45** as a structural H-Phe-Phe-OH mimic in peptides [116]. The resulting conjugate **48** was expected to function as a structural mimic of substance P (Scheme 5.25). However, CD spectroscopic studies indicated that the constraints imposed by the 1,1′-ferrocenophane prohibited the peptide to adopt the characteristic α-helical secondary structure found for native SP in a biomimetic SDS (Sodium *n*-Dodecyl Sulfate) micellar environment. In another interesting paper, Frejd and coworkers prepared a [Leu^5]-Enkephalin mimetic in which the H-Tyr-Gly-Gly-Phe-subunit has been replaced by a Gly-Gly-looped 1,1′-ferrocenyl-*bis*-alanine residue (**49**, Scheme 5.25) [117]. The ferrocene Cp rings of this constrained compound constitute a substitute for the aromatic phenol (Tyr) and phenyl (Phe) rings. NMR-spectroscopic studies revealed this

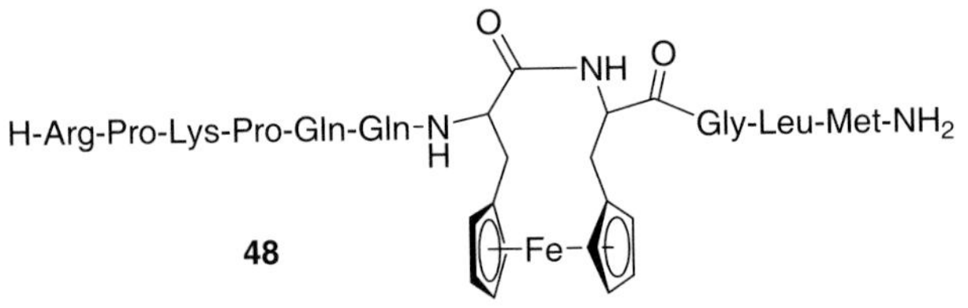

H-Arg-Pro-Lys-Pro-Gln-Gln-N(H)—[...]—NH—[...]—Gly-Leu-Met-NH₂

48

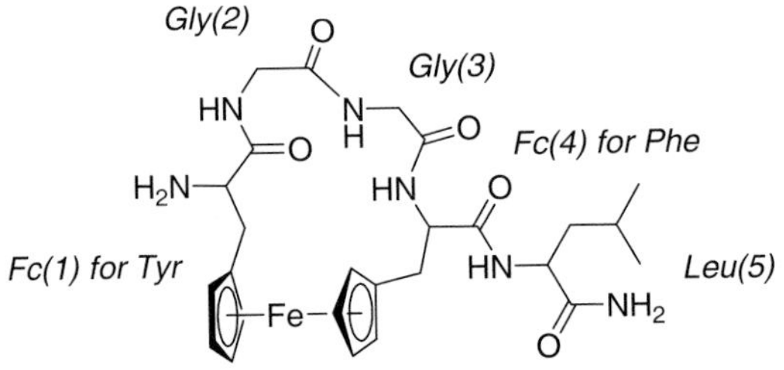

49

Scheme 5.25 Incorporation of derivatives of 1,1′-ferrocenyl-*bis*-alanine **42** as dipeptide mimics into substance P (top, **48**) and enkephalin (bottom, **49**).

conjugate to possess a hydrogen bond between the Fer CO group and the NH moiety of the other Fer ring. This conformation may well serve as a model for native [Leu⁵]-Enkephalin in the single-bend conformation, which is stabilized by a β-turn [119, 120]. Unfortunately, binding studies of this compound to the opioid receptor have not been reported as yet.

5.3.4
Enzymatic Degradation of Organometallic Peptide Derivatives

Two organometallic derivatives of the pentapeptide H-Lys-Gly-Phe-Gln-Gly-OH were prepared and their stability towards hydrolytic degradation by chymotrypsin was investigated [73]. Chymotrypsin belongs, together with trypsin and elastase, to the class of serine proteases and is present in the human pancreas. The high selectivity of the enzyme Chymotrypsin for the scission of peptidic bonds next to the aromatic amino acids Phe, Trp or Tyr has been rationalized based on the X-ray crystal structure of bovine chymotrypsin [121, 122]. The aromatic amino acid Phe was replaced by Fer and Cym to yield the peptides H-Lys-Gly-Fer-Gln-Gly-OH and H-Lys-Gly-Cym-Gln-Gly-OH, respectively. Upon the action of chymotrypsin, the Phe-containing wild-type pentapeptide was hydrolyzed with K_m = 3.85 mM, whereas the Fer-congener was completely stable and not hydrolyzed at all. Interestingly, the analog containing the unnatural amino acid cymantrenylalanine was hydrolyzed, albeit with a lower affinity (K_m = 37 mM, but unchanged V_{max}) than the wild-type Phe derivative [73].

5.4
Conjugates of Organometallics with PNA

5.4.1
Conjugates of PNA Monomers

Peptide Nucleic Acids (PNA) are a class of DNA analogs with very promising properties for applications in molecular biotechnology or medicine. Compared with DNA, the ribose phosphate ester backbone is replaced by a pseudo-peptide backbone in PNA (Scheme 5.26). The nucleobases are linked to this backbone via a carboxymethylene linker [123, 124]. In a chemical sense, PNA resembles peptides more than DNA and is therefore included in this chapter. Topologically however, PNA is very similar to DNA or RNA [125]. PNA binds to complementary DNA or RNA oligomers according to Watson–Crick rules with high stability [126–129]. An increase in stability, as measured by the melting temperature (UV-T_m), between 1 and 3 °C per base pair in a PNA • DNA duplex has been observed compared with a homologous ds-DNA. This value is of course highly dependent on the length and the sequence of the PNA oligomer used. In addition, PNA shows a higher mismatch sensitivity towards non-complementary DNA than is commonly observed in ds-DNA [130, 131]. These favorable properties led to numerous applications for PNA oligomers in biotechnology and as antisense agents [132–136]. Furthermore, a large number of structural variations of the original PNA have been published [137].

Our group published the first organometallic derivatives of PNA monomers [138]. Ferrocene carboxylic acid 3 and benzoic acid chromium tricarbonyl 50 were coupled to the amino group of PNA monomers with different nucleobases (Thymine (T): a and Z-protected Cytosine (Z-C): b) using HBTU as the coupling agent to give compounds 51 and 52, respectively (Scheme 5.27). The activation barrier $\Delta G^{\ddagger}$ for the rotation about the tertiary amide bond has been determined for the ferrocenylated T-PNA monomer 51a to be 75 ± 0.5 kJ mol^{-1}. For the same compound 51a, a T–T self-association constant $K_{TT} = 2.5 \pm 0.2$ M^{-1} has been

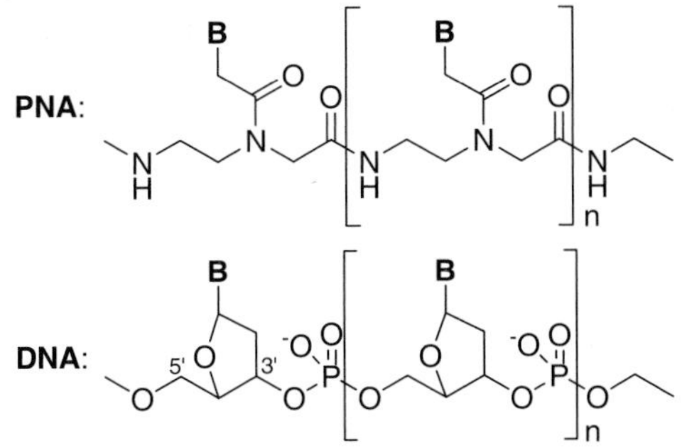

Scheme 5.26 Comparison of PNA (top) and DNA (bottom).
B represents the nucleobases adenine (A), cytosine (C), guanine (G) and thymine (T, DNA) or uracil (U, RNA).

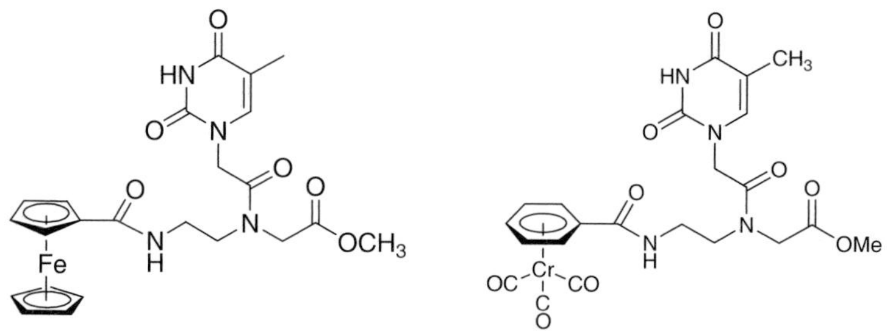

Scheme 5.27 Two examples of organometallic T-PNA monomers. Derivatives of other nucleobases were also prepared, see text.

measured by ^{1}H NMR dilution experiments in CDCl$_3$. In comparison, the association constant K_{AT} for the regular A–T Watson–Crick base pair was found to be 78 ± 9 M^{-1} (in dry CDCl$_3$) by the same experimental technique, using **51a** and tris(isopropyl)adenosine as a soluble adenine derivative. To the best of our knowledge, only these values of association constants for PNA monomers are published in the literature to date [139]. Within the experimental limits, both numbers correspond well to literature values for DNA bases alone, indicating a) that there is no additional stabilization or destabilization to base pairing of the PNA amide backbone and b) that Watson–Crick base pairing is not impaired by the metallocene at the N terminus.

It is also possible to synthesize PNA monomers with an organometallic label at the C terminus. The secondary ferrocene amine derivative that has been used for the preparation of the bimetallic peptide derivatives **5** and **6** (*vide supra*) could be coupled to the C terminus of the *N*-Fmoc-T-PNA monomer to yield **54** (Scheme 5.28). The *N*-Fmoc protecting group can be readily removed under mild conditions and coupling to a second T-PNA monomer is possible, yielding the ferrocenylated PNA dimer **55** [139].

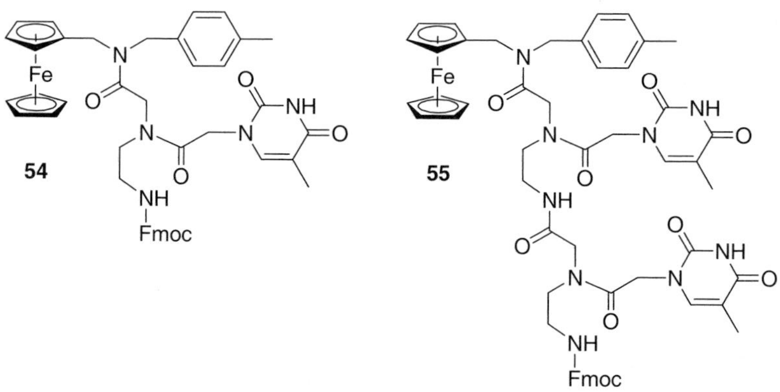

Scheme 5.28 C-terminally labeled PNA derivatives.

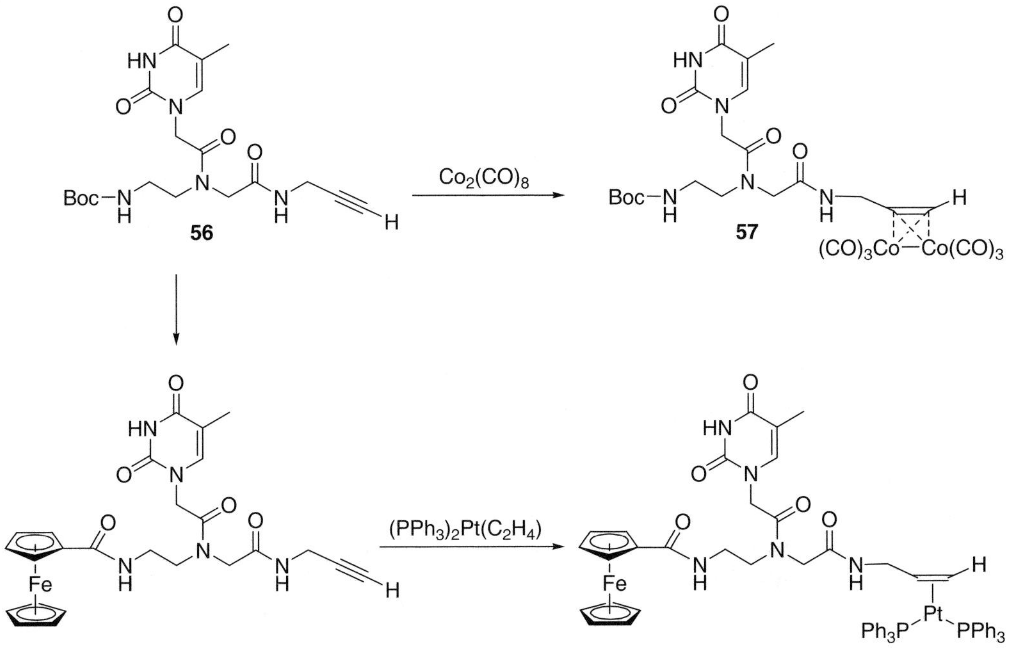

Scheme 5.29 A PNA-alkyne derivative **56**, its $Co_2(CO)_6$ complex **57** (top right), and the first bimetallic PNA monomer **59** (bottom right).

The alkyne group is a versatile binding unit for organometallic fragments, as already shown in Section 5.3.2.1. We have prepared the T-PNA alkyne **56**. Upon addition of $Co_2(CO)_8$, the $Co_2(alkyne)(CO)_6$ complex **57** forms as a dark red solid (Scheme 5.29) [139]. This compound could be characterized unambiguously by NMR and mass spectrometry. Typical bands for the $Co_2(CO)_6$ group can be observed in the IR spectrum of **57** at 2018 cm^{-1} and 2093 cm^{-1}. Starting from **56**, we have also prepared the first bimetallic derivative of a PNA monomer (Scheme 5.29). After Boc deprotection, **56** could be coupled to ferrocene carboxylic acid to yield the ferrocenoyl T-PNA alkyne **58**. Coordination of the Pt(PPh$_3$)$_2$ fragment to **58** yields the bimetallic ferrocene-Pt PNA monomer **59** which could be comprehensively characterized spectroscopically [139]. Most importantly, this compound shows a highly characteristic ^{31}P NMR pattern as a consequence of coupling of the two inequivalent ^{31}P nuclei and the Pt nucleus. In addition, these signals are further split into two because rotation about the central amide bond, which connects the nucleobase to the PNA backbone, is slow and a typical major/minor pattern is observed for all signals.

Interesting organometallic PNA monomers were also reported by Baldoli, Licandro, Maiorana and coworkers [140–142]. An Ugi four-component reaction (U-4CR, see also below in Section 5.5.2.2) was used to assemble an organometallic PNA monomer **60** in a one-pot synthesis in about 20% yield (Scheme 5.30, R = H)

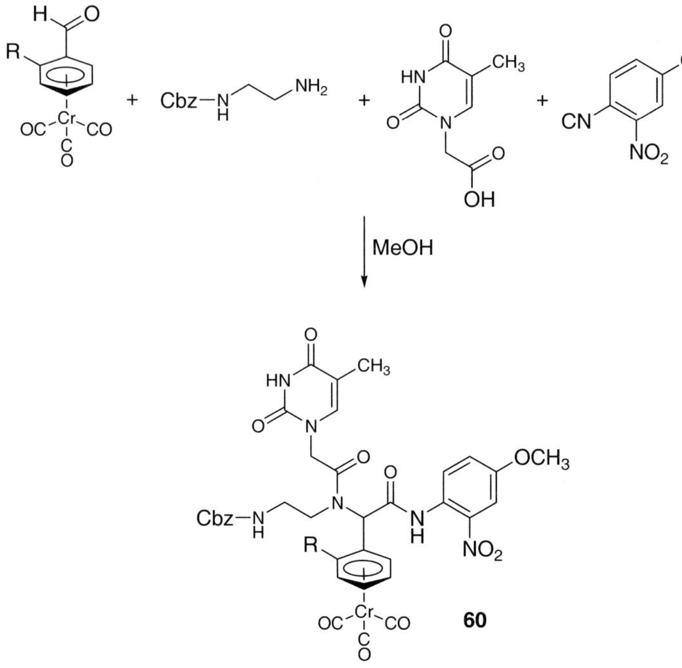

Scheme 5.30 Synthesis of (chiral if R ≠ H) organometallic PNA monomers by Ugi-4CR.

[140]. The yield could be increased to 50% when the benzaldimine from benz-aldehyde chromium tricarbonyl and Cbz-protected ethylenediamine was used directly in the condensation reaction. The Ugi reaction has been used before to assemble PNA monomers [143, 144]. Chiral benzaldehyde chromium tricarbonyl complexes can be easily obtained in enantiomerically pure form and usually lead to high stereoselectivity. It seemed eminently attractive to obtain a chiral organometallic PNA monomer stereocontrolled in a one-pot synthesis by using enantiomerically pure chromium tricarbonyl benzaldehyde derivatives (R ≠ H). Unfortunately, almost equimolar ratios of the two diastereomers were obtained under the experimental conditions, independent of the substitutent R [140]. It is worth noting that the diastereomers could be separated by column chromatogra-phy, thus yielding the first enantiomerically pure organometallic PNA monomers. Very recently, this reaction sequence was extended to yield organometallic PNA dimers like **62** directly from chromium tricarbonyl benzaldimine, thymine acetic acid and an isonitrile PNA monomer **61** in an elegant way (Scheme 5.31) [142]. Although only a tt dimer was prepared originally, this reaction sequence is certainly applicable to combinatorial chemistry to yield all conceivable combinations.

The same group has recently reported the use of Ru-catalyzed cross-metathesis reactions to obtain PNA monomers labeled with a Fischer-type carbene complex (Scheme 5.32) [141]. A suitable PNA monomer **63** was prepared from racemic allylglycine and then subjected to the metal catalyzed metathesis reaction with

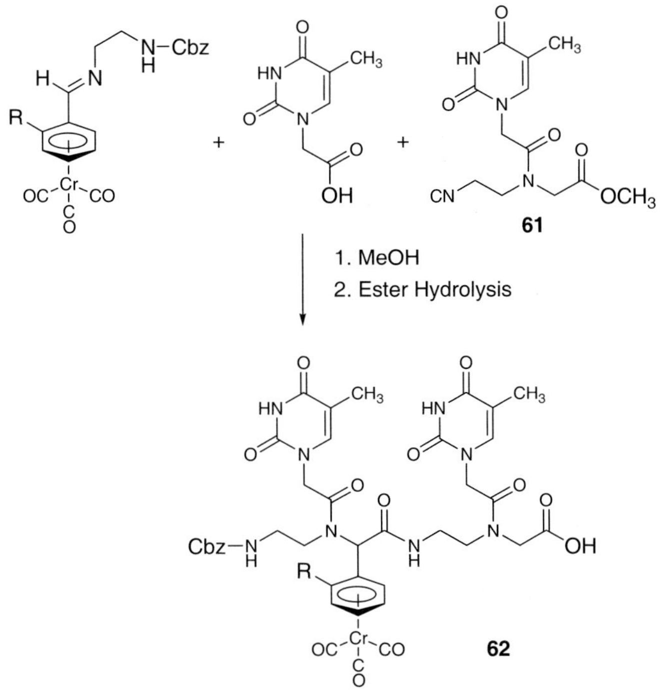

Scheme 5.31 Direct synthesis of organometallic PNA dimers by Ugi-4CR.

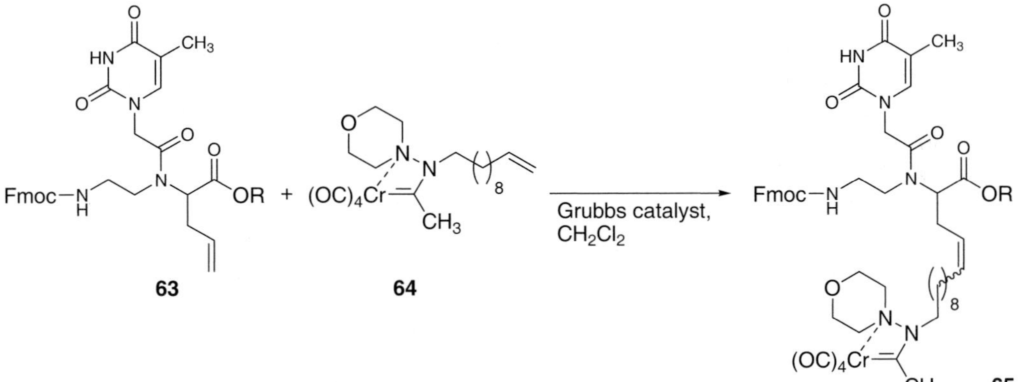

Scheme 5.32 Use of alkene metathesis for the formation of PNA monomers with a chromium carbene moiety.

the Cr(CO)$_4$ carbene **64**. The target PNA monomer **65** was obtained in 40% yield. Since both enantiomers of allylglycine are commercially available, this could be another entry to enantiomerically pure chiral organometallic PNA monomers. It has been observed for purely organic PNA oligomers that the chirality of the PNA backbone may indeed influence the stability and structure of PNA • DNA duplices [145–147].

5.4.2
Conjugates of PNA Oligomers

Our group has prepared the first organometallic derivatives of PNA oligomers that we are aware of [148, 149]. Originally, ferrocene carboxylic acid and chromium tricarbonyl benzoic acid were incorporated into PNA oligomers by solid phase peptide synthesis techniques [148]. The PNA heptamer H-tgg atc g-Gly [Note: According to common convention, the same four letter code is used for PNA as for DNA, small letters however indicate PNA oligomers] was prepared on solid support by standard Fmoc PNA synthesis methods. After deprotection of the last Fmoc group, the heptamer was reacted with activated ferrocene carboxylic acid or chromium tricarbonyl benzoic acid *on the resin*. Cleavage from the resin was achieved by methanolic ammonia with simultaneous removal of the exocyclic protecting groups. Only one main product was observed in reverse phase HPLC of the crude reaction product (> 90%). Whereas the conjugate Fc-CO-tgg atc g-Gly-NH$_2$ **66** was purified by preparative HPLC and shown to have the correct mass by MS, the chromium tricarbonyl derivative could not be obtained in a pure form.

Compared with a tris(bipyridyl)Ru and an acetyl derivative of the same sequence [148], **66** was by far the least hydrophilic derivative and consequently not very soluble in aqueous solvents. Since then, we have also studied the incorporation of cobaltocenium carboxylic acid into PNA oligomers by solid phase synthesis [149]. Cobaltocenium and ferrocene are structurally almost identical. The former, however, has a positive charge and it is thus expected that cobaltocenium PNA conjugates are more soluble in water than the homologous ferrocene derivatives. Following initial studies on smaller oligomers to optimize coupling and deprotection conditions, the decamer CpCo$^+$C$_5$H$_4$-CO-acc ctg tta t-Lys-OH **67** was synthesized by solid phase synthesis techniques. The conjugate could be purified by preparative HPLC and was characterized by MALDI-TOF mass spectrometry. A square wave voltammogram clearly indicated the presence of the cobaltocenium group by a reversible wave at –1.51 V *vs.* FcH/FcH$^+$. This conjugate was also studied in the interaction with complementary DNA. A melting temperature of 46.2 °C was measured with a DNA 18mer containing the complementary sequence. Compared with acetylated PNA of the same length and sequence, a slight stabilization of about 1 °C is observed for the cobaltocenium conjugate **67** which might be attributed to an attractive force between the positively charged cobaltocenium group and the negatively charged DNA strand [149].

5.5
Applications

5.5.1
Organometallic Protecting Groups for Peptide Synthesis

5.5.1.1 Ferrocene-derived Protecting Groups

Eckert and coworkers introduced the ferrocenylmethyl (Fem) group as a lipophilic substituent to mask the fairly polar glycine amide moiety during peptide synthesis in solution [150–152]. Fem-protected glycine (see Scheme 5.33) can be synthesized via imine formation of glycine with ferrocenecarbaldehyde, followed by reduction. Peptide coupling of this modified glycine residue to other amino acids or peptide fragments proceeds readily by the standard carbodiimide method used for peptide synthesis in solution [153, 154]. Advantages that were claimed for this masked glycine derivative for peptide synthesis in solution are: first, the intermediates and products show increased solubility in apolar organic solvents as a consequence of the lipophilic Fem group. This results in complete reactions and improved yields. In addition, the chromatographic purification over silica gel is expedited since the products can be eluted with less polar solvent mixtures. Second, the ferrocene chromophore of the intermediates and products facilitates product identification and isolation during chromatographic purification. Third, the growing peptide is more stable against racemization during the coupling steps, when it is assembled in stepwise fashion starting from the C terminus. Cleavage of the Fem group is postponed until the desired amino acid sequence has been obtained. Removal of the Fem group is accomplished by treatment with TFA/β-thionaphtol in CH_2Cl_2 at room temperature for 2–4 h. Under these conditions, Boc and tert.-butyl protecting groups are simultaneously cleaved [150–152]. The usefulness of Fem-Gly derivatives for peptide synthesis in solution is illustrated by the successful synthesis of [Leu^5]-Enkephalin and H-(Gly)$_6$-OH in solution [152]. In a similar manner, the Fem group has also been used as a protecting group for Asn [151] and Cys [155, 156] residues in peptide synthesis.

Kane-Maguire and coworkers proposed the iron tricarbonyl cyclohexadienyl cation ([(Fe(CO)$_3$(C$_6$H$_7$)]$^+$, Fed) as an alternative to the Fem protecting group [157]. The Fed group can be used as a masking group for the terminal amino group in peptide couplings, thus facilitating purification of the peptide [157]. It is rapidly and quantitatively removed from the peptide with recovery of the [(Fe(CO)$_3$(C$_6$H$_7$)]$^+$ reagent by treatment with TFA. The Fed group is readily introduced into amino

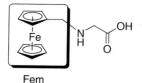

Fem

Scheme 5.33 Ferrocenylmethyl glycine, the ferrocenylmethyl (Fem) group is highlighted.

acids by reaction of the cation with nucleophilic substrates such as the terminal amino group as described above. The order of reactivity, however, is –SH (Cys) > –N= (His) > CO_2^- (Glu) ≫ NH$_2$ (α-NH$_2$ for any amino acid or ε-NH$_2$ for Lys), -OH (Ser), -CONH (Asp) [158]. It is therefore most suited as a selective labeling or protecting group for the side chains of Cys or His. For instance, Fed forms an adduct with the tripeptide Glutathione (γ-Gly-Cys-Glu) at pH 3.1 (isoelectric point of Glutathione) within minutes which precipitates in pure form from the solution. A competition experiment with the Cys-11 derivative of luteinizing hormone-releasing hormone (Glp-His-Trp-Ser-Tyr-Gly-Leu-Arg-Pro-Gly-Cys-OH, see Section 5.3.2.4) underlines the selectivity of Fed. NMR experiments confirm the clean reaction with only the C-terminal Cys sulfhydryl group, and no reaction with the iminidazole ring of His [158]. The mechanism of the addition of this reagent to amino acid esters has been studied in detail [159].

5.5.1.2 Aminocarbene-derived Protecting Groups

A number of groups proposed organometallic groups as amino protecting groups. The tricobalt nonacarbonyl acylium cation [(OC)$_9$Co$_3$CCO]$^+$ reacts with the amino group of amino acids and peptides. In such a reaction, the tricobalt carbonyl derivative **68** is formed (Scheme 5.34) [160]. The metal cluster in **68** can either be seen as an amino protecting group, or as an N-terminal metal carbonyl label. Unfortunately, this chemistry has not been further explored. Lukehart and coworkers proposed the use of rhena-β-ketoimines as heavy atom labels for protein crystallography [161]. One such compound was obtained from the reaction of rhena-β-diketone with glycine ethyl ester.

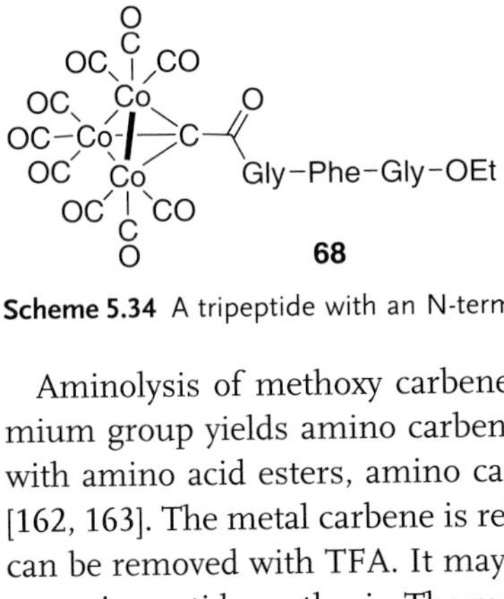

Scheme 5.34 A tripeptide with an N-terminal tricobalt-octacarbonyl labeling group.

Aminolysis of methoxy carbene metal pentacarbonyl complexes of the chromium group yields amino carbene compounds. If the aminolysis is performed with amino acid esters, amino carbene derivatives of amino acids are obtained [162, 163]. The metal carbene is relatively stable under a variety of conditions but can be removed with TFA. It may therefore be regarded as an amino protecting group in peptide synthesis. The metal amino carbene amino acid may be activated and coupled to other amino acid esters. Following this idea, Weiss and Fischer prepared various dipeptides, the tripeptide (OC)$_5$Cr(Ph)-Ala-Ala-Ala-OMe **69** [162] and the tetrapeptide (OC)$_5$Cr(Ph)-Gly-Gly-Pro-Gly-OMe **70** (Scheme 5.35) [163]. Treatment with conc. TFA at 20 °C for 10 min. removes the metal carbene group and furnishes the free peptide esters along with Cr(CO)$_6$. Removal of the metal carbene is also possible with 80% acetic acid (80 °C, 30 min.), which leaves Boc

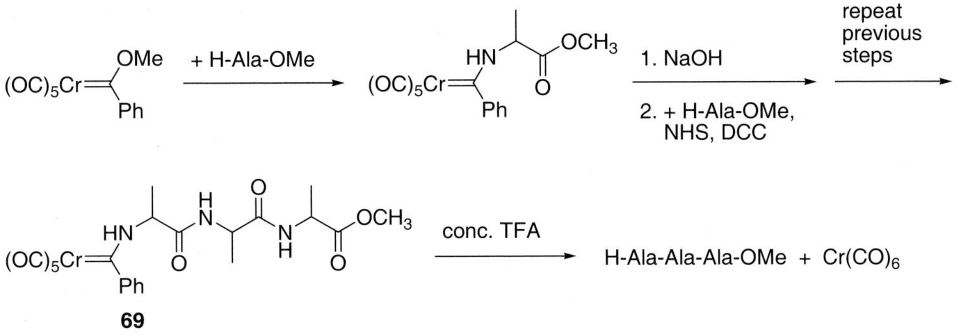

Scheme 5.35 Peptide synthesis in solution using metal carbene protecting groups, as exemplified for the tripeptide Ala-Ala-Ala-OMe.

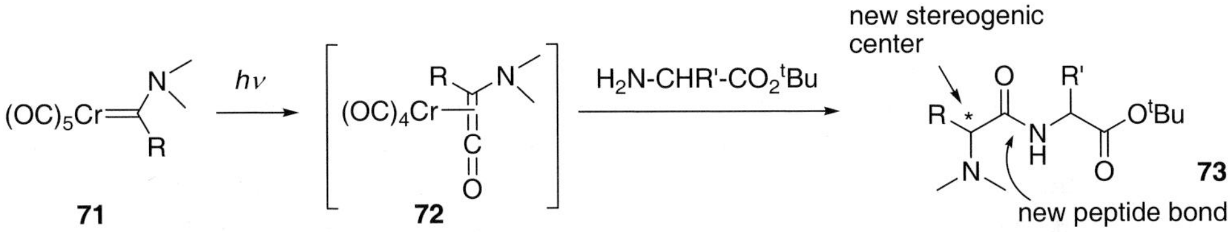

Scheme 5.36 Stereocontrolled dipeptide formation by chromium amino carbene photolysis.

protecting groups in place. Finally, treatment of tungsten amino carbene pentacarbonyls with BBr$_3$ also releases the free amino acid or peptide esters along with *trans*-bromotetracarbonyl-phenylcarbyne-tungsten $(OC)_4W(Br)CPh$ [163].

More recently, Hegedus and coworkers extended this chemistry into a unique route to unnatural amino acids [164]. Photolysis of a chromium pentacarbonyl amino carbene **71** produces a reactive metal-complexed amino ketene **72**. This intermediate **72** can be trapped with an amino acid ester to yield a dipeptide **73**. In this reaction, both a new peptide bond and the stereogenic centre of a new N-terminal amino acid in **73** are produced at the same time. Scheme 5.36 illustrates the principle of this reaction. Use of an appropriate chiral auxiliary on the carbene amino atom generally gives enantiomeric purity of > 95% for the new stereogenic carbon atom. It was possible to perform this chemistry on the resin during SPPS. However, technical difficulties with the photochemistry on the solid support make this process more suitable for solution chemistry. The method is particularly useful for the formation of unnatural amino acids, which are elegantly formed during peptide synthesis. According to Hegedus, the ideal use for this methodology appears to be in segment condensation reactions of polypeptides. Chromium carbene complex chemistry is used to synthesize small, unusual peptide fragments in solution, which are then incorporated into larger peptides by standard SPPS [164]. One successful example of such an approach is the octapeptide Boc-Gly-Ala-**D-homoPhe-D-Ala-Phe**-Val-Leu-Gly-OMe (homoPhe: homo-phenylalanine) [165]. In this compound, the central tripeptide D-homoPhe-D-Ala-Phe (marked boldface) was synthesized in solution by chromium carbene photochemistry. It was added to the tripeptide Val-Leu-Gly on the resin, and the synthesis was completed by addition of the two N-terminal amino acids by SPPS. The crude product was obtained in 72% yield based on the chromium carbene tripeptide and could be readily purified and characterized.

5.5.2
Peptide Synthesis

In almost all sections above, peptides were synthesized by the reaction of an activated carboxylic acid with an amine. In many cases, the carboxylic acid was actually incorporated as part of an organometallic species, thus yielding the organometallic peptide or PNA conjugate. This reaction can also be performed on a solid support and is then part of a solid phase peptide synthesis scheme. However, the formation of peptides is also conceivable by entirely different reactions. In the following sections, three different strategies will be discussed. They are certainly not as well developed as SPPS methods but they represent conceptually appealing alternatives. In all cases, organometallic compounds play a crucial role either as catalysts or chiral auxiliaries. Another common feature is that organometallic intermediates have been isolated and characterized at least in some cases and we can thus truly speak of organometallic peptide conjugates. Numerous other methods exist for the synthesis of amino acids or simple dipeptides with organometallic compounds as catalysts or auxilliaries. As outlined in the introduction, these are not covered herein.

5.5.2.1 Template Synthesis of Peptides with Organometallics
Peptide condensation reactions in the coordination sphere of metals were proposed as early as 1967 [166–168]. Although it seems an attractive alternative at least to standard peptide synthesis in solution, this method never gained widespread attention. Almost 30 years later, Beck et al. discovered a sequence-specific peptide synthesis method that uses half-sandwich complexes of Rh, Ir or Ru [169–175]. In this reaction, N,N'-coordinated peptides are elongated at the N terminus by condensation with amino acid esters. Scheme 5.37 shows the postulated reaction sequence [171]. It is worth noting that formation of the peptide bond does not require activating reagents or protecting groups. The key step is nucleophilic attack of the deprotonated terminal amino group at the ester carbonyl group. Repeated addition of amino acid esters will yield a growing peptide in the coordination sphere of the metal complex. Finally, the peptide can be released from the metal template by methanolic HCl. Several intermediates have been isolated and characterized spectroscopically as well as by X-ray crystallography [170, 171]. Figure 5.6 shows the X-ray single crystal structure of the tetrapeptide Ala-Gly-Gly-Gly-OMe coordinated to the (C_6Me_6)Ru fragment. All reactions proceed under relatively mild conditions in the coordination sphere of the chiral metal half sandwich fragment and no racemization has been observed. In one paper, the catalytic formation of up to a nonapeptide H-(Gly)$_9$-OMe on a {(p-cymene)RuCl} fragment is described [172]. By a similar reaction, the formation of cyclic tetrapeptides on a metal template is reported [173].

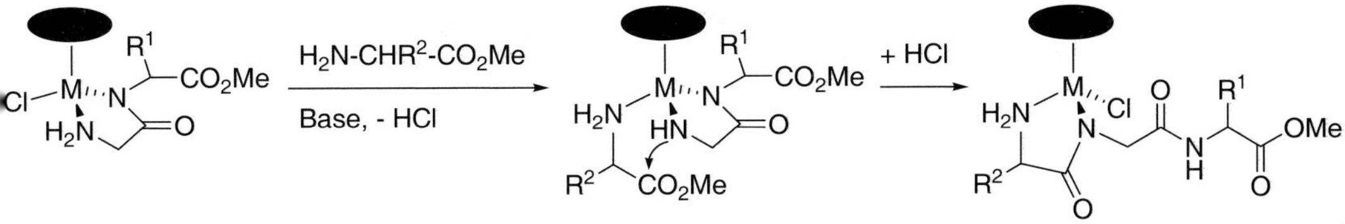

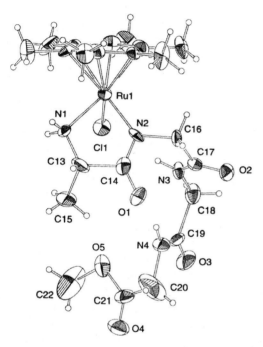

Scheme 5.37 Sequence specific peptide synthesis in the coordination sphere of chiral half sandwich complexes (M = Rh, Ir, π-donor = Cp*; M = Ru, π-donor = C_6Me_6 or cymene).

Fig. 5.6 X-ray single crystal structure of the tetrapeptide Val-Gly-Gly-Gly coordinated to a (C_6Me_6)Ru half-sandwich fragment. (Reprinted from [171]).

5.5.2.2 Ugi Four–component Reaction

The four-component condensation of isocyanides (also called isonitriles), discovered by Ugi in the late 1950s (Ugi four component reaction, Ugi-4CR) [176], provides another alternative to the classical methods of peptide synthesis in solution [177–180]. The Ugi-4CR has been extensively studied in every conceivable aspect and several variants of the general scheme give different products. In one variant that is particularly relevant to the topic of this article, an N-protected peptide or amino acid, a chiral primary amine, an aldehyde and an isonitrile peptide fragment are combined to yield, via a spontaneous rearrangement of an unstable α-adduct, a compound **74** in which a new peptide segment has formed (Scheme 5.38, see also Section 5.4.1 above for an application of Ugi-4CR to PNA synthesis).

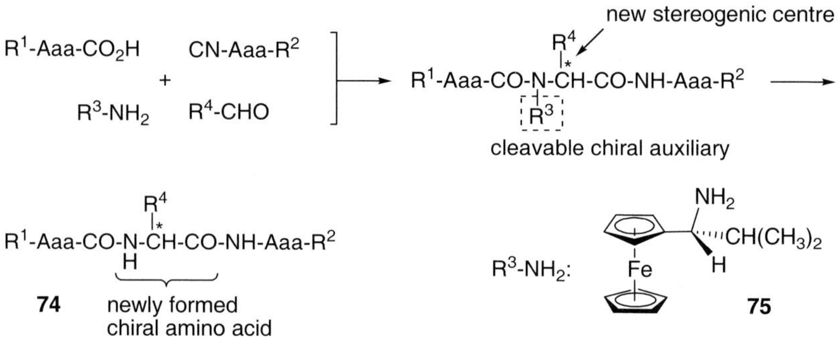

new stereogenic centre

$$R^1\text{-Aaa-CO}_2H \quad CN\text{-Aaa-}R^2$$

$$+$$

$$R^3\text{-NH}_2 \quad R^4\text{-CHO}$$

$$\longrightarrow \quad R^1\text{-Aaa-CO-N-CH-CO-NH-Aaa-}R^2 \quad \longrightarrow$$

cleavable chiral auxiliary

$$R^1\text{-Aaa-CO-N-CH-CO-NH-Aaa-}R^2$$

74 newly formed chiral amino acid

$$R^3\text{-NH}_2:$$

Fe, NH$_2$, CH(CH$_3$)$_2$, H

75

Scheme 5.38 Ugi-4CR for peptide assembly.
Compound **75** ((R)-1-ferrocenyl-2-methyl-propanamine) is a powerful chiral auxiliary R^3-NH$_2$.

If the newly formed peptide segment should consist of a chiral amino acid, it is necessary to use a chiral primary amine as the chiral auxilliary. Chiral α-ferrocenyl alkyl amines, in particular (R)-1-ferrocenyl-2-methyl-propanamine **75** shown in Scheme 5.38, have found widespread application in this stereoselective four component condensations [179, 181, 182–186]. This is because **75** was shown to have strong asymmetric inducing power and it is easily cleaved under acidolytic conditions. In fact, the chiral primary amine can be readily regenerated without racemization, making this reaction very economical [183, 187, 188]. As a proof of concept, tetravaline and Glutathione were synthesized with high enantiomeric purity by stereoselective Ugi-4CR with the help of this chiral ferrocenylalkylamine auxilliary [182, 186, 189].

5.5.2.3 Ruthenium-mediated Coupling of Aryl Ethers for the Synthesis of Cyclic Peptides

Chloroarene compounds can be activated for nucleophilic substitution by complexation to transition metals. [CpRu]$^+$ fragments have been particularly useful for this S$_N$Ar reaction from a synthetic point of view because they can be used for the formation of arylethers as shown in Scheme 5.39 [175]. There is no catalytic variant of this reaction yet. The [CpRu(arene)]$^+$ complexes are stable in air and water and can be isolated, purified and characterized. Indeed, metal decomplexation is achieved by photolysis of the compounds in CH$_3$CN. It is worth noting that this reaction is the exact reverse of the arene complexation reaction mentioned in Section 5.3.2.5. The Ru-assisted arylether formation is the key step in recent total syntheses of several natural products, for example the protease inhibitors K-13 (**76**) and OF-4949-III (**77**, Scheme 5.40). Model compounds for several antibiotics such as Vancomycin have also been synthesized via this route.

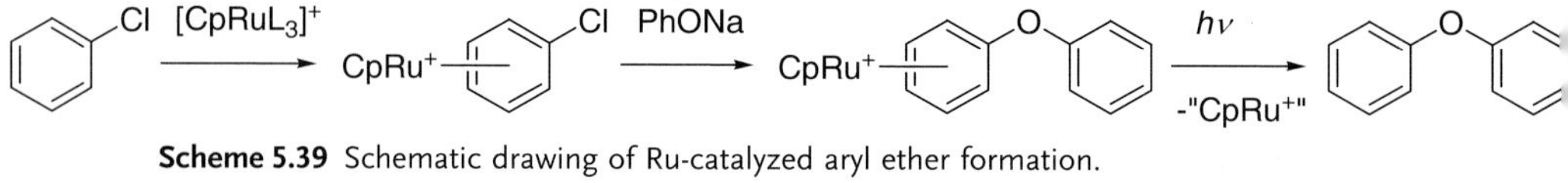

$$\xrightarrow{[CpRuL_3]^+} CpRu^+ \xrightarrow{PhONa} CpRu^+ \xrightarrow[-"CpRu^+"]{h\nu}$$

Scheme 5.39 Schematic drawing of Ru-catalyzed aryl ether formation.

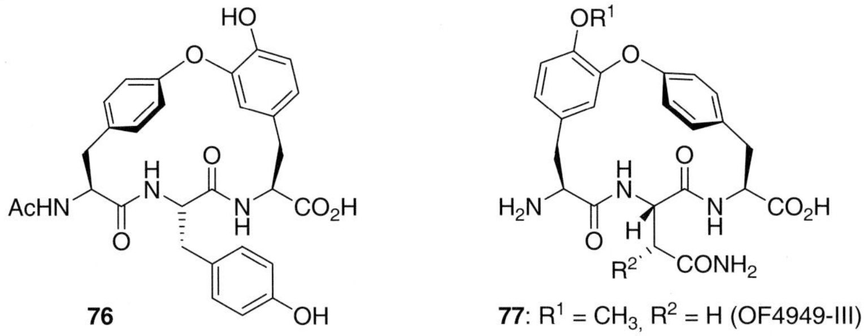

Scheme 5.40 Structural formulae of the cyclic aryl ether peptides K-13 (left) and OF4949-III (top right). Other members of the OF4949 family differ in residues R^1 and R^2.

Both K-13 and OF-4949 are macrocyclic tripeptides which contain the biaryl ether diamino diacid isodityrosine. K-13 is isolated from the culture broth of *Micromonospora halophytica* and is a potent inhibitor of the angiotensin converting enzyme (ACE) [190]. OF-4949 is actually a family of four competitive inhibitors of aminopeptidase B (OF-4949 I-IV) [191]. Both compounds lose their activity completely if the aryl ether bond is opened. Although superficially related, the two compounds differ when viewed from the N-terminal to C-terminal direction in that K-13 has a *para-meta* substitution of the two aromatic rings, whereas OF-4949 has a *meta-para* pattern of substitution. Formal total syntheses for both compounds have been proposed by Pearson and coworkers [192, 193], subsequently convergent syntheses have been published by Janetka and Rich

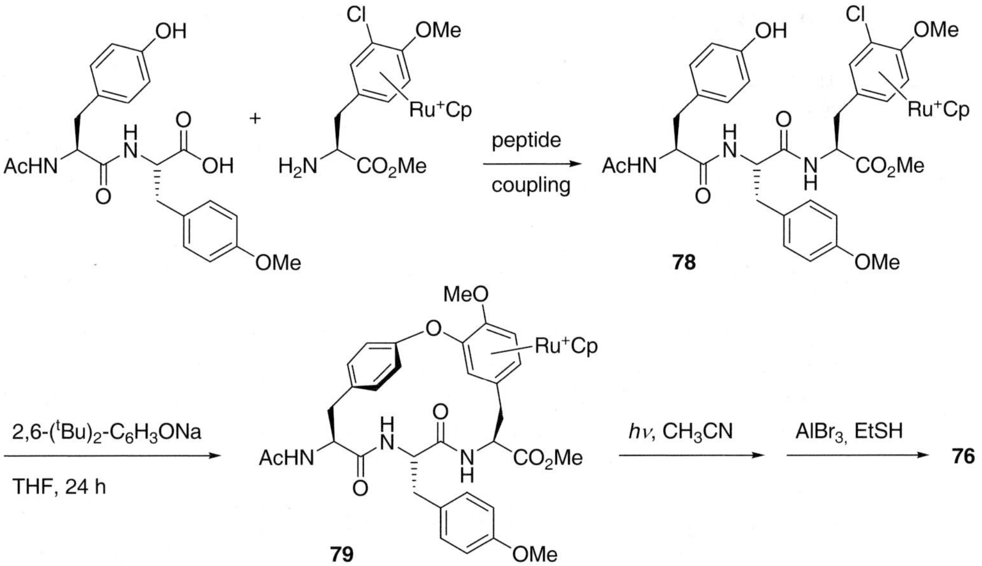

Scheme 5.41 Key steps in the total synthesis of KF-13 (**76**).

[194, 195]. The key steps of the total synthesis of K-13 are summarized in Scheme 5.41 as an example of this method. The CpRu-complexed intermediates, such as **78** and **79**, could be isolated by column chromatography and characterized in all cases. They are used in stochiometric amounts for the closure of the macrocyclic ring.

Lindel and coworkers have recently demonstrated the use of the ruthenium-mediated aryl ether coupling reaction for the synthesis of peptide-like compounds ("peptoids") with alternating peptide and diaryl ether bonds. Their approach involves four different elongation steps and is in principle amendable to automated solid phase synthesis and combinatorial chemistry [196].

5.5.3
Labeling of Peptides

5.5.3.1 HPLC with Electrochemical Detection (HPLC-ECD)

In two papers and a patent, Eckert and Koller described a variety of ferrocene compounds for derivatization of the N terminus of amino acids and the lysine amino groups of proteins, for example BSA [197–199]. The objective was to investigate which of the reagents was best suited for labeling purposes and subsequent electrochemical detection of the conjugates by HPLC-ECD (High Performance Liquid Chromatography with Electrochemical Detection). Among the naturally occurring amino acids, only methionine, tyrosine and tryptophane show an electrochemical activity, but their redox potential is too high for good selectivity (> +0.5 V *vs.* FcH/FcH$^+$). In contrast, ferrocene has a lower redox potential of at least a few hundred mV, depending on the functional groups on both Cp rings. This decrease of the redox potential improves the selectivity.

Reagents for HPLC-ECD derivatization that were investigated by Eckert and Koller include for example Fc-SO$_2$-Cl, Fc-CO-Cl, ferrocene carboxylic acid anhydride, ferrocene propionic acid anhydride, and ferrocenylmethyl isocyanate. Of these derivatives, ferrocenyl propionic acid anhydride was found to be most convenient not only for the labeling itself but also for electrochemical detection of the conjugates [198]. Koppang et al. regard Fc-SO$_2$-Cl and Fc-CO-Cl also to be good derivatization reagents for HPLC-ECD, in contrast to conclusions by Eckert and Koller [200]. In a related paper, Shimada et al. suggest the use of ferrocenylethyl isothiocyanate for the determination of 4-aminobutyric acid in biological samples by HPLC-ECD [201].

Most compounds treated so far are suitable for the derivatization of amino groups in peptides and amino acids. In addition, selective labels for the sulfhydryl group of cysteine have been reported [202–204]. For the labeling of gluthathione **80**, both ferrocenyl-ethylmaleimide **81** and ferrocenyl-iodoacetamide **82** (Scheme 5.42) turned out to be very effective to yield **83** and **84**, respectively [202, 204]. A related maleimide derivative has also been used to label the hexapeptide Ac-Arg-Arg-Ala-Ser-Leu-Cys-OH. This labeled hexapeptide was applied for the detection of serine phosphorylation by protein kinase A via electrochemical methods [203].

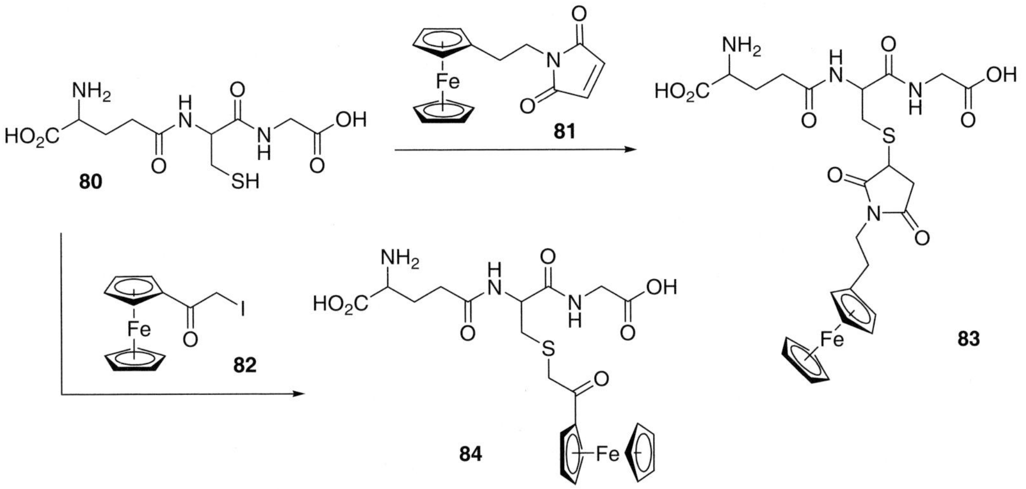

Scheme 5.42 Labeling of glutathione **80** by ferrocenyl-ethylmaleimide **81** and ferrocenyl-iodacetamide **82**.

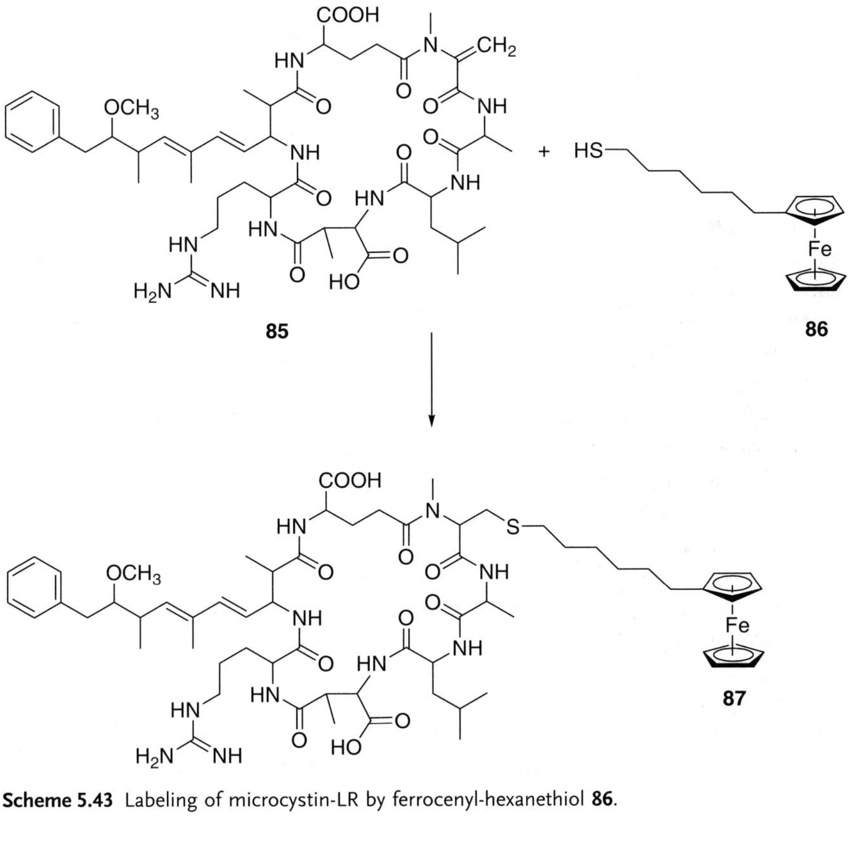

Scheme 5.43 Labeling of microcystin-LR by ferrocenyl-hexanethiol **86**.

Finally, an interesting labeling reaction of a ferrocene thiol with a double bond in a peptide has been reported recently. Microcystins are a class of low-molecular weight cyclic peptide hepatoxins. These microcystins are produced in eutropic lakes and drinking water reservoirs, particularly in hot and relatively dry areas such as China, Australia and South Africa [205]. Because these substances are hazardous to cattle, wildlife, and humans, there is an obvious need for rapid and inexpensive detection of these toxins. Lo, Lam and coworkers developed an elegant method that makes the detection of a particular microcystin, namely microcystin-LR, possible via a quick electrochemical measurement [206]. This method is based on the specific derivatization of the exocyclic double bond of the α, β-unsaturated carbonyl moiety of microcystin-LR **85** with ferrocenyl-hexanethiol **86** (Scheme 5.43), followed by electrochemical detection of the ferrocene tag in the conjugate **87**. The limit of detection for **87** was found to be *ca.* 18 ng. Much lower levels of microcystins can be detected with more expensive methods, such as fluorescence spectroscopy and HPLC-MS. On the other hand, the sensitivity of this new electrochemical detection is in the range of other common detection methods, such as thin-layer chromatography and HPLC with UV-detection. In addition, the electrochemical method is quick, specific and inexpensive.

5.5.3.2 Radioactive Labels

The use of organometallic radiopharmaceuticals for imaging purposes is covered elsewhere in this book, including the $Tc(CO)_3$ fragment which has become very popular in recent years. Therefore, the discussion in this chapter will be rather brief and will only highlight recent applications of organometallic radiolabels for peptides.

Because its properties are very favorable for radio-imaging, ^{99m}Tc is the preferred isotope and a large number of conjugates, also of peptides [207], have been prepared and used. Mostly, the Tc atom is in high oxidation states and coordinated to N and S ligands. The low-spin d^6-$[Tc(CO)_3]^+$ fragment presents an interesting alternative [208]. It can be easily prepared from TcO_4^-, which is readily available in many clinics from a ^{99}Mo generator. This fragment has been developed for clinical applications by Alberto and Schubiger [209]. It has been used for the radiolabeling of serotonin receptors [210] and single-chain antibody fragments [211] as described elsewhere in this book. The same group has also reported the use of histidinyl acetic acid as a chelating group for $[Tc(CO)_3]^+$ in a truncated neuropeptide, namely the 6 amino acid fragment from Neurotensin (NT 8–13) [212]. Beck-Sickinger and coworkers synthesized analogs of Neuropeptide Y (NPY) with a ligand capable of coordinating $[Tc(CO)_3]^+$ [213]. Neuropeptide Y is 36 amino acid peptide [214]. It is formed in the peripheral as well as in the central nervous system and is probably one of the most abundant neuropeptides in the brain. It has various different physiological functions, and its receptors are overexpressed in a number of neuroblastoma, making it an attractive target for tumour scintigraphy [215]. Ala^{26}-NPY was prepared by SPPS, as well as a centrally truncated derivative. Both peptides had the PADA (2-picolylamine-*N*,*N*-diacetic acid) ligand attached to Lys^4. After cleavage from the resin and purification, reaction with $[Tc(CO)_3(H_2O)_3]^+$

yielded the stable conjugates. Of the two conjugates, only the truncated peptide was specifically internalized into receptor expressing cells. Biodistribution studies in mice revealed that even this smaller peptide does not cross the blood–brain barrier, although it has good biological stability.

Katzenellenbogen and coworkers have used a novel double ligand transfer reaction for the labeling of octreotide with Cyclopentadienyl-^{99m}Tc(CO)$_3$ [216]. Octreotide is a physiologically stabilized analog of the cyclic tetradecapeptide somatostatin. Somatostatin receptors are frequently overexpressed in certain type of cancers, and an ^{111}In-derived octreotide is already in use as a somatostatin receptor imaging agent (OctreoScan®). The radiolabeled ^{99m}Tc(CO)$_3$(C$_5$H$_4$-CO$_2$H) is prepared by metal exchange from 1,1′-ferrocene dicarboxylic acid dimethyl ester, followed by ester hydrolysis. The acid is then activated and coupled to octreotide to give the conjugate **88** shown in Scheme 5.44 in 15% decay-corrected yield. The conjugate **88** showed receptor mediated uptake into pancreas and adrenal glands in rats.

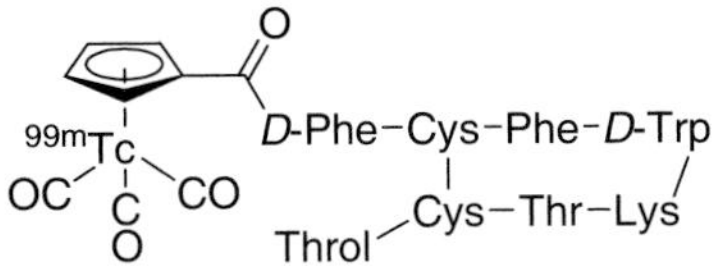

88 (Throl: Threoninol)

Scheme 5.44 Labeling of octreotide by a radioactive Cp^{99m}Tc(CO)$_3$ tag.

Most previous work relies on metal-substitution at the N terminus with suitable chelating ligands. Very recently, van Staveren et al. reported the use of N^ε-amino-propyl-substituted histidine as a ligand for derivatization of peptides at the C terminus. This ligand was coupled to biotin and vitamin B$_{12}$, in addition to the pentapeptide enkephalin. The peptide was prepared by SPPS with a super-labile linker, cleaved from the resin in protected form and purified. Activation and reaction of the protected peptide with the appropriate aminopropyl-His derivative yielded the C-terminally substituted conjugate, which was fully deprotected by addition of strong acid. Reaction with Re(CO)$_3$ or radiolabeled ^{99m}Tc(CO)$_3$ in solution yielded conjugates **89** (Scheme 5.45) [217].

Very interesting work was also reported by Piwnica-Worms and coworkers. They prepared Tat-peptide derivatives like **90** dual-labeled with ^{99m}Tc and fluorescein for scintigraphy and fluorescence microscopy (Scheme 5.45) [218]. Tat-peptide basic regions have been intensely studied because of their ability to translocate across cell membranes, seemingly independent of receptor-mediated endocytosis. The peptides were prepared by SPPS with an N-terminal His ligand, to which Tc(CO)$_3$ or Re(CO)$_3$ is coordinated in solution after cleavage and purification. As an alternative to histidine, diethylenetriamine-pentaacetic acid (DTPA) was also used as an N-terminal metal chelator. The C-terminal cysteine was reacted with fluorescein-5-maleimide already on the resin, thus providing both labels in the same peptide, as shown for **90** in Scheme 5.45. Cellular localization of these

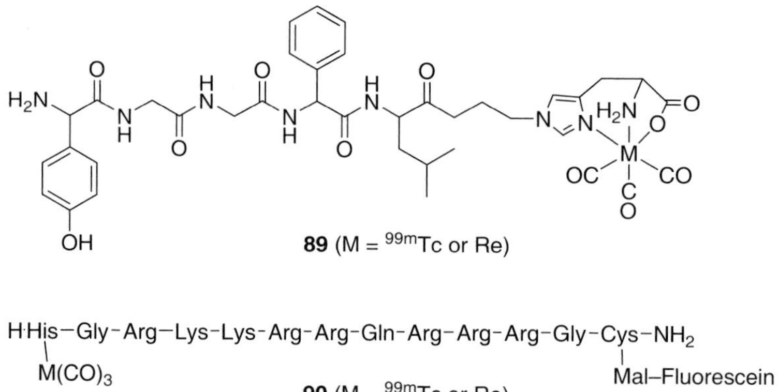

89 (M = ^{99m}Tc or Re)

HHis—Gly—Arg—Lys—Lys—Arg—Arg—Gln—Arg—Arg—Arg—Gly—Cys—NH$_2$
   | |
   M(CO)$_3$ Mal—Fluorescein

90 (M = ^{99m}Tc or Re)

Scheme 5.45 A C-terminally labeled enkephalin derivative **89** peptide and the dual-labeled Tat-peptide derivative **90**.

peptides was confirmed by fluorescence microscopy. The biodistribution in Balb/c mice was studied by scintigraphy. Radiometric and fluorescence microscopy data were in good agreement. For instance, little brain penetration across the blood–brain barrier (BBB) was observed for these conjugates.

5.5.4
Host–guest Chemistry and Biosensors

Han et al. prepared macrocyclic compounds from 1,1′-ferrocene-dicarboxylic acid chloride and pseudo-cystine derivatives with various linkers between the sulfur atoms [219]. The synthesis yielded 1,1′-ferrocenophane derivatives **91** as well as cyclodimers (Scheme 5.46). The redox potential for the $Fe(C_5H_4\text{-}R)_2/Fe(C_5H_4\text{-}R)_2^+$ couple of the 1,1′-ferrocenophane derivatives is dependent on the nature of the anion. In particular, the addition of fluoride resulted in a pronounced redox potential shift of about 120–150 mV to more positive values. From the occurrence of this shift of the redox potential, it is anticipated that halide ions occupy a position in the binding pocket of the 1,1′-ferrocenophane. The binding is stabilized by hydrogen bonds to the two NH moieties and, in the case of the ferrocenium redox state, by an additional electrostatic interaction between the halide and the iron atom. In a subsequent report, cystine and cystine di- and tripeptides $((Aaa)_n\text{-}Cys)_2$ (Aaa = Gly, Ala, Leu, Met, Pro; $n = 0, 1, 2$) were synthesized and reacted with 1,1′-ferrocene-dicarboxylic acid chloride to form ferrocenophanes with a cavity of different size [220]. Into this cavity, alkaline and alkaline earth cations could bind. CD spectroscopy and cyclic voltammetry were used to determine binding constants. For one derivative, high binding affinity for Mg^{2+} and Ca^{2+}, along with a high preference for Ca^{2+} over K^+, was observed which was even better than the values for the natural ionophore valinomycin. In another example, Toniolo, Maggini and coworkers prepared a nonapeptide with a 3_{10}-helical conformation, in which two tyrosine phenol rings were esterified with ferrocene carboxylic acid.

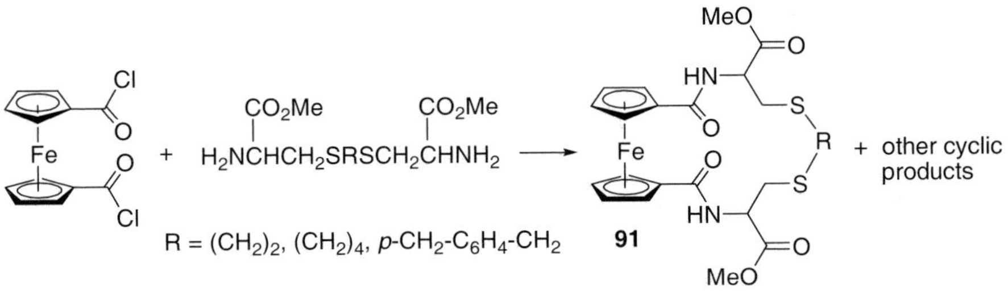

Scheme 5.46 Synthesis of a peptide 1,1'-ferrocenophane **91**, which serves as a host for halide anion binding.

These synthetic helical peptides were shown to have a hydrophobic binding cavity, which can serve as an efficient host for [60] fullerene [221].

Another interesting sensor for small molecules like SO_2 and iodine was reported by van Koten's group. Derivatives of a terdentate organoplatinum(II) compound with a "pincer" NCN-type ligand can be easily prepared by oxidative addition of Pt(0) tetrakis-phosphine into the carbon-halide bond of the NCN ligand [222]. The resulting organo-platinum species are surprisingly stable in air and water. The Pt centre may coordinate small molecules like SO_2 and I_2, which results in a colour change to dark red or brown [223, 224]. In addition, the [195]Pt NMR spectrum show a significant shift upon small molecule coordination. The valine derivative **92** of the Pt-NCN complex (Scheme 5.47) could be coupled to amino acids and peptides [222–224]. It was also coupled to the N terminus of a library of small peptides by solid phase synthesis methods [225]. Successful coupling of the Pt complex to the peptide can be easily verified by addition of I_2 and monitoring the resulting colour change, which can be reversed by washing with DMF/NEt_3. Apart from the fact that this procedures presents a genuine organometallic oxidative addition reaction for the introduction of the organometallic label, it also offers an interesting bioorganometallic marker as an alternative to existing purely organic dyes used in solid-phase screening assays [226].

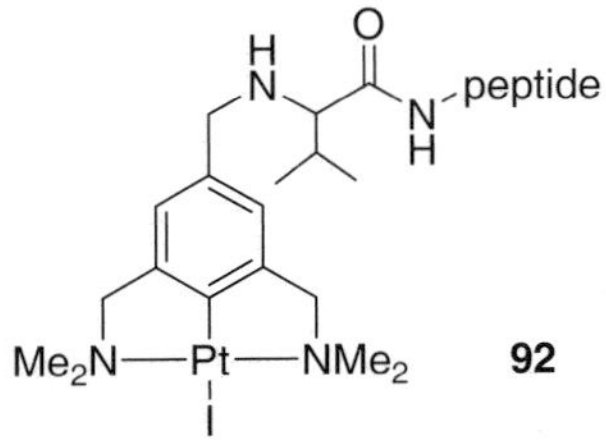

Scheme 5.47 The organoplatinum(II) valine derivative **92** suitable for N-terminal labeling of amino acids and peptides.

Acknowledgements

I am grateful for stimulating discussions with many colleagues, who also freely shared preliminary information that is included in this chapter. Names I would like to mention include: Roger Alberto, Heinz-Bernhard Kraatz, Torbjörn Frejd, Emanuela Licandro, Stefano Maiorana, Vladimir Rapic, Kay Severin, Dave van Staveren. Discussions with Kay Severin were also a great help at a final stage of this manuscript. Most of my group is or was involved in projects described herein, and I would like to thank all coworkers and visiting scientists for their contributions and enthusiam. Finally, I am indebted to Mrs. Karin Weiss for her invaluable help, not only for her support in preparing this review but also for keeping our group running happily. Above all, my family deserves credit for giving me time and room to finish this (and other) projects at night and weekends.

Abbreviations

Aaa	Any amino acid
Aib	Amino iso-butyric acid
AT	Angiotensin
BBB	Blood-brain barrier
BK	Bradykinin
bpa	bis(picolyl)amine
Boc	tert.-Butoxy carbonyl
BSA	Bovine serum albumin
Car	Carboranylalanine
CNS	Central nervous system
Cym	Cymantrenylalanine
Enk	Enkephalin
ESI-MS	Electrospray ionization mass spectrometry
Fac	Ferrocene 1-amino-1'-carboxylic acid
Fc	Ferrocenyl ($CpFeC_5H_4$)
Fc-CO-	Ferrocenoyl ($CpFeC_5H_4$-CO-)
Fed	Iron tricarbonyl cyclohexadienyl cation $[(Fe(CO)_3(C_6H_7)]^+$
Fem	Ferrocenylmethyl
Fer	Ferrocenylalanine
Fmoc	Fluorenyl methoxy carbonyl
GnRH	Gonadotropin-releasing hormone (same as LHRH)
Glp	pyro glutamic acid (also abbreviated pGlu)
GSH	Glutathione (γ-glutamylcysteinylglycine)
GST	Glutathione-(S)-transferase
HPLC	High pressure (or performance) liquid chromatography
HPLC-ECD	HPLC with electrochemical detection
LHRH	Luteinizing hormone-releasing hormone (same as GnRH)
MALDI-TOF	Matrix-assisted laser desorption time-of-flight (mass spectrometry)

NBD	Norbornadiene
NKA	Neurokinin A
NLS	Nuclear localization signal
NT	Neurotensin
NPY	Neuropeptide Y
Oic	Octahydroindole-2-carboxylic acid
Phg	Phenylglycine
Pps	Diphenylphosphine serine
Pps(S)	Diphenylphosphino serine sulfide
Sar	Sarcosine
SP	Substance P
SPPS	Solid Phase Peptide Synthesis
TFA	Trifluoro acetic acid
Ugi-4CR	Ugi four component reaction
Xps	Bis(3,5-dimethylphenyl)phosphino serine

Standard three letter codes for amino acids are used throughout. Unless specifically noted, stereochemistry implies pure L amino acids. Peptides are consistently written from N to C terminus in standard peptide nomenclature. In a peptide, "Gly" corresponds the fragment "HN-CH$_2$-CO". For example, H-Gly-NH$_2$ is the carboxamide of glycine (H$_2$N-CH$_2$-CONH$_2$), Ac-Gly-Ala-OH is *N*-acetylated glycyl alanine.

References

1 K. Schlögl, *Monatsh. Chem.* **1957**, *88*, 601–621.

2 K. Severin, R. Bergs, W. Beck, *Angew. Chem.* **1998**, *110*, 1722–1743; *Angew. Chem. Int. Ed. Engl.* **1998**, *37*, 1634–1654.

3 N. Metzler-Nolte, *Angew. Chem. Int. Ed. Engl.* **2001**, *40*, 1040–1044; *Angew. Chem.* **2001**, *113*, 1072–1076.

4 T. J. Kealy, P. L. Pauson, *Nature* **1951**, *168*, 1039–1040.

5 S. A. Miller, J. A. Tebboth, J. F. Tremaine, *J. Chem. Soc.* **1952**, 632–635.

6 G. Wilkinson, M. Rosenblum, M. C. Whiting, R. B. Woodward, *J. Am. Chem. Soc.* **1952**, *74*, 2125–2126.

7 E. O. Fischer, W. Pfab, *Z. Naturforsch.* **1952**, *7b*, 377–379.

8 R. B. Woodward, M. Rosenblum, M. C. Whiting, *J. Am. Chem. Soc.* **1952**, *74*, 3458–3459.

9 D. R. van Staveren, N. Metzler-Nolte, *Chem. Rev.* **2004**, 5931–5985.

10 H.-B. Kraatz, M. Galka, *Metal Ions in Biological Systems* **2001**, *38*, 385–409.

11 M. Pietraszkiewicz, A. Wieckowska, R. Bilewicz, A. Misicka, L. Piela, K. Bajdor, *Mat. Science Eng. C* **2001**, *18*, 121–124.

12 A. Wieckowska, R. Bilewicz, A. Misicka, M. Pietraszkiewicz, K. Bajdor, L. Piela, *Chem. Phys. Lett.* **2001**, *350*, 447–458.

13 S. Iijima, F. Mizutani, S. Yabuki, Y. Tanaka, M. Asai, T. Katsura, S. Hosaka, M. Ibonai, *Anal. Chim. Acta* **1993**, *281*, 483–487.

14 A. Hess, J. Sehnert, T. Weyhermüller, N. Metzler-Nolte, *Inorg. Chem.* **2000**, *39*, 5437–5443.

15 J. Sehnert, A. Hess, N. Metzler-Nolte, *J. Organomet. Chem.* **2001**, *637–639*, 349–355.

16 R. S. Herrick, R. M. Jarret, T. P. Curran, D. R. Dragoli, M. B. Flaherty, S. E. Lindyberg,

R. A. SLATE, L. C. THORNTON, *Tetrahedron Lett.* **1996**, *37*, 5289–5292.

17 D. P. FAIRLIE, M. L. WEST, A. K. WONG, *Curr. Med. Chem.* **1998**, *5*, 29–62.

18 A. NOMOTO, T. MORIUCHI, S. YAMAZAKI, A. OGAWA, T. HIRAO, *J. Chem. Soc., Chem. Commun.* **1998**, 1963–1964.

19 T. MORIUCHI, A. NOMOTO, K. YOSHIDA, T. HIRAO, *J. Organomet. Chem.* **1999**, *589*, 50–58.

20 T. MORIUCHI, A. NOMOTO, K. YOSHIDA, A. OGAWA, T. HIRAO, *J. Am. Chem. Soc.* **2001**, *123*, 68–75.

21 N. J. LONG, *Metallocenes*, Blackwell Science, Oxford, **1998**.

22 T. MORIUCHI, A. NOMOTO, K. YOSHIDA, T. HIRAO, *Organometallics* **2001**, *20*, 1008–1013.

23 D. R. VAN STAVEREN, T. WEYHERMÜLLER, N. METZLER-NOLTE, *J. Chem. Soc., Dalton Trans.* **2003**, 210–220.

24 E. G. HUTCHINSON, J. M. THORNTON, *Protein Sci.* **1994**, *3*, 2207–2216.

25 C. M. VENKATACHALAM, *Biopolym.* **1968**, *6*, 1425–1434.

26 Y. XU, P. SAWECZKO, H.-B. KRAATZ, *J. Organomet. Chem.* **2001**, *637–639*, 335–342.

27 H. B. KRAATZ, D. M. LEEK, A. HOUMAM, G. D. ENRIGHT, J. LUSZTYK, D. D. M. WAYNER, *J. Organomet. Chem.* **1999**, *589*, 38–49.

28 Y. XU, H.-B. KRAATZ, *Tetrahedron Lett.* **2001**, *42*, 2601–2603.

29 H. UMEZAWA, T. AOYAGI, H. MORISHIMA, M. MATSUZAKI, M. HAMADA, T. TAKEUCHI, *J. Antibiot.* **1970**, *23*, 259–262.

30 I. KATOH, T. YASUNAGA, Y. IKAWA, Y. YOSHINAKA, *Nature* **1987**, *329*, 654–656.

31 R. E. BABINE, S. L. BENDER, *Chem. Rev.* **1997**, *97*, 1359–1472.

32 T. MORIUCHI, K. YOSHIDA, T. HIRAO, *Organometallics* **2001**, *20*, 3101–3105.

33 T. MORIUCHI, K. YOSHIDA, T. HIRAO, *J. Organomet. Chem.* **2001**, *637–639*, 75–79.

34 T. MORIUCHI, K. YOSHIDA, T. HIRAO, *J. Organomet. Chem.* **2003**, *668*, 31–34.

35 I. R. BUTLER, S. C. QUAYLE, *J. Organomet. Chem.* **1998**, *552*, 63–68.

36 T. OKAMURA, K. SAKAUYE, N. UEYAMA, A. NAKAMURA, *Inorg. Chem.* **1998**, *37*, 6731–6736.

37 L. BARISIC, V. RAPIC, V. KOVAC, *Croat. Chem. Acta* **2002**, *75*, 199–210.

38 K. HEINZE, M. SCHLENKER, *Eur. J. Inorg. Chem.* **2004**, 2974–2988.

39 L. BARISIC, M. DROPUCIC, V. RAPIC, H. PRITZKOW, S. KIRIN, N. METZLER-NOLTE, *J. Chem. Soc., Chem. Commun.* **2004**, 2004–2005.

40 K. SEVERIN, W. BECK, G. TROJANDT, K. POLBORN, W. STEGLICH, *Angew. Chem.* **1995**, *107*, 1570–1572; *Angew. Chem. Int. Ed. Engl.* **1995**, *34*, 1449–1451.

41 G. TROJANDT, U. HERR, K. POLBORN, W. STEGLICH, *Chem. Eur. J.* **1997**, *3*, 1254–1268.

42 S. R. GILBERTSON, G. CHEN, M. MCLOUGHLIN, *J. Am. Chem. Soc.* **1994**, *116*, 4481–4482.

43 S. R. GILBERTSON, X. WANG, G. S. HOGE, C. A. KLUG, J. SCHAEFER, *Organometallics* **1996**, *15*, 4678–4680.

44 S. R. GILBERTSON, X. WANG, *J. Org. Chem.* **1996**, *61*, 434–435.

45 S. R. GILBERTSON, G. H. CHEN, J. KAO, A. BEATTY, C. F. CAMPANA, *J. Org. Chem.* **1997**, *62*, 5557–5566.

46 S. R. GILBERTSON, X. F. WANG, *Tetrahedron* **1999**, *55*, 11609–11618.

47 S. R. GILBERTSON, S. E. COLLIBEE, A. AGARKOV, *J. Am. Chem. Soc.* **2000**, *122*, 6522–6523.

48 S. GREENFIELD, A. AGARKOV, S. R. GILBERTSON, *Org. Lett.* **2003**, *5*, 3069–3072.

49 A. AGARKOV, E. W. UFFMANN, S. R. GILBERTSON, *Org. Lett.* **2003**, *5*, 2091–2094.

50 S. R. GILBERTSON, G. W. STARKEY, *J. Org. Chem.* **1996**, *61*, 2922–2923.

51 M. TEPPER, O. STELZER, T. HÄUSLER, W. S. SHELDRICK, *Tetrahedron Lett.* **1997**, *38*, 2257–2258.

52 J. WONNEMANN, M. OBERHOFF, G. ERKER, R. FROHLICH, K. BERGANDER, *Eur. J. Inorg. Chem.* **1999**, 1111–1120.

53 M. OBERHOFF, G. ERKER, R. FROHLICH, *Chem. Eur. J.* **1997**, *3*, 1521–1525.

54 D. HARMSEN, G. ERKER, R. FROHLICH, G. KEHR, *Eur. J. Inorg. Chem.* **2002**, 3156–3171.

55 B. KAYSER, K. POLBORN, W. STEGLICH, W. BECK, *Chem. Ber.* **1997**, *130*, 171–177.

56 J. POSPISEK, S. TOMA, I. FRIC, K. BLÁHA, *Coll. Czech. Chem. Commun.* **1980**, *45*, 435–441.

57 M. KIRA, T. MATSUBARA, H. SHINOHARA, M. SISIDO, *Chem. Lett.* **1997**, 89–90.

58 J. HUGHES, T. W. SMITH, H. W. KOSTERLITZ, L. A. FOTHERGILL, B. A. MORGAN, H. R. MORRIS, *Nature* **1975**, *258*, 577–579.

59 V. CLEMENT-JONES, G. M. BESSER, in S. UDENFRIED, J. MEIENHOFER (Eds.), *The Peptides: Analysis, Synthesis, Biology*, Academic Press, New York, **1984**, Vol. 6, 323–389.

60 S. J. PATERSON, L. E. ROBSON, H. W. KOSTERLITZ, in S. UDENFRIED, J. MEIENHOFER (Eds.), *The Peptides: Analysis, Synthesis, Biology*, Academic Press, New York, **1984**, Vol. 6, 147–189.

61 U. HOFFMANNS, M. DORIA, N. METZLER-NOLTE, unpublished data **2003**.

62 D. R. VAN STAVEREN, N. METZLER-NOLTE, *J. Chem. Soc., Chem. Commun.* **2002**, 1406–1407.

63 D. R. VAN STAVEREN, E. BOTHE, T. WEYHERMÜLLER, N. METZLER-NOLTE, *Eur. J. Inorg. Chem.* **2002**, 1518–1529.

64 D. R. VAN STAVEREN, E. BOTHE, T. WEYHERMÜLLER, N. METZLER-NOLTE, *J. Chem. Soc., Chem. Commun.* **2001**, 131–132.

65 D. R. VAN STAVEREN, E. BILL, E. BOTHE, M. BÜHL, T. WEYHERMÜLLER, N. METZLER-NOLTE, *Chem. Eur. J.* **2002**, *8*, 1649–1662.

66 D. R. VAN STAVEREN, T. WEYHERMÜLLER, N. METZLER-NOLTE, **2004**, unpublished data.

67 H. BRUNNER, W. KÖNIG, B. NUBER, *Tetrahedron Asym.* **1993**, *4*, 699–707.

68 R. F. W. JACKSON, D. TURNER, M. H. BLOCK, *Synlett* **1996**, 862–864.

69 A. RICOUART, P. MAES, T. BATTMANN, B. KERDELHUE, A. TARTAR, C. SERGHERAERT, *Int. J. Pept. Protein Res.* **1988**, *32*, 56–63.

70 E. CUIGNET, C. SERGHERAERT, A. TARTAR, M. DAUTREVAUX, *J. Organomet. Chem.* **1980**, *195*, 325–329.

71 R. EPTON, G. MARR, G. A. WILLMORE, D. HUDSON, P. H. SNELL, C. R. SNELL, *Int. J. Biol. Macromol.* **1981**, *3*, 395–396.

72 E. CUIGNET, M. DAUTREVAUX, C. SERGHERAERT, A. TARTAR, B. ATTALI, J. CROS, *Eur. J. Med. Chem.* **1982**, *17*, 203–206.

73 J. C. BRUNET, E. CUIGNET, M. DAUTREVAUX, A. DEMARLY, H. GRAS, P. MARCINCAL, C. SERGHERAERT, A. TARTAR, J. C. VANVOORDE, M. VANPOUCKE, in K. BRUNFELDT (Ed.), *Peptides 1980. Proceedings of the 16th European Peptide Symposium*, Scriptor, Copenhagen, **1981**, 603–607.

74 U. S. VON EULER, J. H. GADDUM, *J. Physiol.* **1931**, *72*, 74–87.

75 M. M. CHANG, S. E. LEEMAN, H. D. NIALL, *Nature New Biol.* **1971**, *232*, 86–89.

76 G. W. TREGEAR, H. D. NIALL, J. T. POTTS JUN., S. E. LEEMAN, M. M. CHANG, *Nature New Biol.* **1971**, *232*, 87–89.

77 M. OTSUKA, K. YOSHIOKA, *Physiol. Rev.* **1993**, *73*, 229–308.

78 P. HUBLAU, C. SERGHERAERT, L. BALLESTER, M. DAUTREVAUX, *Eur. J. Med. Chem.* **1983**, *18*, 131–133.

79 W. G. SLY, *J. Am. Chem. Soc.* **1959**, *81*, 18–20.

80 N. A. SASAKI, P. POTIER, M. SAVIGNAC, G. JAOUEN, *Tetrahedron Lett.* **1988**, *29*, 5759–5762.

81 F. LE BORGNE, J. P. BEAUCOURT, *Tetrahedron Lett.* **1988**, *29*, 5649–5652.

82 M. JUNG, D. E. KERR, P. D. SENTER, *Arch. Pharmazie* **1997**, *330*, 173–176.

83 T. ROTH, C. ECKERT, H.-H. FIEBIG, M. JUNG, *Anticancer Res.* **2002**, *22*, 2281–2284.

84 I. OTT, B. KIRCHER, R. GUST, *J. Inorg. Biochem.* **2004**, *98*, 485–489.

85 A. TARTAR, A. DEMARLY, C. SERGHERAERT, E. ESCHER, in V. J. HRUBY, D. H. RICH (Eds.), *Proceedings of the 8th American Peptide Symposium*, Pierce Chemical, Rockford, Ill., **1983**, 377–380.

86 P. MAES, A. RICOUART, E. ESCHER, A. TARTAR, C. SERGHERAERT, *Coll. Czech. Chem. Commun.* **1988**, *53*, 2914–2919.

87 J. M. SAAVEDRA, *Endocrine Rev.* **1992**, *13*, 329–380.

88 J. E. TAYLOR, F. V. DEFEUDIS, J. P. MOREAU, *Drug Dev. Res.* **1989**, *16*, 1–11.

89 N.-E. Rhaleb, G. Drapeau, S. Dion, D. Jukic, N. Rouissi, D. Regoli, *British J. Pharmacol.* **1990**, *99*, 445–448.

90 R. Couture, J.-N. Drouin, O. Leukart, D. Regoli, *Can. J. Physiol. Pharmacol.* **1979**, *57*, 1437–1442.

91 R. Couture, A. Fournier, J. Magnan, S. St-Pierre, D. Regoli, *Can. J. Physiol. Pharmacol.* **1979**, *57*, 1427–1436.

92 D. M. Coy, J. A. Vilchez-Martinez, E. J. Coy, A. V. Schally, *J. Med. Chem.* **1976**, *19*, 423–432.

93 R. M. Moriarty, Y. Y. Ku, U. S. Gill, *J. Organomet. Chem.* **1989**, *362*, 187–191.

94 W. S. Sheldrick, S. Heeb, *J. Organomet. Chem.* **1989**, *377*, 357–366.

95 W. S. Sheldrick, A. J. Gleichmann, *J. Organomet. Chem.* **1994**, *470*, 183–187.

96 A. J. Gleichmann, J. M. Wolff, W. S. Sheldrick, *J. Chem. Soc., Dalton Trans.* **1995**, 1549–1554.

97 D. B. Grotjahn, C. Joubran, D. Combs, D. C. Brune, *J. Am. Chem. Soc.* **1998**, *120*, 11814–11815.

98 D. B. Grotjahn, *Coord. Chem. Rev.* **1999**, *192*, 1125–1141.

99 F. Noor, A. Wüstholz, R. Kinscherf, N. Metzler-Nolte, *Angew. Chem. Int. Ed.* **2005**, *44*, 2429–2432; *Angew. Chem.* **2005**, *117*, 2481–2485.

100 B. Rudolf, J. Zakrzewski, M. Salmain, G. Jaouen, *New J. Chem.* **1998**, 813–818.

101 M. Salmain, G. Jaouen, B. Rudolf, J. Zakrzewski, *J. Organomet. Chem.* **1999**, *589*, 98–102.

102 D. Yamamoto, K. Matsumoto, H. Ohishi, T. Ishida, M. Inoue, K. Kitamura, H. Mizuna, *J. Biol. Chem.* **1991**, *266*, 14771–14777.

103 K. Mastumoto, M. Murata, S. Sumiya, K. Kitamura, T. Ishida, *Biochim. Biophys. Acta* **1994**, *1208*, 268–276.

104 K. Mastumoto, M. Murata, S. Sumiya, K. Mizoue, K. Kitamura, T. Ishida, *Biochim. Biophys. Acta* **1998**, *1383*, 93–100.

105 M. O. Funk, Y. Nakagawa, J. Skochdopole, E. T. Kaiser, *Int. J. Pept. Protein Res.* **1979**, *13*, 296–303.

106 K. Plumb, H.-B. Kraatz, *Bioconjugate Chem.* **2003**, *14*, 601–606.

107 M. M. Galka, H.-B. Kraatz, *ChemPhysChem* **2002**, 356–359.

108 J. D. Schmitt, M. S. P. Sansom, I. D. Kerr, G. G. Lunt, R. Eisenthal, *Biochemistry* **1997**, *36*, 1115–1122.

109 R. Latorre, O. Alvarez, *Physiol. Rev.* **1981**, *61*, 77–150.

110 R. O. Fox Jr., F. M. Richards, *Nature* **1982**, *300*, 325–330.

111 G. A. Woolley, B. A. Wallace, *Biochemistry* **1993**, *32*, 9819–9825.

112 S. Kaluz, S. Toma, *Coll. Czech. Chem. Commun.* **1988**, *53*, 638–642.

113 A.-S. Carlström, T. Frejd, *Synthesis* **1989**, 414–418.

114 A.-S. Carlström, T. Frejd, *J. Org. Chem.* **1990**, *55*, 4175–4180.

115 B. Basu, S. K. Chattopadhyay, A. Ritzen, T. Frejd, *Tetrahedron Asym.* **1997**, *8*, 1841–1846.

116 S. Maricic, A. Ritzen, U. Berg, T. Frejd, *Tetrahedron* **2001**, *57*, 6523–6529.

117 S. Maricic, U. Berg, T. Frejd, *Tetrahedron* **2002**, *58*, 3085–3093.

118 S. Maricic, T. Frejd, *J. Org. Chem.* **2002**, *67*, 7600–7606.

119 G. D. Smith, J. F. Griffin, *Science* **1978**, *199*, 1214–1216.

120 D. Picone, A. D'Ursi, A. Motta, T. Tancredi, P. A. Temussi, *Eur. J. Biochem.* **1990**, *192*, 433–439.

121 T. A. Steitz, R. Henderson, D. M. Blow, *J. Mol. Biol.* **1969**, *46*, 337–348.

122 D. M. Blow, *Acc. Chem. Res.* **1976**, *9*, 145–152.

123 P. E. Nielsen, M. Egholm, R. H. Berg, O. Buchardt, *Science* **1991**, *254*, 1497–1500.

124 K. L. Dueholm, M. Engholm, C. Behrens, L. Christensen, H. F. Hansen, T. Vulpius, P. K. H., R. H. Berg, P. E. Nielsen, O. Buchardt, *J. Org. Chem.* **1994**, *59*, 5767–5773.

125 P. E. Nielsen, *Acc. Chem. Res.* **1999**, *32*, 624–630.

126 M. Egholm, O. Buchardt, L. Christensen, C. Behrens, S. M. Freier, D. A. Driver, R. H. Berg, S. K. Kim, B. Norden, P. E. Nielsen, *Nature* **1993**, *365*, 566–568.

127 M. Leijon, A. Gräslund, P. E. Nielsen, O. Buchardt, B. Nordén, S. M. Kristensen, M. Eriksson, *Biochemistry* **1994**, *33*, 9820–9825.

128 M. Eriksson, P. E. Nielsen, *Nature Struct. Biol.* **1996**, *3*, 410–413.

129 V. Menchise, G. D. Simone, T. Tedeschi, R. Corradini, S. Sforza, R. Marchelli, D. Capasso, M. Saviano, C. Pedone, *Proc. Natl. Acad. Sci. USA* **2003**, *100*, 12021–12026.

130 T. Ratilainen, A. Holmén, E. Tuite, G. Haaima, L. Christensen, P. E. Nielsen, B. Nordén, *Biochemistry* **1998**, *31*, 12331–12342.

131 T. Ratilainen, A. Holmén, E. Tuite, P. E. Nielsen, B. Nordén, *Biochemistry* **2000**, *39*, 7781–7791.

132 P. E. Nielsen, *Perspectives in Drug Discovery and Design* **1996**, *4*, 76–84.

133 L. Good, P. E. Nielsen, *Antisense Nucl. Acid Drug Dev.* **1997**, *7*, 431–437.

134 H. Knudsen, P. E. Nielsen, *Anti-Cancer Drugs* **1997**, *8*, 113–118.

135 H. J. Larsen, T. Bentin, P. E. Nielsen, *Biochim. Biophys. Acta* **1999**, *1489*, 159–166.

136 P. E. Nielsen, *Curr. Opin. Biotechnol.* **1999**, *10*, 71–75.

137 E. Uhlmann, A. Peyman, G. Breipohl, D. W. Will, *Angew. Chem.* **1998**, *110*, 2954–2983; *Angew. Chem. Int. Ed. Engl.* **1998**, *37*, 2796–2823.

138 A. Hess, N. Metzler-Nolte, *J. Chem. Soc., Chem. Commun.* **1999**, 885–886.

139 A. Hess, *PhD Thesis*, Ruhr-Universität, Bochum, **1999**.

140 C. Baldoli, S. Maiorana, E. Licandro, G. Zinzalla, D. Perdicchia, *Org. Lett.* **2002**, *4*, 4341–4344.

141 S. Maiorana, E. Licandro, D. Perdicchia, C. Baldoli, B. Vandoni, C. Giannini, M. Salmain, *J. Mol. Catal. A* **2003**, *204–205*, 165–175.

142 C. Baldoli, C. Giannini, E. Licandro, S. Maiorana, G. Zinzalla, *Synlett.* **2004**, 1044–1048.

143 A. Dömling, K. Z. Chi, M. Barrère, *Bioorg. Med. Chem. Lett.* **1999**, *9*, 2871–2874.

144 W. Maison, I. Schlemminger, O. Westerhoff, J. Martens, *Bioorg. Med. Chem. Lett.* **1999**, *9*, 581–584.

145 K. L. Dueholm, K. H. Petersen, D. K. Jensen, M. Egholm, P. E. Nielsen, O. Buchardt, *Bioorg. Med. Chem. Lett.* **1994**, *4*, 1077–1080.

146 S. Sforza, G. Haaima, R. Marchelli, P. E. Nielsen, *Eur. J. Org. Chem.* **1999**, 197–204.

147 G. Haaima, A. Lohse, O. Buchardt, P. E. Nielsen, *Angew. Chem.* **1996**, *108*, 2068–2070; *Angew. Chem. Int. Ed. Engl.* **1996**, *35*, 1939–1942.

148 J. C. Verheijen, G. A. van der Marel, J. H. van Boom, N. Metzler-Nolte, *Bioconjugate Chem.* **2000**, *11*, 741–743.

149 A. Maurer, H.-B. Krautz, N. Metzler-Nolte, *Eur. J. Inorg. Chem.* **2005**, published as Early View article, POI: 10.1002/ejic.200500316.

150 H. Eckert, C. Seidel, *Angew. Chem.* **1986**, *98*, 168–170; *Angew. Chem. Int. Ed. Engl.* **1986**, *25*, 159–161.

151 H. Eckert, Y. Kiesel, C. Seidel, C. Kaulberg, H. Brinkmann, in W. Voelter, E. Bayer, Y. A. Ovchinnikov, T. Ivanov (Eds.), *Chemistry of Peptides and Proteins*, Walter de Gruyter, Berlin, New York, **1986**, Vol. 3, 19–28.

152 H. Eckert, B. Forster, C. Seidel, *Z. Naturforsch. B* **1991**, *46*, 339–352.

153 M. Bodanszky, A. Bodanszky, *The Practice of Peptide Synthesis*, Springer-Verlag, Berlin, **1984**.

154 J. Jones, *The Chemical Synthesis of Peptides*, Clarendon Press, Oxford, **1991**.

155 C. N. C. Drey, A. S. J. Stewart, in D. Theodoropoulos (Ed.), *Peptides 1986*, Walter de Gruyter, Berlin, **1987**, 65–68.

156 A. S. J. Stewart, C. N. C. Drey, *J. Chem. Soc. Perkin Trans. I* **1990**, 1753–1756.

157 L. A. P. Kane-Maguire, R. Kanitz, *J. Organomet. Chem.* **1988**, *353*, C33–C34.

158 J. A. Carver, B. Fates, L. A. P. Kane-Maguire, *J. Chem. Soc., Chem. Commun.* **1993**, 928–929.

159 L. A. P. Kane-Maguire, R. Kanitz, P. Jones, P. A. Williams, *J. Organomet. Chem.* **1994**, *464*, 203–213.

160 J. E. Hallgren, C. S. Eschbach, D. Seyferth, *J. Am. Chem. Soc.* **1972**, *94*, 2547–2549.

161 A. J. Baskar, C. M. Lukehart, K. Srinivasan, *J. Am. Chem. Soc.* **1981**, *103*, 1467–1472.

162 K. Weiss, E. O. Fischer, *Chem. Ber.* **1973**, *106*, 1277–1284.

163 K. Weiss, E. O. Fischer, *Chem. Ber.* **1976**, *109*, 1868–1886.

164 L. S. Hegedus, *Acc. Chem. Res.* **1995**, *28*, 299–305.

165 S. R. Pulley, L. S. Hegedus, *J. Am. Chem. Soc.* **1993**, *115*, 9037–9047.

166 D. A. Buckingham, L. G. Marzilli, A. M. Sargeson, *J. Am. Chem. Soc.* **1967**, *89*, 2772–2773.

167 J. P. Collman, E. Kimura, *J. Am. Chem. Soc.* **1967**, *89*, 6096–6103.

168 R. J. Browne, D. A. Buckingham, C. R. Clark, P. A. Sutton, *Adv. Inorg. Chem.* **2000**, *49*, 307.

169 W. Beck, R. Krämer, *Angew. Chem.* **1991**, *103*, 1492–1493; *Angew. Chem. Int. Ed. Engl.* **1991**, *30*, 1467–1468.

170 R. Krämer, M. Maurus, R. Bergs, K. Polborn, K. Sünkel, B. Wagner, W. Beck, *Chem. Ber.* **1993**, *126*, 1969–1980.

171 R. Krämer, M. Maurus, K. Polborn, K. Sünkel, C. Robl, W. Beck, *Chem. Eur. J.* **1996**, *2*, 1518–1526.

172 W. Hoffmüller, M. Maurus, K. Severin, W. Beck, *Eur. J. Inorg. Chem.* **1998**, 729–731.

173 K. Haas, W. Ponikwar, H. Nöth, W. Beck, *Angew. Chem.* **1998**, *110*, 1200–1203; *Angew. Chem. Int. Ed.* **1998**, *37*, 1086–1089.

174 K. Haas, E. M. Ehrenstorfer-Schafers, K. Polborn, W. Beck, *Eur. J. Inorg. Chem.* **1999**, 465–469.

175 R. Krämer, *Angew. Chem.* **1996**, *108*, 1287–1289; *Angew. Chem. Int. Ed. Engl.* **1996**, *35*, 1197–1199.

176 I. Ugi, *Angew. Chem.* **1959**, *71*, 386.

177 I. Ugi, in W. Foerst (Ed.), *Neuere Methoden der präparativen organischen Chemie*, Verlag Chemie, Weinheim, **1966**, 1.

178 G. Gokel, P. Hoffmann, H. Kleimann, H. Klusacek, G. Lüdke, D. Marquarding, I. Ugi, in I. Ugi (Ed.), *Isonitrile Chemistry*, Academic Press, New York, **1971**, 201–215.

179 R. Urban, I. Ugi, *Angew. Chem.* **1975**, *87*, 67–69; *Angew. Chem. Int. Ed. Engl.* **1975**, *14*, 492–493.

180 I. Ugi, D. Marquarding, R. Urban, in B. Weinstein (Ed.), *Chemistry and Biochemistry of Amino acids, Peptides and Proteins*, Marcel Dekker, New York, **1982**, Vol. 6, 245–289.

181 D. Marquarding, P. Hoffmann, H. Heitzer, I. Ugi, *J. Am. Chem. Soc.* **1970**, *92*, 1969–1971.

182 R. Urban, G. Eberle, D. Marquarding, D. Rehn, H. Rehn, I. Ugi, *Angew. Chem.* **1976**, *88*, 644–646; *Angew. Chem. Int. Ed. Engl.* **1976**, *15*, 627–629.

183 G. Eberle, I. Ugi, *Angew. Chem.* **1976**, *88*, 509–510; *Angew. Chem. Int. Ed. Engl.* **1976**, *15*, 492–493.

184 R. Herrmann, G. Hübener, F. Siglmüller, I. Ugi, *Liebigs Ann.* **1986**, 251–268.

185 F. Siglmüller, R. Herrmann, I. Ugi, *Tetrahedron* **1986**, *42*, 5931–5940.

186 R. Urban, D. Marquarding, I. Ugi, *Hoppe-Seyler's Z. Physiol. Chem.* **1978**, *359*, 1541–1552.

187 A. Ratajczak, B. Misterkiewicz, *J. Organomet. Chem.* **1975**, *91*, 73–79.

188 A. Ratajczak, A. Czech, *Bull. Acad. Pol. Sci., Serie sci. chim.* **1979**, *27*, 661–664.

189 R. Urban, *Tetrahedron* **1979**, *35*, 1841–1843.

190 H. Kase, M. Kaneko, K. Yamada, *J. Antibiot.* **1987**, *40*, 450.

191 S. Sano, K. Ikai, Y. Yoshikawa, T. Nakamura, A. Obayashi, *J. Antibiot.* **1987**, *40*, 512.

192 A. J. Pearson, K. Lee, *J. Org. Chem.* **1994**, *59*, 2304–2313.

193 A. J. Pearson, P. Zhang, K. Lee, *J. Org. Chem.* **1996**, *61*, 6581–6586.

194 J. W. Janetka, D. H. Rich, *J. Am. Chem. Soc.* **1995**, *117*, 10585–10586.

195 J. W. Janetka, D. H. Rich, *J. Am. Chem. Soc.* **1997**, *119*, 6488–6495.

196 A. Schmid, T. Lindel, *Angew. Chem. Int. Ed.* **2004**, *43*, 1581–1583; *Angew. Chem.* **2004**, *116*, 1607–1609.

197 H. Eckert, M. Koller, *Z. Naturforsch. B* **1990**, *45*, 1709–1714.

198 H. Eckert, M. Koller, *J. Liq. Chromatogr.* **1990**, *13*, 3399–3414.

199 German Patent 39193178-A1, **1990**; *Chem. Abstr.* **1991**, *115*, 25536.

200 R. L. Cox, T. W. Schneider, M. D. Koppang, *Anal. Chim. Acta* **1992**, *262*, 145–159.

201 K. Shimada, Y. Kawai, T. Oe, T. Nambara, *J. Liq. Chromatogr.* **1989**, *12*, 359–371.

202 K. Di Gleria, H. A. O. Hill, L. L. Wong, *FEBS Lett.* **1996**, *390*, 142–144.

203 Y. Katayama, S. Sakakihara, M. Maeda, *Anal. Sci.* **2001**, *17*, 17–19.

204 K. K.-W. Lo, J. S.-Y. Lau, D. C.-M. Ng, N. Zhu, *Dalton Transactions* **2002**, 1753–1756.

205 T. N. Duy, P. K. S. Lam, G. R. Shaw, D. W. Connell, *Rev. Environ. Contam. Toxicol.* **2000**, *163*, 113–186.

206 K. K.-W. Lo, D. C.-M. Ng, J. S.-Y. Lau, R. S.-S. Wu, P. K.-S. Lam, *New J. Chem.* **2003**, *27*, 274–279.

207 S. Liu, D. S. Edwards, J. A. Barrett, *Bioconjugate Chem.* **1997**, *8*, 621–636.

208 R. Alberto, R. Schibli, R. Waibel, U. Abram, A. P. Schubiger, *Coord. Chem. Rev.* **1999**, *192*, 901–919.

209 R. Alberto, R. Schibli, A. Egli, A. P. Schubiger, U. Abram, T. A. Kaden, *J. Am. Chem. Soc.* **1998**, *120*, 7987–7988.

210 R. Alberto, R. Schibli, A. P. Schubiger, U. Abram, H. J. Pietzsch, B. Johannsen, *J. Am. Chem. Soc.* **1999**, *121*, 6076–6077.

211 R. Waibel, R. Alberto, J. Willuda, R. Finnern, R. Schibli, A. Stichel-berger, A. Egli, U. Abram, J. P. Mach, A. Pluckthun, P. A. Schubiger, *Nature Biotechnol.* **1999**, *17*, 897–901.

212 A. Egli, R. Alberto, L. Tannahill, R. Schibli, U. Abram, A. Schaffland, R. Waibel, D. Tourwe, L. Jeannin, K. Iterbeke, P. A. Schubiger, *J. Nucl. Med.* **1999**, *40*, 1913–1917.

213 M. Langer, R. La Bella, E. Garcia-Garayoa, A. G. Beck-Sickinger, *Bioconjugate Chem.* **2001**, *2001*, 1028–1034.

214 K. Tatemoto, *Proc. Natl. Acad. Sci. USA* **1982**, *79*, 5485–5489.

215 L. Grundemar, S. R. Bloom, *Neuropeptide Y and drug developments*, Academic Press, San Diego, **1997**.

216 T. W. Spradau, W. B. Edwards, C. J. Anderson, M. J. Welch, J. A. Katzenellenbogen, *Nucl. Med. Biol.* **1999**, *26*, 1–7.

217 D. R. van Staveren, S. Mundwiler, U. Hoffmanns, J. Kyoung Pak, B. Spingler, N. Metzler-Nolte, R. Alberto, *Org. Biomol. Chem.* **2004**, *2*, 2593–2603.

218 K. E. Bullok, M. Dyszlewski, J. L. Prior, C. M. Pica, V. Sharma, D. Piwnica-Worms, *Bioconjugate Chem.* **2002**, *13*, 1226–1237.

219 Q.-W. Han, X.-Q. Zhu, X.-B. Hu, J.-P. Cheng, *Chem. J. Chin. Univ.* **2002**, *23*, 2076–2079.

220 H. Huang, L. Mu, J. He, J.-P. Cheng, *J. Org. Chem.* **2003**, *68*, 7605–7611.

221 A. Bianco, C. Corvaja, M. Crisma, D. M. Guldi, M. Maggini, E. Sartori, C. Toniolo, *Chem. Eur. J.* **2002**, *8*, 1544–1553.

222 M. Albrecht, G. Rodriguez, J. Schoenmaker, G. van Koten, *Org. Lett.* **2000**, *2*, 3461–3464.

223 G. Guillena, G. Rodriguez, G. van Koten, *Tetrahedron Lett.* **2002**, *43*, 3895–3898.

224 G. Guillena, G. Rodriguez, M. Albrecht, G. Van Koten, *Chem. Eur. J.* **2002**, *8*, 5368–5376.

225 G. Guillena, K. M. Halkes, G. Rodriguez, G. D. Batema, G. van Koten, J. P. Kamerling, *Org. Lett.* **2003**, *5*, 2021–2024.

226 M. Albrecht, G. van Koten, *Angew. Chem. Int. Ed.* **2001**, *40*, 3750–3781; *Angew. Chem.* **2001**, *113*, 3866–3898.

6
Labeling of Proteins with Organometallic Complexes: Strategies and Applications

Michèle Salmain

6.1
Introduction

The question of the possible relevance of organometallic π-complexes to biological chemistry was put forward at the end of the 1960s, some 20 years after the discovery of ferrocene [1]. Indeed the first example of the introduction of a transition organometallic complex in a biological molecule was described in 1966. Conjugation of ferrocene to a synthetic polypeptide was achieved by reaction of ferrocenylmethyl isothiocyanate **1**, with a view to using this organometallic complex as radioactive tracer [2].

1

The first example of reaction of a half-sandwich metallocene with a protein was reported a few years later. Tetramethylcyclobutadiene nickel(II) dichloride **2** was shown to react with the enzyme horse liver alcohol dehydrogenase to afford a modified enzyme devoid of zinc and enzymatic activity and containing 8 mol nickel per mol protein. The mode of binding of this organometallic complex had not been fully elucidated at that time but was thought to occur by substitution of the chloride ligands by nucleophilic residues [3].

$Cl \diagdown \overset{Ni}{} \diagup Cl$

2

The versatile chemistry of transition organometallic complexes, their proven biological stability and their sometimes unique physical properties have since

Bioorganometallics: Biomolecules, Labeling, Medicine. Edited by Gérard Jaouen
Copyright © 2006 Wiley-VCH Verlag GmbH & Co. KGaA, Weinheim
ISBN: 3-527-30990-X

opened the way to a number of innovative applications in different areas of biochemistry. We will review in this Chapter the various applications of transition organometallic complexes in biochemistry where these compounds are chemically conjugated to proteins and the various synthetic strategies involved. The strategies of conjugation specifically developed to label peptides and proteins with radioactive organometallic complexes are mentioned in Chapter 4, Section 4.4 and Chapter 5, Section 5.5.3.2.

6.2
Redox Probes

6.2.1
Amperometric Biosensors

Most of the work regarding the covalent labeling of proteins with organometallic redox probes has dealt with the development of enzyme and antibody ampero-metric biosensors. Briefly, amperometric biosensors are analytical devices combining biochemical molecular recognition with enzymes or antibodies and amperometric transduction, i.e. monitoring of the current associated with oxidation or reduction of an electroactive species [4]. Unfortunately, for the great majority of redox enzymes, the redox center(s) (the cofactor(s)) is (are) located sufficiently far from the outermost surface of the protein to be electrically inaccessible. Therefore, direct electron transfer (ET) between the adsorbed enzyme and the electrode is not possible. This hurdle was circumvented by associating with the enzyme a redox mediator, that is an electroactive species able to rapidly exchange electrons with the electrode. Hence, three configurations for mediated enzyme amperometric biosensors have been designed (Fig. 6.1) [5].

In configuration A, ET from the enzyme redox center to the electrode is made possible by addition of a diffusional electron mediator that rapidly shuttles the electron(s) to and from the electrode. In configuration B, redox labels are covalently tethered to the enzyme surface and relay the electrons to and from the electrode.

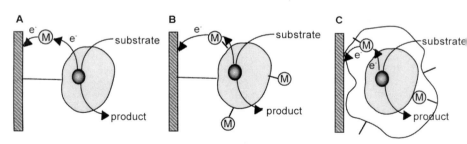

Fig. 6.1 Major amperometric biosensor configurations.
(A) in the presence of a diffusional mediator;
(B) covalent tethering of mediator to the biological recognition element;
(C) immobilization of the biological recognition element in a redox polymer.

In configuration C, the enzyme is immobilized in a cross-linked redox polymer adsorbed onto the electrode surface.

Not surprisingly, ferrocenyl (Fc) derivatives have been widely used either as diffusional or non-diffusional charge transfer mediators because of their almost ideal electrochemical properties, i.e. fast ET rate, low redox potentials and high chemical stability of both the oxidized and the reduced forms. Examples of protein labeling with ferrocene redox probes in the field of amperometric biosensors are reviewed below.

6.2.1.1 Diffusional Mediators

Charge transfer mediation between the redox enzyme glucose oxidase (GOx) with its two buried cofactors FAD and the electrode has been achieved by using the diffusing redox couple 1,1′-dimethylferrocene/1,1′-dimethylferricinium [6]. This finding formed the basis of an oxygen-insensitive glucose amperometric biosensor that is based on the following oxido-reduction equations.

$$\text{Glucose} + \text{GOx-FAD} \rightarrow \text{gluconolactone} + \text{GOx-FADH}_2 \tag{6.1}$$

$$\text{GOx-FADH}_2 + 2\ \text{Med(ox)} \rightarrow \text{GOx-FAD} + 2\ \text{Med(red)} \tag{6.2}$$

$$2\ \text{Med(red)} \rightarrow 2\ \text{Med(ox)} + 2\ \text{e}^- \text{ (to the electrode)} \tag{6.3}$$

This work led to the first mass-marketed single-use glucose enzyme biosensor device called the ExacTECH^{TM} pen (Medisense, Abington, UK) [7].

However, slow leakage of the soluble form of the redox mediator (the ferricinium ion) may lead to biosensor deterioration and precludes its use for on-line monitoring. The use of soluble high molecular weight diffusing mediators should in principle limit this phenomenon. For this purpose, the protein BSA was labeled with Fc groups by reductive amination with ferrocene carboxaldehyde **3** in the presence of sodium borohydride (Scheme 6.1) and the labeled protein was shown to behave as a redox mediator in the electrocatalysis of glucose by GOx [8].

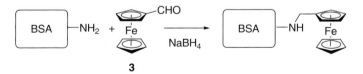

3

Scheme 6.1 Labeling of BSA by reductive amination with **3**.

BSA was chemically modified with ferrocenoylpropionic acid **4** in the presence of the water-soluble carbodiimide EDAC (Scheme 6.2) to afford conjugates with 5 or 23 Fc units covalently bound to the protein.

This redox protein was co-adsorbed with the enzyme fructose deshydrogenase on an electrode and an amperometric biosensor for fructose was elaborated [9].

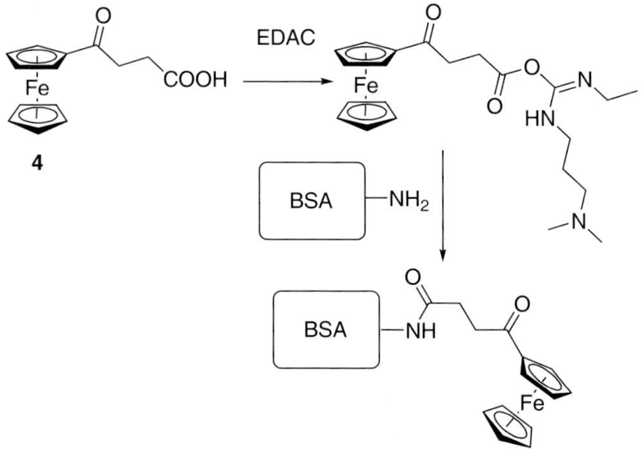

Scheme 6.2 Labeling of BSA by acylation with **4**.

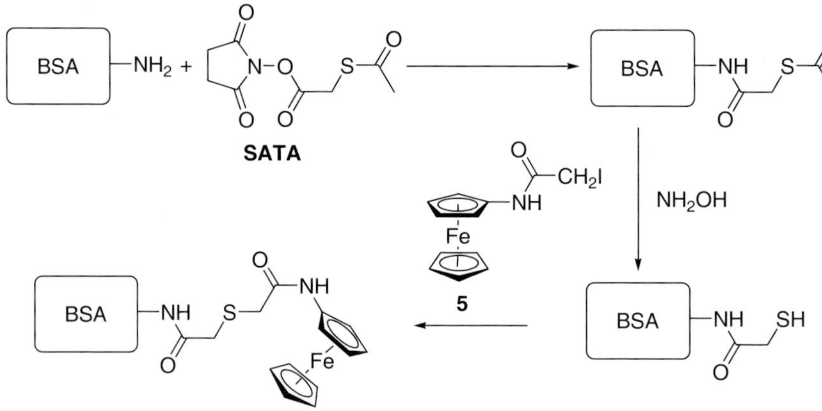

Scheme 6.3 Thiolation of BSA by SATA and labeling with *N*-(ferrocenyl)iodoacetamide **5**.

The sulfhydryl-targeted ferrocenyl derivative *N*-(ferrocenyl)iodoacetamide **5** was synthesized from aminoferrocene and used to label sulfhydryl-modified BSA as illustrated in Scheme 6.3. The Fc-to-BSA ratio for the resulting protein conjugate was estimated to be ca. 40 : 1.

This conjugate displayed reversible electrochemistry but appeared unable to electro-catalyze the oxidation of glucose by glucose oxidase at a glassy carbon electrode. This behavior was explained by the large molecular size of the conjugate [10].

6.2.1.2 Electron Relays

Covalent binding of Fc redox groups to oxido-reductases with buried redox centers such as GOx was first introduced by Heller et al. to introduce so-called "electron relays" randomly at the surface of these enzymes (Fig. 6.1, configuration B) [11].

As seen above, GOx forms the basis of amperometric glucose biosensors which are very useful devices to control the glucose blood level of patients suffering from diabetes. Chemical modification of GOx with Fc entities was performed by reacting the enzyme with the ferrocene-containing reagents **6–12** (Fig. 6.2) under various reaction conditions. The labeling methods and the enzymatic activity of the ferrocene-labeled GOx conjugates are summarized in Table 6.1.

Initially, labeling of GOx was achieved by activation of ferrocene carboxylic acid **6** or ferrocene acetic acid **7** by EDAC and coupling to GOx under partially denaturing conditions. Enzyme conjugates with 12–13 tethered Fc groups and displaying 60% of enzymatic activity with respect to the native enzyme were claimed to be obtained [12, 13]. Later on, a much improved procedure including NHS and omitting the denaturant enabled to obtain mode stable conjugates where Fc groups were exclusively bound to the enzyme lysine residues. A maximum amount of 12 Fc units were chemically bound to GOx under optimized reaction conditions. Interestingly, it was found that the location of a few Fc groups with respect to the redox center and the protein surface is more important for electro-catalytic activity than the number of Fc groups loaded onto the protein [14].

Labeling of GOx with Fc groups by reaction of its lysine residues was also performed by reductive amination with **3** [15, 16]. A lower Fc loading was reached with this procedure (between 1 and 6 Fc/GOx). This loading could be increased by applying high pressure during the reaction [17]. A partial loss of enzymatic activity was observed concomitantly.

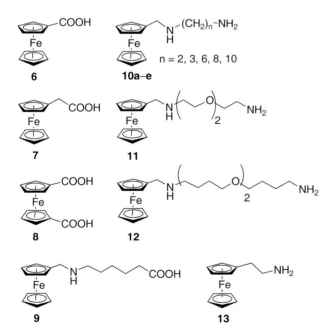

Fig. 6.2 Ferrocenyl complexes **6–13**.

Table 6.1 Labeling of GOx with ferrocenyl moieties.

Ferrocenyl derivative	Co-reagent(s)	Protein binding site	Fc/GOx	Remarks	Reference
6	3 M urea, EDAC	Lysines ?	12 ± 1	60% activity vs. GOx	[11, 12]
7	3 M urea, EDAC		13 ± 1	60% activity vs. GOx	
10a–e	$NaIO_4$, $NaBH_4$	Polysaccharide shell	1–1.5	27–49% activity vs. GOx	[17]
3	$NaBH_4$	Lysines	4.2	Activity not determined	[16]
	$NaBH_4$, 400 MPa		27.5	Reversible denaturation of GOx under pressure	
	$NaBH_4$, 3 M urea		8.1		
6	NHS, EDAC	Lysines	12	64% activity vs. GOx	[13]
7	NHS, EDAC		7.5	69% activity vs. GOx	
8	NHS, EDAC		10	60% activity vs. GOx	
3	$NaBH_4$	Lysines	1–6	30% activity vs. GOx	[14]
6	EDAC	Lysines?			
11	$NaIO_4$, then $NaBH_4$ and glucose	Polysaccharide shell	n.d.	21% activity vs. GOx	[18]
11	EDAC and glucose	Aspartates, glutamates	n.d.	11% activity vs. GOx	
12	EDAC and glucose		n.d.	30% activity vs. GOx	
9	NHS, DCC	Lysines	21	Activity not determined	[20]
	NHS, EDAC, 1 M urea		8		
6	EDAC, 2 M urea	Lysines	n.d.	90% activity vs. GOx Genetically modified GOx	[21]

Another labeling strategy consisted in targeting the oligosaccharide shell of GOx which represents approximately 12% of its weight. Optimal oxidation of these sugar residues with sodium periodate, resulted in the generation of 6.4 aldehyde groups per enzyme molecule. Ferrocenyl amines **10a–e** with spacer arms of different lengths linking the Fc group to the amine were conjugated to oxidized GOx to form Schiff bases that were subsequently reduced by NaBH$_4$ to the secondary amines. An average of 1 to 1.5 Fc units were bound per enzyme molecule. It was shown that the increase of the spacer length resulted in an increase of the electro-catalytic glucose oxidation current by an intramolecular ET process [18]. Ferrocenyl amines **11** and **12** with more flexible hydrophilic spacer arms were synthesized and used to label GOx either through its carboxylic acid residues (aspartates and glutamates) or sugar residues and the resulting conjugates were shown to be even more efficient [19]. A derived labeling strategy was put forward by Nagata et al. [20]. The peripheral sugar chains of GOx were oxidized by periodate to generate aldehyde groups. These groups were converted into hydrazides by reaction with adipyl hydrazide (Scheme 6.4). The modified enzyme was then treated a mixture of **7** and EDAC. This resulted in the binding of 20 redox groups.

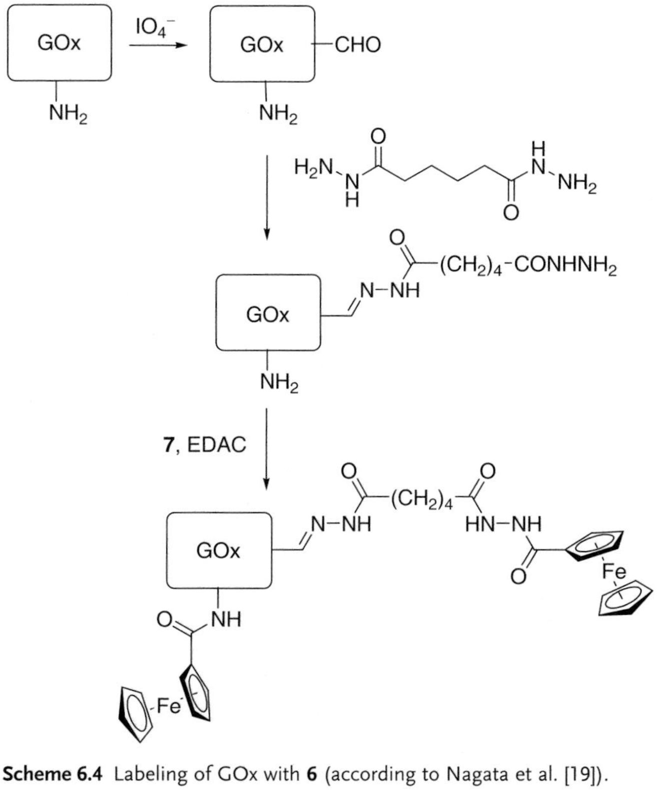

Scheme 6.4 Labeling of GOx with **6** (according to Nagata et al. [19]).

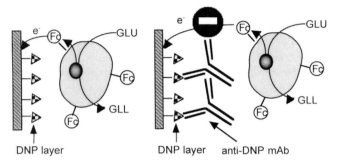

DNP layer   DNP layer   anti-DNP mAb

Fig. 6.3 Principle of amperometric immunosensor for DNP.
Inhibition of amperometric detection of glucose in the presence of
specific anti-DNP antibody.

The carboxylic acid derivative of ferrocene N-(ferrocenylmethyl)caproic acid **9** after activation with the carbodiimide DCC and NHS was also covalently coupled to GOx though its lysines. The resulting ferrocene-tethered enzyme was used as part of a model anti-DNP antibody immunosensor device based on the insulation of electrical contact between Fc-tethered GOx and the electrode by antigen–antibody association as depicted in Fig. 6.3 [21].

GOx was genetically modified so that a polylysine chain was appended at the C terminus of the enzyme mutant. Coupling of ferrocenyl groups was performed by reaction of **6** and EDAC in the presence of 2 M urea, and the resulting enzyme was shown to retain 90% of enzymatic activity as compared with native GOx [22].

6.2.1.3 Electrical Wiring

Immobilization of a redox enzyme in a conductive polymer network is another means of enabling electrical communication between the enzyme and the electrode (see Fig. 6.1, configuration C). A non-organized multilayer of ferrocene-tethered GOx was constructed on the surface of a gold electrode by cross-linking of GOx and 2-aminoethyl ferrocene **13** with glutaraldehyde (Fig. 6.4). The molar ratio between ferrocene and GOx in the film was found around 10 and the sensitivity of glucose response was proportional to the amount of GOx [23].

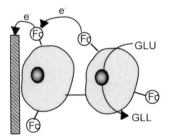

Fig. 6.4 Multilayer of ferrocenyl-tethered GOx built by cross-linking of GOx and **13** in the presence of glutaraldehyde on the surface of a gold electrode.

A multilayered network of GOx was assembled by a stepwise synthesis onto a gold electrode initially covered with a SAM of cystamine. Ferrocene groups were introduced into this network by reaction of **9** in the presence of EDAC, NHS and 1 M urea. The average number of Fc units per GOx molecule in the film was found around 8. The amperometric response to glucose was found to be controlled by the number of protein layers [24].

The protein analog poly(amidoamine) dendrimer of fourth generation (PAMAM G4) was labeled with ferrocene units by reductive amination with compound **4**. A reagentless glucose biosensor was built by alternative layer-by-layer deposition of periodate-activated GOx and PAMAM G4 ferrocene conjugate on a gold electrode initially covered with a SAM of cystamine. The glucose assay sensitivity was correlated to the number of Fc units tethered to PAMAM G4 and the number of deposited layers [25].

PAMAM G4 labeled by ferrocenyl groups was also immobilized on the surface of a gold electrode covered by a monolayer of DTSP [26] or MUA activated by EDAC and pentafluorophenol [27]. This PAMAM G4 ferrocene conjugate monolayer was biotinylated. The bioelectrocatalytic oxidation of glucose by GOx appeared to be inhibited by addition of avidin or by a monoclonal antibiotin antibody because of surface shielding (Fig. 6.5).

Avidin was reacted with compound **9** in the presence of NHS and EDAC to yield a conjugate with 16 Fc units bound per protein molecule. This Fc-tethered conjugate was immobilized in a covalent manner on a SAM of DTSP on a gold electrode. This resulting redox layer was in turn used as a platform to immobilize, by a bioaffinity process, several biotinylated enzymes, including microperoxidase 11 [28], GOx and lactate oxidase [29]. The resulting sensing platforms may be used to measure H_2O_2, glucose and lactate, respectively. The same research group also described the labeling of streptavidin by ferrocenyl groups and examined the charge transport properties of electrodes coated by multilayers of Fc-streptavidin conjugate [30].

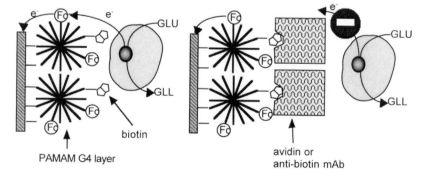

Fig. 6.5 Principle of amperometric biosensor for biotin. Inhibition of amperometric detection of glucose in the presence of specific antibiotin antibody or avidin.

6.2.2
HPLC and Immunoassays

Ferrocene can also be considered as a redox probe directly detectable by an appropriate electrochemical technique and has been employed as such in homogeneous (i.e. separation-free) or heterogeneous immunoassays and for the electrochemical detection of compounds separated by HPLC.

The series of ferrocenyl derivatives **14–21** (Fig. 6.6) was used to chemically derivatize the protein BSA with the intention of determining peptides and proteins in biological fluids by HPLC coupled to ECD. The most efficient labeling agent was anhydride **21** because it combines slow hydrolysis rate and optimal distance between the redox group and the protein core [31].

Ferrocene carboxaldehyde **3** and periodate-oxidized digoxigenin were covalently coupled to BSA lysine residues by reductive amination to yield a conjugate behaving as a multivalent antigen. Binding of antidigoxigenin antibody led to a concentration-related decrease of the oxidation current of the ferrocenyl groups [32].

Monoclonal antibodies anti-HCG or antihistamine were labeled with ferrocene carboxylic acid **7** in the presence of EDAC alone [33, 34] or in the presence of EDAC and sulfo-NHS [35] to yield immunoconjugates with acceptable immunological activity. They were used in amperometric immunoassays of HCG and histamine, respectively (see Chapter 8, Sections 8.4.1 and 8.5.3).

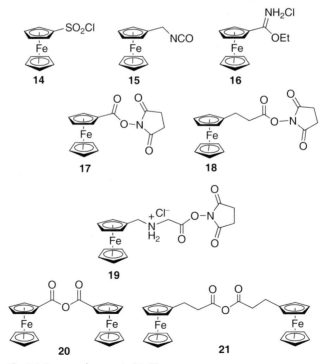

Fig. 6.6 Ferrocenyl reagents **14–21**.

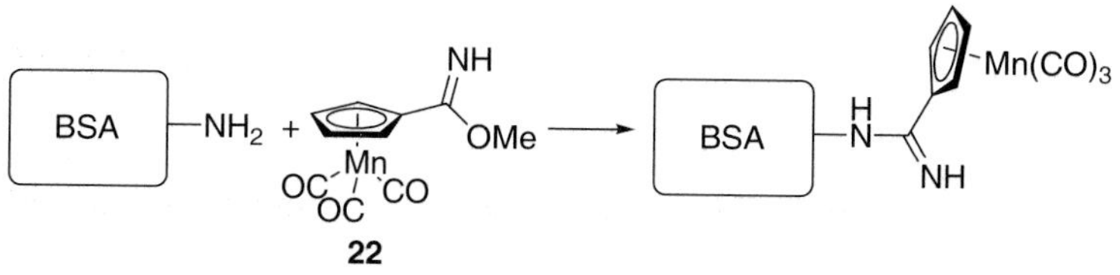

Scheme 6.5 Labeling of BSA by amidination with **22**.

An anti-mouse IgG antibody was labeled by ferrocene carboxylic acid **6** in the presence of EDAC and used in a heterogeneous electrochemical immunoassay with separation of the immunocomplexes by electrophoresis (see Chapter 8, Section 8.5.3) [36].

We recently described the reaction of BSA with cymantrenyl methyl imidate **22** that yielded protein conjugates by reaction of some of BSA lysine residues (Scheme 6.5).

Interestingly, while compound **21** was irreversibly reduced in acidic aqueous medium ($E_p = -1.1$ V versus Ag/AgCl), the probe once covalently bound to the protein undertook a reversible and fast redox process ($E_p = -1.7$ V versus Ag/AgCl) as a consequence of the protection exerted by the protein environment. Voltammetric methods using the direct current as analytical signal turned out to be unsuitable because of concomitant protein catalyzed hydrogen evolution occurring at this potential. The best results were obtained by choosing phase-sensitive AC voltammetry with accumulation of the redox labeled protein at a suitably chosen electrode potential. The detection limit for the BSA-cymantrene conjugate was equal to 200 nM in aqueous medium [37]. Addition of increasing amounts of anti-BSA antibody covalently linked to agarose beads induced a steady decrease of the reduction wave intensity, while it remained constant with an unrelated antibody. We have thus been able to detect the formation of immunocomplexes by association of an organometallic probe with electrochemical detection (Hromadova, Pospisil, Salmain, Jaouen, manuscript in preparation).

6.2.3
Enzyme Structural Studies

Papain, an enzyme of the cysteine proteases family, was rapidly, specifically and irreversibly inactivated by chloroacetyl ferrocene **23** [38]. This property was attributed to the non covalent binding of the organometallic complex at the catalytic site followed by selective alkylation of cysteine 25 as illustrated in Scheme 6.6.

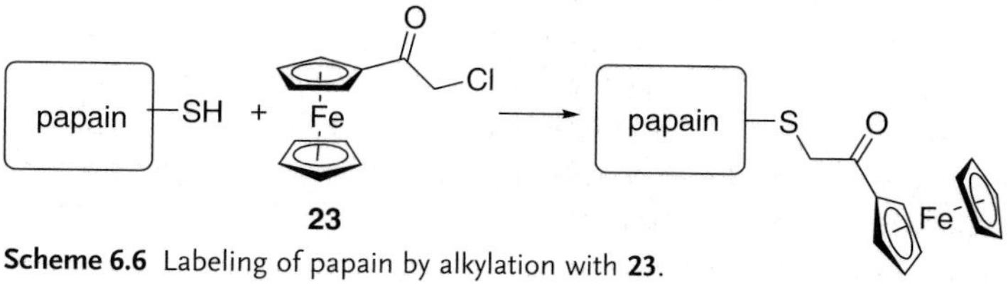

Scheme 6.6 Labeling of papain by alkylation with **23**.

The cysteine-specific electroactive label N-(2-ferrocenemethyl)maleimide **24** was synthesized [39] and applied to label several enzymes containing cysteine residues either in their native sequence or introduced by site-directed mutagenesis.

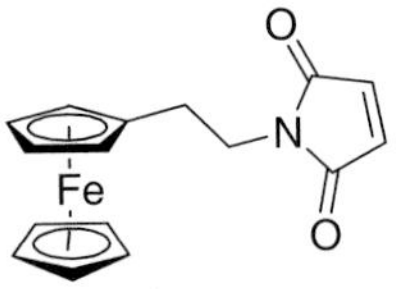

24

For example, treatment of the monooxygenase cytochrome $P450_{cam}$ with **24** resulted in the fast incorporation of Fc redox units as shown by cyclic voltammetry [39]. A mutant of the non-electroactive enzyme β-lactamase where residue Tyr105 located close the active site was replaced by a cysteine by site-directed mutagenesis was successfully labeled with **24** and the redox potential of the probe was seen to shift by 20 mV in the presence of a suicide substrate in the enzyme active site [40]. Labeling with **24** of a cytochrome $P450_{cam}$ mutant, where the exposed Cys334 had been replaced by an alanine, yielded an enzyme adduct with two covalently bound ferrocenyl groups, one at Cys85 and one at Cys136 as revealed by X-ray crystallography [41]. The presence of an Fc group at Cys85 led to dramatic changes in the electrochemical and biochemical behavior of the enzyme, because it is located close to the heme group. Two possible mechanisms for the binding of ferrocene moiety at the enzyme active site were postulated. One is based on the affinity of the active site for ferrocene while the other would request an "internalization" of the Fc moiety attached to the thiol of Cys85.

A cytochrome $P450_{cam}$ mutant genetically engineered so as to introduce a surface cysteine residue at position 344 was labeled with the related compound N-ferrocenyl maleimide **25**.

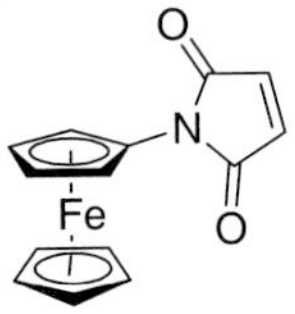

25

Cyclic voltammetry showed that there may be interactions between the iron center of the heme and the ferrocene reporter group [42].

6.2.4
Other applications

The linear poly(ethylene glycol) derivative **26** carrying an Fc group at one end and an N-hydroxysuccinimide ester group at the other end was covalently coupled to mouse anti-goat and goat anti-mouse IgGs.

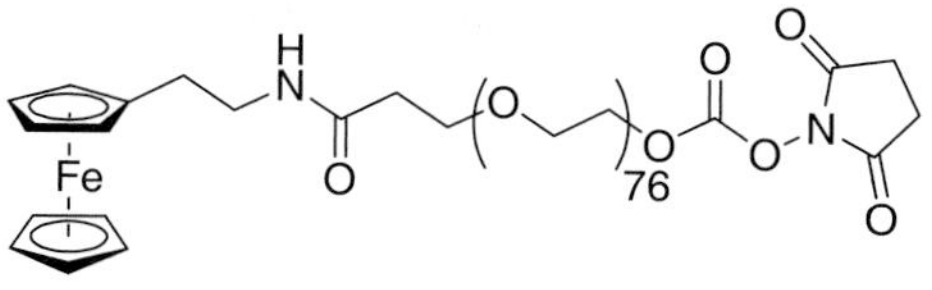

26

Multilayers of alternating goat anti-mouse and mouse anti-goat IgG were built by stepwise immobilization of both antibodies on a glassy carbon electrode. The labeled IgG conjugates were immobilized on top of these molecular layers of variable thickness. Quantitative analysis of the dynamics of the redox probe grafted to the self-assembled construction via the linear and flexible PEG chain was extracted from cyclic voltammetric measurements [43].

6.3
Luminescent Probes

6.3.1
Long-lived Probes

Rhenium (I) complexes of the general structure depicted in Fig. 6.7 have been shown to display remarkable photoluminescent properties since the 1970s [44, 45].

These emission features are associated with the lowest lying MLCT excited state involving a metal-centered $d\pi$-type orbital and a π^* orbital centered on the α-diimine ligand. These transitions appear in the visible range of the spectrum. Noticeably, characteristics of excited states such as emission quantum efficiency, lifetime and, redox potential are greatly affected by the ancillary ligands, solvent polarity and other environmental factors [46].

These photophysical properties have been exploited to design chemical sensors for O_2 [47], cations [46] and anions [48]. Several rhenium(I) polypyridine biotin compounds were shown to display up to a 3-fold enhancement of their emission intensity upon binding to the protein avidin [49].

Interest was recently directed to the conjugation of this class of complexes with proteins, in order to benefit from their exceptional photophysical properties.

N   N = α-diimine
L = pyridine, nitrile, isonitrile

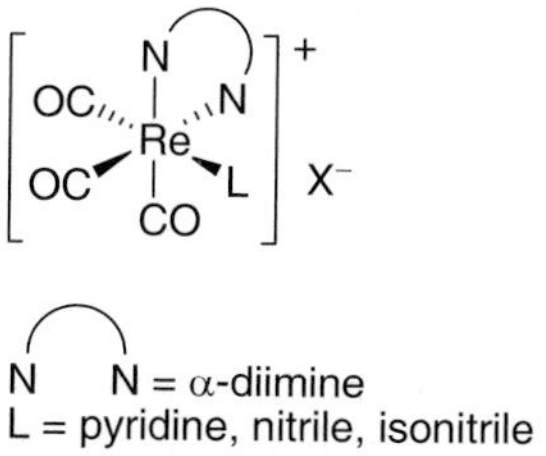

Fig. 6.7 General formula of long-lived, highly luminescent rhenium(I) complexes.

Lakowicz et al. described the first protein conjugable luminescent rhenium(I) complex 27 that was covalently linked to the proteins BSA and bovine IgG by reaction of some of their lysine residues after activation of its carboxylic acid function by a mixture of DCC and NHS (see Chapter 8, Section 8.4.7) [50].

The starting material displayed a large Stokes' shift (100 nm), thereby likely to minimize self-quenching effects, a lifetime over 100 μs and a quantum yield of 0.27 in air-equilibrated glycerol. This complex also displayed highly polarized emission (with a maximum polarization near 0.4 and maximum anisotropy near 0.3) in the absence of rotational diffusion. Once covalently bound to a protein (BSA, IgG, HSA), the photoluminescence lifetime of the luminophore was still around 2.7 μs in air-equilibrated buffer. This property should open the way to fluorescence polarization immunoassays of high molecular weight antigens (see Chapter 8, Section 8.5.4) [51].

The two sulfydryl-targeting rhenium(I) polypyridine complexes 28 and 29 were also recently synthesized and conjugated to cysteine-containing proteins.

Complex 28 conjugated at a dye-to-protein ratio of 0.7 displayed a photoluminescence emission lifetime of 1.1 μs and an anisotropy of 0.34. This long decay time makes it appropriate for measuring correlation times of moderate to large sized proteins [52]. Compound 29 was also conjugated to several proteins and exhibited long luminescence lifetime and poor emission quenching by O_2 owing to lower solvent exposure of the luminophore when bound to the proteins [53].

6.3.2
Electron Tunneling Studies

Electron transfer in biological systems where the electron donor and acceptor are separated by a long molecular distance is encountered in very important processes such as photosynthesis and respiration [54]. As natural systems are not appropriate for such studies, Gray et al. have employed proteins chemically labeled with transition metal complexes to measure ET rates in metalloproteins. In particular, they have shown that long-lived luminescent probes enabled a wider range of ET measurements than is possible with non-luminescent complexes [55]. The blue copper protein azurin is a convenient model for the study of ET in β-sheet proteins. In fact this metalloprotein has a β-barrel tertiary structure and the central Cu atom adopts a trigonal planar structure by coordination of Cys112(S), His117(N) and His46(N) [56].

Wild-type azurin including a single surface histidine at position 83 and Gln107His/His83Gln azurin mutant were chemically modified by the luminescent rhenium(I) phenantroline complex **30**.

30

It is reported that this compound covalently binds to surface histidine residues by substitution of the labile water ligand. However, reaction of **30** is not selective the sole surface histidine since binding to other unidentified residues was noticed together with non-covalent binding. Purification by affinity chromatography followed by cation exchange chromatography afforded the rhenium (I) containing Cu(I) azurins. Zn(II) azurins were obtained in the same manner [57].

Thanks to the favorable photophysical properties of the rhenium(I) phenanthroline moiety, photo-oxidation of $[Re(CO)_3(phen)(His107)]^+$ AzCu$^+$ was carried out in the presence of $[Ru(NH_3)_6]^{3+}$ to generate the $[Re(CO)_3(phen)(His107)]^{2+}$ AzCu$^+$. Instead of direct Cu$^+$ → Re^{2+} ET, multistep ET tunneling was observed, involving the neighboring Tyr108 residue that was rapidly oxidized to Tyr108$^{+\cdot}$, followed by rate-limiting Cu$^+$ → Tyr108$^{+\cdot}$ tunneling ($k = 2 \cdot 10^{4 \, -1}$) (Scheme 6.7) [58].

Phototriggered oxidation of $[Re(CO)_3(phen)(His107)]^+$ AzZn^{2+} and $[Re(CO)_3(phen)(His83)]^+$ AzZn^{2+} by $[Co(NH_3)_5Cl]^{2+}$ generated the aromatic amino acids radicals Tyr108$^{+\cdot}$ and Trp48$^{+\cdot}$, respectively. These species were rapidly deprotonated by neighboring residues to yield the radicals Tyr108$^\cdot$ and Trp48$^\cdot$ as deduced from EPR measurements [57].

The X-ray crystal structure of $[Re(CO)_3(phen)(His83)]^+$ AzCu^{2+} was solved at high resolution. The rate of oxidation of Cu$^+$ by photoexcited Re$^+$ was measured on the protein crystal by transient absorption spectroscopy. It was found to be in

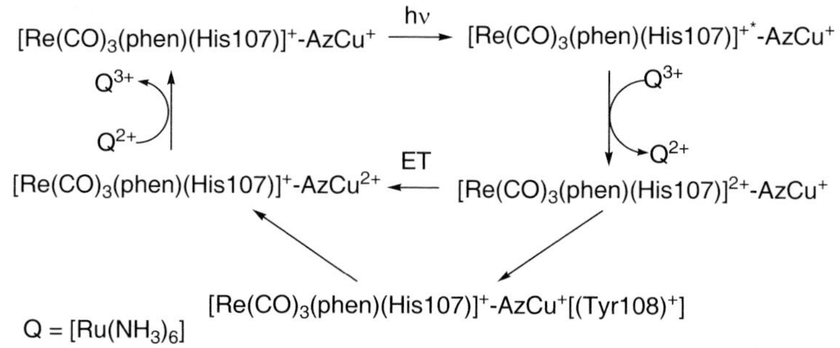

$$[Re(CO)_3(phen)(His107)]^+\text{-AzCu}^+ \xrightarrow{\text{h}\nu} [Re(CO)_3(phen)(His107)]^{+*}\text{-AzCu}^+$$

$$[Re(CO)_3(phen)(His107)]^+\text{-AzCu}^{2+} \xleftarrow{\text{ET}} [Re(CO)_3(phen)(His107)]^{2+}\text{-AzCu}^+$$

Q^{3+} Q^{2+} Q^{3+} Q^{2+}

$$[Re(CO)_3(phen)(His107)]^+\text{-AzCu}^+[(Tyr108)^+]$$

$Q = [Ru(NH_3)_6]$

Scheme 6.7 Mechanism of $Cu^+ \rightarrow Re^{2+}$ ET in $[Re(CO)_3(phen)(His107)]^+$ AzCu$^+$ by multistep tunneling via Tyr108.

the same order of magnitude with respect to ET rate measured in solution. It was concluded that surrounding water apparently plays a minor role in azurin ET reactions [59].

More recently the Gln107/His/Trp48Phe/Tyr42Phe/His83Gln/Tyr108Trp azurin M(II)) mutant (M = Cu or Zn) was labeled with compound **30** to yield after purification $[Re(CO)_3(phen)(His107)]AzM^{2+}$ containing a single Trp at position 108. The Trp108˙ radical generated by an irreversible flash–quench method was shown to be exceptional stable by EPR and optical spectroscopy [60].

6.4
Heavy Metal Probes

6.4.1
Structural Analysis of Proteins by X-ray Crystallography

X-ray crystallography is currently the most powerful analytical method by which three-dimensional structure information on biological macromolecules may be obtained at high resolution. Its application is however limited first by the preparation of single crystals suitable for X-ray diffraction and second by the so-called "phase problem", that is the calculation of phases of diffraction data. Several approaches are available in order to circumvent this latter problem. The most commonly used methods are the multiple and single isomorphous replacement (MIR, SIR). These methods, as well as multiple anomalous diffraction (MAD), require the preparation of heavy atom derivatives, usually by the introduction of electron-dense atoms at distinct locations of the crystal lattice. This is usually done by crystal soaking experiments.

To be useful for the MIR, SIR and MAD phasing methods, the heavy atom derivatives should firstly be isomorphous with the native crystals, that is no alteration of the packing and no change of the unit cell dimensions should occur after binding. Secondly, the number of heavy atom binding sites should be low to facilitate the determination of their position on the Patterson difference maps.

Finally, the fractional increase in intensity of the diffracted beams caused by the binding of the heavy atoms(s) should be much higher that the error margin of the measured intensities. Crick and Magdoff have shown that the fractional change in intensity ΔI can be expressed as:

$$\Delta I = \sqrt{2} \cdot \sqrt{\frac{N_E}{N_P}} \cdot \frac{Z_E}{Z_P} \qquad (6.4)$$

with N_E the number of atoms with atomic number Z_E in a unit cell containing N_P protein atoms with a mean atomic number Z_P [61].The ideal situation occurs when a single heavy atom is bound to every protein molecule at the same site in the crystal. In this case, the site occupancy factor is equal to one. If this occupancy factor is lower than one, Z_E has to be reduced in proportion to the number of protein molecules having a heavy atom associated to them.

A wide range of reagents is available for the preparation of heavy atom derivatives [62, 63]. These reagents are gathered in a databank accessible on the WEB (http://www.sbg.bio.ie.ac.uk/had) [64]. Among them are found a group of transition organometallic compounds, either coordination complexes of Pd(II), Pt(II), Pd(IV), Pt(IV), Au(I) and Au(III) with cyanide ligands, or organomercurials. Because of its high thermodynamic stability $[Pt(CN)_4]^{2-}$ was found to be inert to nucleophilic ligand substitution. This complex remains dianionic under all conditions and binds to positively charged protein residues by electrostatic interaction [63]. This compound cannot therefore be considered as an organometallic protein labeling reagent. The behavior of the anionic complex $[Au(CN)_4]^-$ was however found to be different. Indeed the reaction of the protein β-lactoglobulin (β-LG) with $[Au(CN)_4]^-$ was shown to proceed by substitution of one cyanide ligand by the cysteine residue at position 121 to form a stable adduct (Scheme 6.8).

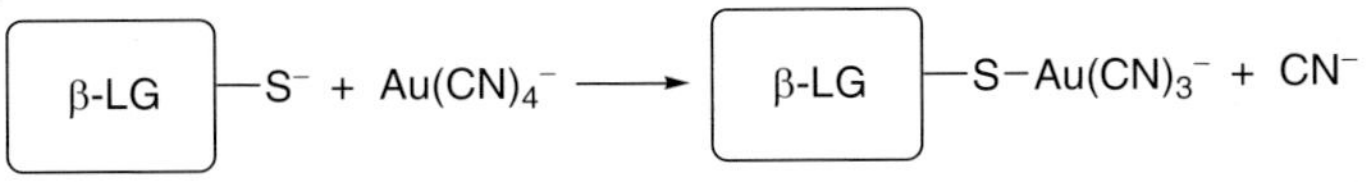

β-LG ─S⁻ + Au(CN)₄⁻ ⟶ β-LG ─S−Au(CN)₃⁻ + CN⁻

Scheme 6.8 Reaction of tetracyanoaurate(III) with β-lactoglobulin.

Unfortunately β-LG crystals treated with $[Au(CN)_4]^-$ were not isomorphous with the native crystals [65].

Organomercurials are supposed to show a preference for cysteine and histidine residues. Small alkyl substituents enable the organomercurial to react with buried protein sites whereas larger aromatic substituents display greater steric selectivity. Anyway, the selectivity of these reagents is still low and often uncontrollable. For example, *p*-chloromercuribenzene sulfonate was shown to bind to the enzyme hen egg white lysozyme exclusively at the positively charged Arg68 residue by interaction with the sulfonate group although this protein has one accessible histidine residue [66].

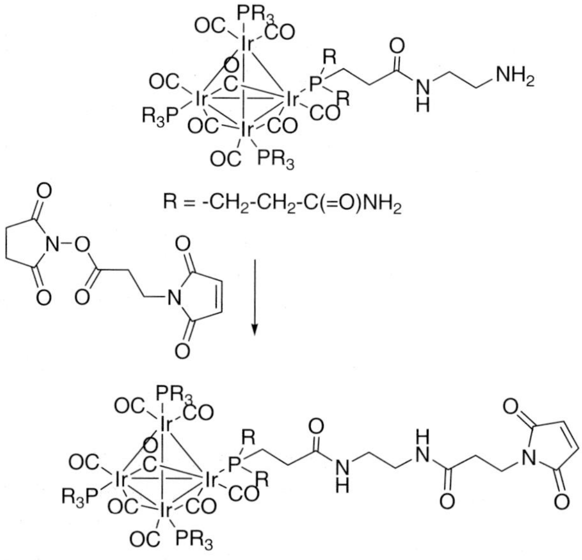

R = -CH$_2$-CH$_2$-C(=O)NH$_2$

31

Scheme 6.9 Synthesis of tetra-iridium cluster **31**.

The advent of molecular engineering has prompted new developments in the area of protein structural analysis with the preparation of protein mutants by site-directed mutagenesis. In this manner, a reactive amino acid residue, generally a cysteine, may be introduced at a well chosen location, providing a heavy atom binding site. There is thus a need to provide heavy atom reagents that will specifically target selected protein residues.

The first transition organometallic complex designed to specifically target cysteine residues was described in 1989 [67]. This monofunctional tetra-iridium cluster **31** synthesized as depicted in Scheme 6.9 was aimed at the derivatization of ribosomal particles. As these biological systems are extremely large, the use of extremely dense and heavy groups for phasing was necessary (see Eq. 6.4).

Labeling of one of the proteins of the 50S ribosome was quantitatively achieved by reaction of **31** in solution. Crystals of the reconstituted ribosome particle were found to be isomorphous with those of the native particle [68]. In order to enable reliable assignments of the various components of the small ribosomal subunit T30S, this particle was labeled with **31**. A 7.2 Å MIR map was constructed from sets of X-ray diffraction data recorded for the native and derivatized ribosome subunit crystals and one major and one minor binding site were located. This iridium derivative was thus helpful to construct a medium-resolution MIR map for a biological particle [69].

We have designed and/or studied several families of mono- and polynuclear organometallic complexes aimed at the selective and covalent chemical modification of proteins through their lysine residues. Chronologically, the first reagents included an N-hydroxysuccinimide ester function that is known from classical

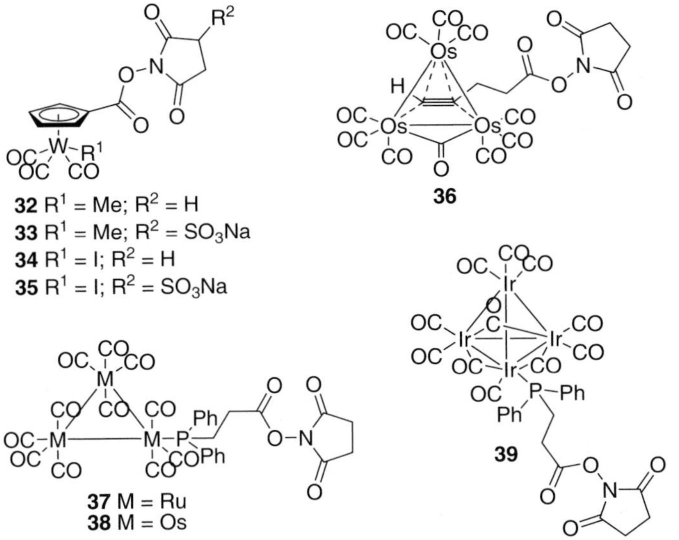

Fig. 6.8 Lysine-targeted transition organometallic reagents for the covalent labeling of proteins with heavy atoms.

protein chemistry to specifically acylate primary amines of peptides and proteins (see Scheme 6.13). Compounds **32–39** (Fig. 6.8) were indeed found to covalently bind to the protein BSA in solution by acylation of some of its lysine residues as evidenced by UV-visible and IR spectroscopy [70–72]. Labeling of the readily crystallizable enzyme lysozyme with **32** and **34** was also achieved in good yield [73].

The tungsten alkoxycarbene complexes **40–41** ("Fischer-type" metallocarbenes, Fig. 6.8) are known to possess a marked electrophilic character at the carbenic carbon. Facile aminolysis reaction occurs with primary and secondary amines, affording stable aminocarbenes (Scheme 6.10) [74].

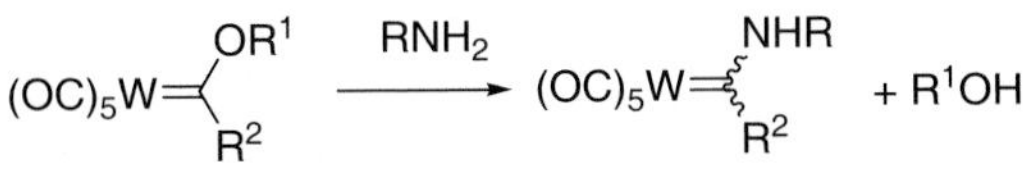

40 $R^1 = R^2 = Me$
41 $R^1 = Et; R^2 = Ph$

Scheme 6.10 Reaction of Fischer-type metallo–alkoxycarbene complexes with primary amines.

These compounds thus appeared to be good candidates for protein heavy atom labeling because they may target particular protein residues and form stable covalent adducts. Moreover their rate of aminolysis was shown to be much higher that their rate of hydrolysis [75]. Reaction of **40** with a series of α-aminoesters always afforded the expected aminocarbene adducts even in the presence of a competing nucleophile on the side chain. Reaction of **40** or **41** with the protein BSA in solution yielded conjugates with spectral characteristics similar to those of aminocarbenes [76].

Tungsten aminocarbene adducts of lysozyme prepared in solution and isolated by preparative RP-HPLC, were further analyzed by MALDI-TOF MS. Unfortunately, the bound organometallic adduct did not withstand the conditions of mass analysis so that lower mass peaks than expected were observed on the spectra. To determine which of the six lysine residues of lysozyme had reacted with **40**, a tungsten–aminocarbene conjugate was proteolyzed with trypsin and the resulting peptides separated by RP-HPLC. The peptides carrying the aminocarbene moiety were identified spectroscopically and submitted to MALDI-TOF. Lysines 13, 33, 97 and 116 were identified as binding sites for the organometallic complex in solution [77].

The $[W(CO)_5(C\text{-}Me)(Lys13)]$lysozyme adduct which is the most abundant species formed in solution, was purified by preparative RP-HPLC and crystallization trials were performed by the hanging drop method. Unfortunately, all the attempts to obtain single crystals of this protein have been unsuccessful to date, yielding only amorphous precipitates (unpublished results). This confirms that crystal soaking is the most successful approach to preparing heavy atom derivatives for X-ray crystallography. However water insolubility of both the organometallic *N*-hydroxysuccinimide esters and the Fischer carbenes complexes precluded lysozyme crystal soaking experiments.

We directed our interest to water-soluble organometallic protein labeling reagents. Pyrylium salts were identified as a promising class of reagents. These compounds (at least the 2,6-dimethyl derivatives) are water-soluble and react with protein lysine residues to yield pyridinium adducts (see Section 6.5). We performed lysozyme crystal-soaking experiments with the 4-benchrotrenyl pyrylium salts **42** and **43** and the 4-ruthenocenyl pyrylium salt **44**. IR spectroscopy (for the benchrotrenyl derivatives) and MALDI-TOF MS (for the ruthenocenyl derivative) provided spectroscopic evidence that these complexes react with accessible lysine residues inside the crystal lattice to afford the pyridinium adducts [78, 79]. Trypsin proteolysis combined with MALDI-TOF analysis of the peptide mixture performed on a crystal of lysozyme treated with **44** showed that lysines 1, 13, 33, 96 and 97 were involved in the reaction, in agreement with the high surface accessibility of these residues [79].

Thus, in our attempts to design targeted organometallic reagents to introduce heavy metals into proteins, we have identified the pyrylium salts substituted by organometallic entities as the most promising series of reagents because they are water-soluble and may be used in crystal soaking experiments. However, no protein crystal structure has so far been solved by this new generation of reagents. Problems may also arise from the high number of surface lysines borne by most proteins that may lead to multiple binding sites. The design of cysteine-targeted heavy organometallic reagents may be envisaged as an alternative to lysine-targeted reagents.

Very recently, Sadler et al. have prepared a half-sandwich arene ruthenium(II) derivative of crystallized lysozyme by slow diffusion of the complex [(η^6-*p*-cymene)-RuCl$_2$(H$_2$O)] in tetragonal lysozyme crystals [80]. High resolution X-ray crystallography of the modified crystals unexpectedly revealed that a single arene ruthenium entity happened to be coordinated to the enzyme by interaction with the imidazole ring borne by the HIS15 residue. This is the first example of this kind of protein structure reported in the literature.

6.4.2
Cryo-electron Microscopy

The monofunctional tetra-iridium cluster **45** synthesized according to Scheme 6.11 was used to label the capsid of HBV assembled from a mutant protein where the cysteines of the wild-type protein had been replaced by alanine and an additional cysteine appended at the C terminus.

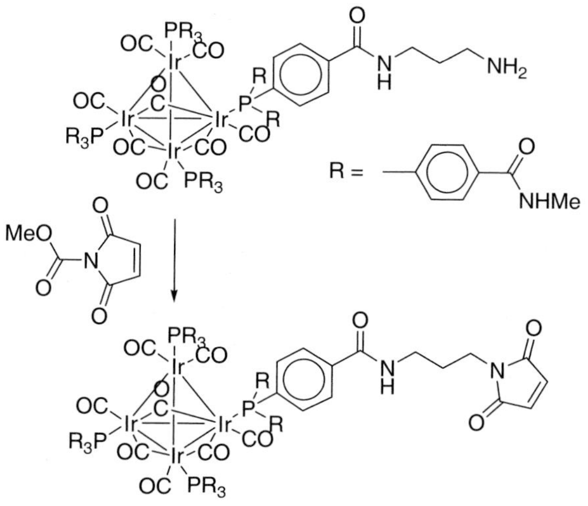

Scheme 6.11 Synthesis of tetra-iridium cluster **45**.

A labeling efficiency of 11% was estimated from spectroscopic measurements. The tetra-iridium cluster was clearly visualized in three-dimensional density maps at good resolution despite the relatively low site occupancy. Interestingly, cysteine residues inaccessible to undecagold reagent were reached with the much smaller tetra-iridium complex **45** [81].

6.4.3
Pharmacological Studies

The reaction of the gold(III) organometallic complex **46** with BSA has been studied by a variety of techniques.

It was shown that this complex is tightly coordinated to the protein and could only displaced by addition of excess cyanide ion. It was speculated that binding of this complex to BSA occurred by coordination of the imidazole group of one histidine residue and loss of the water ligand. Interestingly, both the starting material and the BSA adduct displayed significant cytotoxic activities against several human tumor cell lines [82].

46

6.5
Metallo–carbonyl Probes for Infrared Spectroscopy

Transition metal complexes with carbonyl ligands ("metallo–carbonyl" complexes) have long been known to display typical absorption bands in the mid-IR range. These features are associated with the stretching vibrations v_{CO} of the carbon–oxygen triple bonds (see Chapter 7, Section 7.4). They appear in a spectral region where only few other organic and inorganic vibrators absorb. The band intensities are also remarkable, that is 4 to 10 times more intense than the usual organic group bands.

The first example of the introduction of a metallo–carbonyl group onto a protein was described in a note published in 1972. The reagent used was maleic anhydride iron tetracarbonyl **47** where the double bond of maleic anhydride is complexed with the $Fe(CO)_4$ unit (Scheme 6.12). Reaction of this complex with bovine pancreatic ribonuclease A (RNase), followed by gel filtration chromatography afforded a yellow derivative. The RNase derivative containing 4 g-atoms of Fe/mol of enzyme was shown to exhibit only 2–6% of the activity of the native enzyme while the binding of 2.4 g-atoms of Fe per mole of RNase decreased its catalytic activity by 70%. Interestingly, the IR spectrum of labeled RNase showed a characteristic set of v_{CO} bands in the 1900–2200 cm^{-1} area. The number and position of these absorption bands was very different from the v_{CO} bands measured for compound **47**. At that time, no effort had been put into the determination of the metallo–carbonyl complex binding site(s) [83]. Later on, it was shown that reaction of RNase with both complex **47** and the uncomplexed reagent maleic anhydride involved some of the ten lysine residues of RNase as depicted in

Scheme 6.12. Clearly, reaction of **47** with RNase was not specific of one particular lysine residue but rather involved several lysines. Selective introduction of the metallo–carbonyl group at the RNase active site residue Lys41 was achieved in two steps. First, reaction of RNase with maleic anhydride in the presence of a competitive enzyme inhibitor as a means of protection of the catalytic site led to an enzyme derivative with 3.7 blocked lysines. Removal of the competitive inhibitor by dialysis followed by reaction with compound **47** led to a new enzyme derivative exhibiting 1.1 g-atom of Fe/mol of RNase but exhibited only 29% of the catalytic activity of the unmodified protein [84].

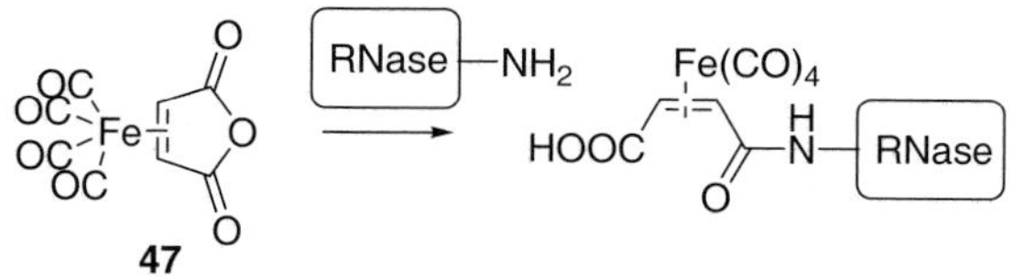

Scheme 6.12 Reaction of RNase with **47**.

We have synthesized and/or studied the reactivity of three series of lysine- and cysteine-targeted metallo–carbonyl complexes with several proteins including BSA, a monoclonal anti-TSH antibody and avidin. Their properties and the IR characteristics of the labeled proteins are gathered in Table 6.2.

The first series of metallo–carbonyl protein labeling reagents was designed from the popular Bolton–Hunter radio-iodination reagent **48** intended for antibody radiolabeling [85].

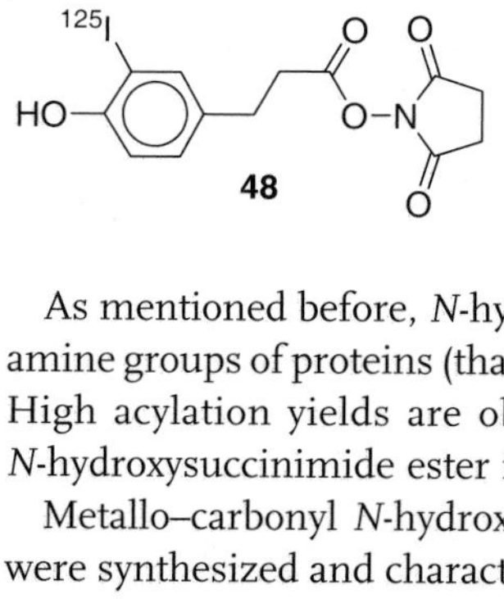

As mentioned before, *N*-hydroxysuccinimide esters readily acylate the primary amine groups of proteins (that is the lysine side chain and the N-terminal residue). High acylation yields are obtained at basic pH, providing aminolysis of the *N*-hydroxysuccinimide ester is faster than its hydrolysis (Scheme 6.13).

Metallo–carbonyl *N*-hydroxysuccinimide esters **49** [86], **50** [87] and **51–53** [88] were synthesized and characterized by classical spectroscopic methods (Fig. 6.9).

Labeling of several proteins was achieved with these complexes [88–91]. Each time, FT-IR analysis of the protein conjugates revealed the presence of several ν_{CO} bands typical of the metallo–carbonyl unit. No significant shift of their position was observed related to the starting reagent because chemical reaction occurs far away from the metallo–carbonyl group. Marginal loss of immunoreactivity of anti-TSH antibody was observed upon modification with alkyne cobalt hexacarbonyl [89] and cyclopentadienyl rhenium tricarbonyl [88] moieties. This point is particularly important when antibodies labeled in this manner are planned to be

Table 6.2 Labeling of proteins with metallo–carbonyl moieties.

Complex	Protein binding site	v_{CO}, cm^{-1}	Reference
49	Lysines	2095, 2054, 2023	[82] (synthesis) [85] (labelling of BSA and anti-TSH mAb) [86] (labelling of avidin)
50	Lysines	2053, 2000	[83] (synthesis) [87] (labelling of BSA)
51–52 53	Lysines Lysines	2025, 1931 2020, 1978	[84] (synthesis, labelling of BSA and anti-TSH mAb)
56	Lysines	2048, 2002	[89] (synthesis, labelling of BSA)
59–60	Lysines	n.d.	[98] (synthesis) [100] (labelling of BSA)
61	Lysines	n.d.	[101] (synthesis, labelling of BSA)
63	Cysteines (neutral pH, r.t.) Cysteines and histidines (neutral pH, 35°C) Cysteines, histidines and lysines (basic pH, 35°C)	2050, 2000	[102] (synthesis) [87] (labelling of BSA) [90, 115] (labelling of dendrimer PAMAM G4)
64	Lysines and/or histidines	2058, 1993	[104] (labelling of chymotrypsin, RNase, lipase, alkaline phosphatase)
65	Histidine	2050, 1978	[103] (labelling of lysozyme)

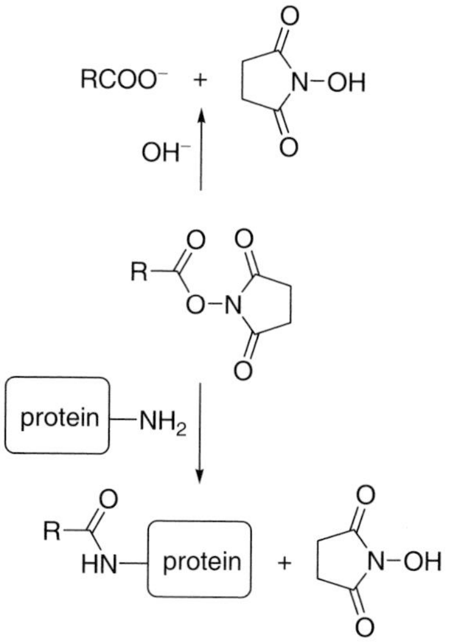

Scheme 6.13 Hydrolysis and aminolysis of N-hydroxysuccinimide esters.

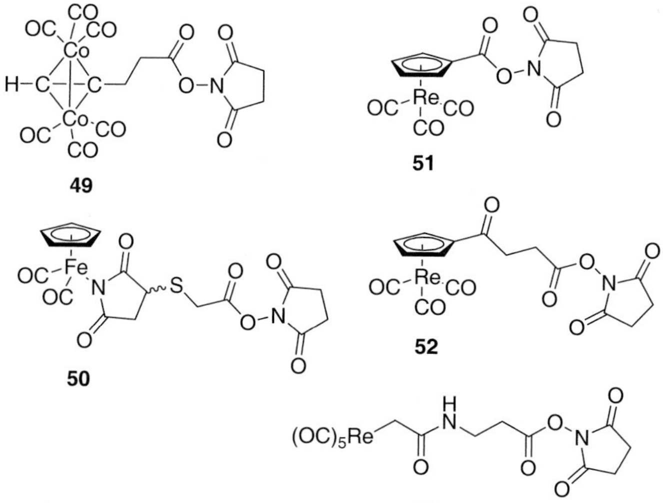

Fig. 6.9 Lysine- and cysteine-targeted transition organometallic reagents
for the labeling of proteins with metallo–carbonyl probes.

used as analytical reagents in immunoassays or as vectors to transport radioisotopes (^{99m}Tc, ^{186}Re or ^{188}Re) to a biological target *in vivo*.

Another popular series of organic reagents for protein chemical modification is the isothiocyanates. They have found numerous applications in peptide and protein chemistry [92]. The best known examples are phenyl isothiocyanate **54** (Edman's sequencing reagent) and fluorescein isothiocyanate **55**.

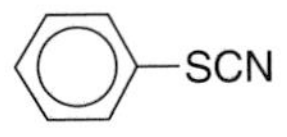

54 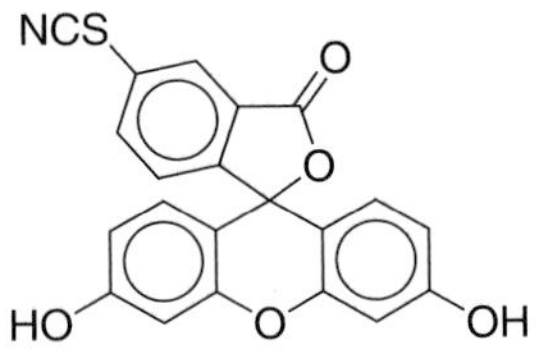   **55**

In this line, compound **56** was synthesized and was shown to label BSA at basic pH in good yield by reaction of some of its amine groups to yield the thiourea derivative (Scheme 6.14) [93].

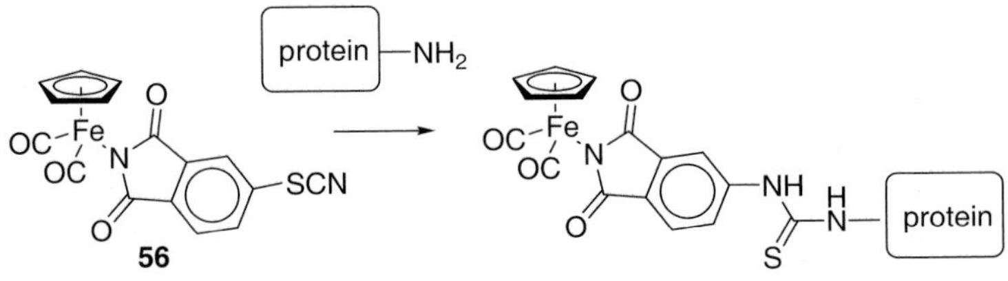

Scheme 6.14 Reaction of BSA with **56**.

Pyrylium salts are aromatic heterocycles with an oxygen atom that display a high electrophilic character at the ortho and para positions of the ring. Di- and tri-substituted compounds are fairly stable and have shown to react with primary amines to yield pyridinium salts (Scheme 6.15) [74]. In aqueous solution, the pyrylium form is in equilibrium with the open form (pseudobase").

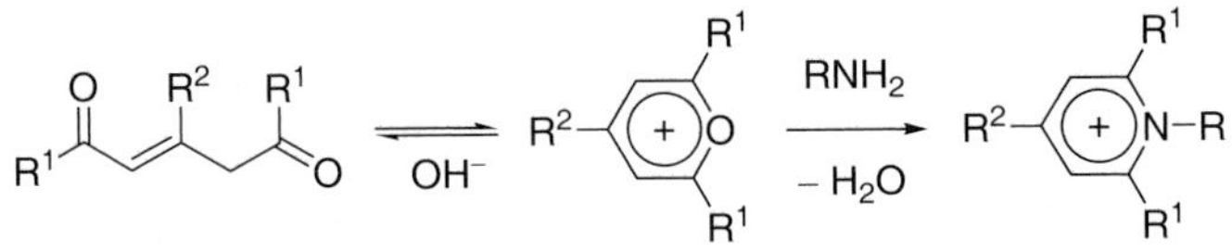

Scheme 6.15 Reaction of pyrylium ions with primary amines and hydroxide anion.

In the 1970s, O'Leary and Samberg described the reaction of 2,4,6-trimethyl-pyrylium perchlorate **57** with α-chymotrypsin and observed the formation of a protein–pyridinium conjugate [94]. Later on, the reaction of the tri-substituted pyrylium salt **58** with the proteins gelatin and α-chymotrypsin [95], glycophorin [96, 97] and the Na^+/glucose cotransporter protein [98] was reported. In all cases, the reaction happened to exclusively involve the proteins lysine residues and formation of the corresponding pyridinium adducts were observed.

Pyrylium salts bearing various metallo–carbonyl moieties at the para position of the heterocycle have been synthesized [99–103]. The manganese and rhenium pyrylium salts **59–60** [104] together with the chromium pyrylium ions **42**, **43** and **61** [105] successfully labeled the protein BSA to afford the corresponding pyridinium adducts as evidenced by the comparison with the spectroscopic data of model pyridinium derivatives. Interestingly, the reaction can readily be monitored by UV-visible spectroscopy.

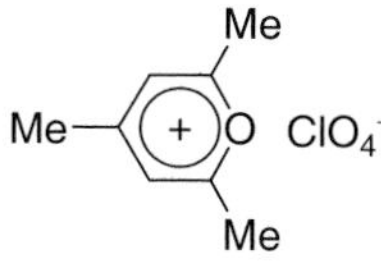

57 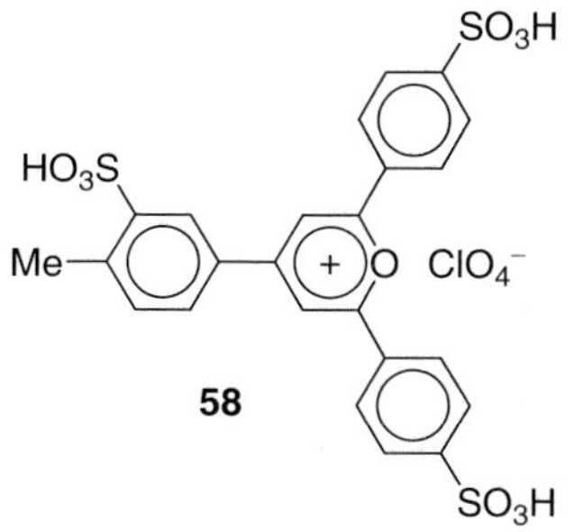 **58**

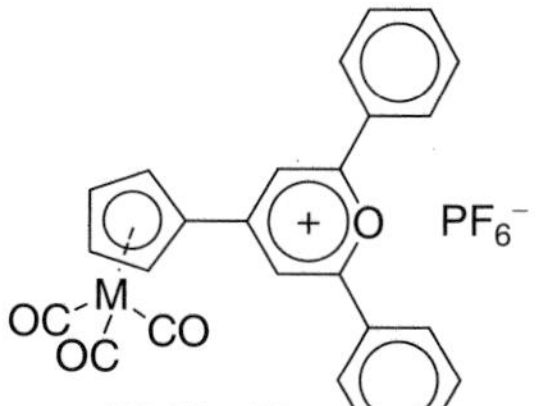

59 M = Mn
60 M = Re

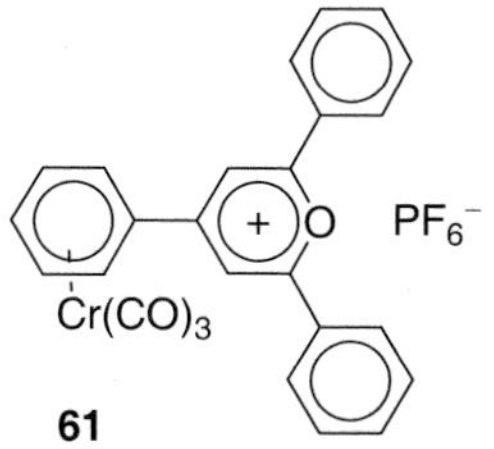

61

N-substituted maleimides are known to be alkylating agents that react with thiols to form stable thioethers, at neutral pH (Scheme 6.16). Hydrolysis can also occur but its rate depends on the nature of the substituent [106].

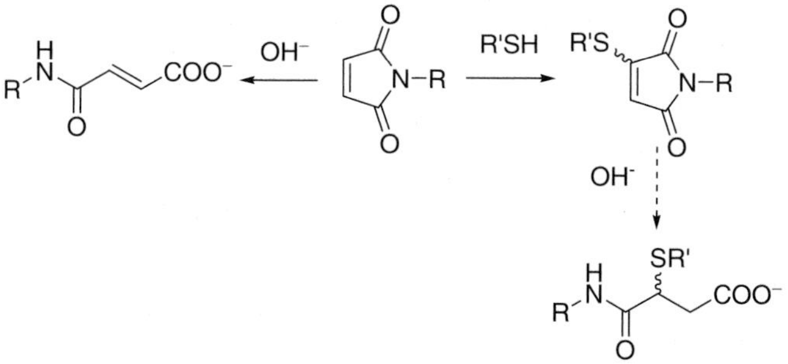

Scheme 6.16 Hydrolysis and alkylation and of N-substituted maleimides by thiols.

For example N-ethylmaleimide **62** is a popular reagent used to permanently block cysteine residues [107]. N-substituted maleimide bearing a cyclopentadienyl iron dicarbonyl moiety **63** was synthesized by Rudolf and Zakrzewski in 1994 [108].

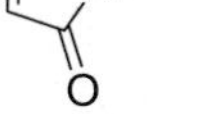

62 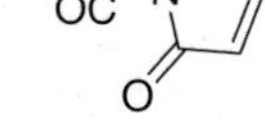 **63**

The reactivity of **63** towards BSA was compared with that of compound **62** under different reactional conditions. It appeared that, except when the reaction was performed at room temperature and neutral pH, the number of metallo–carbonyl units bound per protein always exceeded the initial number of thiols determined by the Ellman's method. In certain conditions, alkylation of histidines and lysines was also proved to occur [91].

The electrophilic character of the dienyl iron tricarbonyl complex **65** as such [109] or masked in the form of the labile pyridinium complex **64** [110] was exploited to label several enzymes in aqueous medium. Nucleophilic addition of the imidazole of histidine residues and/or the primary amine of lysine residues (depending on their accessibility) was shown to occur by spectroscopic methods (Scheme 6.17).

Formation of the protonated or unprotonated forms of the histidine and lysine adducts provided information on the pK_a of the individual protein residues (see Chapter 7, Section 7.7).

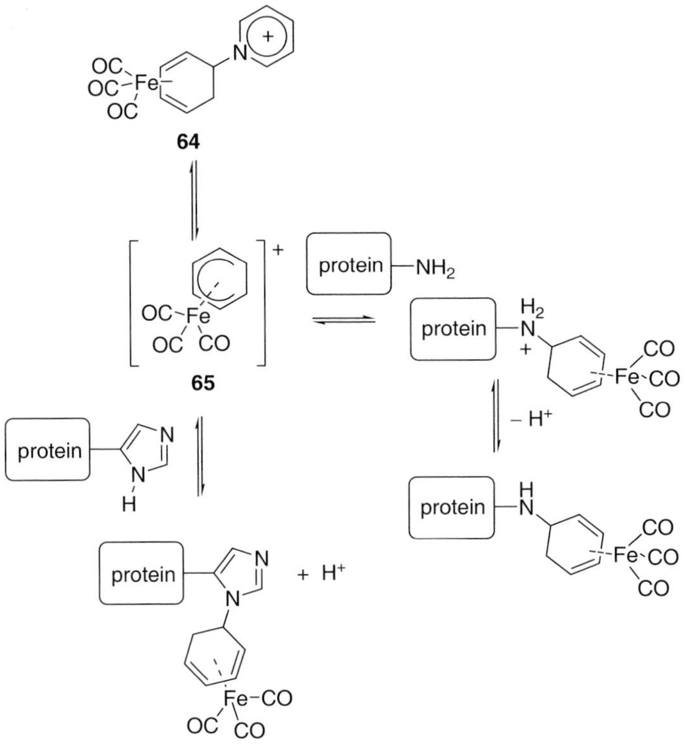

Scheme 6.17 Reaction of compounds **64** and **65** with proteins.

6.6
Conclusions and Outlook

There is now a wide range of synthetic methods available to produce transition organometallic derivatives of proteins. Side-chain selective covalent labeling appears to be achieved only by using organometallic reagents where one of the organic ligands carries a functional group known to specifically target a given residue. In all cases, the aim was to take advantage of the unique *physical* properties (mostly spectroscopic and electrochemical) of these particular metal complexes. Most of the transition organometallic complexes appeared to be fully compatible with the presence of water as solvent and were stable in the long term.

In addition to these probing properties, organometallic complexes are known to display very attractive *chemical* properties. Some of them are in particular able to catalyze a large array of organic reaction processes (including polymerization reactions). The construction *de novo* of metalloenzymes including a metal complex moiety covalently [111] or non-covalently [112, 113] linked to wild-type or genetically engineered proteins and able to catalyze a range of chemical reactions has been the subject of several publications. Combination of directed evolution of enzymes and covalent modification with ligands to form hybrid catalysts has recently been proposed [114]. No doubt these concepts open new perspectives in the preparation of more efficient catalysts to produce new molecules and materials.

Acknowledgements

The French Ministry of Research, the Centre National de la Recherche Scientifique (CNRS), and the European COST (action D8 "Metals in medicine") are gratefully acknowledged for their financial support.

Abbreviations

AC	Alternative current
Az	Azurin
BSA	Bovine serum albumin
DCC	N,N'-dicyclohexylcarbodiimide
DNP	Dinitrophenol
DTSP	3,3'-dithiodipropionic acid di(N-hydroxysuccimide ester)
ECD	Electrochemical detector
EDAC	N-(3-dimethylaminopropyl)-N'-ethylcarbodiimide
ET	Electron transfer
FAD	Flavin adenine dinucleotide
Fc	Ferrocenyl
GLL	Gluconolactone
GLU	Glucose

GOx   Glucose oxidase
HBV   Hepatitis B virus
HCG   Human chorionic gonadotropin
HSA   Human serum albumin
IgG   Immunoglobulin G
MAD   Multiple anomalous diffraction
MIR   Multiple isomorphous replacement
MLCT   Metal-to-ligand charge transfer
MUA   11-mercaptoundecanoic acid
NHS   *N*-hydroxysuccinimide
PAMAM G4   Fourth generation poly(amidoamine)
RP-HPLC   Reverse phase high pressure liquid chromatography
SAM   Self-assembled monolayer
SATA   S-acetylthioacetate
SIR   Single isomorphous replacement

References

1 W. A. Kornicker, B. L. Vallee, *Ann. N. Y. Acad. Sci.* **1969**, *153*, 689–705.

2 T. J. Gill III, L. T. Mann Jr, *J. Immunol.* **1966**, 96, 906–912.

3 R. W. Giese, W. Kornicker, *Biochim. Biophys. Acta* **1983**, *746*, 97–100.

4 J. Wang, *J. Pharm. Biomed. Anal.* **1999**, *19*, 47–53.

5 I. I. Willner, E. Katz, *Angew. Chem. Int. Ed.* **2000**, *39*, 1180–1218.

6 A. E. G. Cass, G. Davies, G. D. Francis, H. A. O. Hill, W. J. Aston, I. J. Higgins, E. V. Plotkin, L. D. L. Scott, A. P. F. Turner, *Anal. Chem.* **1984**, *56*, 667–671.

7 J. E. Pearson, A. Gill, P. Vadgama, *Ann. Clin. Biochem.* **2000**, *37*, 119–145.

8 F. Mizutani, M. Asai, *Denki Kagaku* **1988**, *56*, 1100–1101.

9 H. Shinohara, T. Kusaka, E. Yokota, R. Monden, M. Sisido, *Sens. Actuators B* **2000**, *65*, 144–146.

10 K. K.-W. Lo, J. S.-Y. Lau, D. C.-M. Ng, N. Zhu, *J. Chem. Soc., Dalton Trans.* **2002**, 1753–1756.

11 A. Heller, *Acc. Chem. Res.* **1990**, *23*, 128–134.

12 Y. Degani, A. Heller, *J. Phys. Chem.* **1987**, *91*, 1285–1289.

13 Y. Degani, A. Heller, *J. Am. Chem. Soc.* **1988**, *110*, 2615–2620.

14 A. Badia, R. Carlini, A. Fernandez, F. Battaglini, S. R. Mikkelsen, A. M. English, *J. Am. Chem. Soc.* **1993**, *115*, 7053–7060.

15 T. Suzawa, Y. Ikariyama, M. Aizawa, *Anal. Chem.* **1994**, *66*, 3889–3894.

16 M. Imamura, T. Haruyama, E. Kobatake, Y. Ikariyama, M. Aizawa, *Sens. Actuators B* **1995**, *24*, 113–116.

17 S. Kunugi, Y. Murakami, K. Ikeda, N. Itoh, *Int. J. Biol. Macromol.* **1992**, *14*, 210–214.

18 W. Schuhmann, T. J. Ohara, H. L. Schmidt, A. Heller, *J. Am. Chem. Soc.* **1991**, *113*, 1394–1397.

19 W. Schuhmann, *Biosens. Bioelectron.* **1995**, *10*, 181–193.

20 R. Nagata, S. A. Clark, K. Yokoyama, E. Tamiya, I. Karube, *Anal. Chim. Acta* **1995**, *304*, 157–164.

21 R. Blonder, E. Katz, Y. Cohen, N. Itzhak, A. Riklin, I. Willner, *Anal. Chem.* **1996**, *68*, 3151–3157.

22 L.-Q. Chen, X.-E. Zhang, W.-H. Xie, Y.-F. Zhou, Z.-P. Zhang, A. E. G. Cass, *Biosens. Bioelectron.* **2002**, *17*, 851–857.

23 S. Kuwabata, T. Okamoto, Y. Kajiya, H. Yoneyama, *Anal. Chem.* **1995**, *67*, 1684–1690.

24 A. Riklin, I. Willner, *Anal. Chem.* **1995**, *67*, 4118–4126.

25 H. C. Yoon, M. Y. Hong, H. S. Kim, *Anal. Chem.* **2000**, *72*, 4420–4427.

26 H. C. Yoon, M. Y. Hong, H. S. Kim, *Anal. Biochem.* **2000**, *282*, 121–128.

27 H. C. Yoon, D. Lee, H.-S. Kim, *Anal. Chim. Acta* **2002**, *456*, 209–218.

28 C. Padeste, A. Grubelnik, L. Tiefenauer, *Biosens. Bioelectron.* **2000**, *15*, 431–438.

29 C. Padeste, B. Steiger, A. Grubelnik, L. Tiefenauer, *Biosens. Bioelectron.* **2003**, *19*, 239–247.

30 B. Steiger, C. Padeste, A. Grubelnik, L. Tiefenauer, *Electrochim. Acta* **2003**, *48*, 761–769.

31 H. Eckert, M. Koller, *J. Liq. Chromatog.* **1990**, *13*, 3399–3414.

32 T. Suzawa, Y. Ikariyama, M. Aizawa, *Bull. Soc. Chem. Jpn.* **1995**, *68*, 167.

33 T. K. Lim, T. Matsunaga, *Biosens. Bioelectron.* **2001**, *16*, 1063–1069.

34 T. K. Lim, S. Imai, T. Matsunaga, *Biotechnol. Bioeng.* **2002**, *77*, 758–763.

35 T. K. Lim, H. Ohta, T. Matsunaga, *Anal. Chem.* **2003**, *75*, 2984–2989.

36 J. Wang, A. Ibanez, M. P. Chatrathi, *Electrophoresis* **2002**, *23*, 3744–3749.

37 M. Hromadova, M. Salmain, R. Sokolova, L. Pospisil, G. Jaouen, *J. Organomet. Chem.* **2003**, *668*, 17–24.

38 K. T. Douglas, O. S. Ejim, K. Taylor, *J. Inhibition* **1992**, *6*, 233–242.

39 K. Di Gleria, H. A. Hill, L. L. Wong, *FEBS Lett.* **1996**, *390*, 142–144.

40 K. Di Gleria, C. M. Halliwell, C. Jacob, H. A. Hill, *FEBS Lett.* **1997**, *400*, 155–157.

41 K. Di Gleria, D. P. Nickerson, H. A. O. Hill, L.-L. Wong, V. Fulop, *J. Am. Chem. Soc.* **1998**, *120*, 46–52.

42 K. K.-W. Lo, L. L. Wong, H. A. Hill, *FEBS Lett.* **1999**, *451*, 342–346.

43 A. Anne, C. Demaille, J. Moiroux, *J. Am. Chem. Soc.* **2001**, *123*, 4817–4825.

44 S. M. Fredericks, J. C. Luong, M. S. Wrighton, *J. Am. Chem. Soc.* **1979**, *101*, 7415–7417.

45 L. Sacksteder, M. Lee, J. N. Demas, B. A. DeGraff, *J. Am. Chem. Soc.* **1993**, *115*, 8230–8238.

46 Y. Chen, B. P. Sullivan, *J. Chem. Educ.* **1997**, *74*, 685–689.

47 L. Sacksteder, J. N. Demas, B. A. DeGraff, *Anal. Chem.* **1993**, *65*, 3480–3483.

48 S. S. Sun, A. J. Lees, P. Y. Zavalij, *Inorg. Chem.* **2003**, *42*, 3445–3453.

49 K. K. Lo, W. K. Hui, D. C. Ng, *J. Am. Chem. Soc.* **2002**, *124*, 9344–9345.

50 X. Q. Guo, F. N. Castellano, L. Li, H. Szmacinski, J. R. Lakowicz, J. Sipior, *Anal. Biochem.* **1997**, *254*, 179–186.

51 X.-Q. Guo, F. N. Castellano, L. Li, J. R. Lakowicz, *Anal. Chem.* **1998**, *70*, 632–637.

52 J. D. Dattelbaum, O. O. Abugo, J. R. Lakowicz, *Bioconjugate Chem.* **2000**, *11*, 533–536.

53 K. K. Lo, W. K. Hui, D. C. Ng, K. K. Cheung, *Inorg. Chem.* **2002**, *41*, 40–46.

54 H. B. Gray, J. R. Winkler, *Annu. Rev. Biochem.* **1996**, *65*, 537–561.

55 I. J. Chang, H. B. Gray, J. R. Winkler, *J. Am. Chem. Soc.* **1991**, *113*, 7056–7057.

56 E. T. Adman, L. H. Jensen, *Isr. J. Chem.* **1981**, *21*, 8.

57 A. J. Di Bilio, B. R. Crane, W. A. Wehbi, C. N. Kiser, M. M. Abu-Omar, R. M. Carlos, J. H. Richards, J. R. Winkler, H. B. Gray, *J. Am. Chem. Soc.* **2001**, *123*, 3181–3182.

58 J. R. Winkler, A. J. Di Bilio, N. A. Rarrow, J. H. Richards, H. B. Gray, *Pure & Appl. Chem.* **1999**, *71*, 1753–1764.

59 B. R. Crane, A. J. Di Bilio, J. R. Winkler, H. B. Gray, *J. Am. Chem. Soc.* **2001**, *123*, 11623–11631.

60 J. E. Miller, C. Gradinaru, B. R. Crane, A. J. Di Bilio, W. A. Wehbi, S. Un, J. R. Winkler, H. B. Gray, *J. Am. Chem. Soc.* **2003**, *125*, 14220–14221.

61 F. C. H. Crick, B. S. Magdoff, *Acta Cryst.* **1956**, *9*, 901–908.

62 T. L. Blundell, L. N. Johnson, *Protein crystallography*, Academic Press, London, **1976**.

63 G. A. Petsko, *Methods Enzymol.* **1985**, *114*, 147–156.

64 S. A. Islam, D. Carvin, M. J. Sternberg, T. L. Blundell, *Acta Crystallogr. D Biol. Crystallogr.* **1998**, *54*, 1199–1206.

65 L. Sawyer, D. W. Green, *Biochim. Biophys. Acta* **1979**, *579*, 234–239.

66 C. C. F. Blake, *Adv. Protein Chem.* **1968**, *23*, 59–120.

67 W. Jahn, *Z. Naturforsch.* **1989**, *44b*, 79–82.

68 S. Weinstein, W. Jahn, H. Hansen, H. G. Wittmann, A. Yonath, *J. Biol. Chem.* **1989**, *264*, 19138–19142.

69 S. Weinstein, W. Jahn, C. Glotz, F. Schlunzen, I. Levin, D. Janell, J. Harms, I. Kolln, H. A. Hansen, M. Gluhmann, W. S. Bennett, H. Bartels, A. Bashan, I. Agmon, M. Kessler, M. Pioletti, H. Avila, K. Anagnostopoulos, M. Peretz, T. Auerbach, F. Franceschi, A. Yonath, *J. Struct. Biol.* **1999**, *127*, 141–151.

70 D. Osella, P. Pollone, M. Ravera, M. Salmain, G. Jaouen, *Bioconjugate Chem.* **1999**, *10*, 607–612.

71 D. Osella, M. Ravera, M. Vicenti, M. Salmain, G. Jaouen, *Organometallics* **1996**, *15*, 3037–3041.

72 A. Gorfti, M. Salmain, G. Jaouen, M. J. McGlinchey, A. Bennouna, A. Mousser, *Organometallics* **1996**, *15*, 142–151.

73 M. Salmain, A. Gorfti, G. Jaouen, *Eur. J. Biochem.* **1998**, *258*, 192–199.

74 A. T. Balaban, G. W. Fischer, A. Dinulescu, A. V. Koblik, C. N. Dorofeenko, V. V. Mezhritski, W. Schroth, in A. R. Katritzky (Ed.), *Advances in heterocyclic chemistry, Suppl. II* Academic Press, New York, **1982**, pp. 114–127.

75 C. F. Bernasconi, M. W. Stronach, *J. Am. Chem. Soc.* **1993**, *115*, 1341–1346.

76 M. Salmain, E. Licandro, S. Maiorana, H. Tran-Huy, G. Jaouen, *J. Organomet. Chem.* **2001**, *617–618*, 376–382.

77 M. Salmain, J. C. Blais, H. Tran Huy, C. Compain, G. Jaouen, *Eur. J. Biochem.* **2001**, *268*, 5479–5487.

78 D. P. Egan, M. Salmain, P. McArdle, G. Jaouen, B. Caro, *Spectrochim. Acta A* **2002**, *58*, 941–951.

79 M. Salmain, B. Caro, F. Le Guen-Robin, J.-C. Blais, G. Jaouen, *ChemBioChem* **2004**, *5*, 99–109.

80 I. W. McNae, K. Fishburne, A. Habtemariam, T. M. Hunter, M. Melchart, F. Wang, M. D. Walkinshaw, P. J. Sadler, *Chem. Commun.* **2004**, 1786–1787.

81 N. Cheng, J. F. Conway, N. R. Watts, J. F. Hainfeld, V. Joshi, R. D. Powell, S. J. Stahl, P. E. Wingfield, A. C. Steven, *J. Struct. Biol.* **1999**, *127*, 169–179.

82 G. Marcon, L. Messori, P. Orioli, M. A. Cinellu, G. Minghetti, *Eur. J. Biochem.* **2003**, *270*, 4655–4661.

83 R. W. Giese, B. L. Vallee, *J. Am. Chem. Soc.* **1972**, *94*, 6199–6200.

84 R. W. Giese, *J. Inorg. Biochem.* **1983**, *18*, 301–311.

85 A. E. Bolton, W. M. Hunter, *Biochem. J.* **1973**, *133*, 529–539.

86 M. Salmain, A. Vessières, I. S. Butler, G. Jaouen, *Bioconjugate Chem.* **1991**, *2*, 13–15.

87 B. Rudolf, J. Zakrzewski, M. Salmain, G. Jaouen, *Tetrahedron Lett.* **1998**, *39*, 4281–4282.

88 M. Salmain, M. Gunn, A. Gorfti, S. Top, G. Jaouen, *Bioconjugate Chem.* **1993**, *4*, 425–433.

89 A. Varenne, M. Salmain, C. Brisson, G. Jaouen, *Bioconjugate Chem.* **1992**, *3*, 471–476.

90 M. Salmain, N. Fischer-Durand, L. Cavalier, B. Rudolf, J. Zakrzewski, G. Jaouen, *Bioconjugate Chem.* **2002**, *13*, 693–698.

91 B. Rudolf, J. Zakrzewski, M. Salmain, G. Jaouen, *New J. Chem.* **1998**, 813–818.

92 S. S. Wong, *Chemistry of protein conjugation and cross-linking*, CRC Press, Boca Raton (Fl), **1991**.

93 A. Kazimierczak, J. Zakrzewski, M. Salmain, G. Jaouen, *Bioconjugate Chem.* **1997**, *8*, 489–494.

94 M. H. O'Leary, G. A. Samberg, *J. Am. Chem. Soc.* **1971**, *93*, 3530.

95 A. R. Katritzky, J. L. Mokrosz, M. L. Lopez-Rodriguez, *J. Chem. Soc., Perkin Trans. 2* **1984**, 875–878.

96 K. Dill, S. Hu, A. R. Katritzky, M. Sutharchanadevi, *J. Biochem. Biophys. Methods* **1988**, *17*, 75–78.

97 K. Dill, S. H. Hu, M. Sutharchanadevi, A. R. Katritzky, *J. Protein Chem.* **1988**, *7*, 341–348.

98 A. Fernandez, A. R. Katritzky, M. Sutharchanadevi, B. R. Stevens, *Biochem. Biophys. Res. Commun.* **1989**, *163*, 1356–1363.

99 B. Caro, D. Sénéchal, M.-C. Sénéchal-Tocquer, P. Marrec, J.-Y. Saillard, C. Trikli, S. Kahlal, *Tetrahedron Lett.* **1993**, *34*, 7259–7262.

100 B. Caro, F. Robin-Le Guen, M.-C. Sénéchal-Tocquer, V. Prat, J. Vaissermann, *J. Organomet. Chem.* **1997**, *543*, 87–92.

101 P. Haquette, M. Salmain, Y. Marchal, Y. Marsac, G. Jaouen, D. Cunningham, D. P. Egan, P. McArdle, *Chem. Commun.* **2001**, 1504–1505.

102 K. L. Malisza, S. Top, J. Vaissermann, M.-C. Sénéchal-Tocquer, D. Sénéchal, J. Y. Saillard, S. Triki, S. Kahlal, J. F. Britten, M. J. McGlinchey, G. Jaouen, *Organometallics* **1995**, *14*, 5273–5280.

103 A. G. Milaev, O. Y. Okhlobystin, *Khim. Geterosikl. Soedin* **1985**, *5*, 593–597.

104 M. Salmain, K. L. Malisza, S. Top, G. Jaouen, M.-C. Sénéchal-Tocquer, D. Sénéchal, B. Caro, *Bioconjugate Chem.* **1994**, *5*, 655–659.

105 B. Caro, F. Le Guen-Robin, M. Salmain, G. Jaouen, *Tetrahedron* **2000**, *56*, 257–263.

106 G. T. Hermanson, *Bioconjugate techniques*, Academic Press, New York, **1996**.

107 D. G. Smyth, A. Nagamatsu, J. S. Fruton, *Biochem. J.* **1960**, *91*, 589.

108 B. Rudolf, J. Zakrzewski, *Tetrahedron Lett.* **1994**, *35*, 9611–9612.

109 J. G. Carver, B. Fates, L. A. P. Kane-Maguire, *Chem. Commun.* **1993**, 928.

110 C. E. Anson, C. S. Creaser, O. Egyed, G. R. Stephenson, *Spectrochim. Acta A* **1997**, *53*, 1867–1877.

111 D. Qi, C.-M. Tann, D. Haring, M. D. Di Stefano, *Chem. Rev.* **2001**, *101*, 3081–3111.

112 J. Collot, J. Gradinaru, N. Humbert, M. Skander, A. Zocchi, T. R. Ward, *J. Am. Chem. Soc.* **2003**, *125*, 9030–9031.

113 M. E. Wilson, G. M. Whitesides, *J. Am. Chem. Soc.* **1978**, *100*, 306–307.

114 M. T. Reetz, *Proc. Natl. Acad. Sci. USA* **2004**, *101*, 5716–5722.

115 N. Fischer-Durand, M. Salmain, B. Rudolf, A. Vessières, J. Zakrzewski, G. Jaouen, *ChemBioChem* **2004**, *5*, 519–525.

7
Organometallic Bioprobes

G. Richard Stephenson

7.1
Introduction

The interplay between chemistry and biology is still of growing importance. At the molecular scale, organic compounds can interact with natural molecular, biopolymer and suprabiomolecular structures with great precision. The design and study of these interactions has over the last decade been an important priority area of research within the field of molecular recognition, and many significant advances have been made. Indeed, the importance in biological systems of molecular recognition is one of the crucial reasons that this topic has received so much attention; recognition events between small biomolecules, between small molecules and biological macromolecules, and between macromolecules themselves, hold the key to the functioning of natural systems, and so to the healthy or diseased growth of cells and organisms. The study of biological recognition events can be motivated by the desire to control them, as is the case in the search for new classes of drugs and antibiotics, or research to provide tools to explore in intimate detail the exact workings of nature.

7.2
The Definition of the Terms Bioprobes and Molecular Bioprobes

In general, "bioprobes" are defined [1] as follows: "Bioprobes are functional molecules or devices that provide information about biological systems. They can function on a macroscopic, microscopic, or submicroscopic scale (bionanometrology and molecular-scale observation of biological systems)." The term "bioprobe" is used in this chapter, however, to describe molecular structures that can function as probes to provide information about biological systems. These are *molecular* bioprobes, as distinct, for example, from microelectrodes or other macroscopic

Bioorganometallics: Biomolecules, Labeling, Medicine. Edited by Gérard Jaouen
Copyright © 2006 Wiley-VCH Verlag GmbH & Co. KGaA, Weinheim
ISBN: 3-527-30990-X

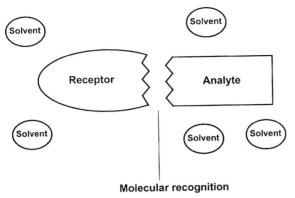

Molecular recognition

Fig. 7.1 Cartoon to illustrate molecular recognition in host–guest chemistry.

devices. This chapter is about *functional molecules*, and how organometallic chemistry can offer new designs for functional molecules that address biological questions. This same distinction occurs in other aspects of analytical science. For example, sensors can be macroscopic devices ("lab on a chip" devices [2] would be a good illustration) or can be functional molecules which "sense" at the molecular scale, and the sometimes mis-used or mis-understood term "*molecular sensor*" is most properly reserved for the definition of molecular-scale devices that respond and give information about analytes or other external measurables (the more general terms "chemosensor" or "chemical sensor" [3] adequately fill the need to describe sensors that respond to chemical substances). In the same way, the "*molecular bioprobes*" that form the topic of this chapter are not merely devices that respond to biological molecules – they are functional molecular scale materials that explore the workings of biological systems.

It is also important to establish that the chemistry of molecular bioprobes is not a sub-discipline of molecular recognition [i.e. "host–guest" chemistry [4] (Fig. 7.1) in which a "guest molecule" binds in an artificial receptor site (the "host") which through selective interactions provides the "recognition" effect]; rather, it is an *application* of molecular recognition.

Bioprobes must do more than simply recognize biologically relevant molecular features. They must also incorporate a method to read out the information that is available from the molecular recognition event. In this sense, the design of molecular bioprobes is analogous to the now extensively reviewed design issues for molecular sensors [5]. Specifically, they typically contain three components (Fig. 7.2): a binding site for an analyte molecule; a signal transduction portion of the molecule; and a responsive portion of the molecule which provides a read-out of the information.

Recognition, transduction and response are thus all important for the correct functioning of the bioprobe, but the performance of such designs is normally limited by the selectivity of the receptor. Where similar analytes compete for binding in the receptor, there is often a lack of discriminating power in the sensor, and even at its present advanced stage of development, the optimization of the

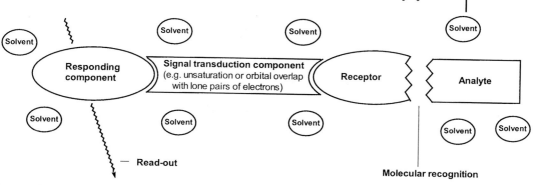

Fig. 7.2 Cartoon to illustrate designs for molecular sensors in which the presence of a signal transduction component allows a spectroscopically active group to respond to the binding of an analyte molecule at the receptor.

host–guest chemistry of the receptor for a specific analyte is far from straight-forward. Furthermore, this approach requires the design and synthesis of an optimized receptor for each analyte, a time-consuming process and one that does not provide a single sensor that can be applied to each of a range of similar analytes. The approach to the design of bioprobes discussed in this chapter is based on a novel strategy for exploiting multiple complementary responses to gain additional selectivity. For precise applications to study biological problems, great selectivity will be needed.

7.3
Response Strategies for the Read-out of Information

Typical and currently popular approaches to signal transduction and read-out of information employ spectroscopic methods based on UV, fluorescence, and electrochemistry. In its simplest form, the binding of a guest molecule in the receptor changes the spectroscopic properties of a chromophore and so the UV/visible spectrum of the molecule changes [6]. Measuring the UV spectrum in the presence and absence of the analyte, and comparing the results, can give the basis for an analytical measurement, and calibration curves can be defined. For well-defined systems, it can suffice to measure the ratio of two signals (the absorption maxima, for example, of the empty host molecule, and the host–guest complex) and construct a calibration graph from the ratio of signals and analyte (guest molecule) concentration. In more challenging applications, method validation can seek out wavelengths where interference from absorptions of other chromophores is minimal, and ratioing at these wavelengths can provide the calibration. The use of first derivative spectra [7] and the ratio-spectra method [8] can offer further improvements. Fluorescence spectroscopy can be used in a similar way [9] to UV/visible methods and has the advantage of greater sensitivity, and although there are biological molecules that fluoresce, interference from background fluorescence is far less likely than from competing absorbance by natural chromophores in the biological sample when UV/visible spectra are used.

Fluorescence thus offers improved sensitivity and lower levels of interference, but the molecular designs are often more complicated. The example shown in Fig. 7.2 would be typical of a sensor exploiting fluorescence as the read-out method. A fluorophore (the "responding component") would be covalently linked to the molecular receptor, so that the design of the binding component, and the read-out method can be separated. Signal transduction between these two portions of the molecular sensor is important if it is to function correctly, and often conjugated π-systems provide the best answer. For both UV/visible and fluorescence-based systems, however the underpinning spectroscopy has as its basis electronic excitation between full and empty π/π^* orbitals (or n/π^* orbitals). Changes in the molecule to optimize signal transduction can thus also change the spectroscopic and fluorescence properties of the molecule. Despite this potential drawback, both UV/visible and fluorometric read-out methods are popular and many successful and elegant applications have been developed. Furthermore, in the case of fluorescence, sophisticated designs in which fluorescence quenching [10] is employed to give large changes to the spectroscopic responses upon binding of the analyte, can truly offer unique advantages as a signal transduction and read-out strategy. However, when applied to biological systems, the sample must be transparent to the exciting/emitted radiation.

Read-out methods based on electrochemistry [11] have also been extensively studied. Here, the measurement of changes in redox potentials provides the mechanism to identify the binding of the guest analyte. The redox-active part of the system can be chosen to avoid competition from natural background electrochemical phenomena, and designs are often analogous to those needed for fluorescence-based systems, as few receptor molecules will also be naturally redox-active. It is again better to separate the two functions, linking them by a signal transduction component. An advantage of the redox chemistry approach is that many subtle molecular factors can influence the electrochemical responses of the electrophore, and a variety of signal transduction strategies can be used. Organometallic structures often offer rich and well-defined electrochemical responses, so for redox-based methods (far more so than for UV/visible/fluorescence-based approaches) it has been seen as worthwhile to introduce an organometallic structure as the responsive section of the molecular sensor. A typical case would be that of ferrocene [12], which combines high levels of chemical stability with a well-defined range of synthetic procedures for derivatization and useful and characteristic electrochemical responses, but many organometallic compounds can provide useful systems. Lithium and sodium binding in an azacrown, for example, have been studied by cyclic voltammetry of cobalticinium complexes [13]. Efficient bioprobes have been developed using ferrocene to provide electrochemical read-out. The most typical cases would be studies of electron-transfer pathways in redox-active proteins by the surface attachment of ferrocenyl derivatives [14]. A tricarbonyl(cyclopentadienyl)manganese derivative of bovine serum albumin has also been used in electrochemical studies [15]. Ferrocene derivatives of oligonucleotides have been used as gene sensors [16] using differential pulse voltammograms as the read-out method.

Conventionally, IR spectroscopy is not commonly used in molecular sensors as the read-out method. This is because spectroscopic responses from other components of the analytical sample are normally regarded as too great and too diverse. In the study of biological systems, in particular, background absorbance from water makes large sections of the IR spectrum opaque, and so unusable in a spectroscopic read-out strategy. However, in a small region around 2000 cm^{-1}, there is a window in the water background spectrum, and excellent transmission of IR radiation is possible [17]. The natural IR absorbances of organic functional groups do not fall within this window, so competition from background spectra of biological samples (e.g. protein, nucleic acid, and even whole cell preparations) will not interfere with the spectroscopic read-out through the window (it should be noted, however, that carbonylmetal derivatives of metalloenzymes [18, 19] produce spectroscopic signals in the window). Since IR spectroscopy is vibrational spectroscopy, the spectroscopic transitions are between vibrational states of functional groups within the molecule, and although such effects can be coupled (vibrational coupling), by and large, separate functional groups respond separately in IR spectroscopy. Thus the molecular features introduced into a bioprobe to achieve good signal transduction between the receptor site and the responding group are less likely to interfere when vibrational spectroscopy provides the read-out process, than for example would be the case when π-systems link the receptor and responding chromophores for read-out by UV/visible spectroscopy. The conceptual advantages of IR are thus minimal background absorbances from biological functional groups in the window in the water background spectrum, and a freedom from problems with compromises in the read-out method forced from molecular design changes introduced to achieve efficient signal transduction from the binding of the analyte to the receptor portion of molecular sensors and bioprobes. Also, IR spectroscopy operates on a very fast timescale, and so has the potential to resolve details of rapidly changing events.

A final method of read-out deserves a mention, though its use is more specialized, and the spectrometers required are more expensive. Electron spin provides an extremely sophisticated and sensitive tool to probe biological systems [20], and read-out by EPR (electron paramagnetic resonance) spectroscopy is a powerful technique. The conceptual advantages identified for IR-based methods in the preceding paragraph are also true of EPR-based read-out, in the sense that low levels of background signals can be expected from typical biological samples, especially when the spectroscopic resolution of EPR is taken into account, and the spectroscopic transitions that are exploited (transitions between electron spin states) are distinct from the general properties of most organic functional groups. Specialized organic molecules (stable organic radicals) are typically employed in this work, and just as ferrocene-based electrochemistry has been employed with the study of electron transfer pathways between surfaces of proteins and redox-active groups within the protein structure (see above), so too can electron-spin based methods provide detailed information, employing surface-mounted stable free radicals. The sensitivity and freedom from background signals possible for EPR makes this more attractive than other spin-based methods (for example, nuclear magnetic resonance) but in general, signal transduction between a receptor and an electron spin based read-out system will be complicated by interactions of the electron with features in the transduction and receptor components. What is a powerful technique for the study of electron transfer in proteins may not be so easily generalized to a design for bioprobes for less specialized applications. Similarly, NMR-based methods with "unnatural" nuclei (e.g. ^{19}F NMR of fluoro-carbon groups [21]), while avoiding conceptual difficulties with spin diffusion into the signal transduction and receptor components, lack the great sensitivity of other methods. Because of its sensitivity, EPR is more widely used. In the case of EPR, as with electrochemical methods, organometallic structures can play an important role, providing stable paramagnetic organometallic components for use in the same fashion as stable organic radicals.

In conclusion, for the development of molecular bioprobes, several factors are important. High sensitivity in the spectroscopic response is crucial, as the greater the sensitivity, the wider the range of biological issues that can be addressed. Similarly, low competition from background spectroscopic transitions is important. Practical sensitivity of response is not just the fundamental sensitivity of the spectroscopic method, but the ability to identify changing signals from the read-out system in the presence of the background spectra of the biological sample. More specialized read-out methods, however, lead to molecular designs which have separate sections of the molecule for achieving recognition, and creating the spectroscopic response, and it is advantageous to be able to optimize these two sections of the design independently. Consequently, complications arise when changes in the structures of the host molecules in the host–guest chemistry of the receptor, or the signal transduction component that relays information to the responding group, also create intrinsic changes in the spectroscopic responses of the system. Independent read-out methods, which are sensitive to the difference in properties between the host and the host–guest complex, but not to other aspects of the molecular architectures of the host and signal transduction components, will provide the most straightforward applications. In this sense, the IR-based method shows great promise, as it meets all the criteria.

7.4
Organometallic Components for Organometallic Bioprobes – Opening up the Advantages of IR-based Read-out Methods

The very factor that makes IR spectra, recorded through the window in the water background spectrum, exceptionally free from interference from signals from the sample, in principle constitutes a problem in the choice of functional groups for development in read-out techniques. Organic functional groups do not exhibit vibrational bands in the 1900–2100 cm^{-1} range. Only very weak overtone bands of bending modes of organic functional groups lie in this part of the spectrum.

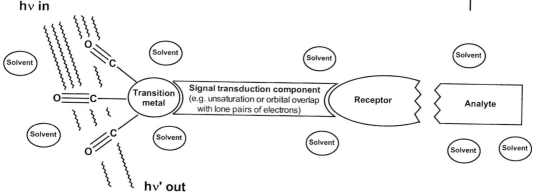

Fig. 7.3 Cartoon to illustrate an organometalcarbonyl-based sensor in which binding of an analyte at the receptor influences the vibrational modes of carbonylmetal groups by changing the distribution of charge density in the organic ligand, allowing the response to be measured by FTIR spectroscopy.

Fortunately (and uniquely) organometallic chemistry holds the answer to this problem, as the vibrational stretching mode of carbon monoxide (centered on 2143 cm^{-1}) is moved to lower wavenumbers in carbonylmetal transition metal complexes. Most carbonylmetal complexes, whether primary carbonyl complexes or organometallic complexes including organic ligands, show intense vibrational stretching modes in the window in the water background spectrum, so when connected to a receptor function by a signal transduction component (Fig. 7.3) a suitable design for a bioprobe exploiting IR as the read-out method can be determined.

The highly developed chemistry of the organic ligands provides good prospects for joining the responsive carbonylmetal component, the signal transduction section, and the receptor moiety. Furthermore, the complicated vibration/rotation band structure of carbon monoxide gas is simplified in carbonylmetal complexes to a single intense vibrational mode (the CO stretching mode of an M–CO complex) in the 1800–2100 cm^{-1} region.

The cause of the shift of the vibrational frequency of free CO into the 1800–2100 cm^{-1} region in transition metal complexes is well understood [22, 23]. The bonding (Fig. 7.4) between the carbon of the CO molecule and the metal comprises a σ bond (from the overlap of a symmetric metal orbital with the sp hybrid orbital on carbon, which in free CO [24] holds a non-bonded pair of electrons at carbon), and a π bond formed from overlap of a π* orbital of the CO triple bond with a suitable symmetry (antisymmetric) metal centered atomic orbital. Although this π* orbital of the CO triple bond is antibonding with respect to C≡O, it forms a bonding interaction between the carbonyl ligand and the metal. The pair of electrons associated initially with the lone pair of the carbon atom of CO becomes part of the filled bonding system of the metal complex, so the formation of this bond is often referred to as σ-donation. Similarly, as the π* antibonding orbital is empty in the structure of CO itself, it can be regarded as fulfilling a π-acceptor function as the metal complex forms [25].

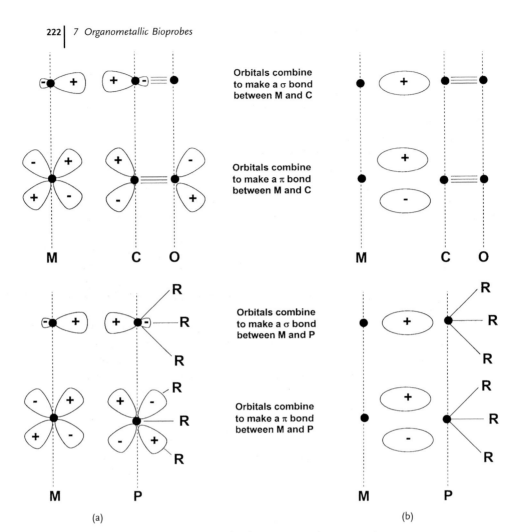

Fig. 7.4 Orbital overlap in the bonding between (a) carbon monoxide, and (b) phosphine ligands and a transition metal centre (based on [23]).

Although the true picture is a question of balance between the magnitudes of orbital coefficients of the σ and π bonding orbitals at M and C in the M–CO structure, the concepts of σ-donation and π-acceptor properties can provide a useful qualitative model to understand the effect of complexation on the strength of the CO bond and the relative shifts in the vibrational frequency of its stretching mode [ν(CO)]. In general, cationic carbonylmetal complexes have ν(CO) at higher wavenumbers than neutral complexes, which similarly have ν(CO) at higher wavenumbers than anionic complexes. Anionic complexes have greater electron density on the metal than a comparable neutral structure, and π-acceptor properties of ligands provide a means for charge delocalization. Greater transfer of electron density into an orbital which involves the π* antibonding orbital of the carbonyl ligand weakens the C≡O bond and so lowers the vibrational frequency of the C≡O

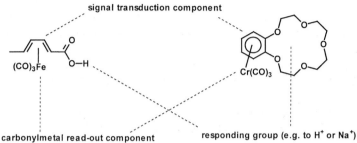

shifts vibrational response to lower wavenumbers

shifts vibrational response to higher wavenumbers

signal transduction component

carbonylmetal read-out component

responding group (e.g. to H⁺ or Na⁺)

Fig. 7.5 Examples of carbonylmetal structures that show changes in their IR spectra in response to changes in H⁺ and alkali metal ion concentrations.

stretching mode. A comparable argument applies to cationic structures, where lower electron density at the metal corresponds to a reduced tendency to transfer electron density into the π-acceptor orbitals, so there is less weakening of the C≡O bond, and a ν(CO) stretching mode at higher energy (higher wavenumbers). The same principles apply when partial charges are considered, and offer an explanation for the expected function of bioprobe structures of the type illustrated in Fig. 7.3. Binding of the analyte at the receptor can change the charge density in the receptor, a change that is conveyed by the signal transduction component to result in a corresponding change in charge density at the transition metal and so a change in the balance between σ-donation and π-acceptor properties of the CO ligand. Examples of carbonylmetal structures [26–29] that should respond in this way to simple chemical changes are shown in Fig. 7.5.

Figure 7.4 also shows the similarity in bonding in phosphine (PR₃) complexes and carbonyl complexes. Both these ligands are π-acceptor ligands, and form part of a series of isolobal [30] ligands (including carbenes, isonitiles, nitriles) that draw electron density away from the metal when present in transition metal complexes. When these ligands are bound to the metal in combinations, they affect the positions of the vibrational signals of the carbonyl groups. For example, replacing a carbonyl ligand by a phosphine ligand moves the vibrational frequency of any remaining carbonyl ligands to lower frequency because phosphines are weaker π acceptors than carbon monoxide.

When several carbonyl ligands of the same type are present, vibrational coupling [23] will produce a more complicated IR spectrum in the 1900–2100 cm⁻¹ region. This is illustrated in Fig. 7.6.

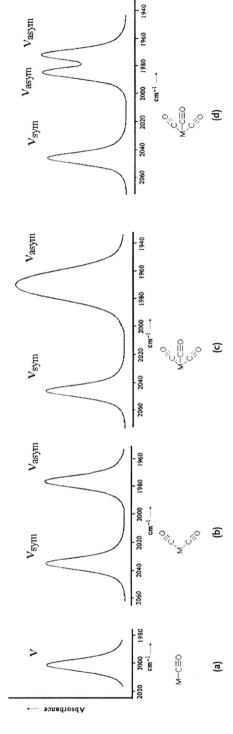

Fig. 7.6 Idealized representations of typical IR spectra of $\nu_{sym}(CO)$ and $\nu_{asym}(CO)$ vibrational stretching modes for carbonylmetal groups $[M(CO)_n]$: (a) $n = 1$; (b) $n = 2$; (c, d) $n = 3$].

While an isolated M(CO) group (M–C≡O) produces (Fig. 7.6a) a single sharp, narrow vibrational band for the v(CO) stretching mode, an $M(CO)_2$ group always produces (Fig. 7.6b) two distinct vibrationally coupled signals. It is incorrect to think of these signals as arising each from one of the individual carbonyl groups. The higher energy vibrational band corresponds to a symmetric vibrational mode of the entire $M(CO)_2$ group, and the lower energy (lower frequency) vibrational band corresponds to an alternative asymmetric vibrational mode. The separation of the two bands [v_{sym}(CO) and v_{asym}(CO)] depends on the efficiency of vibrational coupling, so in bioprobes containing an $M(CO)_2$ group as the responding group, both the position and the separation of the v_{sym}(CO) and v_{asym}(CO) bands can convey the information. While the position (strictly speaking, the position of the center point between the two band maxima) is strongly influenced by the charge on the metal, the vibrational coupling (and hence the separation) is often influenced strongly by changes in the geometry of the complex. There is more information available from an $M(CO)_2$ responding group than from an M(CO) responding group. When the responding group is $M(CO)_3$, there is still a unique symmetric stretching mode which again produces the highest wavenumber vibrational band. With $M(CO)_3$, however, there are two possibilities for asymmetric stretching modes and these are often degenerate (i.e. have the same energy for the vibrational transition). The signals from these degenerate vibrations therefore appear at exactly the same place in the IR spectrum and the result (Fig. 7.6c) is a small v_{sym}(CO) band followed at lower frequency by a single v_{asym}(CO) band of roughly twice the intensity (the degree of vibrational coupling also influences the intensity ratio [32, 33]). The number of vibrational modes for an $LM(CO)_n$ structure can be determined [44] from the symmetry of the complex and the number of CO ligands (n). For example, when the symmetry is low, the degeneracy of the asymmetric v_{asym}(CO) bands is lost. In practice, this often amounts only to a slight broadening (and lowering) of the v_{asym}(CO) band envelope, but in nonpolar solvents such as cyclohexane, the v_{asym}(CO) bands can sometimes be partially resolved (Fig. 7.6d). Even the overlapping v_{asym}(CO) bands, however, are valuable in bioprobe applications, since loss of degeneracy and the degree of separation of the two contributions to the band envelope are a consequence of subtle changes in the structure of the complex. In the IR spectrum, the position of the whole group of signals on the frequency scale (a center point based on a weighted average of the positions of contributing bands), the separation of v_{sym}(CO) and v_{asym}(CO) bands, and the exact form of the signals from the vibrationally coupled v_{asym}(CO) vibrational modes, all convey information [35], and are influenced to different extents by changes in bonding, geometry and charge distribution in the molecule. There is more information available from the IR spectra of $LM(CO)_n$ structures than might at first be thought possible, and the most information is available from complexes with larger values of n.

It can be seen from this discussion that the IR based read-out of information from organometallic bioprobes with carbonyl ligands by analysis of the vibrational band envelopes viewed through the window of the water background spectrum has a number of unusual features. The v(CO) signals are intense and narrow, can be adjusted in terms of their position on the wavenumber scale by changing other ligands in the complexes, and respond in a differential fashion to a variety and charge/structural effects in the complex. A significant advantage of the IR spectroscopy of the carbonyl ligands of organometallic bioprobes is that the signals of several organometallic complexes can be determined independently but in a single spectroscopic experiment. The vibrational bands are sufficiently narrow that combinations of complexes can be chosen to give signals that do not overlap, i.e. a "multi-carbonylmetal" method [26, 36]. In the study of biological systems, where many competing influences need to be resolved, this factor may prove to be crucial. Multi-carbonylmetal systems can be envisaged which respond to the same analytical feature of interest (this will improve the precision of the measurement, or help to distinguish competing influences from the true responses to the feature of interest). Alternatively, complexes with different receptors can be used together for the simultaneous determination of two factors. The multi-carbonyl-metal approach is an important conceptual advantage of the IR-based methods, benefits greatly in terms of practicality from the very narrow form of the $\nu(CO)$ signals, and offers a solution to limitations in receptor selectivity when bioprobes are built from receptor components, $M(CO)_n$ responding groups, and η^n organic ligands with extended π-systems as the signal transduction component (Fig. 7.3). The challenge in developing practical methods based on this approach is to seek out complementary pairs (or ultimately higher combinations: see Section 7.8.4) of complexes which respond very differently (see Section 7.5.1, Eq. 7.2) to a key factor of interest, but are similarly influenced by other competing effects, so that by comparison the contribution of the desired effect can be differentiated. Methods to quantify differential responses of pairs (and later, multiple combinations) of metal complexes will be an important tool in the optimization of bioprobe designs. So far, these methods have only been demonstrated for solvent effects (see Section 7.5.1), but current work in Norwich aims to extend this in a study of different sugar/disaccharide/oligosaccharide analytes [37].

7.5
Selectivity of Responses in IR-based Read-out Methods

Now that the capabilities of the IR-read-out from $LM(CO)_n$ responding groups have been introduced, more careful consideration can be made of the issues for bioprobe designs presented in Fig. 7.3. It has been established for some time [22, 38] that change in the solvation environment significantly influences the $\nu(CO)$ signals (Fig. 7.7a).

Changing from non-polar to polar solvents both shifts and broadens [33, 39] the signals. This may in part be due to changes in charge distribution within the structure (induced by the greater solvation of charged functionality by more polar solvents), but also has a contribution from the direct consequence of increased dipole–dipole interactions between the vibrating dipole of the read-out group, and the dipoles of the solvent molecule. Both effects are significant, and changes to vibrational spectra through the influence of different solvation environments can be quite large (of the order of several wavenumbers). Solvent-induced shifts in tetraphenylporphyrin carbonylruthenium complexes have been described and found to correlate with the carbonyl vibrational lifetime [40]. The width of the vibrational bands has also been shown to correlate with vibrational frequency [33] and a study has been made of the solvent effect on bandwidths [41]. Effects of these types can be quantified [42] and related to intermolecular interactions [43] and transition dipole moment effects [44].

7.5.1
Solvent-induced Effects

Our initial study [45] of the environment sensitivity of IR-active carbonylmetal complexes used Bellamy plots [46] of $v_{sym}(CO)$ against $v_{asym}(CO)$ vibrational modes [22] to demonstrate that solvents of different polarities produce spectra in which the positions of the centers of the vibrational bands are shifted substantially relative to one another, but correlate to a linear plot in which the responses to each solvent are spread out along the diagonal between hexane and DMA [data for tricarbonyl(1-methoxycyclohexadiene)iron (**6**): Fig. 7.7b]. The degree of differentiation of environments demonstrated by each complex was judged in our original study [45] by calculation of the diagonal separation (δ) of the extreme points on the plot, corresponding to polar solvents such as dimethylformamide (DMF) and dimethylacetamide (DMA) and non-polar solvents such as hexane and cyclohexane. The parameter (δ) was defined (Eq. 7.1) as:

$$\delta = \{[v_s(\text{hexane}) - v_s(\text{DMA})]^2 + [v_{as}(\text{hexane}) - v_{as}(\text{DMA})]^2\}^{1/2} \qquad (7.1)$$

in cases where the vibrational signals for measurements in hexane and DMA lie at the extremes of the Bellamy plot.

The objective of our work is to apply these procedures to address biological issues, so the differentiation of environment-induced shifts in the IR spectra is of particular importance. The ends of the diagonals in Bellamy plots correspond to hydrophilic and hydrophobic environments. However, the more subtle factors that influence the relative positions of bands in different solvents will ultimately prove more important to address specific biological questions, so we have examined in more detail [47], the factors that contribute to high discriminatory effects. The simple calculation of Δ (Eq. 7.1) [45] is not on its own an adequate guide, since a large variation of the ordering of the peaks, when the spectra of one complex are compared with another, can be just as powerful as a means of obtaining discrimination in a multi-carbonylmetal procedure.

To achieve this, a reference solvent is chosen (allowing the calculation of Δ_{RS} [47]), and for each combination of complex and solvent, a normalized separation ($\Delta_{N(solvent)(complex)}$) is calculated taking into account the diagonal displacement relative to the reference solvent, and the spread of values contributing to the whole of the Bellamy plot. For two complexes ("a" and "b") and two solvents ("m" and

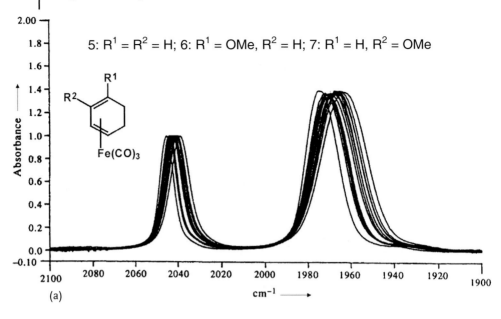

(a)

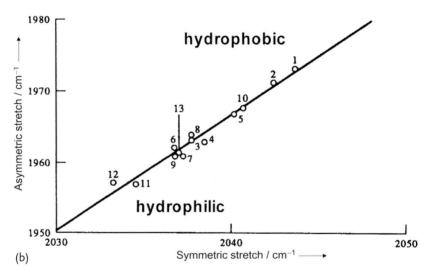

(b)

Fig. 7.7 (a) FTIR Spectra of the tricarbonyliron complex **6** in different solvents (normalized by placing the v_{sym}(CO) band maxima at the same height on the absorbance scale). (b) A Bellamy plot of v_{sym}(CO) and v_{asym}(CO) for data taken from [45]: hydrophobic solvents tend to produce data at the top right of the diagonal (high wavenumbers), and hydrophilic solvents tend to produce data at the bottom left of the diagram (low wavenumbers).

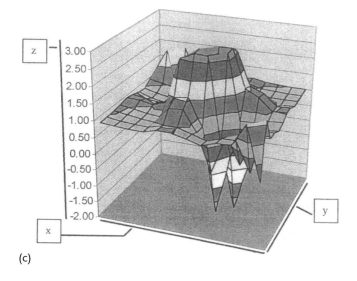

(c)

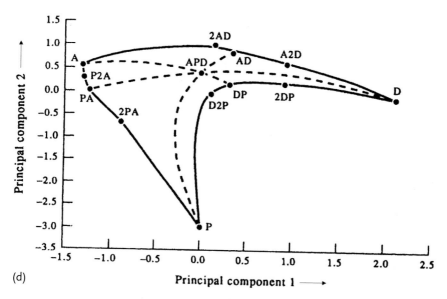

(d)

Fig. 7.7 (c) A two dimensional plot comparing responses from different solvents by complexes **6** and **7** (large positive or negative values in the z direction identify pairs of solvents (read from the x, y axes [47]) that will be well-differentiated by comparing spectra from complexes **6** and **7**. (d) An example of principal component analysis plotting "factor scores" ("principal components" 1 and 2) generated by statistical analysis of data from complex **5** in mixtures of three different solvents ("A": ethyl acetate; "D": dimethyl acetamide; "P": propan-2-ol). Data from pure solvents is placed at the corners of the plot, and data from binary mixtures lies along the edges.

"n"), the ratio of Δ_Ns has been used [47] as a measure of differential solvent discrimination (Eq. 7.2 and Fig. 7.7c; data from complexes **6** and **7**).

$$\frac{\Delta_{N \text{ (complex a)}}}{\Delta_{N \text{ (complex b)}}} = \frac{\Delta_{R \text{ (solvent n) (complex a)}} - \Delta_{R \text{ (solvent m) (complex a)}}}{\Delta_{R \text{ (solvent n) (complex b)}} - \Delta_{R \text{ (solvent m) (complex b)}}} \times \frac{\Delta_{R \text{ (DMA) (complex b)}}}{\Delta_{R \text{ (DMA) (complex a)}}} \quad (7.2)$$

Principal Component Analysis (PCA) has proved a powerful statistical method to distiguish the contributions of multiple effects on a particular complex. The data for tricarbonyl(cyclohexadiene)iron (**5**) produces the PCA plot shown in Fig. 7.7d, in which values for pure solvents lie at the corners, data for binary mixtures, define the edges, and three-solvent mixtures give points that lie inside the boundaries [45]. Although so far demonstrated for single complexes and up to three solvents, the combination of data from two differentially responding complexes (see Fig. 7.7c) in a multi-carbonylmetal approach should give exceptionally clear measures of properties of hydrophobic and hydrophilic environments when exploited in bioprobe applications.

7.5.2
Responses to pH

The principles on which carbonylmetal-based IR read-out of changes in pH are based [26] have already been introduced in Section 7.4. The hexadienoate complex **1** (Fig. 7.5) provides a typical example. Both the symmetric and antisymmetric modes respond to changes in pH, but the $v_{\text{sym}}(CO)$ band provides the simplest case, because $v_{\text{asym}}(CO)$ in this case comprises an overlapping pair of antisymmetric modes. The position of $v_{\text{sym}}(CO)$, however, can simply be measured from the peak top. In Fig. 7.8a, the data represented by filled circles (read against the scale on the left hand vertical axis) shows the position of the $v_{\text{sym}}(CO)$ band of **1/2** in a pH range between 3 and 8.6 [26].

The multi-carbonylmetal method has been used in this case by preparing the $Fe(CO)_2PPh_3$ analog of **1**. The change from CO to PPh_3 shifts the position of the vibrational bands (see Section 7.4) and also influences the pK_a of the acid (PPh_3 is a weaker π-acceptor than CO, so $Fe(CO)_2PPh_3$ complexes are less effective anion stabilizers than their $Fe(CO)_3$ counterparts). The open circles in Fig. 7.8a (read against the right-hand axis) give the data for $v_{\text{asym}}(CO)$ of the phosphine complex. The two together give both range and precision in a single experiment. Other carbonylmetal complexes behave similarly and can be compared (Fig. 7.9b) by plotting normalized values calculated by correcting for the overall change in band positions at high and low pH. The spectra in Fig. 7.9a provide an interesting comparison to the spectra shown earlier in Fig. 7.7a. Whereas different solvents/ solvent mixtures (Fig. 7.7a) produce progressive shifts of the whole band envelope, the pH response from complexes containing an ionizable group show by the clear isobestic points the interconversion of two discrete species (the associated and dissociated forms). The intensities of the IR bands can be used to assess the position of equilibrium, and give straight line log plots against pH (Fig. 7.9b) in the regions where association/dissociation is incomplete.

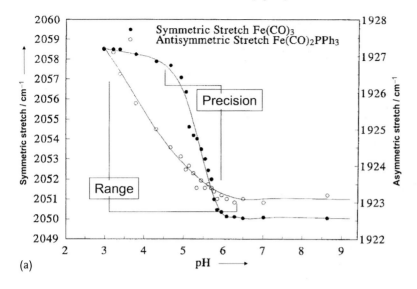

(a)

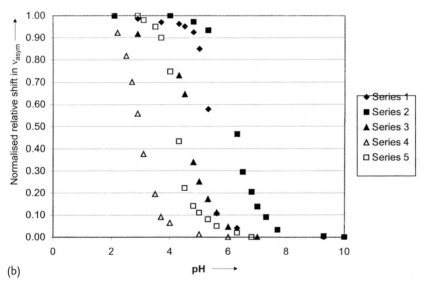

(b)

Fig. 7.8 (a) Plot of IR spectra from tricarbonyl(η^4-2,4-hexadienoic acid)iron **1**
(higher wavenumber scale) and dicarbonyl((η^4-2,4-hexadienoic
acid)(triphenylphosphine)iron (lower wavenumber scale) at pH 3–9.
(b) pH titration curves for (methyl 4-hydroxybenzoate)Cr(CO)$_3$ (series 1);
(4-hydroxytoluene)Cr(CO)$_3$ (series 2); (o-phthallic acid)Cr(CO)$_3$ (series 3);
complex **1** (series 4); (benzoic acid)Cr(CO)$_3$ (series 5) (data taken from [26]).
Relative shifts are calculated for each spectrum by subtracting the limiting
value at high pH from the measured values, and this data is normalized by
dividing by the separation in wavenumbers between the limiting values at
high and low pH.

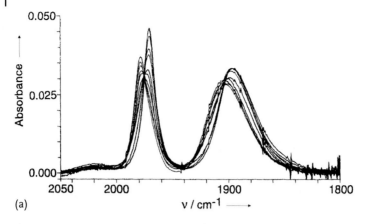

(a)

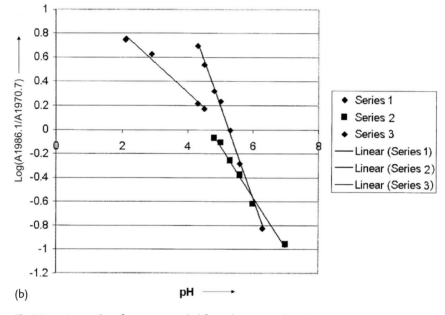

(b)

Fig. 7.9 (a) An overlay of spectra recorded from (benzoic acid)Cr(CO)$_3$ in buffers ranging from pH 2.1–6.0 (corresponding to series 5 in Fig. 7.8b). (b) Log plots of the ratios of absorbances of the associated and dissociated forms against pH [Series 1: first dissociation of (o-phthallic acid)Cr(CO)$_3$; Series 2: second dissociation of (o-phthallic acid)Cr(CO)$_3$; Series 3: (methyl 4-hydroxybenzoate)Cr(CO)$_3$].

7.5.3
Responses to Alkali Metal Ion Concentrations

The general design discussed earlier (Figs. 7.3 and 7.5), works [28] for alkali metal ion-sensing by employing classic [48] crown ether structures to bind the ions. As with pH, interconversion of two discrete species (**3** and **4**) is observed, and plots of band position against $[M^+]$ show a rising curve to a limiting value at which the crown is saturated.

It is important to note that a good fit (e.g. **8**) between the metal and the crown does not necessarily provide a good sensor, since although it gives accuracy at low concentrations (0–3 mM: Fig. 7.10a), the crown becomes saturated too soon to give a good range of response. The smaller crown, which fits the large potassium ion less well, gives a lower binding constant and a wider range of response (Fig. 7.10b). Complex **3**, for example, has been shown to respond well to Rb^+, Cs^+ and Ba^+ as well as Li^+, Na^+, and K^+ [49].

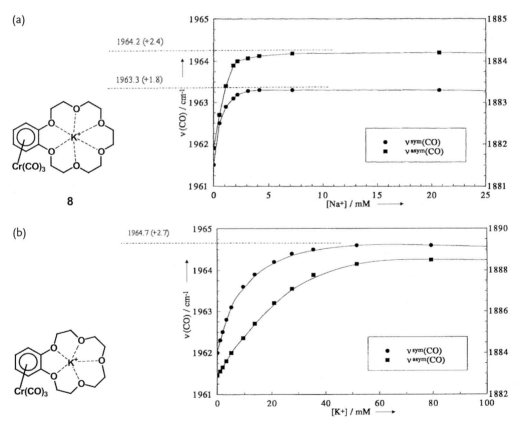

Fig. 7.10 Plots of positions of symmetric (high wavenumber) and antisymmetric (low wavenumber) vibrational modes of tricarbonylchromium complexes of benzo-18-crown-6 (a) and benzo-15-crown-5 (b) at different concentrations of alkali metal ions.

The use of non-selective receptors discussed above for the Li^+, Na^+, K^+, Rb^+, Cs^+ and Ba^+ case simplifies the sensor design, eases the synthetic challenge of the construction, and provides a sensor that is capable of general applicability, but might be regarded as a problem in biological conditions for the measurement of Li^+, Na^+, and K^+ because of interference when these ions are present together in solution. In fact, however, because of the advantages of carbonylmetal/IR-based methods discussed in Section 7.4, this is not the case, and a non-selective receptor can provide information that distinguishes the responses to the different analytes. Once again, it is signal transduction, not receptor selectivity, that is the key to optimization of response in carbonylmetal-based methods. In fact, the measurement of alkali metal ions as mixtures provides the earliest, and still one of the best, examples [28] of PCA methods to discriminate differential responses of IR band profiles. The procedure is analogous to the method described for solvent polarity in Section 7.5.1, but in this case relies on differences in the nature of the interaction of the metal in the crown ether. As described above, Na^+ seems to be particularly effectively bound in the crown-5 structure, and Li^+ (which fits inside the crown) and K^+ (which is displaced further out from the center of the crown) bind differently to Na^+, and to each other. Figure 7.11 shows the PCA factor score plot that can be produced by the analysis of mixtures of Li^+, Na^+, and K^+ ions [28].

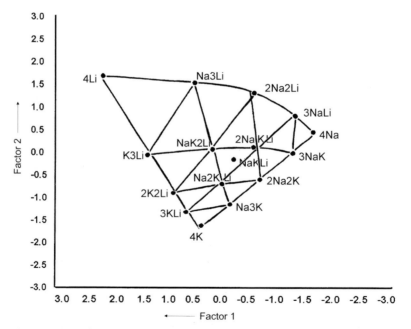

Fig. 7.11 Principal component analysis plot of factor scores calculated from vibrational bands of complex **3** with different ratios of alkali metal ions. Data from Li^+, Na^+ and K^+ define the corners of the plot, and binary mixtures define the edges (the relative amounts of the elements are indicated beside each point).

At the present stage of development, the method is not truly quantitative, since ratios of two ions in the presence of large concentrations of the third, would be hard to distinguish with accuracy, but by combining a multi-carbonylmetal design with the PCA method, and with an additional factor score to generate a 3D plot (with pure solvent as the forth corner), a considerably more accurate version of the method should be possible. Even the present procedure should be suitable for use in bioprobe applications, measuring $Na^+:K^+$ ratios for example, or $[Li^+]$ in biological media in the presence of other ions.

7.5.4
Responses to π-Stacking Interactions between Organic Structures

Because of concerns about carcinogenicity, benzene was not used as a solvent in the original systematic studies of solvent responses of tricarbonyliron and tricarbonylchromium complexes (see Section 7.5.1), and in the later work on differential solvent-induced shifts where toluene was used, although the data was found to conform to the diagonal of Bellamy plots, measurements with toluene were only performed with tricarbonyliron complexes. As attention turned to bioprobe applications of carbonylmetal-based systems, a search for probes for π-stacking effects was begun [50], because of the importance of π–π interactions with aromatic side-chains in proteins and between adjacent bases in nucleic acids. Arene complexes should be suitable for this purpose, because the aromatic ring remains flat after η^6-complexation, and the empty antibonding orbitals of the M–C bonding interactions lie in the region above the complexed arene, on the side opposite to the metal. These should be correctly positioned to form charge transfer interactions with electron rich aromatic rings. Indeed, it was already known that the vibrational bands of tricarbonylchromium complexes are shifted to lower wavenumbers when measured in benzene (compared with cyclohexane), a shift that has been ascribed, at least in part, to the consequences of π-stacking [51]. To study such effects it seems sensible to add electron-donating substituents to the uncomplexed arene, as the increase in electron density should help to promote charge transfer complexation between the arene and the complexed η^6 ligand. Indeed, the opposite effect had already been reported in the literature, as 1,3,5-trinitrobenzene has been shown to interact with benzene(tricarbonyl)chromium (**9**) via the carbonyl ligands [52].

In our work, computer-based methods were used to identify the individual contributions to overlapping vibrational bands for spectra (Fig. 7.12a) recorded in mixtures of dimethoxybenzene in cyclohexane. Fig. 7.12b shows that while the position of the higher wavenumber $\nu_{sym}(CO)$ and $\nu_{asym}(CO)$ bands remains essentially constant, the new pair of $\nu_{sym}(CO)$ and $\nu_{asym}(CO)$ bands shift further to lower wavenumbers as the contribution of the π-stacking increases [50]. This is consistent with either additional π-stacking to form taller stacks [53], or shifts due to random solvent effects in the mixed cyclohexane/arene solvent environment. The new pair of $\nu_{sym}(CO)$ and $\nu_{asym}(CO)$ bands also broaden more as they shift more, an effect that can be caused by progressively adding a more polar solvent to an apolar solvent such as cyclohexane.

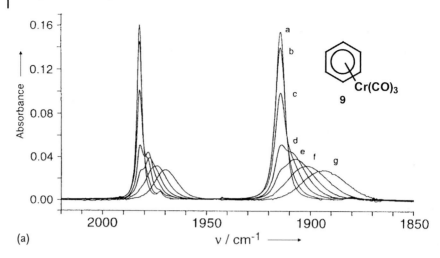

(a)

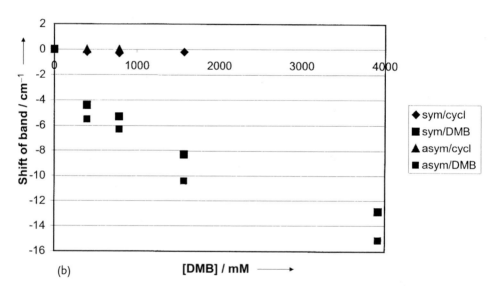

(b)

Fig. 7.12 (a) An overlay of spectra of (benzene)Cr(CO)$_3$ in cyclohexane (trace a) and in the presence of 1,2-dimethoxybenzene (trace b: 39 mM; trace c: 79 mM; trace d: 392 mM; trace e: 785 mM; trace f: 1.57 M; trace g: 3.92 M). (b) Based on data from [50] in which computer analysis of band envelopes (shown in a) identifies contributions from cyclohexane and 1,2-dimethoxybenzene

(DMB) environments. Shifts in band positions are plotted against concentration of 1,2-dimethoxybenzene "sym/cycl": ν_{sym}(CO), cyclohexane environment; "sym/DMB": ν_{sym}(CO), 1,2-dimethoxy-benzene environment; "asym/cycl": ν_{asym}(CO), cyclohexane environment; "asym/DMB": ν_{asym}(CO), 1,2-dimethoxy-benzene environment.

The information available so far clearly demonstrates that arenes complexed by $Cr(CO)_3$ are suitable candidates for probes for π-stacking effects. Solvated and π-stacked structures show different IR band positions, and progressive inter-conversion of the two forms occurs as the tendency to π-stack is increased. This behaviour is more similar to the effects of interconversion of associated and dissociated forms in pH measurements (Section 7.5.2 and Fig. 7.9a) or empty and occupied crowns in metal ion measurement (Section 7.5.3), than to the shifting of bands that occurs when different solvents are mixed (Section 7.5.1 and Fig. 7.7a). This in itself is evidence for a more specific type of interaction in cases where arenes can approach the uncomplexed face of the (arene)$Cr(CO)_3$ complex. Furthermore, direct evidence for face-to-face interactions between free and complexed arenes is available from studies of cyclophane complexes, as in the solid state, the separation between the arene rings of cyclophanes has been shown by X-ray crystallography to decrease when one of the rings is complexed to $Cr(CO)_3$ [54].

7.6
Examples of Organometalcarbonyl Bioprobe Structures

Two distinct approaches have been used for the development of organometallic bioprobe structures. The organometallic component can be attached to a small molecule that interacts with a biological receptor system (Fig. 7.13), or directly to a biological macromolecular structure.

The original studies on oestrogens by Jaouen and Vessières [55] is typical of the first approach, in which, for example, steroidal compounds were derivatized at different positions in search of examples that retained biological activity. This work provided the basis for the later development of the CMIA method (metallo-immunoassay: [56]; carbonylmetalloimmunoassay: [57, 58]; see Section 8.5.2). Typical examples are **10** [59, 60], **11** [61] and **12** [62]. The anticipated mode of action is of **10** and **11** is different, since **10** should bind competitively to the receptor, but **11** has the capability of covalent attachment by generation [63] of a metal stabilized electrophile by displacement of OH from the propargylic position. The larger cluster **12** provides additional CO ligands in order to enhance detectability at very low concentrations. Detection of femtomole quantities of carbonylmetal complexes has been achieved [64] using specialized FTIR instruments. The cyclo-pentadienylrhenium complex **13** has been shown by modelling [65] to fit well in the hormone binding domain [66] of the oestradiol receptor. Derivatives of zeranol [67] and the natural mycotoxin zearalenone, such as **14** and **15** [68] have been prepared to seek a system for the biological recognition of these structures. A carbonylmanganese derivative **16** [69] of phenobarbital has also been examined as a tracer compound [70], and has been shown to have a detection limit (64 scans at 4 cm^{-1} resolution) of 10 pmol in a microcavity cell [71]. Derivatives **17** and **18** of carbamazepine and phenytoin are also detectable [72] in the 10–100 pmol range. Compounds such as **16, 17,** and **18** have non-overlapping IR bands, and

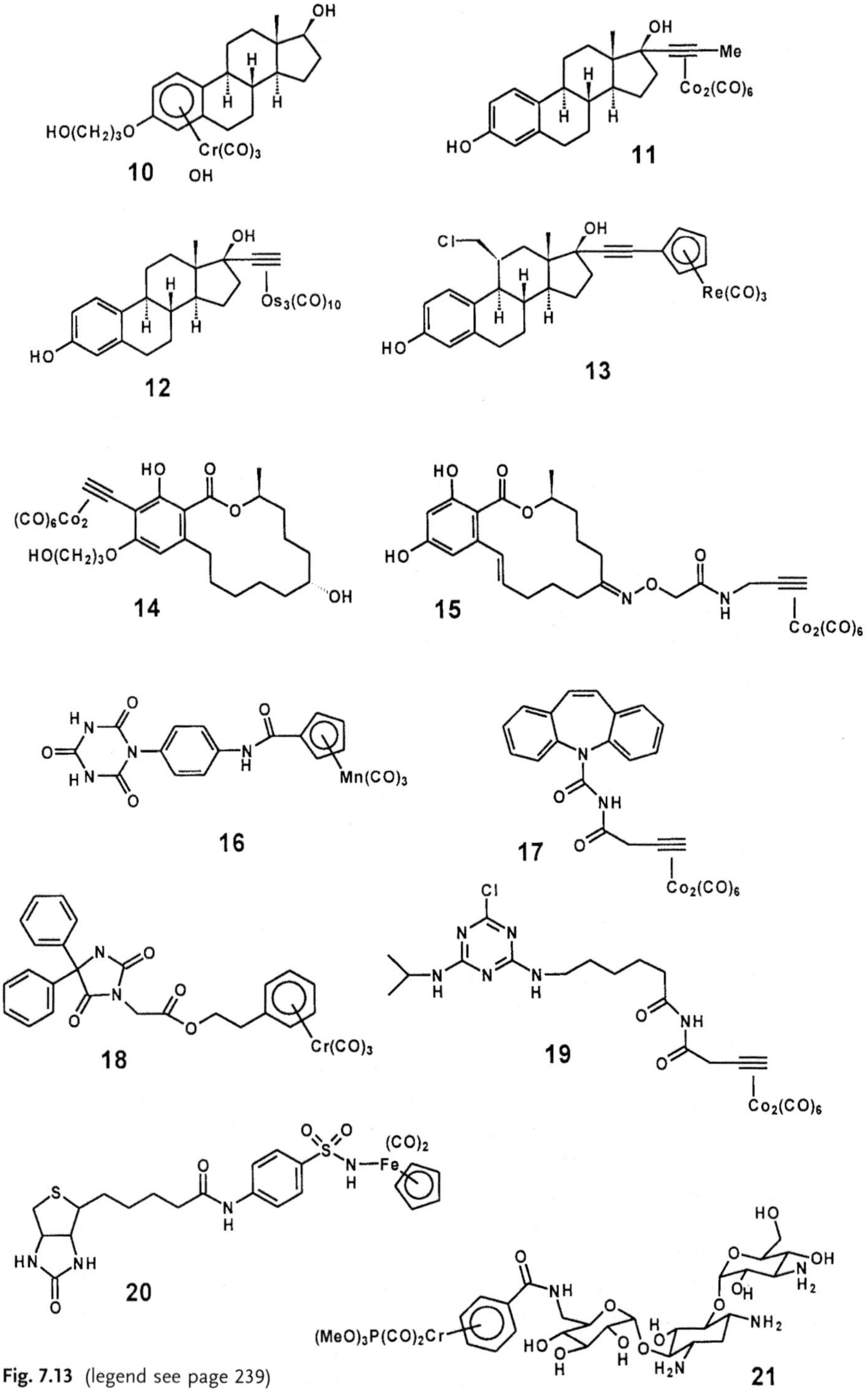

Fig. 7.13 (legend see page 239)

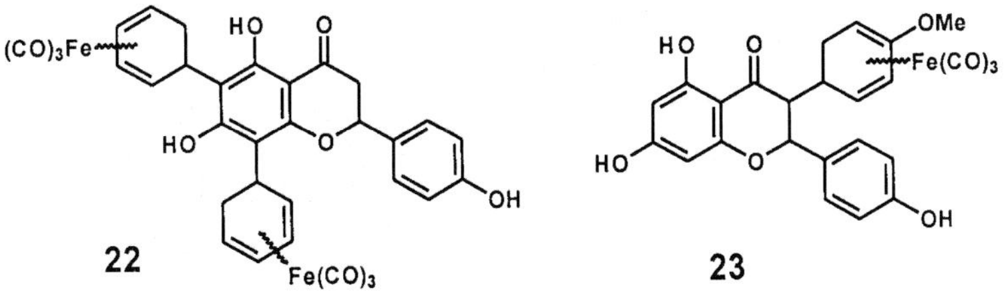

Fig. 7.13 Examples of organometallic derivatives of biologically active molecules prepared for use in competition assay and bioprobe applications.

so can be measured concurrently in a single spectroscopic experiment. The carbonylcobalt complex **19** is an organometallic tracer derived from the herbicide atrazine [73]. Biotin has been derivatized [74] as the iron complex **20** for use in binding studies with avidin. Tricarbonylchromium and dicarbonylchromium phosphite (e.g. **21**) derivatives of kanamycin A have been prepared from the aminosugar antibiotic using the activated ester method developed (see next paragraph) for proteins. Derivatives of tobramycin have been similarly prepared [75]. Compounds of these types have been synthesised for use in competition assays, where the IR absorbance is used in the same way as a radioisotope emission in conventional immunoassays. The same type of approach in which a biologically active substance is converted into an active organometallic derivative has been used to prepare the flavonoid-based bioprobes discussed in Section 7.8. The intended principle of application of the complexes, however, is different, as the bioprobe approach seeks changes in spectroscopic response upon binding to the target, whereas the competition assay work aims at establishing binding affinities by quantifying attached and removable material. In both cases, however, highly efficient spectrometers and organometallic compounds that still exhibit biological activity are required, so a good screen for biological activity is also needed. The flavonoids shown in Fig. 7.13 are examples of an extensive set of compounds [76, 77] that were prepared to study the possibility (see Section 7.8) of binding to a protein thought to be a regulatory protein for the induction of genes associated with root nodulation that leads to nitrogen fixation in the roots of legumes. Structure **22** [76] was inactive [78], but **23** [77] was still able [78] to bring about the induction of gene expression.

The alternative to the small molecule approach is to attach the carbonylmetal component to a biological macromolecule, and this has been examined by several research groups. As early as 1988, Le Borgne and Beaucourt, and Sasaki, Potier, Savignac and Jaouen had attached [79, 80] carbonylcobalt complexes to short peptides, employing an alkyne-containing amino acid. Cyclopentadienylmanganese carbonyl complexes have also been used [80]. There is a good selection of carbonylmetal-containing amino acid derivatives available (e.g. **24–28**, Fig. 7.14), including a $Cr(CO)_3$ η^6-complex of phenylalanine (**26**) [81], and related η^5 $Mn(CO)_3$ and η^4 $Fe(CO)_3$ structures (**27** and **28**) [82].

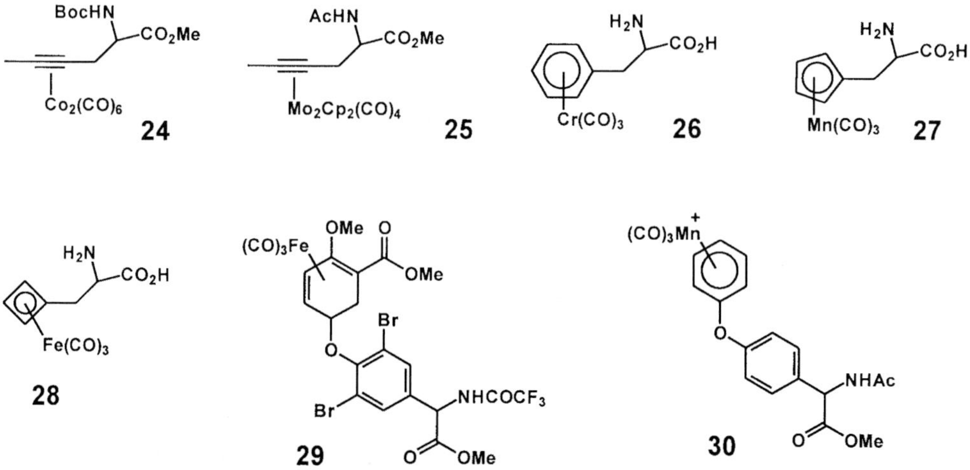

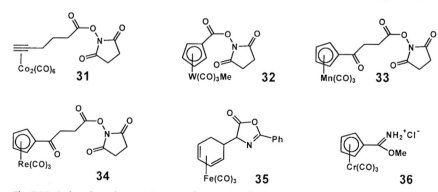

Fig. 7.14 Examples of carbonylmetal derivatives of amino acids.

More specialized examples are the Fe(CO)$_3$ derivative **29** of the thyroxine analog SK&F L-94901 [83] and the cationic tyrosine derivative **30** [84]. Later, the Jaouen group generalized this chemistry by adapting (Fig. 7.15) a succinimidyl-based delivery system [85] to attach carbonylmetal compounds to intact proteins. First cobalt [86] and later manganese, rhenium [87] and tungsten [88] complexes (**31–34**) were examined. Similarly, oxazolinone [89] and imidoester [90] approaches have been used to deliver carbonyliron and carbonylchromium complexes, respectively from **35** and **36**. Fischer carbene complexes have also been employed [91] in the derivatization of proteins. Synthesis of amino acid structures exploiting electrophilic organometallic chemistry [92] provides alternative access to carbonyl-metal derivatives that have potential themselves for use in biological studies. The electrophilicity of organometallic structures has been used itself as a means to promote direct attachment of carbonylmetal complexes by reaction with the nucleophilic surface groups of peptides and proteins (see Section 7.7) [93–97].

Fig. 7.15 Carbonylmetal-containing transfer reagents for use in the derivatization of proteins with organometalcarbonyl groups.

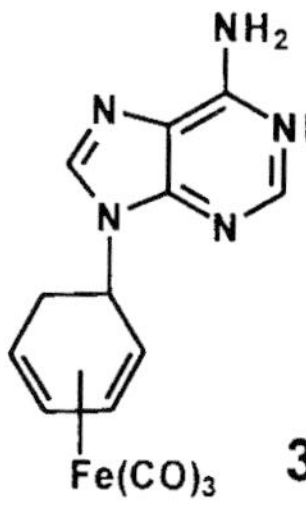

Fig. 7.16 Examples of carbonylmetal derivatives of nucleotides and nucleotide bases.

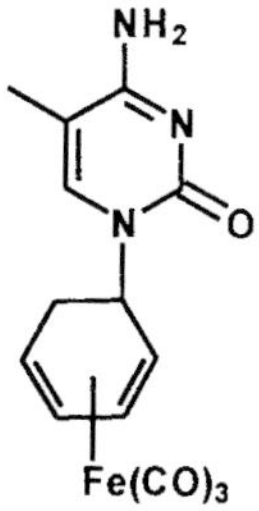

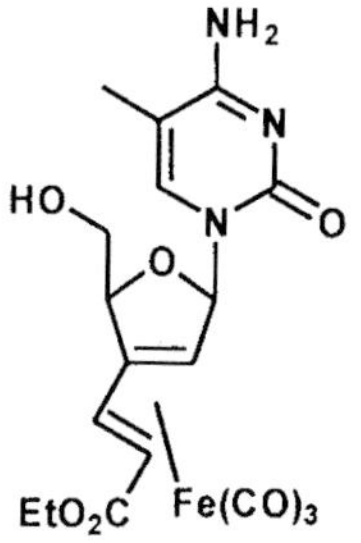

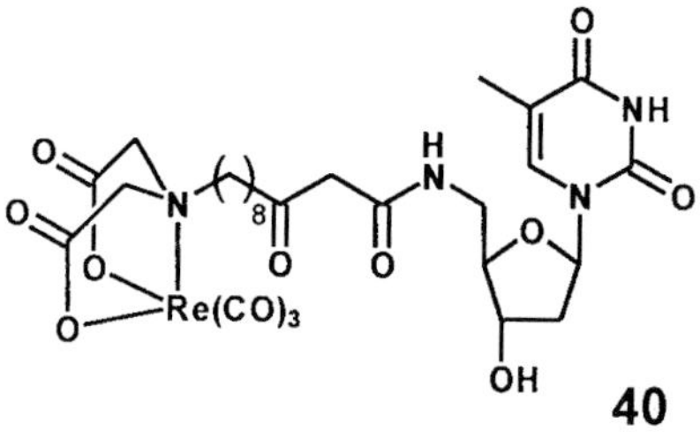

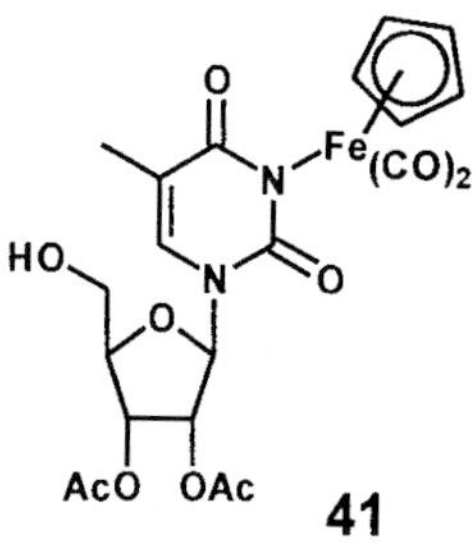

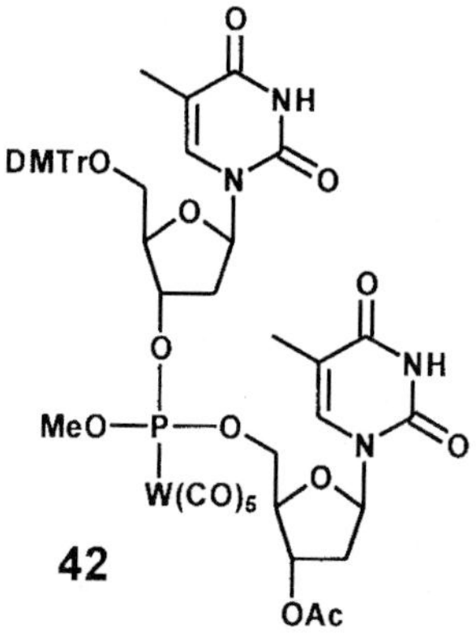

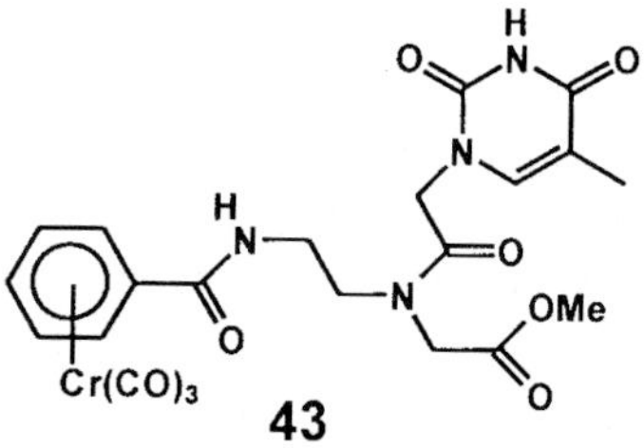

Nucleotide bases (Fig. 7.16) have been derivatized by use of electrophilic organoiron complexes. The carbonyliron adducts **37** and **38** are typical examples [98]. A recent development in this field has led to the synthesis [99] of an extensive set of structures, including **39** which more closely resembles a nucleoside. The thymidine derivative **40** is an inhibitor of human thymidine kinase [100]. In the case of **41**, the carbonylmetal component is attached at the heterocyclic base [101]. The carbonylmetal component can also be attached to dinucleotides by co-ordination at a phosphite, placed (as in **42**) where the phosphate link would normally join the ribose rings [102]. The attachment of carbonylmetal labels to oligonucleotides for analysis by FTIR has also been described. The succini-midyl-based delivery strategy is the same as that used with proteins, as a

$H_2N(CH_2)_6OP(O)_2O$ linker was added to a universal primer oligonucleotide sequence (GTTTTCCCAGTCACGAC) [103]. This was employed with *Bacillus stearotherophilus* (BST) DNA polymerase. The carbonylchromium complex **43** has been prepared in a study of carbonylmetal labels on PNA (peptide nucleic acid) monomers [104].

7.7
Use of Organometallic Bioprobes with Proteins

Work in Norwich on attaching carbonylmetal bioprobes to the surfaces of proteins [95, 105] has exploited the electrophilicity of tricarbonyl(cyclohexadienyl)iron(1+) salts in reactions (see Section 7.6) with nucleophilic groups on the protein, for example surface lysine NH_2 groups (Fig. 7.17).

In the original investigations of this approach [93], a pyridinium derivative dissolved in methanol was used to introduce the organometallic electrophile to aqueous preparations of the enzymes, but the true reactive species is likely to be the cyclohexadienyl cation **44**. Indeed, at the same time in Australia, Kane-Maguire's group [94] was using **44** in direct reactions with proteins, and this is probably the most straightforward procedure. In the Norwich study [95, 105], reactions of the electrophile with the protein α-chymotrypsin was examined first. Monitoring by IR, bands at 2051.3 and 1979.5 cm^{-1} (assigned as two overlapping $\nu_{asym}(CO)$ vibrations at 1981.5 and 1973.8 cm^{-1}) gradually decreased in intensity as higher frequency shoulders appeared at about 2058 and 1993 cm^{-1}. Over a period of time (100 minutes) the new signals at 2058 and 1993 cm^{-1} increased in intensity (Fig. 7.18). This was interpreted as evidence for addition of the tricarbonyliron complex to surface lysine groups to form ammonium ions by alkylation of the NH_2 groups. Increasing the concentration of the enzyme should increase the

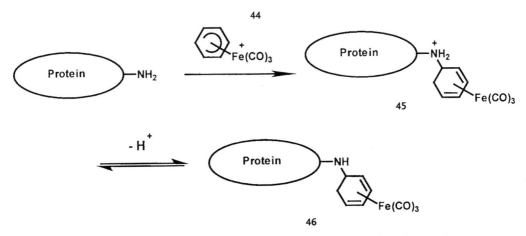

Fig. 7.17 Use of electrophilic organometallic reagents (e.g. **44**) to derivatize proteins.

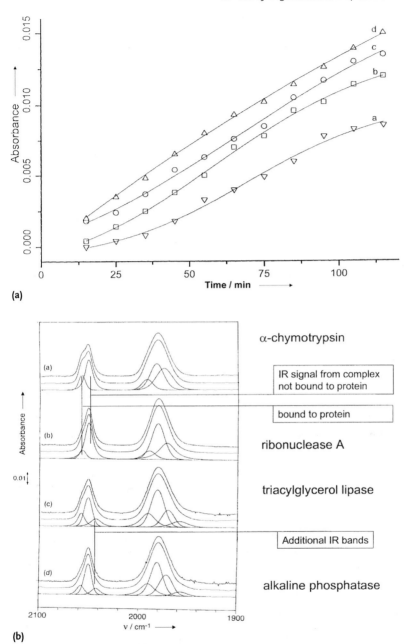

Fig. 7.18 FTIR spectral data from organometalcarbonyl derivatives of proteins. (a) Plots of intensity of ν_{sym}(CO) at 2058 cm^{-1} against reaction time at different concentrations of α-chymotrypsin (trace a: 1.96 mM; trace b: 2.52 mM; trace c: 3.32 mM; trace d: 4.82 mM). (b) IR bands recorded for α-chymotrypsin, ribonuclease A, triacylglycerol lipase, and alkaline phosphatase (30–120-mg samples dissolved in 1 ml 0.2 M phosphate buffer).

rate of appearance of the new signals, and this was found to be the case. Figure 7.18a shows the rise in intensity of the $v_{sym}(CO)$ signal at 2058 cm^{-1} at four different concentrations of α-chymotrypsin (1.96, 2.52, 3.32, and 4.82 mM). The highest line in the plot corresponds to the greatest contribution of the new band at 2058 cm^{-1} and the highest concentration of chymotrypsin. In this way, it was confirmed that binding to chymotrypsin was the cause of the new signals in the IR spectra. [105].

Similar results were obtained with ribonuclease A, alkaline phosphatase and triacylglycerol lipase (Fig. 7.18b). In the last two examples, however, a second pair of new $v_{sym}(CO)$ and $v_{asym}(CO)$ signals appeared at lower wavenumbers. This process was fast, and reached a limiting value while the high wavenumber bands continued to grow in intensity as the slow reaction continued. The explanation for this behaviour is that in triacylglycerol lipase and alkaline phosphatase, some surface lysines are in environments that cause them to be substantially in the NH$_2$ form at the start of the experiment, and so readily available as nucleophiles. The remaining lysine side-chain groups are largely in the protonated NH$_3^+$ form, and so react more slowly, as dissociation is necessary prior to the nucleophile addition step. In the fast-reaction case, the ammonium ion **45** formed upon nucleophile addition becomes immediately deprotonated to form **46**. This is consistent with the time dependent behaviour of the low wavenumber signals, which appeared rapidly at the start of the reaction, and then did not change with time. A consequence of this observation is that it should, in these cases, be possible to place two different types of organometallic bioprobe on the protein surface, by selectively alkylating the fast-reacting groups with a small quantity of the tricarbonyliron electrophile and a short reaction time, followed by addition of a second, spectroscopically complementary complex [e.g. dicarbonyl(triphenyl-phosphine)iron]. Such multi-carbonylmetal procedures will make available novel multiply labelled proteins for spectroscopic study.

The reactions discussed above also indicate the possibility of monitoring reactions on the protein surface by the IR responses of the probe groups, and this has already been demonstrated [105] in the case of α-chymotrypsin for acid/base equilibria (i.e. interconversion of **45** and **46**). On addition of sufficient 0.2 M sodium hydroxide solution to bring the pH to 10, the high wavenumber signal is substantially depleted, and the band envelope was broadened and distorted towards lower wavenumbers. This change is reversible, and on returning to pH 7.5 by addition of ethanoic acid, the original high frequency shoulder reappears (the change in intensities is a consequence of the dilution caused by this procedure).

The work performed to date with tricarbonyl(cyclohexadienyl)iron adducts of enzymes has established the principles that IR-based experiments can be used to study the reactivity and properties of enzyme surface functional groups, but there is much left to be done in this area. Electrospray MS results (see for example ref. [97]) are needed to count the number of surface groups that have been derivatized, and establish whether the gradual increase in intensity of the signal from the adduct is a consequence of random derivatization of all the available surface lysines, and reflects a progressive order of kinetic reactivity of lysines in different environments. As yet, only fast- and slow-reacting environments have been identified, and these have been shown to have different pK_a-related properties. Degradation experiments, and MS determination of which amino acid side-chains have been alkylated, also remains to be examined. However, in order to establish with reasonable certainty that surface lysines are indeed the target of the derivatization reactions, a number of IR-based experiments were performed. A selection of individual amino acids were derivatized by the same procedure used in the enzyme studies, and the IR band positions established [105]. All these substrates have available the amino groups of the amino acid itself, and in almost all cases (cysteine and histidine were the exceptions) IR bands around 2058–9 and 1991–2 cm^{-1} [$\nu_{sym}(CO)/\nu_{asym}(CO)$, respectively] were observed. In addition, aspartic acid, glutamic acid and serine showed signals at 2043–5 and 1956–60 cm^{-1}, assigned to C–O bond formation at the acid and OH groups. In the case of histidine, signals at 2057.1 and 1990.6 cm^{-1} were observed. These are close to the values expected for reactions of NH_2 groups, so the experiment was repeated using N-acetyl histidine in which the NH_2 group was acylated, and so much less nucleophilic. Similar bands at 2057.9 and 1990.0 were again observed, and these have been assigned to adducts of the imidazole in the histidine side chain. These experiments have made available a set of diagnostic IR signatures (Fig. 7.19) for the amino acid side-chain groups in the tricarbonyliron series of complexes.

The conclusions were confirmed for CO_2H, NH_2 and guanidino groups by examination of the derivatization of polymers of glutamic acid, lysine, and arginine, where the effect of the N-terminal NH_2 group would be insignificant compared with the repeated side-chain functionality. Polylysine gave the typical signals at 2058.2 and 1994.2 cm^{-1}, again similar to the values assigned to addition to surface lysine $RR'NH_2^+$ groups on the protein surfaces. The characteristic $\nu_{sym}(CO)$ vibrational frequences for acid and guanidino groups are also shown in Fig. 7.19. Data for cysteine SH groups are not presented in Fig. 7.19, because the assignment is not fully validated, but reaction of cysteine gives only new IR signals at 2043.4/ 1955.9 cm^{-1} [$\nu_{sym}(CO)/\nu_{asym}(CO)$, respectively]. This is clearly not a result of C–N bond formation at the amino group of the amino acid unit, and in view of the low reactivity of carboxylic acid groups (see below) it also seems unlikely that the carboxylic acid should be alkylated faster than the amine in this case. On this basis, the signals at 2043.4 and 1955.9 cm^{-1} should correspond to the RSR' adduct from the SH side-chain group of cysteine. Comparing serine (OH) and cysteine (SH), the NH_2 group dominated the competition for the nucleophile in the serine case, and the SH group dominated the reaction with cysteine. In the tricarbonyliron series, secondary amines, ethers, thioethers, and esters (RR'NH, ROR', RSR' and RCO_2R') are too similar to distinguish from vibrational band positions alone, and similarly imidazoyl and guanidinyl adducts, show similar properties that are also close to the values for $RR'NH_2^+$. Further experiments with other complementary organometallic electrophiles will be needed to complete the library of diagnostic tools. However, the rate of reaction with the electrophile also gives information, and comparing RR'NH and RCO_2R', for example, reaction with amine NH groups appears fast, compared with the corresponding reaction with carboxylic acid COOH

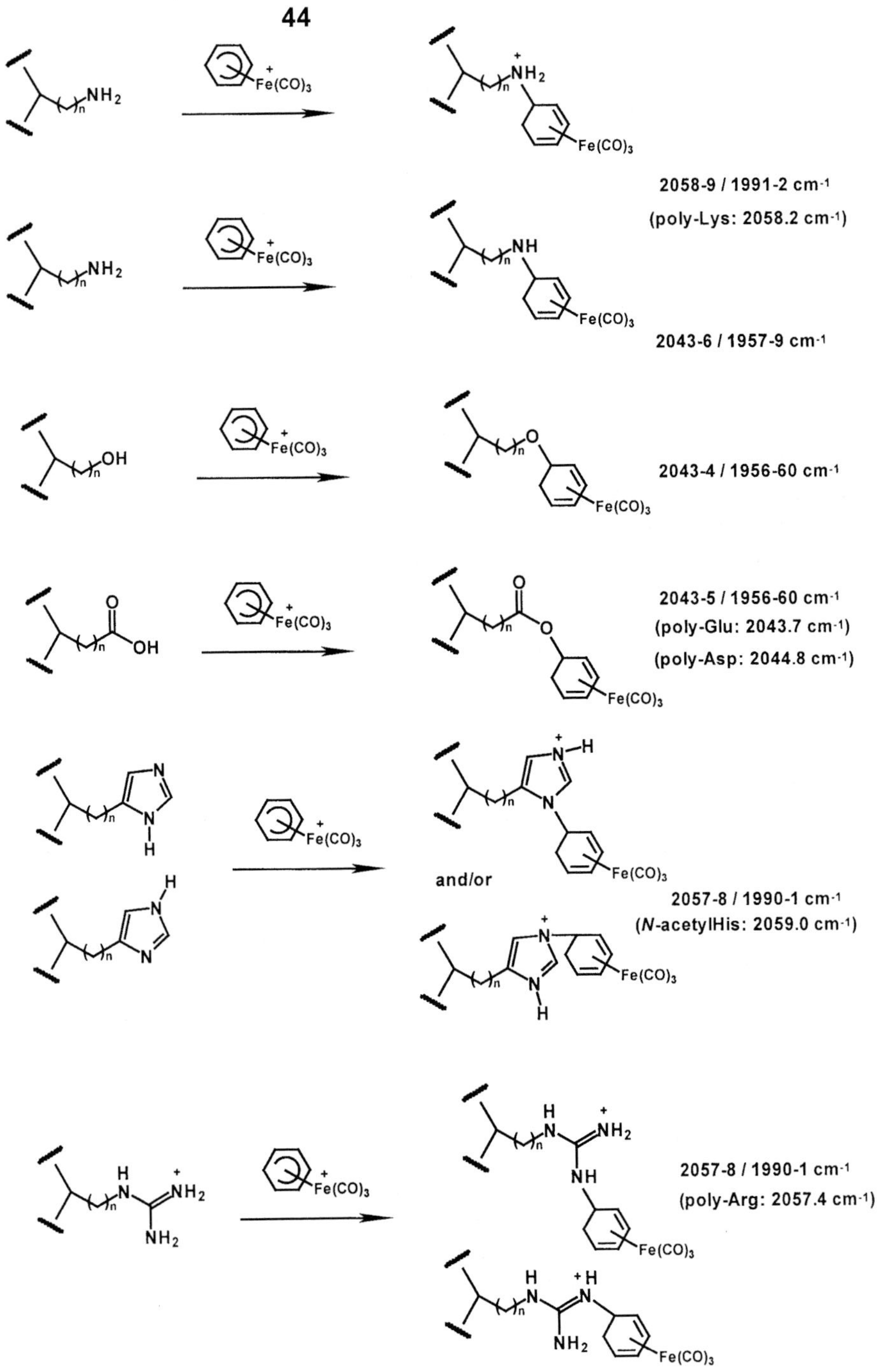

44

2058-9 / 1991-2 cm⁻¹

(poly-Lys: 2058.2 cm⁻¹)

2043-6 / 1957-9 cm⁻¹

2043-4 / 1956-60 cm⁻¹

2043-5 / 1956-60 cm⁻¹

(poly-Glu: 2043.7 cm⁻¹)

(poly-Asp: 2044.8 cm⁻¹)

2057-8 / 1990-1 cm⁻¹

(*N*-acetylHis: 2059.0 cm⁻¹)

2057-8 / 1990-1 cm⁻¹

(poly-Arg: 2057.4 cm⁻¹)

Fig. 7.19 Representative $\nu_{sym}(CO)/\nu_{asym}(CO)$ bands for amino acids derivatized by reaction with **44**.

groups. This was confirmed by examining the reaction of a co-polymer of alanine, lysine, glutamic acid and tyrosine, which gave bands typical for NH_2^+ adducts [2058.2/1991.6 cm^{-1}], showing kinetic selectivity for reaction at the lysines.

The pH titration experiments with the surface lysine groups described above were also validated by model studies. For this purpose, isopropylamine was added to the cyclohexadienyliron electrophile to afford the corresponding ammonium salt. Again, characteristic high wavenumber bands for the ammonium ion were observed, and in this simple case, the reaction could be taken to completion. This characteristic $\nu_{sym}(CO)/\nu_{asym}(CO)$ signature for this isopropylammonium ion was 2058.7/1988.6 cm^{-1}. The pH of the solution was gradually raised to 10.5, to produce a new spectrum in which the 2058.7/1988.6 cm^{-1} bands were largely replaced by new signals at 2046.2 and 1974.8 cm^{-1}. In this case, good isobestic points show the clean interconversion of the charged and neutral species, further evidence for the assignment of the structures with IR vibrational frequencies about 2046 and 1975 cm^{-1} as the neutral NH form of the lysine adduct, both in the model study, and the experiments with α-chymotypsin.

Although there is much left to do in the study and characterization of carbonyl-metal adducts of amino acids and polyamino acids, and enzyme surface groups, the general principles have been well established. It is possible to deliver electrophilic cyclohexadienyliron complexes to protein surfaces, monitor their attachment to the surface side-chain functional groups, and gain information about the types of groups participating in the reaction, by consideration of the vibrational signatures of the adducts and the rate of adduct formation. The practicality of detailed IR studies of the protein derivatives in realistic aqueous conditions such as phosphate buffers has also been established [94–96, 105]. Furthermore, detection of the bound probe in the presence of the unbound probe, has been demonstrated [105] in this work (a capability that will be important in applications of the type discussed in Section 7.8.2). Reactions of functional groups on the protein surface can be followed by IR, and the data available at this stage is suitable to study changes in local pH on the protein surface, or the acidity/basicity of the biological environment in which the protein is located. It is also possible to detect differences in active sites. Differences in ν(CO) for carbonylmetal complexes of different mutant heme proteins have been determined [19] working in 50 : 50 glycerol : phosphate buffer mixtures.

7.8
Power of Genetics in the Design of Bioprobe Experiments

The organometallic flavonoid derivatives **22** and **23** discussed in Section 7.6 (Fig. 7.13) form part of a set of compounds which were designed for a study of biological switching events in root nodulation in legumes [106]. This provides a good example to illustrate how genetics can be important in the design of experiments employing the IR-responsive carbonylmetal bioprobes that are the subject of this chapter. The concepts of "functional and dysfunctional probes", "remote and local response capabilities", and the use of genetics to provide "functional and dysfunctional receptors", can be defined at this stage, even though the final IR experiments envisaged for the nodulation gene study still await more strongly binding probe structures. The definition and discussion of these concepts will complete the chapter, and set out a general approach that should be applicable to a wide selection of biological problems.

7.8.1
Remote and Local Response Capabilities

Sections 7.5.1 to 7.5.4 have provided examples of two distinct types of interactions that influence the vibrational responses of the carbonylmetal groups. The first is solute/solvent or environment-induced effects, which produce a progressive shift of the band positions. Alternatively, specific interactions can be relayed to the carbonylmetal section of the molecule when electronic communication is present. Comparing these effects can provide valuable information to establish the biological interpretation of the spectroscopic results.

Figure 7.20 illustrates the basis of this type of analysis. Examples a and b show illustrations of specific and random effects when the organometallic structure enters a binding site.

For Fig. 7.20a, interactions between OH groups on, for example, an aromatic ring, and features in the receptor, will influence the charge distribution in the molecule, and so cause a change in the vibrational properties of the M–CO groups. The same molecule can sit in many ways near the active site (Fig. 7.20b), so there will be random interactions as well. In cases such as these, the bioprobe is functional (see Section 7.8.2) and interactions of bound probe with the receptor of the types shown in Fig. 7.20a, b and solvent interactions of the free probe and the medium must all be taken into account. Because the organometallic structure is in electronic communication with the binding part of the molecule, the hydrogen-bonding interactions in the active site should have a significant effect by disturbing the charge distribution in the complex. This effect is a consequence of interactions taking place some distance from the carbonylmetal complex, i.e. "remote sensitivity". The same type of interaction is shown in Fig. 20c, except that the conjugation between the binding section and the carbonylmetal complex is interrupted in this case. Now the changes in hydrogen bonding will have far less effect on the vibrational bands of the complex. Only local effects ("local sensitivity") will influence the IR spectra. These will be random interactions with the receptor (Fig. 7.20d) and random interactions of the free probe with the medium. Probes with "remote sensitivity" respond to both remote and local effects, while those described as having "local sensitivity" are responding only to local effects. Comparing the two systems is a powerful approach to identifying and quantifying the remote effects.

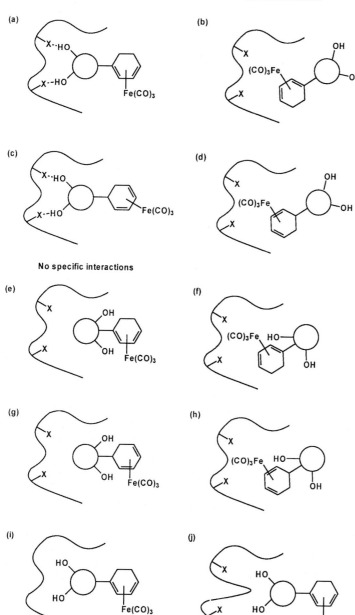

Fig. 7.20 Cartoons to illustrate possibilities for specific and random effects on ν(CO) vibrational modes of carbonylmetal derivatives introduced into biological receptor structures, and organometallic structures that show remote and/or local sensitivity to such environmental effects.

7.8.2
Functional and Dysfunctional Probes

In Fig. 7.20 e and f the bioprobe is no longer suitable to bind to the receptor (for example, the point of attachment of the carbonylmetal complex may sterically block the binding event). These probes are "dysfunctional", but will share some of the random interactions that are exhibited by the "functional" probes shown in Fig. 7.20 a–d. Comparing the spectroscopic responses of a "dysfunctional" probe and a "locally sensitive" probe will give additional information to help identify the true remote response from Fig. 7.20a. To be more precise, the probe shown in Fig. 7.20e is "dysfunctional" *and* "remotely sensitive", so pH changes that might influence the association/dissociation chemistry of the OH groups will still influence the spectroscopy (situations f and b are similar), and if the OH groups are phenolic such effects could be substantial. In d, however, which is only "locally sensitive", changes in pH will be much less significant. Comparing b, d, e, and f and locally sensitive analogs of the dysfunctional structures (g and h) will provide information to distinguish pH changes between the active site and the bulk medium and specific hydrogen bonding interactions.

7.8.3
Functional and Dysfunctional Receptors

The final pair of examples (i and j) shown in Fig. 7.20 correspond to situations where the receptor itself is unable to bind the bioprobe molecule. It could be (case i) that key functionally is deleted (for example, by a mutation in the gene that codes for the enzyme) or blocked (for example, by covalent reaction with an irreversible inhibitor). Alternatively (case j), the shape of the active site might be altered by mutations so that the approach of the bioprobe to the binding site is blocked. In situations i and j, even "functional remotely sensitive" bioprobes will be unable to show the response assigned to the remote interactions. This is the true test for the validity of a method based on these design principles. The specific difference between remote and local sensitivity with functional receptors should be absent when those same structures are used with a dysfunctional receptor.

7.8.4
Functional and Dysfunctional Probes and Receptors to Study the Induction of nod Gene Expression

The principles examined in Sections 7.8.1 to 7.8.3 could apply for enzyme active sites and other biological receptor structures. In both cases, the availability of good test for correct binding in the receptor is an important criterion in selecting a biological problem for study by the IR-responsive bioprobe approach. In the case of enzymes, such assays are usually straight forward, since an ability to measure conversion of the substrate into the product will allow both k_{cat} and k_i to be determined, and so provide the means to identify good examples of functional and dysfunctional probes and receptors. In the case of other biological receptor structures, either a direct measure of the biological consequence of the binding event, or an indirect assay, is needed, and more care with the development/choice of the biological system is required.

A project in Norwich to address the molecular events that lead to gene induction and the onset of root nodulation in legumes, and so to the accumulation of nitrogen-fixing bacteria in the nodules on the legume roots, was chosen to elucidate the methods discussed in this chapter. Collaboration with geneticists was crucial for this purpose, and in the case of these studies was possible because of the close links between the University of East Anglia and the John Innes Centre on the Norwich Research Park. This collaboration provided both the assay for gene induction [107], and scenarios for functional and dysfunctional biological samples. A description of the design of these experiments will conclude this chapter. It is our expectation that only when all these factors are successfully in place can one hope to decode the information from the IR-responsive bioprobe approach, and reach clear conclusions about the molecular recognition events that form the basis of the biological processes. When all the necessary tools are in place, however, because of the precise nature of the comparisons that can be made, the power of PCA-based methods to distinguish similar spectroscopic effects (see Sections 7.5.1 and 7.5.3), and the prospect of multi-carbonylmetal methods with spectroscopically complementary probe groups (see Section 7.5.2), the proposal is that despite the complexity of the system, it will be possible to isolate a specific spectroscopic response, and prove its relationship to the biological problem being addressed.

Figure 7.21 summarizes some of the key points about the early stages of the chemical biology of root nodulation. The roots of legumes exude flavonoid signal molecules [108] and *Rhizobium* soil bacteria respond to them [107, 109–111] to approach and ultimately enter the roots to form nodules [106]. A section of the *Rhizobium* genome is specifically involved in this process, and one of the genes (*nod*D) is constitutive [109] (i.e. it is continuously transcribed, and the protein it codes for is always available). When flavonoids are present, all the genes are transcribed [112–114], and this induction of *nod* gene expression requires the *nod*D gene to be present [*nod*D(–) strains do not respond to flavonoids]. Thus it was proposed that the protein coded by *nod*D was the regulatory protein that binds flavonoids and then switches on the other *nod* genes. It was also known that OH groups on both aromatic rings were necessary [115] for flavonoids to have biological activity [highly oxygenated flavonoids such as naringinin (**47**) and eriodictiol (**48**) are typical examples], so oxygenated flavonoid derivatives with carbonylmetal groups attached were prepared and tested.

In fact, in studies of this type, "dysfunctional probes" are easy to find [76, 78], as the most likely outcome of adding a cyclohexadienyliron complex will be to spoil the specific interactions between flavonoid and the putative receptor (compound **22** in Fig. 7.13 is an example). However, it is worth emphasising that even such negative results are important, and advance the project, since for the final spectroscopic comparisons it will be necessary to choose good examples of

Legume

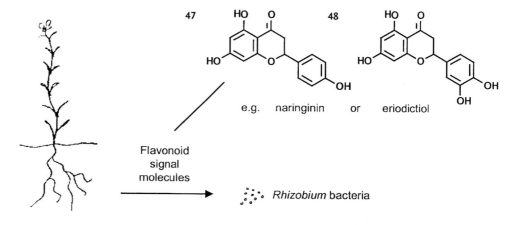

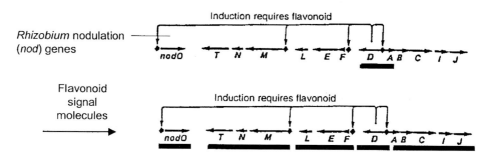

Fig. 7.21 Illustration of the role of flavonoid signal molecules (e.g. 47 and 48) in the root nodulation of legumes.

dysfunctional probes to set up comparisons. The main search, however, is for active organometallic derivatives, and to distinguish active from inactive, a good screen is needed. In this case, an additional gene had been inserted in the nodA-nodB-nodC section of the Rhizobium gene sequence, near to nodD. This additional gene (lacZ) came from E. coli not Rhizobium and codes for β-galacto-sidase. When gene induction occurs, the additional gene is also transcribed and β-galactosidase is made by the Rhizobium mutant [107]. This is easily detected by colorimetric assay [116] for the hydrolysis of p-nitrophenol galactoside. In this example, genetics expertise had been essential so that a suitable screen could be available, and without this capability, the Rhizobium gene induction problem would not have been suitable to target with IR-responsive bioprobes. Complex 23 (Fig. 7.13) is an example of a flavonoid derivative which retained the ability to induce nod gene expression.

The final role of genetics in this project was to set up the functional/dysfunctional comparison [78] in the biological sample. A simple approach for a dysfunctional case is obvious, as it corresponds to the *nodD*(–) mutant discussed above. It would be possible to use the wild type as the functional case, but a more powerful approach is available from an isogenic strain in which *nodD* is overexpressed. When cultured, this will have a far higher concentration of the D protein than the wild type, so improving the chances of detecting the (D protein) · (flavonoid) interaction by spectroscopy. In this case, *Rhizobium leguminosarum* WT(8400pRL1), the *nodD*(–) strain, and the same mutant carrying multiple copies of *nodD* on plasmid pIJ1518, were used. Since the two strains are identical except for the production of protein D, any differences in the IR spectra should be due to interactions with protein D.

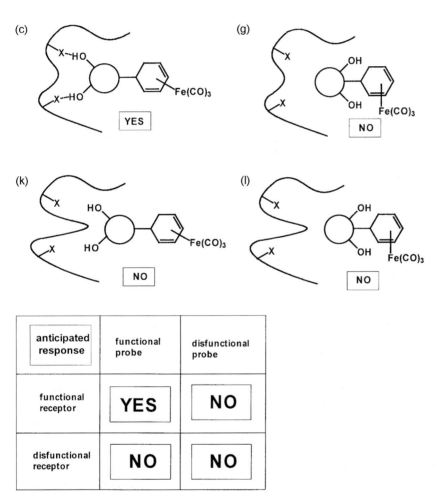

anticipated response	functional probe	disfunctional probe
functional receptor	YES	NO
disfunctional receptor	NO	NO

Fig. 7.22 Comparison of functional and dysfunctional systems [compare (c) and (g) with Fig. 7.20; (k) and (l) are new examples of (j)] to allow the spectroscopic identification of specific effects due to the interaction of a carbonylmetal bioprobe with a biological receptor.

This experiment design effectively has three controls, since it will only be valid to interpret differences between the functional probe and the dysfunctional probe (Fig. 7.22: comparing c and g) as a consequence of the binding of the functional probe in the receptor, if these differences can no longer be detected when the dysfunctional biological sample is used (Fig. 7.22: protein D absent).

Similarly, the significance of differences between the functional and dysfunctional receptors (Fig. 7.22: comparing c and k), can be tested since the same effect should not be visible in either case g or case i. This is not the same as setting a criterion that g, k and l, should all be the same, and c uniquely different. The point is more subtle than that, and the requirement for success is that any observable differences should vanish when a dysfunctional component is put in place of a functional one, and that this should occur for every combination that uses dysfunctional components.

As yet, the full set of comparisons available from "locally responsive" structures such as **23** (Fig. 7.13) and the corresponding "remotely responsive" analogs await further synthetic work, but as a final experiment [78] with the compounds available at this stage, the feasibility of the required type of spectroscopic measurements was tested with **23** at concentrations as low as 50 µM. Even at this low level, and in the presence of the debris from the lysed cells of *Rhizobium leguminosarum* WT(8400pRL1).pIJ1518, recognisable v_{sym}(CO) and v_{asym}(CO) stretching vibrational modes were detected at 2045 and 1965 cm^{-1}. With the synthetic [76, 78], spectroscopic [78], genetic [78, 107] and conceptual (this chapter) techniques now all in place, further progress awaits the synthesis of more strongly binding organometallic derivatives [the problem to be addressed is similar to the challenges facing the Jaouen group in their successful pioneering work addressing [72] oestrogen receptors (see Section 7.6, and Chapter 3)], and remote response (see Section 7.8.1) analogs of the most highly active structures.

7.9
Conclusions

The work described here is part of an ongoing programme for the development of carbonylmetal molecular bioprobes for use in bioanalysis. We have progressed from simple combinations of molecular receptors and the carbonylmetal complexes that provide the IR read-out, to multi-carbonylmetal methods [26, 36] and supported molecular sensing devices [117–119]. When multi-carbonylmetal methods are applied in bioanalysis, several independent sources of information are accessible in the same spectroscopic experiment, so direct effects arising from signal transduction of the host–guest chemistry of the receptor to the vibrational spectroscopy of the carbonylmetal component can be distinguished from general effects arising from random interactions with the diverse range of substances in the immediate environment of the probe group in a biological medium. This is an essential requirement in the application of the IR-responsive bioprobe methodology to the analysis of biological systems.

Acknowledgements

This chapter corresponds to part of a lecture presented at the First International Symposium on Bioorganometallic Chemistry [1], and GRS thanks the organizers for financial support to attend "ISBOMC'02" (for the synthetic chemistry section of the lecture, see references [76, 78]). The work discussed has been conceived and performed entirely in collaboration with Professor Colin S. Creaser, Nottingham Trent University, Nottingham, UK. This collaboration between Norwich and Nottingham has been supported by EPRSC grant (GR/L75719), with earlier work supported by BBSRC grant PG83/547, AFRC grants PG83/503, PG83/519, EPSRC grant GR/J83068, and SERC grant GR/G49852. The genetics part of these investigations relies on the expertise of research groups at the John Innes Centre, Norwich, especially that of Dr J. Allan Downie, whose guidance with conceptual aspects of *Rhizobium* genetics, and help with practical aspects of the work, is gratefully acknowledged. We also thank Karen E. Wilson for assistance with genetics and cell culture experiments used in screening the organometallic flavonoid derivatives. We thank BBSRC ISIS and the Royal Society for support, the EPSRC National Mass Spectrometry Service Centre, University of Wales, Swansea, UK for high resolution mass spectrometric measurements, and J. Varga and J. Lejtovicz (Chemical Research Center of the Hungarian Academy of Sciences) for use of unpublished curve analysis software. An influential review by Severin, Bergs, and Beck [120] covering the bioorganometallic chemistry of amino acid and peptide derivatives to 1998 (see also [121]), and the reviews cited in references [55, 63, 65] are also acknowledged.

References

1 The definitions of the terms bioprobes and bionanometrology were presented at the *First International Symposium on Bioorganometallic Chemistry*, Paris, 18–20 July, **2002**.

2 R. B. M. SCHASFOORT, *Expert Review of Anticancer Therapy* **2004**, *1*, 123–132; B. H. WEIGL, R. L. BARDELL, C. R. CABRERA, *Adv. Drug Delivery Rev.* **2003** 55, 349–377; G. M. WALKER, H. C. ZERINGUE, D. J. BEEBE, *Lab on a Chip* **2004**, *4*, 91–97; E. VERPOORTE, *Lab on a Chip* **2003**, *3*, 60N; V. SRINIVASAN, V. K. PAMULA, R. B. FAIR, *Anal. Chim. Acta* **2004**, *507*, 145–150.

3 J. JANATA, M. JOSOWICZ, P. VANÝSEK, D. M. DeVANEY, *Anal. Chem.* **1998**, *70*, 179R–208R.

4 Reviews: D. J. CRAM, *Science*, **1988**, *240*, 760–7; H.-J. SCHNEIDER, *Angew. Chem.*

Int. Ed. Engl. **1991**, *30*, 1417; D. J. CRAM, *From Design to Discovery*, Oxford University Press, Oxford, **1990**; T. WANG, J. S. BRADSHAW, R. M. IZATT, *J. Heterocycl. Chem.* **1994**, *31*, 1097–1114; examples: E. P. KYBA, R. C. HELGESON, K. MADAN, G. W. GOKEL, T. L. TARNOWSKI, S. S. MOORE, D. J. CRAM, *J. Am. Chem. Soc.* **1977**, *99*, 2564–2571; R. ARNECKE, V. BÖHMER, R. CACCIAPAGLIA, A. D. CORT, L. MANDOLINI, *Tetrahedron* **1997**, *53*, 4901–4908; S. ROELENS, R. TORRITI, *Supramolecular Chem.*, **1999**, *10*, 225–232; S. C. ZIMMERMAN, C. M. VAN ZYL, G. S. HAMILTON, *J. Am. Chem. Soc.*, **1989**, *111*, 1373–1381; G. ARENA, A. CASNATI, A. CONTINO, G. G. LOMBARDO, D. SCIOTTO, R. UNGARO, *Chem. Eur. J.* **1999**, *5*, 738–744.

5 J.-H. SCHNEIDER, *Angew. Chem. Int. Ed. Engl.* **1993**, *32*, 848–850; W. SIMON, *Chimia*, **1990**, *44*, 395; R. MURRAY, R. DESSY, W. HEINEMAN, F. JANTA, W. R. SEITZ (Eds.), *Chemical Sensors and Microinstrumentation, ACS Symposium Series*, **1988**, *403*; O. S. WOLFBEIS, *J. Anal. Chem.* **1990**, *337*, 522–527.

6 Reviews: P. M. SAITO, K. KIKUCHI, *Opt. Rev.* **1997**, *5*, 527; M. N. TAIB, R. NARAYANASWAMY, *Analyst*, **1995**, *120*, 1617–1627; O. S. WOLFBEIS, *Anal. Chem.* **2000**, *72*, 81R–90R; examples: T. HAO, X. XING, C. C. LIU, *Sens. Actuators B*, **1993**, *10*, 155; M. TAKEUCHI, M. TAGUCHI, H. SHINMORI, S. SHINKAI, *Bull. Chem. Soc. Jap.* **1996**, *69*, 2613; C. J. DACIS, P. T. LEWIS, M. E. CARROLL, M. W. READ, R. CUETO, R. M. STRONGIN, *Org. Lett.* **1999**, *1*, 331–334; K. R. A. S. SANDANAYAKE, S. SHINKAI, *Chem. Commun.* **1994**, 1083; C. J. WARD, P. PATEL, P. R. ASHTON, T. D. JAMES, *Chem. Commun.* **2000**, 229–230; L. A. CABELL, M. K. MONAHAN, E. V. ANSLYN, *Tetrahedron Lett.* **1999**, *40*, 7753–7756; J. J. LAVIGNE, E. V. ANSYL, *Angew. Chem. Int. Ed. Engl.* **1999**, *38*, 3666; R. NARAYANASWAMY, *Analyst*, **1993**, *118*, 317–322; T. HAO, X. XING, C. C. LUI, *Sens. Actuators B* **1994**, *10*, 155.

7 B. MORELLI, *J. Pharmaceut. Biomed. Anal.* **2003**, *32*, 257.

8 B. MORELLI, *Anal. Lett.* **1994**, *41*, 673.

9 Reviews: A. W. CZARNIK, *Acc. Chem. Res.* **1994**, *27*, 302; T. D. JAMES, P. LINNANA, S. SHINKAI, *Chem. Commun.* **1996**, 281–288; E. N. BRODY, L. GOLD, *Rev. Mol. Biotechnol.* **2000**, *74*, 5–13; J. HESSELBERTH, M. P. ROBERTSON, S. JHAVERI, A. D. ELLINGTON, *Rev. Mol. Biotechnol.* **2000**, *74*, 15–25; A. W. CZARNIK (Ed.), *Fluorescent Chemosensors for Ion and Molecule Recognition*, American Chemical Society, Washington DC, **1993**, 538; examples: A. W. CZARNIK, *Chem. Biol.*, **1995**, *2*, 423–428; A. MIYAWAKI, J. LLOPIS, R. HELM, J. M. McCAFFERY, J. A. ADAMS, M. IKURA, R. Y. TSIEN, *Nature* **1997**, *388*, 882–887; R. B. THOMPSON,

B. P. MILIWAL, V. L. FELICCIA, C. A. FIERKE, K. McCALL, *Anal. Chem.* **1998**, *70*, 4717–4723; K. YAMANA, T. MITSUI, H. NAKANO, *Tetrahedron* **1999**, *55*, 9143–9150; R. R. FRANCE, I. CUMPSTEY, T. D. BUTTERS, A. J. FAIRBANKS, M. R. WORMALD, *Tetrahedron Asym.* **2000**, *11*, 4985–4994; B. WETZL, M. GRUBER, B. OSWALD, A. DÜRKOP, B. WEIDGANS, M. PROBST, O. S. WOLFEIS, *J. Chromatogr.* **2003**, *793*, 83–92; C. Arnosti, *J. Chromatogr.* **2003**, *793*, 181–191; R. KENOTH, S. S. KOMATH, M. J. SWAMY, *Arch. Biochem. Biophys.* **2003**, *413*, 131–138; G. K. WALKUP, B. IMPERIALI, *J. Am. Chem. Soc.* **1996**, *118*, 3053–3054; S. DEO, H. A. GODWIN, *J. Am. Chem. Soc.* **2000**, *122*, 174–175; H. A. GODWIN, J. M. BERG, *J. Am. Chem. Soc.* **1996**, *118*, 6514–6515; S. D. JAYASENA, *Clin. Chem.* **1999**, *45*, 1628–1650; G. A. SOUKUP, R. R. BREAKER, *Curr. Opin. Struct. Biol.* **2000**, *10*, 318–325; R. R. BREAKER, *Curr. Opin. Struct. Biol.* **2002**, *13*, 31–39; M. J. P. LEINER, *Anal. Chim. Acta* **1990**, *225*, 209.

10 Reviews: A. P. DE SILVA, H. Q. N. GUNARATNE, T. GUNNLAUGSOSN, C. P. McCOY, P. R. S. MAXWELL, J. T. RADEMACHER, T. E. RICE, *Pure Appl. Chem.* **1996**, *68*, 1443; A. P. DE SILVA, T. GUNNLAUGSOSN, C. P. McCOY, *J. Chem. Educ.* **1997**, *74*, 53; R. NARAYANASWAMY, *Chem. Eng. News* **1985**, 204; R. BISSEL, A. P. DE SILVA, H. Q. N. GUNARANTA, P. L. M. LYNCH, G. E. M. MAGUIRE, K. R. A. S. SANANAYAKE, *Chem. Soc. Rev.* **1992**, *21*, 187; examples: M. IRIE, T. YOROZU, K. HAYASHI, *J. Am. Chem. Soc.* **1978**, *100*, 2236; A. P. DE SILVA, T. GUNNLAUGSOSN, T. E. RICE, *Analyst*, **1996**, *121*, 1759; A. P. DE SILVA, H. Q. N. GUNARATNE, T. GUNNLAUGSOSN, P. L. M. LYNCH, *New J. Chem.* **1996**, *20*, 871; A. P. DE SILVA, H. Q. N. GUNARATNE, C. P. McCOY, *J. Am. Chem. Soc.* **1997**, *119*, 7891; T. D. JAMES, K. R. A. S. SANANAYAKE, R. IGUCHI, S. SHINKAI, *J. Am. Chem. Soc.* **1995**, *117*, 8982; M. BIELECKI, H. EGGERT,

J. C. NORRILD, *J. Chem. Soc. Perkin Trans. 2* **1999**, 449; Y. SHIOMI, M. SAISHO, K. TSUKAGOSHI, S. SHINKAI, *Supra. Mol. Chem.* **1993**, *2*, 11; Y. SHIOMI, M. SAISHO, K. TSUKAGOSHI, S. SHINKAI, *J. Chem. Soc., Perkin Trans. 1* **1993**, 2111; A. J. BRIAN, A. P. DE SILVA, S. A. DE SILVA, R. A. D. RUPASINHA, K. R. A. S. SANANAYAKE, *Biosensors* **1989**, *4*, 169; S. A. DE SILVA, A. P. DE SILVA, *Chem. Commun.* **1986**, 1709; T. D. JAMES, K. R. A. S. SANANAYAKE, S. SHINKAI, *Nature* **1995**, *374*, 345; C. R. COOPER, T. D. JAMES, *J. Chem. Soc. Perkin Trans. 1* **2000**, 963–969.

11 Reviews: S. TAKENAKA, *Small Mol. DNA RNA Binders* **2003**, *1*, 234; S. SUBRAHMANYAM, S. A. PILETSKI, A. P. F. TURNER, *Anal. Chem.* **2002**, *74*, 3942; C. N. CAMPBELL, A. HELLER, D. J. CARUANA, D. W. SCHMIDTKE, *Electroanal. Meth. Biol. Mater.* **2002**, 439; R. I. STEFAN, H. Y. ABOUL-ENEIN, J. F. VAN STADEN, *Sens. Update* **2002**, *10*, 123; S. O. KELLEY, *Electroanal. Meth. Biol. Mater.* **2002**, 1; M. FOJA, *Electroanalysis*, **2002**, *14*, 1449; R. P. BUCK, V. COSOFRET, *Pure Appl. Chem.* **1993**, *65*, 1849; examples: R. IYER, V. PAVLOV, I. KATAKIS, L. G. BACHAS, *Anal. Chem.* **2003**, *75*, 3898; C. J. YU, H. YOWANTO, Y. WAN, T. J. MEADE, Y. CHONG, M. STRONG, L. H. DONILON, J. F. KAYYEM, M. GOZIN, G. F. BLACKBURN, *J. Am. Chem. Soc.* **2000**, *122*, 6767; O. W. SCHUHMANN, H.-L. SCHMIDT, *Adv. Biosensors* **1992**, *2*, 79; O. KAZUNORI, R. NAGANAWA, H. RADECKA, M. KATAOKA, E. KIMURA, T. KOIKE, K. TOHDA, H. FURUTA, *Supramol. Chem.* **1994**, *4*, 101; O. KAZUNORI, R. NAGANAWA, H. RADECKA, M. KATAOKA, E. KIMURA, T. KOIKE, K. TOHDA, M. TANGE, H. FURUTA, *Supramol. Chem.* **1994**, *4*, 101.

12 Reviews: A. HELLER, *Acc. Chem. Res.* **1990**, *23*, 128; A. D. RYABOV, *Angew. Chem. Int. Ed. Engl.* **1991**, *23*, 931; A. N. J. MOORE, D. D. M. WAYNER, *Can. J. Chem.* **1999**, *77*, 681; examples: A. E. G. CASS, G. DAVIS, G. D. FRANCIS, H. A. O. HILL, W. J. ASTON,

I. J. HIGGINS, E. O. PLOTKIN, L. D. L. SCOTT, A. P. F. TURNER, *Anal. Chem.* **1984**, *56*, 667; A. E. G. CASS, G. DAVIS, M. J. GREEN, H. A. O. HILL, *Electroanal. Chem.* **1985**, 117; P. D. HALE, L. I. BOGUSLAVSKY, T. INAGAKI, H. I. KARAN, H. S. LEE, T. A. SKOTHEIM, Y. OKAMOTO, *Anal. Chem.* **1991**, *63*, 677; H.-Z. BU, S. R. MIKKELSEN, A. M. ENGLISH, *Anal. Chem.* **1995**, *67*, 4071; L. GORDON, *Electroanalysis* **1995**, *7*, 23; Y. XU, H. B. KRAATZ, *Tetrahedron Lett.* **2001**, *42*, 2601; B. SCHMIDT, *Clin. Chim. Acta* **1997**, *266*, 33; J. A. DUINE, *J. Biosci. Bioeng.* **1999**, *88*, 231; V. LAURINAVICIUS, J. RAZUMIENE, B. KURTINAITIENE, I. LAPENAITE, I. BACHMATOVA, L. MARCINKEVICIENE, R. MESKYS, A. RAMANAVICIUS, *Bioelectrochem.* **2002**, *55*, 29; M. SMOLANDER, H. L. LIVIO, L. RASANEN, *Biosens. Bioelectron.* **1992**, 637.

13 H. PLENIO, D. BURTH, *Organometallics* **1996**, *15*, 1151–1156.

14 A. M. ABEYSEKERA, J. GRIMSHAW, S. D. PERERA, *Chem. Commun.* **1990**, 1797–1800; H. ECKERT, M. KOLLER, *Z. Naturforsch.* **1990**, *45*, 1709–1714; I. LAVASTRE, J. BESANÇON, C. MOÏSE, P. BROSSIER, *Bull. Soc. Chim. Fr.* **1995**, *132*, 188–195; K. DI GLERIA, H. A. O. HILL, L. L. WONG, *FEBS Lett.* **1996**, *390*, 142–144; H. B. KRAATZ, J. LUESTYK, G. D. ENRIGHT, *Inorg. Chem.* **1997**, *36*, 2400–2405.

15 M. HROMADOVÁ, M. SALMAIN, R. SOKOLOVÁ, L. POSPÍSIL, G. JAOUEN, *J. Organomet. Chem.* **2003**, *668*, 17–24.

16 T. IHARA, M. NAKAYAMA, M. MURATA, K. NAKANO, M. MAEDA, *Chem. Commun.* **1997**, 1609.

17 G. JAOUEN, A. VESSIÈRES, S. TOP, A. A. ISMAIL, I. S. BUTLER, *J. Am. Chem. Soc.* **1985**, *107*, 4778–4780.

18 J. O. ALBEN, W. S. CAUGHEY, *Biochemistry* **1968**, *7*, 175–183; J. C. OWRUTSKI, M. LI, B. LOCKE, R. M. HOCHSTRASSER, *J. Phys. Chem.* **1995**, *99*, 4842; J. R. HILL, D. D. DLOTT, M. D. FAYER, K. A. PETERSON, C. W. RELLA, M. M. ROSENBLATT, K. S. SUSLICK, C. J. ZIEGLER, *Chem. Phys. Lett.* **1995**, *244*, 218; J. R. HILL,

D. D. Dlott, C. W. Rella, T. A. Smith,
H. A. Schwettman, K. A. Peterson,
A. Kwok Rector, M. D. Fayer,
Biospectroscopy **1996**, *2*, 277; R. B. Dyer,
K. A. Peterson, P. O. Stoutland,
W. H. Woodruff, *J. Am. Chem. Soc.*
1991, *113*, 6276–6277.

19 J. R. Hill, D. D. Dlott, C. W. Rella,
K. A. Peterson, S. M. Dacatur,
S. G. Boxer, M. D. Fayer, *J. Phys.
Chem.* **1996**, *100*, 12100–12107.

20 Reviews: D. Marsh, L. I. Horváth,
M. J. Swamy, S. Mantripragada,
J. H. Kleinschmidt, *Mol. Membrane
Biol.* **2002**, *19*, 247–255; P. P. Borbat,
A. J. Costa-Filho, K. A. Earle,
J. K. Moscicki, J. H. Freed, *Science*
2001, *291*, 266–269; examples:
M. B. Sankaram, P. J. Brophy,
D. Marsh, *Biochem.* **1989**, *28*, 9699–9707;
D. Marsh, L. I. Horváth, *Biochim.
Biophys. Acta* **1998**, *1376*, 267–296;
A. Spaltenstein, B. H. Robinson,
P. B. Hopkins, *J. Am. Chem. Soc.* **1988**,
110, 1299; S. Burgard; M. Harada;
Y. Kagawa; W. E. Trommer;
P. D. Vogel, *Cell Biochem. Biophys.*
2003, *39*, 175–182; M. J. Swamy,
D. Marsh, *Biochim. Biophys. Acta*, **2001**,
1513, 122–130; M. J. Swamy,
L. I. Horváth, P. J. Brophy, D. Marsh,
Biochemistry **1999**, *38*, 16333–16339;
M. Ramakrishnan, V. Anbazhagan,
T. V. Pratap, D. Marsh, M. J. Swamy,
Biophys. J. **2001**, *81*, 2215–2225;
S.-C. Hung, W. Wang, S. I. Chan,
H. M. Chen, *Biophys. J.* **1999**, *77*,
3120–3133.

21 Reviews: L. Feilding, *Curr. Top. Med.
Chem.* **2003**, *3*, 39–53; J. T. Gerig, *Prog.
Nucl. Mag. Reson. Spectrosc.* **1994**, *26*,
293; M. A. Danielson, J. J. Falke,
Annu. Rev. Biophys. Biomol. Struct. **1996**,
25, 163–195; examples: J. Feeney,
J. E. McCormick, C. J. Bauer,
B. Birdsall, C. M. Moody,
B. A. Starkman, D. W. Young,
P. Franis, R. H. Havin, W. D. Arnold,
E. Oldfield, *J. Am. Chem. Soc.* **1996**,
118, 8700–8706; J. Scheuring, J. Lee,
M. Cushman, H. Patel, D. A. Patrick,
A. Bacher, *Biochemistry* **1994**, *33*,
7634–7640; B. Salopek-Sondi,
M. D. Vaughan, M. C. Skeels,

J. F. Honek, L. A. Luck, *J. Biomol.
Struct. Dyn.* **2003**, *21*, 235–246;
J. T. Gerig, *Prog. NMR Spectrosc.* **1994**,
26, 293–370; J. G. Pearson, B. Montez,
H. Le, E. Oldfield, E. Y. T. Chien,
S. G. Sligar, *Biochemistry* **1997**, *36*,
3590–3599; H. Duewel, E. Daub,
V. Robinson, J. F. Honek, *Biochem.*
1997, *36*, 3404–3416; C. Lian, H. Le,
B. Montez, J. Patterson, S. Harrell,
D. Laws, I. Matsumura, J. Pearson,
E. Oldfield, *Biochemistry* **1994**, *33*,
5238–5245; S. D. Hoeltzli,
C. Frieden, *Biochemistry* **1994**, *33*,
5502–5509; H. S. Duewel, J. F. Honek,
J. Protein Chem. **1998**, *17*, 337.

22 D. A. Brown, H. Sloan, *J. Chem. Soc.*
1962, 3849–3854.

23 F. A. Cotton, R. M. Wing, *Inorg.
Chem.* **1965**, *4*, 314; F. A. Cotton,
Helv. Chim. Acta **1967**, *Fasc. Extraord.*
(A. Werner Commemoration Vol., Vergal)
117; M. B. Hall, R. F. Fenske, *Inorg.
Chem.* **1972**, *11*, 1619; F. A. Cotton,
G. Wilkinson, *Advanced Inorganic
Chemistry*, 2nd edition, Wiley
Interscience, New York, **1966**, 730–735;
F. J. McQuillin, D. G. Parker,
G. R. Stephenson, *Transition Metal
Organometallics for Organic Synthesis*,
Cambridge University Press,
Cambridge, **1991**, 19–20.

24 S. P. Walsh, W. A. Goddard,
J. Am. Chem. Soc. **1976**, *98*, 7908;
J. B. Johnson, W. G. Klemperer,
J. Am. Chem. Soc. **1977**, *99*, 7132.

25 L. Pauling, *Nature of the Chemical
Bond*, 3rd edition, Cornell University
Press, Ithaca, N. Y., **1964**.

26 C. E. Anson, T. J. Baldwin,
C. S. Creaser, G. R. Stephenson,
Organometallics **1996**, *15*, 1451–1456.

27 R. Eiss, *Inorg. Chem.* **1970**, *9*, 1650–1655;
A. Musco, G. Pairo, R. Palumbo,
Chem. Ind. (Milan) **1968**, *50*, 550.

28 C. E. Anson, C. S. Creaser,
G. R. Stephenson, *Chem. Commun.*
1994, 2175–2176.

29 K. J. Odell, E. M. Hyde, B. L. Shaw,
I. Shepherd, *J. Organomet Chem.* **1979**,
168, 103.

30 R. Hoffmann, *Angew. Chem. Int. Ed.
Engl.* **1982**, *21*, 711; F. G. A. Stone,
Angew. Chem. Int. Ed. Engl. **1984**, *23*, 89.

31 P. S. Braterman, *Metal Carbonyl Spectra*, Academic Press, New York, **1975**; E. Maslowsky Jr., *Vibrational Spectra of Organometallic Compounds*, Wiley, London, **1977**; P. S. Braterman, R. Bau, H. D. Kaesz, *Inorg. Chem.* **1967**, *6*, 2097–2102.

32 W. P. Anderson, T. L. Brown, *J. Organomet. Chem.* **1971**, *32*, 343.

33 D. J. Darensbourg, *Inorg. Chem.* **1972**, *11*, 1606–1609.

34 F. A. Cotton, *Chemical Applications of Group Theory*, Wiley-Interscience, New York, 2nd edition, **1971**, Chapter 10; K. Nakamoto, *Infrared Spectra of Inorganic and Coordination Compounds*, Wiley-Interscience, New York, 2nd edition, **1970**.

35 J. K. Burdett, *Inorg. Chem.* **1981**, *20*, 2607–2615.

36 M. Salmain, A. Vessières, A. Varenne, P. Brossier, G. Jaouen, *J. Organomet. Chem.* **1999**, *589*, 92–97; A. Varenne, A. Vessières, M. Salmain, P. Brossier, G. Jaouen, *Immunoanal., Biol. Spec.* **1994**, *9*, 205–214; M. Salmain, A. Vessières, P. Brossier, G. Jaouen, *Anal. Biochem.* **1993**, *208*, 117–120.

37 G. R. Stephenson, S. Sudhakaran-Pillai, unpublished results.

38 C. C. Barraclough, J. Lewis, R. S. Nyholm, *J. Chem. Soc.* **1961**, 2582–2584; W. Beck, K. Lottes, *Z. Naturforsch.* **1964**, *19b*, 987.

39 L. M. Haines, M. H. B. Stiddard, *Adv. Inorg. Chem. Radiochem.* **1970**, *12*, 53.

40 J. R. Hill, C. J. Ziegler, K. S. Suslick, D. D. Dlott, C. W. Rella, M. D. Fayer, *J. Phys. Chem.* **1996**, *100*, 18023–18032.

41 I. P. Clark, M. W. George, J. J. Turner, *J. Phys. Chem. A* **1997**, *101*, 8367–8370.

42 P. C. Painter, G. J. Pehlert, Y. Hu, M. M. Colman, *Macromol.* **1999**, *32*, 2055–2057.

43 E. J. Hutchinson, D. Ben-Amotz, *J. Phys. Chem. B* **1988**, *102*, 3354.

44 T. C. Cheam, S. Krimm, *Chem. Phys. Lett.* **1984**, *107*, 613.

45 C. S. Creaser, M. A. Fey, G. R. Stephenson, *Spectrochim. Acta* **1994**, *50A*, 1295–1299.

46 L. J. Bellamy, R. L. Williams, *J. Chem. Soc.* **1957**, 863; L. J. Bellamy, H. E. Hallam, R. L. Williams, *Trans. Faraday Soc.* **1958**, *54*, 1120; L. J. Bellamy, R. L. Williams, *Trans. Faraday Soc.* **1959**, *55*, 14; A. D. E. Pullin, *Proc. Roy. Soc. A* **1960**, *255*, 39–43; C. Reichert, *Solvent effects in Organic Chemistry*, in H. F. Ebel (Ed.), *Monographs in Modern Chemistry*, Vol. 13, Verlag Chemie, Weinheim, **1979**.

47 C. E. Anson, T. J. Baldwin, N. J. Clayden, C. S. Creaser, O. Egyed, M. A. Fey, W. E. Hutchinson, A. Kavanaugh, G. R. Stephenson, P. I. Walker, *J. Optoelectronics and Adv. Mat.* **2003**, *5*, 533–554.

48 R. M. Izatt, K. Pawlak, J. S. Bradshaw, R. L. Bruening, *Chem. Rev.* **1991**, *91*, 1721–2085.

49 C. E. Anson, C. S. Creaser, G. R. Stephenson, unpublished results.

50 C. E. Anson, C. S. Creaser, G. R. Stephenson, *Spectrochim. Acta*, **1996**, *52A*, 1183–1191.

51 M. J. Aroney, M. K. Cooper, R. K. Pierens, S. J. Pratten, S. W. Filipczuk, *J. Organomet. Chem.* **1982**, *234*, C20.

52 H. Kobayashi, K. Kobayashi, Y. Kaizu, *Inorg. Chem.* **1981**, *20*, 4135; O. L. Carter, A. T. McPhail, G. A. Sim, *J. Chem. Soc. A* **1966**, 822.

53 D. J. Cram, D. I. Wilkinson, *J. Am. Chem. Soc.* **1960**, *82*, 5721; H. Ohno, H. Horita, T. Otsubo, Y. Sakata, S. Misumi, *Tetrahedron Lett.* **1977**, 265; E. Langer, H. Lehner, *J. Organomet. Chem.* **1979**, *173*, 47.

54 Y. Kai, N. Yasuoka, N. Kasai, *Acta Cryst.* **1978**, *B34*, 2840; H. Oike, Y. Kai, K. Miki, N. Tanaka, N. Kasai, *Bull. Chem. Soc. Jpn* **1987**, *60*, 1993–1999; A. de Meijere, O. Reiser, M. Stöbbe, J. Kopf, G. Adiwidjaja, V. Sinnwell, S. I. Khan, *Acta Chem. Scand.* **1988**, *A42*, 611.

55 G. Jaouen, A. Vessières, *Pure Appl. Chem.* **1985**, *57*, 1865–1874.

56 M. Cais, S. Dani, Y. Eden, O. Gandolfi, M. Horn, E. E. Isaacs, Y. Josephi, Y. Saar, E. Slovin, L. Snarsky, *Nature* **1977**, 534–535.

57 A. Vessières, M. Salmain, V. Philomin, G. Jaouen, *Immunoanal. Biol. Spec.* **1992**, *31*, 9.

58 M. Salmain, A. Vessières, I. S. Butler, G. Jaouen, *Bioconjugate Chem.* **1991**, *1*, 59.

59 G. Jaouen, A. Vessières, S. Top, A. A. Ismail, I. S. Butler, *J. Am. Chem. Soc.* **1985**, *107*, 4780–4781.

60 A. Vessières, S. Top, A. A. Ismail, I. S. Butler, M. Louer, G. Jaouen, *Biochemistry* **1988**, *27*, 6659–6666.

61 M. Savignac, G. Jaouen, C. A. Rodger, R. E. Perrier, B. G. Sayer, M. J. McGlinchey, *J. Org. Chem.* **1986**, *51*, 2328–2332.

62 A. Vessières, S. Top, C. Vaillant, D. Osella, J.-P. Mornon, G. Jaouen, *Angew. Chem. Int. Ed. Engl.* **1992**, *31*, 753–755.

63 G. Jaouen, A. Vessières, *Pure Appl. Chem.* **1989**, *61*, 565–572.

64 I. S. Butler, A. Vessières, M. Salmain, P. Brossier, G. Jaouen, *Appl. Spectrosc.* **1989**, *43*, 1497.

65 G. Jaouen, A. Vessières, I. S. Butler, *Acc. Chem. Res.* **1993**, *26*, 361–369.

66 S. Green, P. Waiter, V. Kumar, A. Krust, J. M. Bornert, P. Argos, P. Chambon, *Nature* **1986**, *320*, 134.

67 M. Gruselle, J.-L. Rossignol, A. Vessières, G. Jaouen, *J. Organomet. Chem.* **1987**, *328*, C12–C15.

68 M. Gruselle, P. Deprez, A. Vessières, S. Greenfield, G. Jaouen, J.-P. Larue, D. Thouvenot, *J. Organomet. Chem.* **1989**, *359*, C53–C56.

69 I. Lavastre, J. Besançon, P. Brossier, C. Moïse, *Appl. Organomet. Chem.* **1990**, *4*, 9.

70 M. Salmain, A. Vessières, P. Brossier, I. S. Butler, G. Jaouen, *J. Immunol. Methods* **1992**, *148*, 56–75.

71 M. Salmain, A. Vessières, G. Jaouen, I. S. Butler, *Anal. Chem.* **1991**, *63*, 2323–2329.

72 M. Salmain, A. Vessières, P. Brossier, G. Jaouen, *Anal. Biochem.* **1993**, *208*, 117.

73 N. Fischer-Durand, A. Vessières, J.-M. Heldt, F. Le Bideau, G. Jaouen, *J. Organomet. Chem.* **2003**, *668*, 59–66.

74 B. Rudolf, M. Makowska, A. Domagala, A. Rybarczyk-Pirek, J. Zakrzewski, *J. Organomet. Chem.* **2003**, *668*, 95–100.

75 J. Szymoniak, B. El Mouatassim, J. Besançon, C. Moïse, P. Brossier, *Tetrahedron* **1993**, *49*, 3109–3116.

76 A. V. Malkov, L. Mojovic, G. R. Stephenson, A. T. Turner, C. S. Creaser, *J. Organomet. Chem.* **1999**, *589*, 103–110.

77 C. E. Anson, C. S. Creaser, A. V. Malkov, L. Mojovic, G. R. Stephenson, *J. Organomet. Chem.* **2003**, *668*, 101–122.

78 C. E. Anson, C. S. Creaser, J. A. Downie, O. Egyed, A. V. Malkov, L. Mojovic, G. R. Stephenson, A. T. Turner, K. E. Wilson, *Bioorg. Med. Chem. Lett.* **1998**, *8*, 3549–3554.

79 F. Le Borgne, J. P. Beaucourt, *Tetrahedron Lett.* **1988**, *29*, 5649–5652.

80 N. A. Sasaki, P. Potier, M. Savignac, G. Jaouen, *Tetrahedron Lett.* **1988**, *29*, 5759–5762.

81 C. Sergheraert, J. C. Brunet, A. Tartar, *Chem. Commun.* **1982**, 1417–1418; C. Sergheraert, A. Tartar, *J. Organomet. Chem.* **1982**, *240*, 163–168.

82 J. C. Brunet, E. Cuingnet, H. Gras, P. Marcincal, A. Mocz, C. Sergheraert, A. Tartar, *J. Organomet. Chem.* **1981**, *216*, 73–77.

83 F. Hosner, M. Voyle, *J. Organomet. Chem.* **1988**, *347*, 365.

84 A. J. Pearson, S. H. Lee, F. Gouzoules, *J. Chem. Soc. Perkin Trans. 1* **1990**, 2251–2254.

85 A. E. Bolton, W. M. Hunter, *Biochem. J.*, **1973**, *133*, 529–539.

86 A. Varenne, M. Salmain, C. Brisson, G. Jaouen, *Bioconjugate Chem.* **1992**, *3*, 471–476.

87 M. Salmain, M. Gunn, A. Gorfti, S. Top, G. Jaouen, *Bioconjugate Chem.* **1993**, *4*, 425–433.

88 A. Gorfti, S. Top, G. Jaouen, *Chem. Commun.* **1994**, 433–434.

89 W. Bauer, M. Prem, K. Polborn, K. Sünkel, W. Steglich, W. Beck, *Eur. J. Inorg. Chem.* **1998**, 485–493.

90 S. Blanalt, M. Salmain, B. Malézieux, G. Jaouen, *Tetrahedron Lett.* **1996**, *37*, 6561–6564.

91 M. Salmain, E. Licandro, C. Baldoli, S. Maiorana, H. Tran-Huy, G. Jaouen, *J. Organomet. Chem.* **2001**, *617*, 376–382.

92 M. Chaari, J. P. Lavergne,
P. Vaillefont, *Synth. Commun.* **1989**,
19, 1211–1216; F. Rose-Munch,
K. Aniss, E. Rose, *J. Organomet. Chem.*
1990, *385*, C1–C3; F. Rose-Munch,
K. Aniss, *Tetrahedron Lett.*, **1990**, *31*,
6351–6354; M. Chaari, A. Jenhi,
J. P. Lavergne, P. Vaillefont,
J. Organomet. Chem. **1991**, *401*, C10–C13;
M. Chaari, A. Jenhi, J. P. Lavergne,
P. Vaillefont, *Tetrahedron* **1991**, *47*,
4619–4630; F. Rose-Munch, K. Aniss,
E. Rose, J. Vaisserman, *J. Organomet.
Chem.* **1991**, *415*, 223–255; M. J. Dunn,
R. F. W. Jackson, G. R. Stephenson,
Synlett **1992**, 905–906; J.-P. Genet,
R. D. A. Hudson, W.-D. Meng,
E. Roberts, G. R. Stephenson,
S. Thorimbert, *Synlett* **1994**, 631–634;
R. D. A. Hudson, S. A. Osborne,
G. R. Stephenson, *Synlett* **1996**,
845–846.

93 M. A. Fey, *PhD Thesis*, University of
East Anglia, Norwich, **1992**.

94 J. A. Carver, B. Fates,
L. A. P. Kane-Maguire,
Chem. Commun. **1993**, 928–929.

95 C. E. Anson, C. S. Creaser, O. Egyed,
M. A. Fey, G. R. Stephenson,
Chem. Commun. **1994**, 39–40.

96 L. A. P. Kane-Maguire, R. Kanitz,
P. Jones, P. A. Williams, *J. Organomet.
Chem.* **1994**, *464*, 203–213.

97 K. L. Bennet, J. A. Carver,
D. M. David, L. A. P. Kane-Maguire,
M. M. Sheil, *J. Coord. Chem.* **1995**, *34*,
351–355.

98 F. Franke, I. D. Jenkins, *Aust. J. Chem.*
1978, *31*, 595–603.

99 D. Schlawe, A. Majdalani,
J. Velcicky, E. Hessler, T. Wiedler,
A. Prokop, H.-G. Schalz,
Angew. Chem. Int. Ed. Engl. **2004**, *43*,
1731–1734.

100 S. Schibli, M. Netter, L. Scapozza,
M. Birringer, P. Schelling,
C. Dumas, J. Schoch,
P. A. Schubiger, *J. Organomet. Chem.*
2003, *668*, 67–74.

101 K. Kowalski, J. Zakrzewski,
J. Organomet. Chem. **2003**, *668*, 91–94.

102 J. M. Dalla Riva Toma,
D. E. Bergstrom, *J. Org. Chem.* **1994**,
59, 2418–2422.

103 Z. Wang, B. A. Roe, K. M. Nicholas,
R. L. White, *J. Am. Chem. Soc.* **1993**,
115, 4399–4400.

104 A, Hess, N. Metzler-Nolte, *Chem.
Commun.* **1999**, 885–886.

105 C. E. Anson, C. S. Creaser, O. Egyed,
G. R. Stephenson, *Spectrochim. Acta*
1997, *53A*, 1867–1877.

106 A. W. B. Johnston, *Nodulation*, in
J. Kendrew, E. Lawrence (Eds.),
Encyclopaedia of Molecular Biology **1994**,
741–745; A. W. B. Johnston, *Biological
Nitrogen Fixation*, in J. L. Marx (Ed.),
A Revolution in Biotechnology **1989**,
103–118; J. A. Downie,
A. W. B. Johnston, *Plant Cell and
Environment* **1988**, *11*, 403–412;
L. Rossen, E. O. Davis,
A. W. B. Johnston, *Trends Biochem. Sci.*
1987, *12*, 430–433; L. Rossen,
A. W. B. Johnston, J. A. Downie,
Nucleic Acids Res. **1984**, *12*, 9497–9508;
J. A. Downie, C. D. Knoght,
A. W. B. Johnston, *Mol. Gen. Genet.*
1985, *198*, 225; J. A. Downie,
S. A. Walker, *Curr. Opin. Plant Biol.*
1999, *2*, 483–489; M. R. Bladergroen,
H. P. Spaink, *Curr. Opin. Plant Biol.*
1998, *1*, 353–359; A. W. B. Johnston,
E. O. Davis, A. Economou, J. E. Burn,
J. A. Downie, *The Complexity of the
Regulation of Expression of Nodulation
Genes of* Rhizobium, in G. W. Lycett,
D. Grierson (Eds.), *Genetic Engineering
of Crop Plants* **1990**, pp. 249–258;
J. A. Downie, C. Marie, A.-K. Schieu,
J. L. Firmin, K. E. Wilson,
A. E. Davies, T. M. Cubo,
A. Mavridou, A. W. B. Johnston,
A. Economou, *Genetic and Biochemical
Studies on the Nodulation Genes of*
Rhizobium leguminosarum bv. viciae,
in H. Hennecke, D. P. S. Verma (Eds.),
*Advances in Molecular Genetics of Plant–
Microbe Interactions* **1990**, pp. 134–141;
A. W. B. Johnston, *The Symbiosis
between* Rhizobium *and Legumes*, in
D. A. Hopwood, K. F. Chater (Eds.),
Genetics of Bacterial Diversity **1989**,
pp. 393–414.

107 J. L. Firmin, K. E. Wilson, L. Rossen,
A. W. B. Johnston, *Nature* **1986**, *324*, 90.

108 H. R. M. Schlaman, R. J. H. Okker,
B. J. J. Lugtenberg, *J. Bacteriol.* **1992**,

174, 5177–5182; C. A. Shearman, L. Rossan, A. W. B. Johnston, J. A. Downie, *EMBO J.* **1986**, *5*, 647; K. Recourt, J. Schripsema, J. W. Kijne, A. A. N. van Brussel, B. J. J. Lugtenberg, *Plant Mol. Biol.* **1991**, *16*, 841–852; S. A. J. Zaat, C. A. Wijffelman, I. H. M. Mulders, A. A. N. van Brussel, B. J. J. Lugtenberg, *Plant Physiol.* **1988**, *86*, 1298–1303.

109 L. Rossen, C. A. Shearman, A. W. B. Johnston, J. A. Downie, *EMBO J.* **1985**, *4*, 3369.

110 H. P. Spaink, R. J. H. Okker, C. A. Wijffelman, E. Pees, B. J. J. Lugtenberg, *Plant Mol. Biol.* **1987**, *9*, 27–39.

111 A. O. Ovtsyna, G. J. Rademaker, E. Esser, J. Weinman, B. G. Rolfe, I. A. Tikhonovich, B. J. J. Lugtenberg, J. E. Thomas-Oates, H. P. Spaink, *Mol. Plant Microbe Interact.* **1999**, *12*, 252–258; M. Hungria, A. W. B. Johnston, D. A. Phillips, *Mol. Plant Microbe Interact.* **1992**, *5*, 199–203; M. Schultze, B. Quilet-Sire, E. Kondorosi, H. Virelizier, J. N. Glushka, G. Endre, S. D. Gero, A. Kondorosi, *Proc. Natl Acad. Sci.* **1992**, *89*, 192; P. Lerouge, P. Roche, C. Faucher, F. Maillet, G. Truchet, J. C. Prome, J. Dearie, *Nature* **1990**, *344*, 781.

112 J. Burn, J., L. Rossen, A. W. B. Johnston, *Genes and Development* **1987**, *1*, 456–464; A. A. Begum, S. Leibovitch, P. Migner, F. Zhang, *J. Exp. Bot.* **2001**, *52*, 1537–1543.

113 J. A. Downie, C. D. Knight, A. W. B. Johnston, L. Rossen, *Mol. Gen. Genet.* **1985**, *198*, 255–262; J. E. Burn, W. D. Hamilton,

J. C. Wootton, A. W. B. Johnston, *Mol. Microbiol.* **1989**, *3*, 1567–1577.

114 G.-F. Hong, J. E. Burn, A. W. B. Johnston, *Nucl. Acids Res.* **1987**, *15*, 9677–9691; E. O. Davis, A. W. B. Johnston, *Mol. Microbiol.* **1990**, *4*, 933–941; C. A. Shearman, L. Rossen, A. W. B. Johnston, J. A. Downie, *EMBO J.* **1986**, *5*, 647–652; J. McIver, M. A. Djordjevic, J. J. Weinman, G. L. Bender, B. G. Rolfe, *Mol. Plant Microbe Interact.* **1989**, *2*, 97–106; H. P. Spaink, R. J. H. Okker, C. A. Wijffelman, T. Tak, L. Goosen-deRoo, E. Dees, A. A. N. van Brussel, B. J. J. Lugtenberg, *J. Bacteriol.* **1989**, *171*, 4045–4053.

115 S. A. J. Zaat, J. Scripsema, C. A. Wijffelman, A. A. N. van Brussel, B. J. J. Lugtenberg, *Plant. Mol. Biol.* **1989**, *13*, 175–188.

116 J. H. Miller, *Experiments in Molecular Genetics*, Cold Spring Harbour Laboratory Press, New York, **1972**, *82*, p. 6609.

117 C. S. Creaser, W. E. Hutchinson, G. R. Stephenson, *Appl. Spectrosc.* **2000**, *54*, 1624–1628.

118 C. S. Creaser, W. E. Hutchinson, G. R. Stephenson, *Analyst* **2001**, *126*, 647–651.

119 C. S. Creaser, W. E. Hutchinson, G. R. Stephenson, *Sens. Actuators B* **2002**, *82*, 150–157.

120 K. Severin, R. Bergs, W. Beck, *Angew. Chem. Int. Ed. Engl.*, **1998**, *37*, 1634–1654.

121 G. R. Stephenson, *Transition Metal Organometallic Methods for the Synthesis of Amino Acids*, in Atta-ur-Rahman (Ed.), *Studies in Natural Product Chemistry*, Elsevier, Amsterdam, *16*, **1995**.

8
Organometallic Complexes as Tracers in Non-isotopic Immunoassay

Michèle Salmain and Anne Vessières

8.1
Introduction

The idea of using metal atoms in organometallic or coordination complexes as labeling agents for haptens or proteins (antigens) in immunoassays was first suggested by Cais in 1977 [1]. Cais named this procedure the metalloimmunoassay (MIA), taking his inspiration from the term radioimmunoassay (RIA), coined in 1960 by Yalow and Berson [2] to describe the immunoassay of insulin that was the first to use radioactive tracers and antibodies [3], and was awarded the 1977 Nobel prize for medicine. The use of radioactive β (^{14}C, ^{3}H) or γ (^{131}I) radiation-emitting tracers significantly increased the sensitivity of the immunoassay, allowing quantification of substances present in the body in low (picomolar) concentrations, including circulating (steroid and proteic) hormones, prescribed drugs or illegal substances [4, 5]. Limitations inherent in the use of these radioactive tracers, however, soon became apparent. These include the health risks associated with handling of the radioactive substances involved, especially in the case of gamma emitters, and radioactive waste disposal problems. The consequent strict regulation of radioisotope usage has led to a progressive reduction in their use, to the point where they now represent only a small fraction of the immunoassay market. To address these limitations, various non-isotopic techniques have seen the light of day since the late 1960s [6] most of them using enzymes [7–9], fluorescent compounds [10–12] or luminescent compounds [13–15]. The principle behind these techniques is to combine a tracer with an analytical method sensitive enough to quantify analytes at the picomole level.

Cais's pioneering work played a part in this search for novel tracers that could be used in place of radioactive labels. When this work began in the late 1970s, the structure of ferrocene had been known for 20 years [16, 17], and a range of stable and varied organometallic complexes was thus available. Cais and coworkers took advantage of the new type of metal–carbon π bond discovered with ferrocene to synthesize novel molecules that might be candidates for use as non-isotopic tracers, assuming of course that a sufficiently sensitive analytical method could be found

Bioorganometallics: Biomolecules, Labeling, Medicine. Edited by Gérard Jaouen
Copyright © 2006 Wiley-VCH Verlag GmbH & Co. KGaA, Weinheim
ISBN: 3-527-30990-X

to detect them at very low concentrations. The purpose of this chapter is to summarize current progress in the use of these organometallic complexes in non-isotopic immunoassay, which as we shall see continues to be a subject of ongoing research, recently extended to include environmental monitoring, new detection techniques suitable for miniaturization and the prospect of novel tracers based on organometallic dendrimers [18, 19].

Since the subject of the immunoassay may not be familiar to readers whose interests are in the field of organometallic chemistry, we will begin this chapter with a brief overview of the principle behind the immunoassay method before describing the synthesis of organometallic tracers and reviewing the various detection methods that have facilitated the creation of novel immunoassays.

8.2
Principle of an Immunoassay

An immunoassay is the indirect assay of an analyte whose concentration in a matrix (blood, plasma, urine, drinking water, etc.) is at most several micrograms/litre. It is based on the specific interaction between three components (Fig. 8.1): the analyte (antigen) to be assayed, a specific (polyclonal or monoclonal) antibody of the analyte, plus a tracer consisting of the analyte (antigen) or the antibody, modified by addition of a label that can be detected at the nanomolar (or better, at the picomolar) level by an analytical technique such as radioactivity, enzymatic detection, electrochemical detection, atomic absorption spectroscopy (AAS), Fourier-transform infrared (FT–IR) spectroscopy, or fluorescence polarization (FP).

Substances for analysis can be divided into three categories: (1) haptens, which are molecules with a molecular weight of less than 1 kD. These include steroids, medications, illegal substances, pesticides; (2) proteins such as tumoral markers, antibodies, β-HCG; (3) viruses.

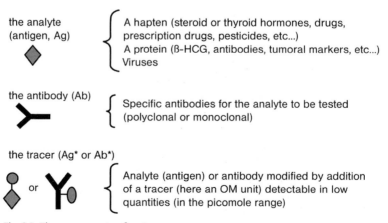

Fig. 8.1 The components of an immunoassay.

Two main types of immunoassay, classed as competitive and non-competitive, share the market. They are shown schematically in Fig. 8.2.

For a competitive immunoassay, a fixed quantity of tracer is incubated in the presence of a fixed (lesser) quantity of antibody and variable quantities of the analyte. After incubation, the free and bound fractions of the tracer are separated. Quantification of one or other of the fractions allows calculation of the amount of analyte present. This technique is particularly suited to hapten assay. This assay method, requiring a separation step, is known as a heterogeneous assay. In a few particular cases, the free and bound fractions of the tracer give different signals, eliminating the need for the separation step. This is known as a homogeneous-phase assay.

In the case of a non-competitive immunoassay, an excess of the specific antibody of the analyte is bound to a solid phase, such as the wells of a microtitration plate, and the solution containing the analyte is added. After incubation and washing, a second, labeled antibody is added to the solution. After a second incubation phase the free fraction of antibody is removed and the bound fraction is quantified. This is directly proportional to the quantity of analyte present. This technique advanced rapidly when monoclonal antibodies became available, since this made it possible to produce unlimited quantities of antibodies with clearly defined characteristics [20]. This second type of assay is also known as a sandwich immunoassay, and is best suited to the assay of proteins rather than haptens. The widespread ELISA (Enzyme Linked Immunosorbent Assay) test belongs to this category [21–24].

In all cases development of a new non-isotopic immunoassay proceeds via the following steps: (1) acquisition of specific polyclonal or monoclonal antibodies for the substance to be assayed, (2) preparation of a tracer, which may be either the analyte or the antibody modified by addition of label that can be detected at low levels, (3) selection of a separation method for the free and bound fractions of the tracer, except in the particular case of a homogeneous assay (discussed below),

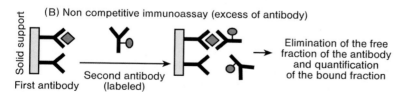

Fig. 8.2 The two types of immunoassays.

(4) selection of a method to quantify the tracer, (5) preparation of the immunoassay itself, leading to the establishment of a calibration curve for quantification of the analyte.

This chapter is concerned exclusively with non-isotopic immunoassays using organometallic tracers, that is, tracers containing a direct metal–carbon bond. Non-isotopic immunoassays using coordination complexes, and particularly chelates, have followed a parallel development and are outside the scope of this chapter [10, 11]. Preparation of a new immunoassay is a multidisciplinary endeavor that requires input from both organometallic and analytical chemistry in order to achieve the immunological analysis. There are always obstacles to be overcome, which perhaps explains the relative paucity of new assays. It is the multidisciplinary challenge, however, that makes the enterprise particularly interesting.

8.3
Obtaining Specific Antibodies

The first step in the preparation of a new immunoassay is to obtain specific antibodies. Since immunoassays using organometallic tracers are most commonly employed to assay hapten-type analytes, i.e. low molecular weight (< 1 kD) molecules, which cannot by themselves induce production of specific antibodies, an immunogen must first be prepared by coupling the hapten to a carrier protein such as bovine serum albumin (BSA). Injection of this immunogen into an animal (rabbit, goat or horse) once a month for several months stimulates production of polyclonal antibodies specific to the substance to be assayed, which are then available in the blood serum of the animal (the antibody-containing serum is known as antiserum). Various preparation methods for immunogens are described in the reference [25]. Most are based on the following principle: the analyte is modified by addition of a function (such as a carboxylic acid function) which then allows coupling via an amide bond to the lysine residues of the carrier protein. In order to obtain the best possible immune response against the hapten, the latter is often separated from the carrier protein by a chain of varied length. Several hapten molecules are attached to the carrier protein, but a number of residues in the 8–25 range is generally considered to give the best results when the carrier protein is BSA. An example of the preparation of an immunogen is shown in Scheme 8.1. This is the method used to obtain the anti-diphenylhydantoin specific antibody [26].

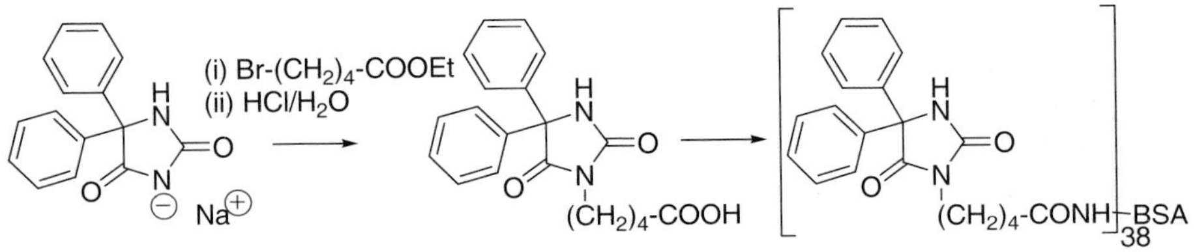

Scheme 8.1 Preparation of the diphenylhydantoin immunogen **2**.

For the immunoassay of haptens, which is of primary interest here, monoclonal antibodies have not been shown to be superior and polyclonal antibodies are still frequently used.

8.4
Synthesis of the Organometallic Tracers

In the great majority of examples of MIA, the organometallic tracers used are molecules derived from the analyte to be assayed (structures shown in Table 8.1) to which an organometallic moiety has been attached. The analyte is normally a hapten (cf. definition above).

The preparation strategy for the tracer must take into account the role it will play in the competitive immunoassay as shown in Scheme 8.2, which is the most commonly used MIA format.

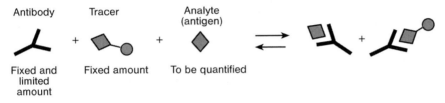

Scheme 8.2 Competitive immunoassay.

In this format, there is competition between the analyte and the tracer to bind to a limited amount of the antibody. This means that the tracer must be recognized by the antibody, but the analyte must also be able to displace the antibody–tracer bond. The affinity of the tracer for the antibody must therefore not be greater than that of the analyte for the same antibody. For this reason the organometallic moiety is normally attached at the same position used when preparing the immunogen, and for optimum sensitivity the assay can be fine-tuned afterwards by varying the nature and length of the carbon arm. Finally, the tracer must be stable in air and biological media, must not require overly-constricting precautions such as degassing of solvents, and must itself possess particular physical and chemical properties. Once all these constraints are taken into account, only a relatively limited number of candidate organometallic moieties remain. These will be reviewed by type of complex.

8.4.1
Preparation of Tracers Labeled with Ferrocene

It is possible to attach a ferrocene entity to a hapten or an antibody via formation of an amide bond between the carboxylic acid of the ferrocene or one of its derivatives and an amine (whether pre-existing or attached to a hapten) or the amino groups of the antibody lysines.

Table 8.1 The analytes, their tracers, their organometallic units (OM unit) and the analytical methods used for their quantification.

Analytes	Tracer; OM unit	Analytical method	Reference
Antidepressant			
Nortriptyline	12; Cy (Scheme 8.8)	AAS	62
Antiepileptics			
Carbamazepine	33; $[Co_2(CO)_6]$ (Scheme 8.15)	FT-IR	45
Phenobarbital	16; Cy (Scheme 8.9)	AAS	60
	idem	FT-IR	71
	40; [Co]$^+$; (Scheme 8.16)	Electrochemistry (SWV)	55
Diphenylhydantoin (DPH)	24; $[Cr(CO)_3]$ (Scheme 8.12)	FT-IR	26
	10; Fp (Scheme 8.7)	FT-IR	32
	37; [Co]$^+$ (Scheme 8.16)	Electrochemistry (SWV)	49
	45; [Fc]$^+$ (Scheme 8.18)	Electrochemistry (SWV)	55

Table 8.1 (continued)

Analytes	Tracer; OM unit	Analytical method	Reference
Drugs (miscellaneous)			
Lidocaine (cardiac arrhythmias)	4; Fc (Scheme 8.3)	Electrochemistry (Cyclic voltammetry)	27
Theophylline (cardiac stimulant)	6; Fc (Scheme 8.4)	Electrochemistry (Cyclic voltammetry)	28
Histamine (allergic reactions)	**Fc-Antibody**: Fc (Scheme 8.6)	Electrochemistry	30
Drug of abuse			
Amphetamine	35; [Co$^+$] (Scheme 8.16)	Electrochemistry (SWV)	48

AAS: Atomic absorption spectroscopy; FT–IR: Fourier-transform infrared spectroscopy; SWV: Square wave voltammetry.

Table 8.1 (continued)

Analytes	Tracer; OM unit	Analytical method	Reference
Hormones			
Cortisol	**27**; $[Co_2(CO)_6]$ (Scheme 8.13)	FT-IR	70
3,3′,5-L-triiodothyronine (T$_3$)	**8**; Fc (Scheme 8.5)	Electrochemistry (Amperometry)	29
Pesticides			
Atrazine (ATZ)	**30**; $[Co_2(CO)_6]$ (Scheme 8.14)	FT-IR	44
Chlortoluron	**19**; Cy (Scheme 8.10)	FT-IR	39
Miscellaneous			
Biotin	**21**; Cy (Scheme 8.11)	AAS	65

8.4.1.1 Preparation of a Tracer for Lidocaine

The ferrocenyl tracer **4**, derived from lidocaine, a drug used in the treatment of cardiac arrhythmias, was obtained by acylation of p-aminolidocaine with 1',3-dimethylferrocene carboxylic acid chloride **3** (Scheme 8.3) [27]. This tracer was used in developing one of the first MIAs [27].

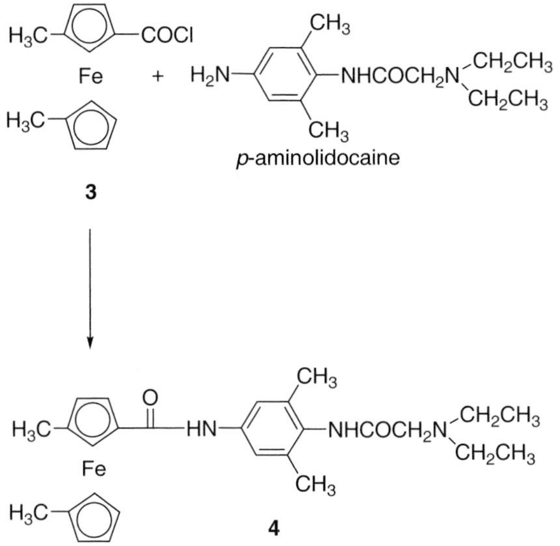

Scheme 8.3 Synthesis of the ferrocenyl tracer of lidocaine **4**.

8.4.1.2 Preparation of a Tracer for Theophylline

Theophylline is used as a cardiac stimulant. Preparation of 8-(ferrocenylmethyl)-theophylline **6**, was carried out by condensation of 5,6-amino-1,3-dimethyluracil with ferrocenylacetic acid **5** in N,N'-dimethylaniline at reflux, followed by cyclization in a basic medium (Scheme 8.4) [28].

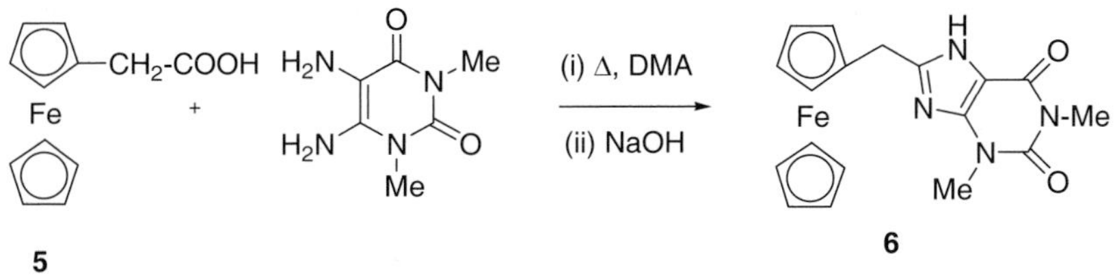

Scheme 8.4 Synthesis of the ferrocenyl tracer of theophylline **6**.

8.4.1.3 Preparation of a Tracer for Triiodothyronine

Synthesis of the tracer **8** derived from triiodothyronine, a natural hormone, was carried out by direct peptide coupling between the amine function of triiodothyronine and ferrocenyl carboxylic acid **7**, previously activated by EDAC (Scheme 8.5) [29].

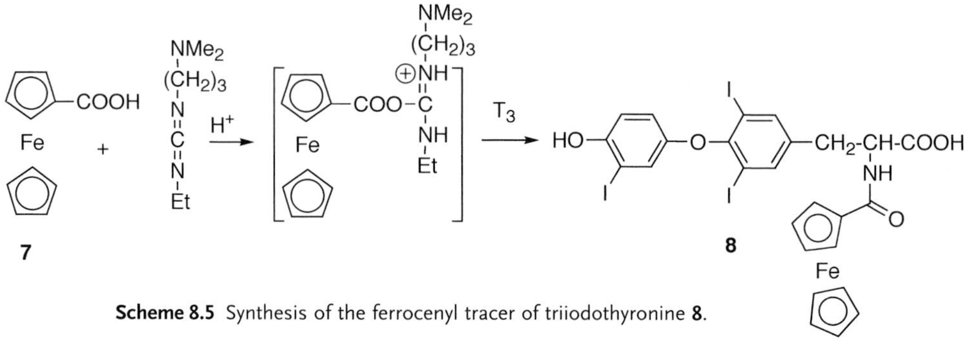

Scheme 8.5 Synthesis of the ferrocenyl tracer of triiodothyronine **8**.

8.4.1.4 Labeling of Antibodies (IgG)

Two monoclonal antibodies (IgG) were labeled with ferrocene, the first specific to histamine [30], a molecule implicated in allergic reactions, and the second specific to the hormone HCG [31], a glycoprotein biosynthesized by placental tissue throughout pregnancy. This procedure, involving a fraction of their lysine residues, is very similar to that used for triiodothyronine, except for the fact that activation of the ferrocenyl carboxylic acid **7** is achieved by reaction of EDAC and NHSS to give an intermediate N-sulfosuccinimidyl ester (Scheme 8.6). The number of Fc residues attached to the antibodies ranges from 4 to 11.

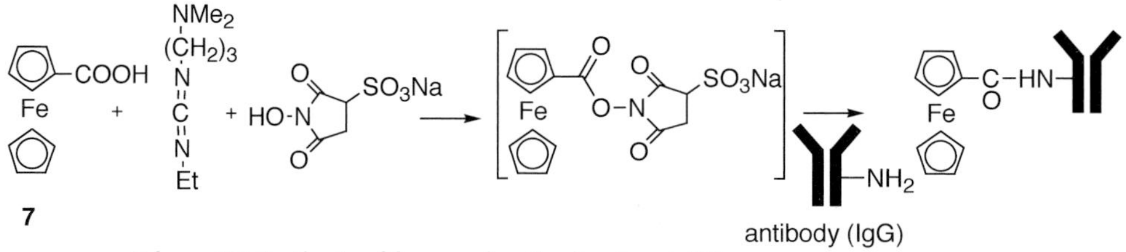

Scheme 8.6 Synthesis of ferrocenyl-antibodies ($n = 4–11$).

8.4.2
Preparation of a Tracer for Diphenylhydantoin Labeled with a (Cyclopentadienyl)dicarbonyl Iron (Fp) Entity

An organometallic tracer of diphenylhydantoin (DPH) **10**, including the Fp unit was prepared in one step (Scheme 8.7) [32] by taking advantage of the facile substitution under sunlight of the iodo ligand of iodo(cyclopentadienyl)dicarbonyl-iron (II) **9** by imidato(–) ligands generated *in situ* using diisopropylamine [33].

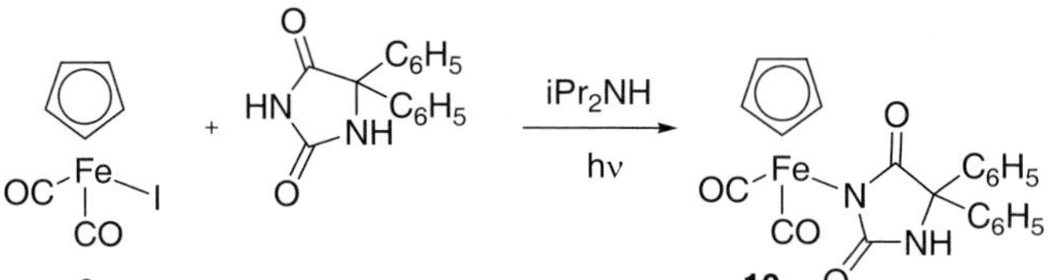

Scheme 8.7 Synthesis of Fp-diphenylhydantoin **10**.

8.4.3
**Synthesis of Tracers Labeled with a Cymantrene
(Cyclopentadienyl Manganese Tricarbonyl) Entity**

8.4.3.1 **Preparation of Tracers for Nortriptyline and Phenobarbital**
Nortriptyline, a drug belonging to the class of the tricyclic antidepressants, was labeled directly in one step by acylation of its secondary amino function by the acid chloride of cymantrene (Scheme 8.8) [34].

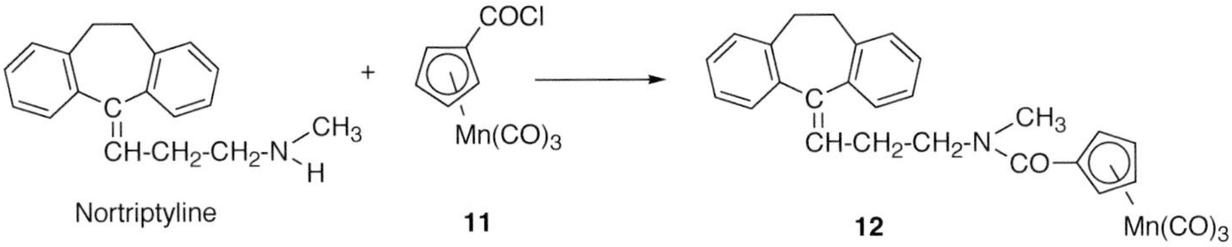

Scheme 8.8 Synthesis of the cymantrenyl tracer of nortriptyline **12**.

Labeling of phenobarbital, a barbiturate used in the treatment of convulsions, was carried out by acylation of p-aminophenobarbital **15** by the acid chloride of cymantrene **11** (Scheme 8.8). Synthesis of **15** was performed by nitration of the benzene ring of phenobarbital [35] to give a mixture of m- and p-nitrophenobarbital **13** and **14** followed by reduction of p-nitrophenobarbital by hydrogenation in the presence of palladium on charcoal [36].

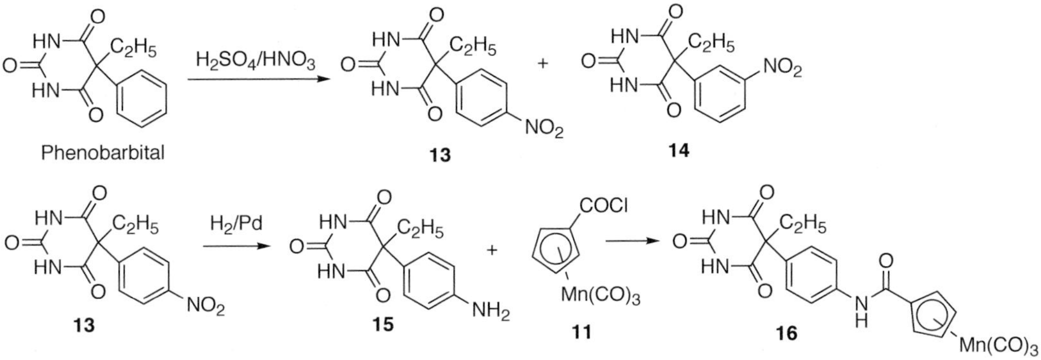

Scheme 8.9 Synthesis of the cymantrenyl tracer of phenobarbital **16**.

8.4.3.2 **Preparation of a Tracer for Chlortoluron**
Chlortoluron is a herbicide extensively used in the control of four types of wild oats and black grass in winter barley [37]. The starting point for the synthesis of a tracer for chlortoluron bearing a cymantrene fragment is the amino derivative 3-chloro-4-methylaniline aminocaproate **17**. This compound is synthesized by peptide coupling between 3-chloro-4-methylaniline and 6-aminocaproic acid, converted to the acid chloride in situ by reaction with thionyl chloride in the

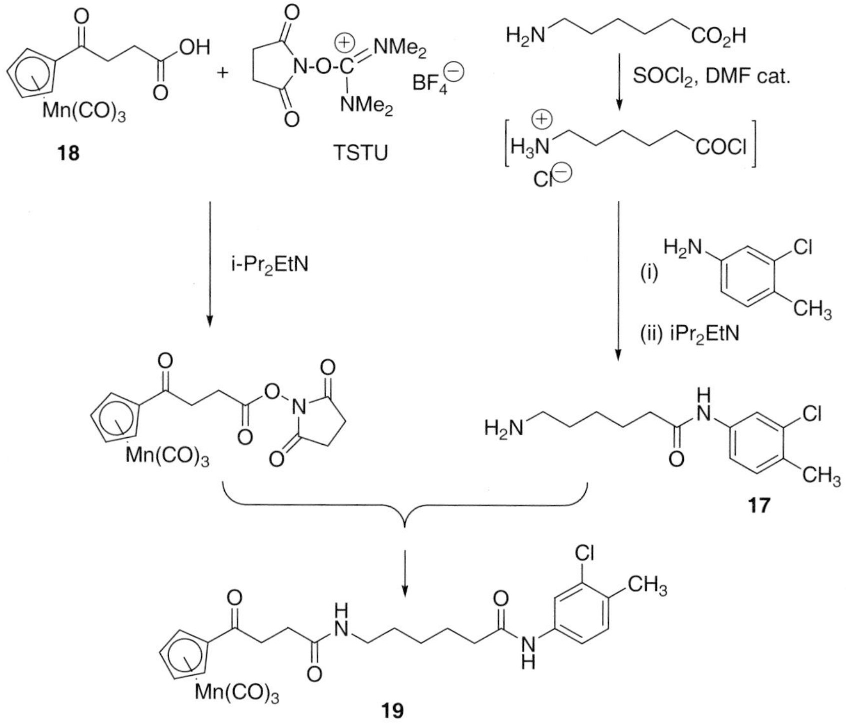

Scheme 8.10 Synthesis of the cymantrenyl derivative of chlortoluron **19**.

presence of DMF. The ammonium salt thus obtained is then deprotonated by action of diisopropylethylamine. Meanwhile, β-cymantrenoyl propionic acid, prepared by the method of Dabard et Le Plouzennec [38], is converted to the N-succinimidyl ester by action of TSTU and diisopropylethylamine. The two compounds are coupled to give the tracer of chlortoluron **19** (Scheme 8.10) [39].

8.4.3.3 Preparation of a Tracer for Biotin

The cymantrenyl-biotin tracer **21** is obtained by coupling of cymantrenyl carboxaldehyde **20** prepared according to the method of Leblanc et al. [40] with biotin hydrazide (Scheme 8.11) [41].

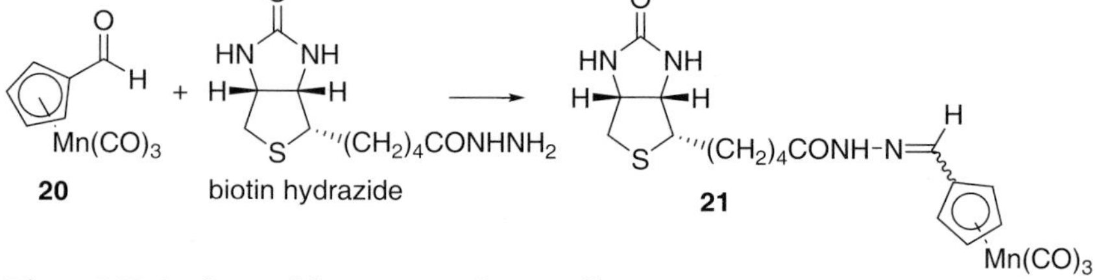

Scheme 8.11 Synthesis of the cymantrenyl tracer of biotin.

8.4.4
Synthesis of Diphenylhydantoin Bearing a Benchrotrene (Benzene chromium Tricarbonyl) Entity

Diphenylhydantoin is a drug used in the treatment of epilepsy. The tracer **24** for this substance was synthesized by esterification of diphenylhydantoin 3-acetic acid **22** by (2-phenylethanol)chromiumtricarbonyl **23** (Scheme 8.12) [26], the latter having been prepared in two steps by alkylation of the sodium salt of diphenyl-hydantoin by ethyl bromoacetate followed by acid hydrolysis of the ethyl ester.

Scheme 8.12 synthesis of the benchrotrenyl tracer of diphenylhydantoin **24**.

8.4.5
Synthesis of Tracers Bearing an Alkyne Dicobalt Hexacarbonyl Moiety

8.4.5.1 Preparation of Tracers for Cortisol and Atrazine
The common principle behind the labeling of these molecules by an alkyne dicobalt hexacarbonyl moiety lies in conjugating the alkyne by peptide coupling between a carboxylic derivative of the analyte and propargylamine, followed by complexation of the alkyne by a dicobalt hexacarbonyl unit.

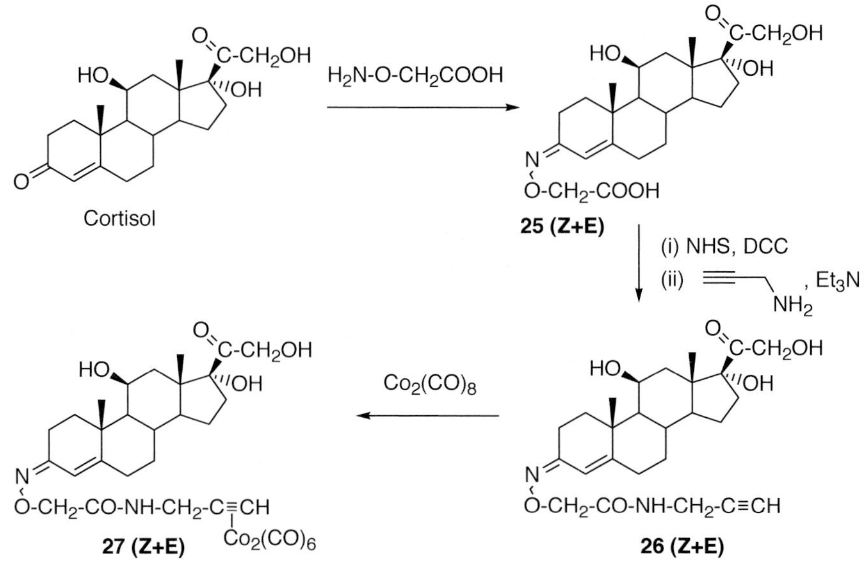

Scheme 8.13 Synthesis of the alkyne dicobalt hexacarbonyl tracer of cortisol **27**.

For cortisol, a natural steroid hormone, position 3 of the ketone was chosen as the site of attachment for the organometallic moiety, since this site had also been used to derive the immunogen responsible for producing the antibodies employed in the assay. The synthetic strategy is summarized in Scheme 8.13.

Oxime **25** carrying a carboxylic acid function was synthesized in a regioselective manner from cortisol and obtained as a mixture of the Z+E isomers. An alkyne was then attached by peptide coupling with propargylamine to yield **26**. Finally, the triple bond of **26** was complexed by action of $Co_2(CO)_8$ to give **27** as a mixture of two separable isomers [42].

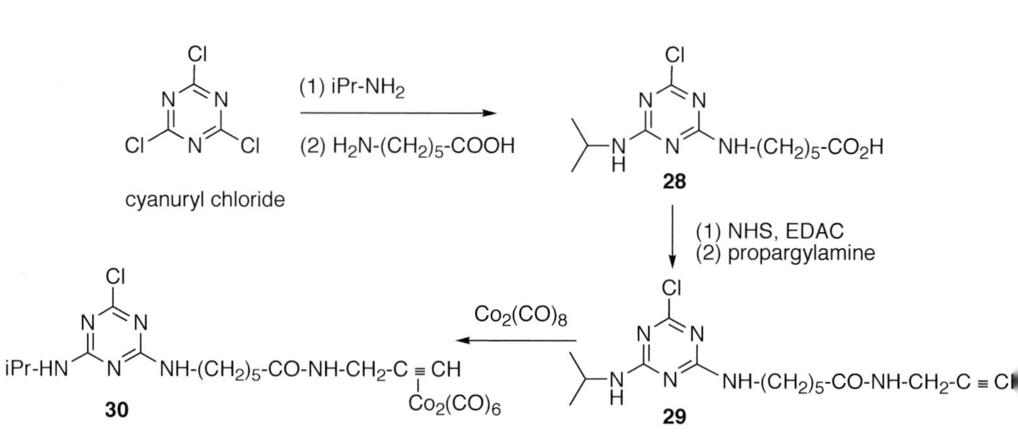

Scheme 8.14 Synthesis of the alkyne dicobalt hexacarbonyl tracer of atrazine **30**.

For atrazine, a herbicide of the triazine family, synthesis of the carboxylic derivative is carried out starting from cyanuryl chloride via double substitution of chlorine by isopropylamine and then 6-aminocaproic acid by the method of [43]. This derivative is then conjugated to propargylamine by peptide coupling in the presence of NHS and EDAC and the triple bond complexed by $Co_2(CO)_8$ to yield the atrazine tracer **30** (Scheme 8.14) [44].

8.4.5.2 Preparation of a Tracer for Carbamazepine

Labeling of carbamazepine, an anti-epileptic, is carried out by conjugating the alkyne dicobalt hexacarbonyl entity to an amino derivative of the analyte **31** via peptide coupling with (*N*-succinimidyl-4-pentynoate) hexacarbonyl dicobalt **32** (Scheme 8.15) [45]. The amino derivative of carbamazepine **31** is prepared in one step from iminostilbene. (*N*-succinimidyl-4-pentynoate) hexacarbonyl dicobalt **32** is synthesized in two steps by esterification of pentynoic acid in the presence of NHS and DCC, followed by complexation of the triple bond by $Co_2(CO)_8$ to give the tracer **33** [46].

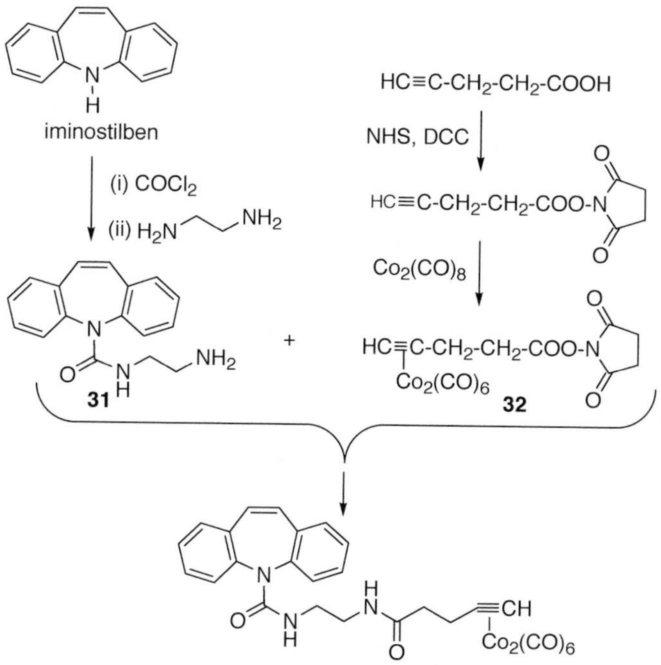

Scheme 8.15 Synthesis of the alkyne dicobalt hexacarbonyl tracer of carbamazepine **33**.

8.4.6
Synthesis of Cationic Tracers

8.4.6.1   **Tracers Labeled with a Cobaltocenium Entity**
Three tracers incorporating the cobaltocenium fragment have been described and utilized in MIA. These are the tracers for amphetamine, a psycho-stimulant, and for diphenylhydantoin and phenobarbital, whose therapeutic activity is outlined above. The general procedure is to use peptide coupling to conjugate an amine, whether pre-existing or added to the analyte, with the acid chloride of cobaltocenium hexafluorophosphate **34** prepared according to the method of Sheats [47] (Scheme 8.16).

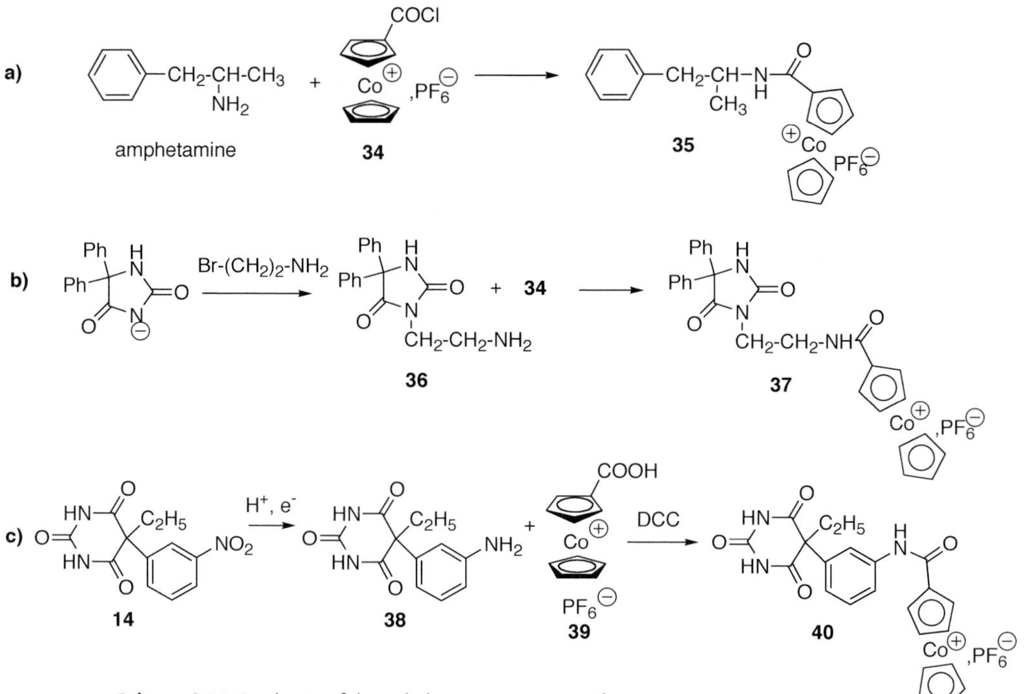

Scheme 8.16 Synthesis of the cobaltocenium tracers of
(a) amphetamine **35**, (b) diphenylhydantoin **37** and (c) phenobarbital **40**.

In the case of amphetamine, the coupling is carried out directly from its primary amine [48]. Labeling of diphenylhydantoin was carried out by coupling onto 3-(2-aminoethyl)-5,5-diphenylhydantoin **36** [49], the latter amino derivative having been prepared by the method of O'Neal [50]. In the case of phenobarbital, labeling is effected by coupling between m-aminophenobarbital **38** and cobaltocenium carboxylic acid hexafluorophosphate **39** in the presence of DCC [51]. Preparation of m-aminophenobarbital **38** was by electrochemical reduction of the nitro derivative **14** [51], whose synthesis is outlined in Scheme 8.9.

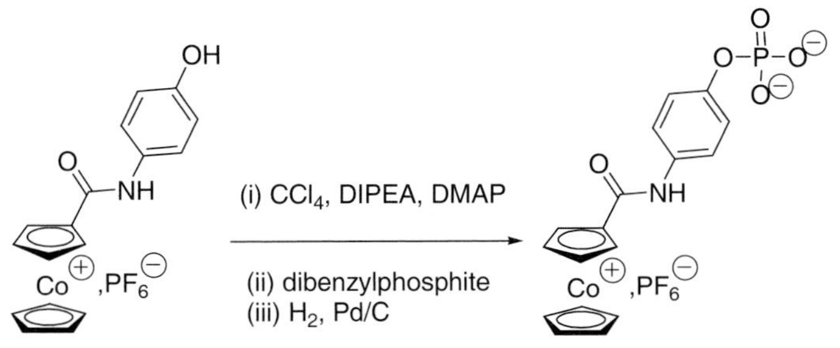

Scheme 8.17 Synthesis of a colbatocenium tracer of a substrate of the alkaline phosphatase.

A cobaltocenium derivative working as a substrate of the enzyme alkaline phosphatase (AP), namely [(4-hydroxyphenyl)amino]-carbonyl]cobaltocenium hexafluorophosphate **42** was prepared starting from [(4-phenylphosphate)amino]-carbonyl]colbatocenium hexafluorophosphate **41** following the general procedure of Silverberg et al. [52]. Compound **41** was obtained as described by Beer et al. [53] as shown in Scheme 8.17 [54].

8.4.6.2 Cationic Tracers Including a Ferrocene Entity

A cationic tracer for diphenylhydantoin including a ferrocene fragment was synthesized in two steps starting from the sodium salt of diphenylhydantoin. A bromopropyl chain was first attached at position 3 by nucleophilic substitution of 1,3-dibromopropane. N,N-dimethylaminomethylferrocene **44** was then alkylated by the brominated derivative of diphenylhydantoin to obtain the tracer **45** in the form of its quaternary ammonium salt (Scheme 8.18) [55].

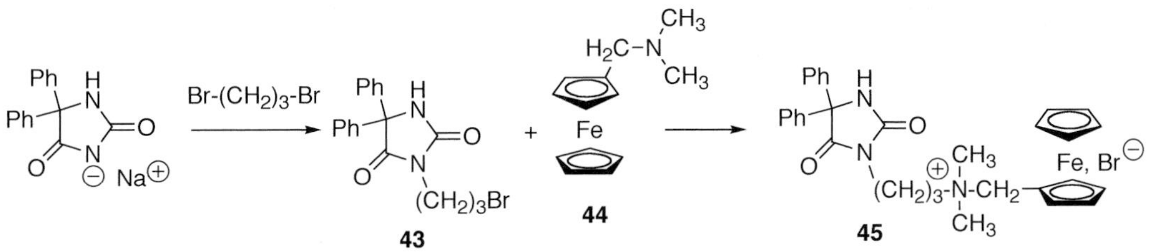

Scheme 8.18 Synthesis of a cationic tracer of diphenylhydantoin **45**.

To increase water solubility of the theophylline tracer and decrease the level of nonspecific interaction between the tracer and antibody, aminomethylation of the substituted cyclopentadienyl ligand was performed via a Mannich reaction in the presence of dimethylaminomethane, after which the tertiary amine was quaternized by reaction with methyl iodide (Scheme 8.19) [28].

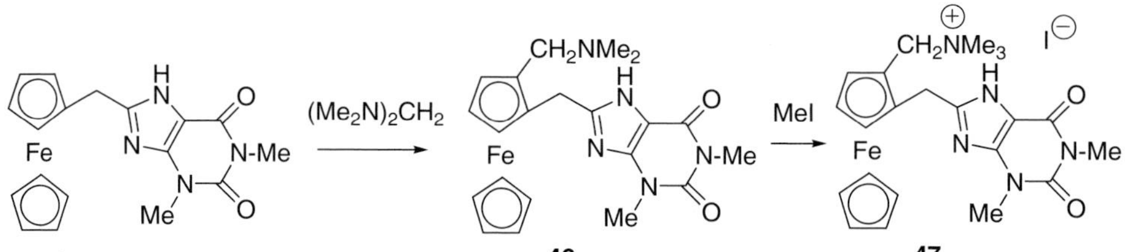

Scheme 8.19 Synthesis of a cationic tracer of theophylline **47**.

8.4.7
Synthesis of a Tracer Bearing a Rhenium Tricarbonyl Fragment

Labeling of human serum albumin (HSA) was carried out by acylation of a fraction of its lysines with the N-succinimidyl ester of the complex [Re(bcp)(CO)$_3$-(4-COOHPy)]ClO$_4^-$ where bcp is 2,9-dimethyl-4,7-diphenyl-1,10-phenantroline, and 4-COOHPy is isonicotinic acid. This ester Re-L **48**, is synthesized in 3 steps starting from bcp, as shown in Scheme 8.20. It is then coupled to HSA to give the complex (Re-L)$_n$-HSA [56]. On average, two Re(CO)$_3$ fragments per molecule of HSA are conjugated.

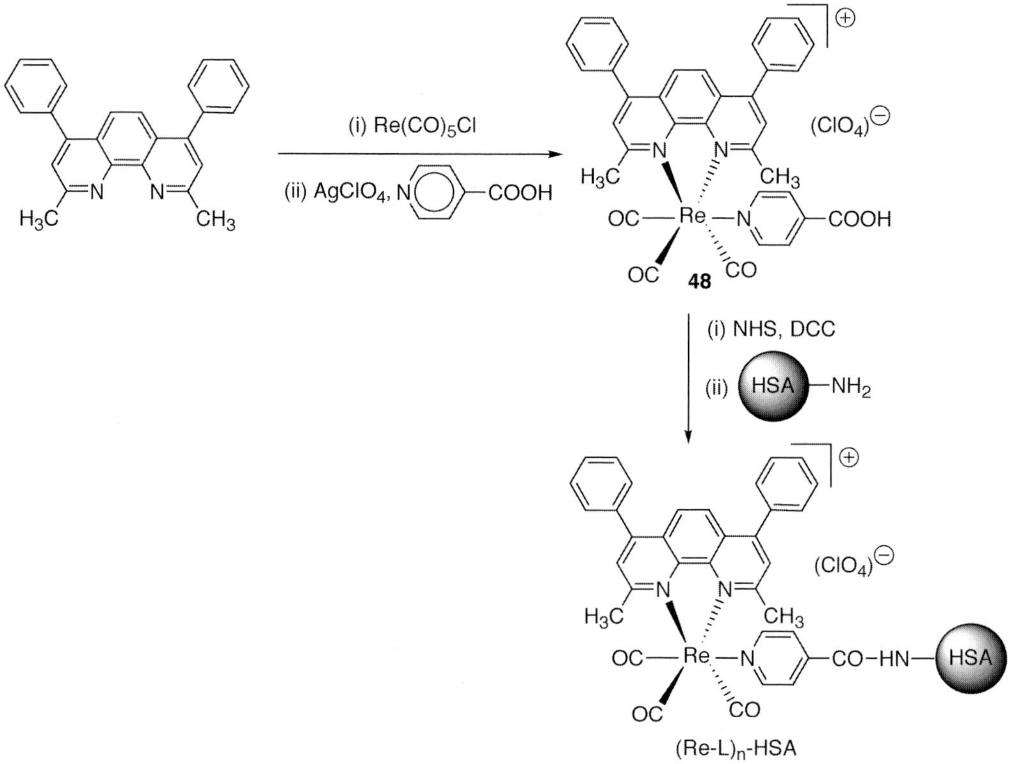

Scheme 8.20 Synthesis of human serum albumin labeled with a rhenium tricarbonyl fragment.

8.5
Examples of Mono- and Multi-metalloimmunoassays (MIA)

Once specific antibodies of the proposed analyte and a tracer have been obtained, an analytical method must be chosen to quantify the tracer. The next step in developing the immunoassay is to establish a dilution curve for the antibodies, obtained by incubation of a fixed quantity of tracer in the presence of variable dilutions of antibodies. From this it is possible to determine the titer of the antibodies (the inverse of the dilution of antibodies binding 50% of the tracer), which gives an indication of the quantity of specific antibodies of the analyte contained in the antiserum. Finally a calibration curve must be established to allow quantification of the analyte. In the case of a competitive immunoassay, this is obtained by incubation of a fixed quantity of antibody (equal to the titer value) in the presence of a fixed quantity of tracer (able to bind ~50% of the antibody) and variable quantities of the analyte to be assayed. *Note:* most of the organometallic complexes reported here are insoluble in water and in the buffers used for immunological analyses, but their solubility in these media can easily be attained by addition of a small percentage of alcohol (~5%).

8.5.1
Metalloimmunoassay Using Atomic Absorption Spectroscopy

Cais, who coined the term metalloimmunoassay, prepared a series of organo-metallic markers (metallohaptens) with the potential for use in immunoassay. These are derivatives of steroids (estradiol, estriol), mood-altering drugs (amphetamine and cocaine) or medications (morphine and barbiturates), labeled with ferrocene, cobaltocenium and cymantrene [57–59]. He then assessed the use of a Sepharose 4B column to separate the free and bound fractions of the tracer and quantification of the tracer by atomic absorption spectrometry [58].

In fact it was P. Brossier who first succeeded in completing all the steps of a MIA of this type, establishing an immunoassay for nortriptyline and for phenobarbital [60–62]. These are non-homogeneous competitive immunoassays using polyclonal antibodies [63]. The metallohaptens used are the cymantrenyl complexes of nortryptiline **12** and phenobarbital **16**, the separation of the free and bound fractions is achieved by selective extraction of the free fraction using an organic solvent [64] and quantification of the metal atom of the tracer (in this case Mn) is performed by Zeemann effect atomic absorption spectroscopy (AAS).

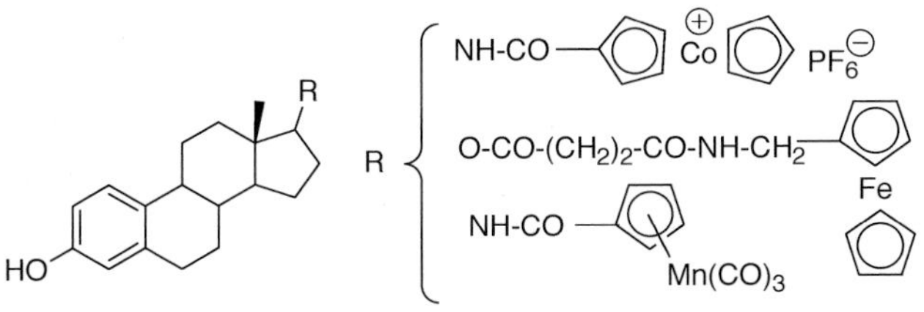

Fig. 8.3 The first examples of metallohaptens.

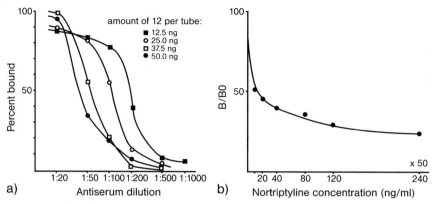

Fig. 8.4 MIA of notriptyline using **12** as tracer and AAS as the method of detection.
(a) Antiserum dilution curves; (b) calibration curve. (Data adapted from [62]).

The first ferrocene-based labels were quickly abandoned, as trace quantities of iron are found in the buffer solutions, and thus perturb the results of quantification by AAS. Cymantrene-based markers were next favored, since manganese is much less likely to occur naturally. Insert Fig. 8.4 shows the MIA of nortriptyline [62]. Antiserum dilution curves were obtained in the presence of 12.5–50.0 ng tracer **12** per tube (Fig. 8.4a).

The feasibility of the assay was demonstrated by the calibration curve shown in Fig. 8.4b. The range of sensitivity for the assay is between 10 and 240 ng mL^{-1} nortriptyline. This is entirely compatible with concentrations of medications in the blood, which are in the microgram mL^{-1} range (cf. Table 8.2). To increase the sensitivity of the assay, signal amplification was obtained using the biotin/streptavidin system. The principle of this 4-step assay is shown in Fig. 8.5. This is

Table 8.2 Range of sensitivity obtained in CMIA.

Analyte	Tracer	Therapeutic range	(B/Bo)$_{50}$	Ref.
Carbamazepine	**33** [Co$_2$(CO)$_6$]-Carbamazepine	7–42 μM	0.2 μM	45
Phenobarbital	**16** [Cy]-Phenobarbital	43–172 μM	0.4 μM	71
Diphenyl-hydantoin	**10** [Fp]-DPH **24** [Cr(CO)$_3$]-DPH	10–200 μM	0.046 μM 0.06 μM	32 26
Cortisol	**27** [Co$_2$(CO)$_6$]-Cortisol	0.14–0.69 μM	0.1 μM	70
Chlortoluron	**19** [Mn(CO)$_3$]-Chlortoluron	MAC[a]: 0.1 μg/L (5 nM)	100 nM	39
Atrazine	**30** [Co$_2$(CO)$_6$]-Atrazine	MAC[a]: 0.1 μg/L (5 nM)	400 nM	44

[a] MAC: Maximum admissible concentration set by the European Community for drinking water.

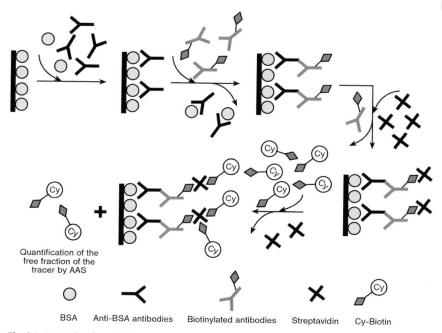

Quantification of the
free fraction of the
tracer by AAS

| BSA | Anti-BSA antibodies | Biotinylated antibodies | Streptavidin | Cy-Biotin |

Fig. 8.5 Principle of a solid-phase streptavidin-cymantrenyl-biotin competitive immunoassay [65].

a solid phase indirect competitive immunoassay of BSA on a solid phase. In the first step there is competition to bind to a fixed quantity of rabbit anti-BSA antibodies between the BSA to be assayed and the BSA attached to the solid phase. In the second step a biotinylated secondary antibody (anti-rabbit IgG) is added, and binds to the primary antibody. In the third step streptavidin is added, which binds to the biotinylated antibody, and in the fourth and final step a fixed quantity of the tracer cymantrenyl-biotin **21** is added, prepared as shown in Scheme 8.11. Each molecule of streptavidin possesses 4 binding sites, thus several tracer molecules can be attached to each antibody, providing signal amplification. Quantification of the tracer is by AAS of the free fraction of cymantrenyl-biotin **21**. In this way a calibration curve was obtained for quantities of BSA ranging between 0 and 200 mg L^{-1}. The detection limit for manganese by AAS is 125 fg per assay of 25 µL (i.e. 92 nM) [65, 66].

8.5.2
Detection by Fourier-transform Infrared Spectroscopy (Carbonyl Metalloimmuno Assay, CMIA)

Carbonyl metalloimmunoassay (CMIA) is a heterogeneous competitive-type metalloimmunoassay using metal carbonyl complexes as tracers, organic solvents (ethyl acetate, isopropyl ether) to separate the free and bound fractions, and Fourier-transform infrared (FT–IR) spectroscopy for quantification. The choice of FT–IR spectrocopy as the analytical method is based on the well-known property of these metal carbonyl complexes to have characteristic mid-IR bands between 1850 and 2200 cm^{-1}, a region of the spectrum where most proteins and organic molecules do not absorb (for details see Chapter 7). In addition, these bands are 8–10 times more intense than all the other bands in the spectrum. Each metal carbonyl complex possesses distinct and characteristic v_{CO} bands, as for example do **33**, the alkyne dicobalt hexacarbonyl tracer of carbamazepine, **24**, the cymantrenyl tracer of phenobarbital and **16**, the benchrotrenyl derivatives of diphenylhydantoin whose infrared spectra are shown in Fig. 8.6. It is this unique property of the M–CO complexes that makes it possible to perform a simultaneous quantitative analysis for two or three of these tracers, thus opening the door to the simultaneous multi-immunoassay to be discussed later in this chapter [67].

The use of infrared spectrometry for quantitive analysis became possible only in the 1980s, when affordable and user-friendly benchtop Fourier-transform spectrometers became available. The sensitivity of the FT–IR spectroscopy was, however, insufficient to meet the requirements of the immunoassay. To address this problem, an instrument equipped with a liquid nitrogen-cooled detector made from a semi-conducting material, for example MCT (mercury–cadmium–telluride) or InSb (indium antimonide), was used to increase sensitivity by a factor of 20 compared with the thermal detector DTGS found in standard FT–IR machines. Use of a "light-pipe" cell with a long optical path (20 mm) for a

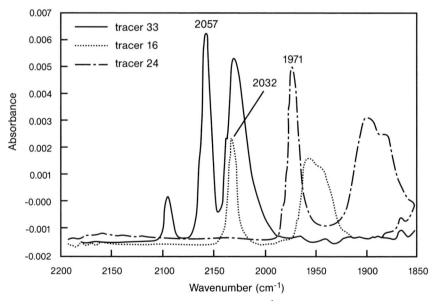

Fig. 8.6 Superimposition of the IR spectra (20 scans, 4 cm^{-1} resolution) of the three metal carbonyl tracers **33**, **24** and **16** in solution in chloroform recorded on a MB100 FT-spectrometer (Bomem) equipped with an InSb detector and a light-pipe cell. Wavenumbers of the analytical bands used for quantitative analysis are shown for each tracer.

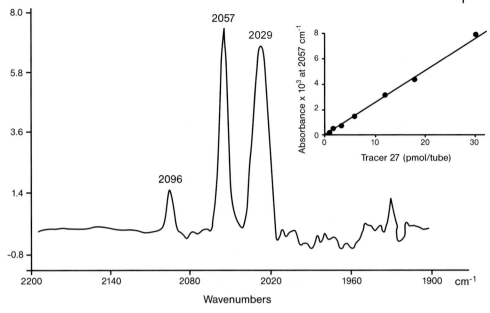

Fig. 8.7 FT–IR spectrum of 30 pmol of **27** in CCl$_4$ (10 scans, 4 cm^{-1} resolution).
Expansion of the v(CO) region between 2200 and 1900 cm^{-1}. Inset Beer's
law plot for **27** ranging from 1.25 to 30 pmol/tube. Absorbance at 2057 cm^{-1}
(r = 0.997). (Data adapted from [70]).

fill volume of only 30 μL also increased the sensitivity of the spectrometer, by a
factor of 200 [68]. The height of the peaks is proportional to the quantity of tracer
which allows their quantification (Fig. 8.7). For each complex, the analytical peak,
i.e. the peak that will give the most sensitive analyses, is identified. For example,
the analytical peak is the one at 2057 cm^{-1} for **33**, at 2032 cm^{-1} for **16**, and at
1971 cm^{-1} for **24**. Cobalt carbonyl tracers give the best detection limit, with a
limit of 300 fmol being attained for an analysis of **33** in 30 μL CCl$_4$ (i.e. 10 nM)
[69].

8.5.2.1 Mono-immunoassays by CMIA
Six CMIA mono-immunoassays were developed (cf. Table 8.1). The first of these
were in the area of clinical biology, with CMIA assays of antiepileptic medications
such as carbamazepine [45], phenobarbital [71], diphenylhydantoin [26, 32], and
a steroid hormone, cortisol [70]. More recently they have been extended into the
environmental area, with assays of pesticides atrazine [44] and chlortoluron [39].
All these assays used polyclonal antibodies. For each of the analytes dilution curves
and standard curves were obtained. The curves obtained for carbamazepine,
representative of the results for all compounds, are shown in Fig. 8.8.

The range of sensitivity of an immunoassay is given by the (B/Bo)$_{50}$ value,
which is the quantity of the analyte which displaces 50% of the quantity of tracer.
Values obtained in CMIA for the antiepileptics and cortisol are listed in Table 8.2.

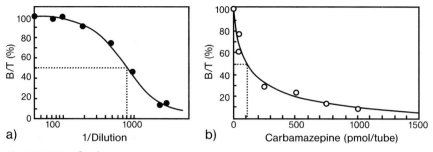

Fig. 8.8 CMIA of carbamazepine.
(a) Antiserum dilution curve; (b) standard curve of carbamazepine.
Both curves were obtained in the presence of 30 pmol of **33**. (Data from [45]).

At the therapeutic range for antiepileptic medications, CMIA requires only a few microlitres of serum from the patient; for cortisol the assay of serum from a patient suffering from hyperandrogeny was performed with 50 µL of serum. Where comparison was possible, it was shown that the sensitivity of the CMIA method is identical to that of a RIA assay using a tracer labeled with ^{14}C.

Two of the principal advantages of CMIA are the absence of non-specific signals in the infrared spectral region (due to the artificial nature of the tracers), and the possibility of performing simultaneous multi-immunoassays.

8.5.2.2 Multi-immunoassay by CMIA

The possibility of performing several immunoassays simultaneously is of obvious economic interest, since it economizes on both time and reagents. It is especially appropriate for making particular diagnoses that involve the assay of several biochemical parameters (endocrine diseases, fertility tests, cardiac markers, thyroid conditions, diagnostic procedures for Down's syndrome) [72]. However, the scarcity of simultaneously analyzable tracers means that the number of multi-immuno-assays based on the simultaneous detection of several tracers remains very limited. The Delfia® commercial systems using rare earth chelates as tracers and time resolved fluorescence as the method of detection permit simultaneous dual-immunoassays [73, 74] and even quadruple-immunoassay using four chelates (Eu(III) Tb(III), Sm(III) and Dy(III)) [75]. This example is unique in the literature. Metal carbonyl labels, with their characteristic ν_{CO} vibration bands that are distinct for each tracer, present another possibility for the simultaneous multi-immuno-assay of several analytes in a single sample. An example of a double immunoassay was reported for the case of a mixture of two antiepileptics, carbamazepine and diphenylhydantoin using a mixture of tracers **33** and **24** [76]. Since the infrared bands of these two tracers have no overlap (cf. Fig. 8.6), quantification is a simple matter of measuring the height of the peak at 2057 cm^{-1} for **33** and the peak at 1971 cm^{-1} for **24**. The simultaneous triple immunoassay of three antiepileptics (carbamazepine, phenobarbital and diphenylhydantoin) has also been reported [77]. In the case of the mixture of three tracers **33**, **16** and **24** there is overlap of the bands (cf. Fig. 8.9) except for the one for the cobalt complex at 2058 cm^{-1}.

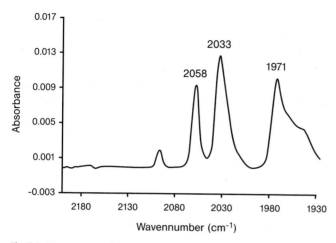

Fig. 8.9 IR spectrum of the mixture of the three tracers **33**, **24** and **16**. (Data from [77]).

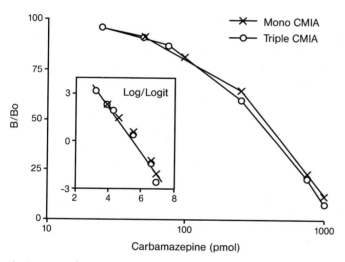

Fig. 8.10 Superimposition of the standard curves of carbamazepine in mono and triple CMIA. Inset: linearization of the curves. (Data from [77]).

Calculation methods such as multivariate method based on least-squares or partial least-squares algorithms [78] or univariate method [77] allow quantification of each tracer in the 10–100 pmol range. The standard curves for the assay of carbamazepine in mono and triple-CMIA are shown in Fig. 8.10.

8.5.2.3 New Developments in the CMIA Method

The sensitivity of the CMIA method is well suited for quantification of drugs in blood serum, but this ceases to be the case when working with pesticides, which require a sensitivity compatible with the European Community regulation

(Maximum Admissible Concentration (MAC) = 0.1 µg/L) for pesticides in drinking water. Recent studies have been aimed at increasing the sensitivity of CMIA by attaching several M–CO probes borne by PAMAM dendrimers to antibodies [19] or to avidin which can then be used in conjugation with biotinylated antibodies [18]. Detection of these labeled antibodies can be carried out on membranes, making it possible to omit the step of extracting the free fraction by solvent.

8.5.3
Electrochemical Detection

Electrochemistry is a cheap analytical method that can be used with cloudy test samples and works well with miniaturization. It was thus adopted early as an ideal analytical method when novel reliable and sensitive non-isotopic immuno-assays were developed [79]. In addition, ferrocene has long been known as an excellent redox mediator, used in many applications some of which are discussed in this book (cf. Chapters 6 and 9). One example is in commercial biosensor devices, such as the hand-held Medisense Precision QID glucose meter for use by diabetics to determine their sugar levels in whole blood [80]. It is therefore unsurprising to find that some of the oldest and best developed MIAs employ electrochemical methods to quantify the tracer.

8.5.3.1 Homogeneous Ferrocene-mediated Amperometric Immunoassay
The potential of ferrocene as an electrochemical label in immunoassays was first explored by Weber and Purdy [81] who prepared a ferrocene–morphine conjugate and demonstrated that its electrochemistry was perturbed on binding to the anti-body so that a homogeneous competitive assay could, in principle, be configured in which separation of free and antibody-bound label was unnecessary. These ideas were extended by McNeil and led to the immunoassay of lidocaine, which became one of the first complete MIAs to be published [27]. The tracer used in this immunoassay is lidocaine labeled with dimethyl-ferrocene **4** (Scheme 8.3) detected by cyclic voltammetry. Labeled lidocaine acts as an electron acceptor for glucose oxidase (GOD) in a non-oxygen dependent manner [82], leading to a substantial amplification of the signal, visible in Fig. 8.11 [28].

Since binding of **4** to a specific antibody produces an electrochemically inert entity, the intensity of the current is proportional to the quantity of free tracer in the medium, and thus inversely related to the quantity of analyte present. This is a competitive homogeneous immunoassay in which separation of the free and bound fractions of the tracer is not required (Fig. 8.12).

A calibration curve was obtained in the 5–50 µM range for lidocaine, which corresponds well to its therapeutic range (6–21 µM). The approach used in this assay was recently used again by Forrow et al. for an assay of theophylline, the tracer in this case being the derivative of theophylline labeled with a quaternary ammo-nium salt of ferrocene **6** (Scheme 8.17) [28]. Use of a cationic tracer increases the solubility of the tracer in biological media and lessens the non-specific hydrophobic binding between the uncharged ferrocene derivatives and the antibodies.

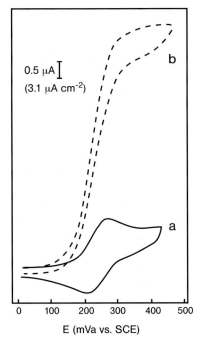

Fig. 8.11 (a) Cyclic voltammogram (at 5 mV s^{-1}) of **6** (0.25 mM) in 0.6 ml of phosphate buffer at a gold electrode (area 0.16 cm^2); (b) as for **a**, but with the addition of 100 mL of stock glucose oxidase solution at a final concentration of 2.1 mM. (Data from [28]).

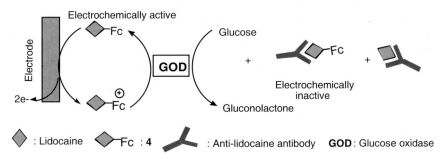

: Lidocaine   Fc : **4**   : Anti-lidocaine antibody   **GOD**: Glucose oxidase

Fig. 8.12 Ferrocene-mediated amperometric homogeneous immunoassay of lidocaine. Only free ferrocenyl lidocaine tracer **4** is electrochemically active [28].

8.5.3.2 Homogeneous Electrochemical Immunoassay (Square Wave Voltammetry)

The principle of this competitive immunoassay without separation, developed by Degrand and Limoges, is based on electrochemical detection of a cationic electroactive label, a cobaltocenium salt and on the specific properties of Nafion™, a polyanionic perfluorosulfonate possessing a strong affinity for small-sized cationic organic molecules. Its effect is to sort the molecules within a mixture according to their size and charge [48].

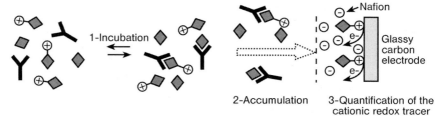

Fig. 8.13 Principle of a homogeneous electrochemical immunoassay using a Nafion coated electrode as the detector for a cationic organometallic tracer.

In the case of an immunoassay (Fig. 8.13) only the free fraction of the cationic tracers is extracted from the mixture and accumulates at the surface of the electrode where it is detected electrochemically. The antibody–cationic tracer complexes are too large to penetrate the polymer and are thus not detected. This is an elegant approach to the separation of free and bound fractions of tracers, permitting a homogeneous-phase like immunoassay.

The tracer is quantified by square wave voltammetry and the intensity of the peak current is proportional to the concentration of free tracer that accumulates at the surface of the electrode, and thus inversely related to the quantity of analyte present in the solution. The amplification of the signal (intensity of the current) obtained by use of Nafion is clearly shown in Fig. 8.14. The cobaltocenium tracers of amphetamine **35** and diphenylhydantoin **37** (Scheme 8.16a and b) were used for the development of the immunoassay of these substances [49, 83]. This method allows analytes to be assayed in the range of 2–4 μM.

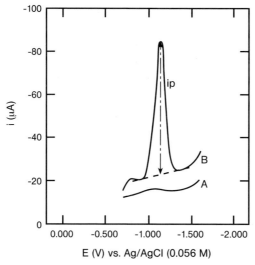

Fig. 8.14 Square wave voltammetry (f = 100 Hz, Esw = 50 mV) of [Co]$^+$ of complex **35** at glassy carbon (GC) electrode (A) and GC/Nafion (B) (film thickness 0.4 μ) for 5 min at 600 rpm and T = 22 °C. (Data from [48]).

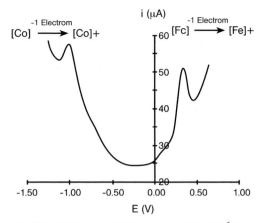

Fig. 8.15 SWV curve of [Co]$^+$ complex **36** (1 x 10^{-6} M) and [Fc]$^+$ complex **39** (2 x 10^{-6} M) at Nafion-loaded CPE (carbon paste electrode) in buffer containing 1% EtOH and 10% rabbit normal serum. (Data adapted from [55]).

One of the advantages of the technique is the potential for performing a simultaneous double immunoassay by combining tracers with 2 different cationic eletroactive markers. The first marker **40** is the cobaltocenium coupled to phenobarbital (Scheme 8.16c) while the second tracer **45** is diphenylhydantoin couple to ferrocene (Scheme 8.18). These two electroactive tracers, possess the unique advantage of having very different standard potentials (E^0 = 0.26 V for the ferricinium/ferrocene pair and E^0 = −1.05 V for the cobaltocenium/cobaltocene pair) thus permitting simultaneous quantification of the two tracers (Fig. 8.15). This allowed development of the double simultaneous immunoassay of the two drugs, phenobarbital and diphenylhydantoin [55]. It was possible to perform the immunoassay of these two antiepileptic drugs at a detection limit of 0.5 μmol L^{-1}.

8.5.3.3 Electrochemical Flow Immunoassay System

Two immunoassays of this type have been described with differing methods to separate the free and bound fractions of the tracer. In the assay described by Matsunaga, the method used is based on the difference in the isoelectric point (pI) values of the free and bound antibodies, permitting separation on an ion exchange capillary column [84–86]. The bound fraction of the antibodies (pI = 5.6) is not retained on the column and is quantified by continuous amperometric flow analysis, while the free fraction (pI = 7) is subsequently eluted using a buffer (pH 6.0 with 50 mM malonate) and can be reused up to 8 times (Fig. 8.16).

Separation based on the isoelectric point difference can also be obtained on glass fiber membrane modified with anion [87] or on a multichanneled matrix column coated with cation-exchange resin [30]. This electrochemical flow immunoassay system was also used for the assay of HCG, the tracer being a monoclonal antibody labeled with ferrocene and prepared as shown in Scheme 8.6 [31, 87]. A linear relationship was observed between 10 and 2000 mI mL^{-1} of HCG, indicating that this immunoassay is suitable for diagnosis of pregnancy and/or gonadal tumors in

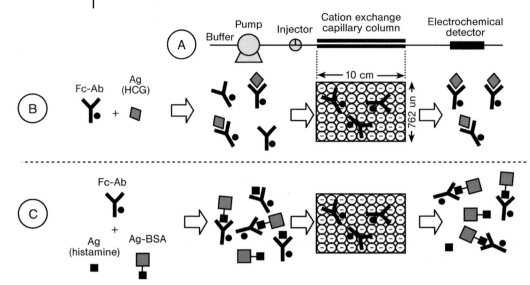

Fig. 8.16 (A) Schematic diagram of the electrochemical capillary flow immunoassay system. (B) Principle of electrochemical capillary flow immunoassay for a protein (HCG), (C) for a hapten (histamine). (Adapted from [31]).

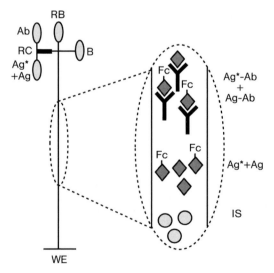

Fig. 8.17 Schematic of the on-chip competitive electrochemical immunoassay protocols described by Wang et al. [29]. (RB) running buffer, (B) unused buffer reservoir, (Ag*) Ferrocene labeled antigen, (Ag) antigen, (Ab) antibody, (RC) reaction chamber, (IS) internal standard, (WE) screen-printed working electrode.

clinical practice. This immunoassay has also been reported for the assay of a hapten, histamine (Fig. 8.16C). In this case there is competition for binding to the labeled (Fc–Ab) antibody between the histamine to be assayed and the histamine bound to BSA. It is the (Fc–Ab)-[BSA-histamine] complex that becomes the tracer and that is quantified by electrochemistry [30]. The assay yields a good relationship between current and histamine concentration in the range of 200–2000 ng mL^{-1} and allows the assay of histamine released in whole blood within 2 min. The entire system has been miniaturized and is available on a credit-card sized chip.

In the second assay type described by Wang, the separation of the free and bound fractions is carried out by capillary electrophoresis (Fig. 8.17) [29]. The feasibility of this assay has been shown for the miniaturized competitive assay of a hapten (triiodothyronine, T_3), the tracer **8** being the hapten labeled with ferrocene was prepared as shown in Scheme 8.5. The free and bound fractions of the tracers are quantified by amperometric detection after leaving the electrophoresis column, using gold-plated screen-printed carbon working electrodes. The detection limit is 1 µg mL^{-1} of T_3.

8.5.4
Detection by Fluorescence Polarization (FP)

The use of fluorescent labels has been very successful for inorganic complexes such as the europium chelates, which are used today in the Delfia® commercial system. In contrast, fluorescence has rarely been employed in the area of organometallic tracers, and only the immunoassay developed by Lakowicz uses such a tracer [88]. This is a homogeneous competitive immunoassay, the tracer being the complex (Re-L)$_n$-HSA, **48** obtained as shown in Scheme 8.20 and the detection method fluorescence polarization (FP). This compound **48** displays highly polarized emission (with a maximum polarization near 0.4 and maximum anisotropy near 0.3) in the absence of rotational diffusion and a long average lifetime (2.7 µs) when bound to proteins in air-equilibrated aqueous solution. The steady-state polarization of (Re-L)$_n$-HSA was sensitive to the binding of anti-HSA, resulting in a significant increase in luminescence polarization. Consequently (Re-L)$_n$-HSA can be used as a tracer in a competitive immunoassay with unlabeled HSA acting as an antigen (Fig. 8.18).

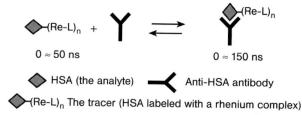

Fig. 8.18 Principle of Fluorescence Polarization Immunoassay (FPIA). The rotational correlation time (θ) is different for the free [(Re-L)$_n$-HSA] and bound [(Re-L)$_n$-HSA]-anti-HSA-antibody fraction of the tracer.

The sensitivity of the assay is in the 100 nM range. Use of this organometallic rhenium tracer has made it possible to extend the range of use of the hapten assay technique in which it is widely employed to include high molecular weight analytes (10^5 to 10^8 D).

8.6
Use of Organometallic Complexes as Substrates or Co-substrates for Enzyme Immunoassay

In the mid 1970s, immunoassay sensitivity was greatly increased with the arrival of the enzymo immunoassay (EIA) [21–24]. In this type of assay, an enzyme attached to the detection antibody (cf. Fig. 8.2B) or to the analyte allows amplification of the signal. These types of assay are currently the most widespread. Derivatives of ferrocene have been used in this type of assay, several of which are discussed below. This is in fact an extension of the term MIA. In fact, the organometallic complex as used here is not the tracer but the substrate or co-substrate (redox mediator) of an enzyme.

8.6.1
Organometallic Complexes Used as Enzyme Substrates

This is an immunoassay in which competition for binding to the specific antibodies, attached to magnetic beads, is between the analyte to be assayed and the analyte attached to an enzyme, alkaline phosphatase (AP) (Fig. 8.19).

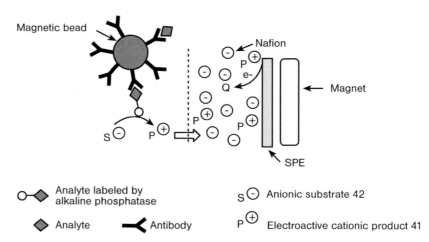

Fig. 8.19 Key step of the magnetic electrochemical assay using an organometallic complex S⁻ as substrate for an enzyme (Alkaline Phosphatase) linked to the analyte and a screen-printed electrode (SPE) coated with Nafion for the electrochemical detection of the cationic product P⁺. (Adapted from [89]).

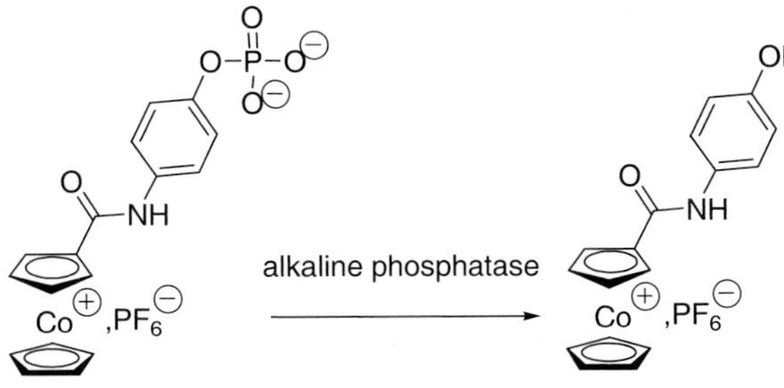

Scheme 8.21 Hydrolysis, catalyzed by alkaline phosphatase, of phosphoric acid ester **42**.

After incubation the magnetic beads cluster at a magnet near the electrode and the activity of the enzyme attached to the beads is measured, using as substrate the organometallic complex **42**, prepared as shown in Scheme 8.17. This anionic substrate **42** is transformed by the alkaline phosphatase in the cationic phenol derivative **41** (Scheme 8.21).

This is then concentrated in the polyanionic Nafion film coating the electrode and quantified by cyclic voltammetry as shown in Fig. 8.19. The feasibility of this immunoassay has been demonstrated for a herbicide, 2,4-dichlorophenoxyacetic acid (2,4-D) with a detection limit of 0.01 μg L^{-1} [89].

8.6.2
Organometallic Complexes Used as Enzyme Co-substrates (Redox Mediators)

8.6.2.1 Flow Injection Immunoassay with Electrochemical Detection
This immunosensor-type assay uses ferrocene methanol as co-substrate (redox mediator) of glucose oxidase (GOD). The various steps of the assay, shown in Fig. 8.20, are as follows: (1) Covalent immobilization of protein A on graphite-polystyrene screen-printed electrodes (SPEs) (2) Addition of the rabbit IgG to be quantified which is captured specifically by protein A (3) Addition of a biotinylated goat anti-rabbit antibody (4) Addition of avidin-GOD conjugate (5) Addition of glucose and ferrocene methanol (6) Measurement of catalytic current by flow injection immunoassay. The voltammetric current corresponds to the one-electron oxidation of the ferrocenyl group to ferricinium. The electrode can subsequently be regenerated up to 30 times. The feasibility of the assay has been demonstrated for the case of monoclonal mouse anti-human prolactin (PL) with a detection limit of 0.02 μg mL^{-1} [90].

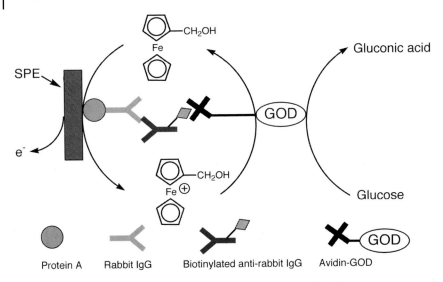

Fig. 8.20 Principle of an enzyme immunoassay using ferrocene methanol as the redox mediator of glucose oxidase (GOD).

8.6.2.2 Dual Enzyme Immunoassay with Amperometric Detection

This is a non-competitive immunoassay using ferrocene carboxylic acid **7** as the redox mediator of horseradish peroxidase (POD), which is itself attached to the detection antibody (Scheme 8.21) [91]. When an anode potential is applied, a catalytic current is generated at the surface of the electrode, with an intensity proportional to the quantity of POD located in close proximity to the surface of the electrodes, and thus proportional to the quantity of captured analyte. Applying several antibodies sequentially on the same electrode permitted a double immunoassay of LH and FSH with detection limits respectively of 2.1 and 1.8 IU L^{-1}.

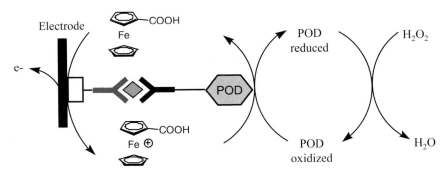

Fig. 8.21 Principle of a non-competitive immunoassay using ferrocene carboxylic acid as the redox mediator of horseradish peroxidase (POD).

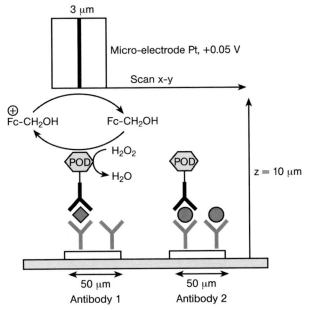

Fig. 8.22 Principle of a dual immunoassay using ferrocene methanol as redox mediator of horseradish peroxidase (POD) and electrochemical microscopy detection. (Adapted from [91].)

8.6.2.3 Dual Enzyme Immunoassay Using Electrochemical Microscopy Detection

This is another non-competitive immunoassay using a detection antibody labeled with POD, and ferrocenyl methanol as a redox mediator, but here the detection method makes use of the electrochemical scanning microscope [91] which, as shown in Fig. 8.22, involves using a micro-electrode of 3 μ in diameter to scan a smooth solid phase on which are distributed distinct microzones (50 μ × 50 μ) of the specific antibodies of the substances to be assayed (here HCG and PL). The micro-electrode acts as a microprobe able to identify and quantify locally the electro-chemical reactions generated at the surface of the solid phase.

Using an x–y scan, a three-dimensional image is obtained of the electrochemical phenomena occurring at the surface. In this way it is possible to obtain a quantitative image of the enzyme conjugate fixed at the surface of the antibody microzones, and thus indirectly to assay the analytes sandwiched between them. Analytical detection limits of 0.1 IU mL^{-1} and 3 ng mL^{-1} were established for HCG and PL respectively.

8.7
Conclusions

When our research began into the use of bioorganometallics in immunological analysis, the photosensitivity of metal carbonyl complexes and their insolubility in aqueous solvents were seen as major obstacles to their use in biological media. Despite this, and with the collaboration of others, notably P. Brossier in Dijon, and C. Degrand and B. Limoges in Clermont–Ferrand, we patiently plowed our furrow and allowed this novel and multidisciplinary endeavor at the intersection of organometallic, analytical and biomedical chemistry to reach its potential, culminating in the development of new non-isotopic immunoassays. A quarter-century later, two tracer/detection methods seem to stand out from the rest. They are (1) ferrocene complexes which, because of their unusual redox behavior, can be detected by electrochemistry, an economical method that lends itself to miniaturization and automation; and (2) metal carbonyl complexes whose $M(CO)$ vibrations can be detected by infrared. The need to optimize the infrared technique to attain the required sensitivity allowed us to reach a detection limit of 300 femtomoles which, to the best of our knowledge, has not been bettered since. This is still insufficient, however, to allow assay of environmental pollutants without preconcentration. Recent developments aimed at amplifying the infrared signal by multiple labeling of the antibodies using organometallic dendrimers may give new impetus to this method, and would also eliminate the need for a separation step. The chief advantages of the CMIA assay method are still the absence of non-specific signals, thanks to the artificial nature of the tracers, and the potential for simultaneous multiple immunoassay, a possibility rarely offered by other techniques.

In Europe immunoassays represented a market of 1.76 billion Euros in 2003 [92]. The DELFIA® system, which uses tracers based on europium chelates, was one of the select few to share this very competitive market. ELISA, thanks to its ease of use, is uncontestibly the most widely-used system, but it is not best adapted for hapten assay, which is a particularly advantageous field for the multiple immunoassay, and there is nothing to prevent bioorganometallic chemistry making an exemplary breakthrough in a new technology like that of microarrays [93]. The likelihood is enhanced by the fact that the sensitivity of FT–IR spectrometers should increase even more when existing technology, in the form of new detectors and beam splitters, eventually reaches the market.

Acknowledgments

We wish particularly to thank Professor M. Cais (Technion; Israel Institute of Technology) who during his sabbatical visit to Rennes in 1978–1979 allowed us to witness the first halting steps of the metalloimmunoassay, Professors I. S. Butler (McGill University) and P. Brossier (Université de Bourgogne) for their major contribution to the development of the CMIA method; Prof. Butler by offering us the opportunity from 1981 onwards of working on a Fourier-transform infrared spectrometer, Prof. Brossier for his enthusiasm, his collaboration and his "Fauves de Bourgogne" rabbits, our source of specific antibodies without which this research would not have been possible. Nor do we wish to forget the students and collaborators who are participating, or who have participated in our research: N. Fischer-Durand, J. M. Heldt, F. Le Bideau, V. Philomin, A. Varenne, and B. McGlinchey for her valuable help in the translation of our articles. We also wish to thank the Association France-Québec, the Région de Bourgogne, the CNRS and MRT for financial support.

Abbreviations

AAS	Atomic absorption spectroscopy
Ab	Antibody
Ag	Antigen
Ag-M	Organometallic antigen
AP	alkaline phosphatase
BSA	Bovine serum albumin
CPE	Carbon paste electrode
CMIA	Carbonylmetalloimmunoassay
Cy	Cymantrenyl
DCC	N,N'-diisopropylcarbodiimide
DMA	N,N-dimethylaniline
DMAP	N,N-dimethylaminopyridine
DPH	Diphenyl hydantoin
EDAC	N-ethyl-N'-(3-dimethylaminopropyl)carbodiimide
EIA	Enzyme immunoassay
ELISA	Enzyme linked immuno sorbent assay
Fc	ferrocene
FIA	Flow injection assay
Fp	cyclopentadienyl dicarbonyl iron
FP	fluorescence polarization
FPIA	fluorescence polarization immunoassay
FT–IR	Fourier-transform infrared
GC	Glassy Carbon
GOD	Glucose oxidase
HCG	human chorionic gonadotrophin
HSA	human serum albumin
IgG	Immunoglobulin G
MIA	MetalloImmunoAssay
NHS	N-hydroxysuccinimide
NHSS	N-hydroxysulfosuccinimide
PL	Prolactin
POD	horseradish peroxidase

RIA   Radioimmunoassay

SWV   Square wave voltammetry

TSTU N,N,N',N'-tetramethyl-O-(N-succinimidyl)uronium tetrafluoroborate

References

1 M. Cais, S. Dani, Y. Eden, O. Gandolfi, M. Horn, E. E. Isaacs, Y. Josephy, Y. Saar, E. Slovin, L. Snarsky, *Nature* **1977**, *270*, 534–535.

2 R. S. Yalow, S. A. Berson, *J. Clin. Invest.* **1960**, *39*, 1157.

3 R. S. Yalow, S. A. Berson, *Nature* **1959**, *84*, 1648.

4 R. Edwards, *Immunoassay*, William Heinemann Medical Books, London, **1985**.

5 T. Chard, *An Introduction to Radioimmunoassay and Related Techniques*, Elsevier/North Holland Biomedical Press, Amsterdam, **1978**.

6 J. P. Gosling, *Clin. Chem.* **1990**, *36*, 1408–1427.

7 G. B. Wisdom, *Ligand Rev.* **1981**, *3*, 44–49.

8 M. Oellerich, *J. Clin. Chem. Clin. Biochem.* **1984**, *22*, 895–904.

9 E. Ishikawa, *Clin. Biochem* **1987**, *20*, 375–385.

10 E. P. Diamandis, *Clin. Biochem* **1988**, *1988*, 139–150.

11 I. Hemmilä, *Clin. Chem.* **1985**, *31*, 359–370.

12 D. S. Smith, M. H. H. Al-Hakiem, J. A. Landon, *Ann. Clin. Biochem.* **1981**, *18*, 253–274.

13 I. Weeks, J. S. Woodhead, *J. Clin. Immunoassay* **1984**, *7*, 82–89.

14 W. R. Seitz, *Clin. Biochem* **1984**, *17*, 120–125.

15 L. J. Kricka, G. H. G. Thorpe, *Ligand Rev.* **1981**, *3*, 17–24.

16 E. O. Fischer, W. Pfab, *Z. Naturforsch* **1952**, *7b*, 377.

17 G. Wilkinson, A. Stone, E. W. Abel, *Comprehensive organometallic chemistry. The synthesis, reactions and structures of organometallics compounds*, Pergamon Press, New York, **1982**.

18 J. M. Heldt, N. Fischer-Durand, M. Salmain, A. Vessières, G. Jaouen, *J. Organomet. Chem.* **2004**, *689*, 4775–4782.

19 N. Fischer-Durand, M. Salmain, B. Rudolf, A. Vessières, J. Zakrzewski, G. Jaouen, *ChemBioChem* **2004**, *5*, 519–525.

20 G. Köhler, C. Milstein, *Nature* **1975**, *256*, 495–497.

21 E. Engvall, P. Perlmann, *Immunochemistry* **1971**, *8*, 871–874.

22 L. Wide, Solid-phase antigen-antibody systems, in K. E. Kirkham, W. M. Hunter (Eds.), *Radioimmunoassay Methods*, pp. 405–412, Churchill Livingstone, Edinburgh, **1971**.

23 L. Bélanger, C. Sylvestre, D. Dufour., *Clin. Chim. Acta* **1973**, *48*, 15–18.

24 R. Maiolini, R. Messeyeff, *J. Immunol. Methods* **1975**, *8*, 223–234.

25 B. F. Erlanger, *Methods Enzymol.* **1980**, *70*, 85–104.

26 A. Varenne, A. Vessières, P. Brossier, G. Jaouen, *Res. Commun. Chem. Pathol. Pharmacol.* **1994**, *84*, 81–92.

27 K. Di Gleria, H. A. O. Hill, C. J. McNeil, *Anal. Chem.* **1986**, *58*, 1203–1205.

28 N. J. Forrow, N. C. Foulds, J. E. Frew, J. T. Law, *Bioconjugate Chem.* **2004**, *15*, 137–144.

29 J. Wang, A. Ibanez, M. P. Chatrathi, *Electrophoresis* **2002**, *23*, 3744–3749.

30 T. K. Lim, H. Ohta, T. Matsunaga, *Anal. Chem.* **2003**, *75*, 3316–3321.

31 T. K. Lim, T. Matsunaga, *Biosens. Bioelectron.* **2001**, *16*, 1063–1069.

32 A. Vessières, K. Kowalski, J. Zakrzewski, A. Stepien, M. Grabowski, G. Jaouen, *Bioconjugate Chem.* **1999**, *10*, 379–385.

33 J. Zakrzewski, *J. Organomet. Chem.* **1989**, *359*, 215–219.

34 I. Lavastre, J. Besançon, P. Brossier, C. Moise, *Appl. Organomet. Chem.* **1990**, *4*, 9–17.

35 E. W. Bousquet, R. Adams, *J. Am. Chem. Soc.* **1930**, *52*, 224.

36 P. Brossier, C. Moise, Exemples d'utilisation d'un complexe organo-métallique comme marqueur de molécules thérapeutiques, in *Radioimmunoassay and related procedures in medicine 1982*, pp. 779–786, International Atomic Energy Agency, Vienna, **1982**.

37 M. Katmeh, G. Frost, W. Aherne, D. Stevenson, *Analyst* **1994**, *119*, 431.

38 R. Dabard, M. Le Plouzennec, *Bull. Soc. Chim.* **1972**, 3594.

39 A. Vessières, N. Fischer-Durand, F. Le Bideau, P. Janvier, J. M. Heldt, S. Ben Rejeb, G. Jaouen, *Appl. Organomet. Chem.* **2002**, *16*, 669–674.

40 J. C. Leblanc, C. Moise, J. Tirouflet, *J. Organomet. Chem.* **1985**, *292*, 225–228.

41 I. Rémy, P. Brossier, I. Lavastre, J. Besançon, C. Moise, *J. Pharm. Biomed. Anal.* **1991**, 965–967.

42 V. Philomin, A. Vessières, M. Gruselle, G. Jaouen, *Bioconjugate Chem.* **1993**, *4*, 419–424.

43 B. Dunbar, B. Riggle, G. Niswender, *J. Agric. Food Chem.* **1990**, *38*, 433.

44 N. Fischer-Durand, A. Vessières, J. M. Heldt, F. Le Bideau, G. Jaouen, *J. Organomet. Chem.* **2003**, *668*, 59–66.

45 A. Varenne, A. Vessières, M. Salmain, P. Brossier, G. Jaouen, *J. Immunol. Methods* **1995**, *186*, 195–204.

46 M. Salmain, A. Vessières, I. S. Butler, G. Jaouen, *Bioconjugate Chem.* **1991**, *2*, 13–15.

47 J. E. Sheats, M. D. Rausch, *J. Org. Chem.* **1970**, *35*, 3245.

48 B. Limoges, C. Degrand, P. Brossier, R. L. Blankespoor, *Anal. Chem.* **1993**, *65*, 1054–1060.

49 A. Le Gal La Salle, B. Limoges, J. Y. Anizon, C. Degrand, P. Brossier, *J. Electroanal. Chem.* **1993**, *350*, 329–335.

50 J. S. O'Neal, S. G. Schulman, *Anal. Chem.* **1984**, *56*, 2588.

51 A. L. Bordes, B. Schöllhorn, B. Limoges, C. Degrand, *Appl. Organomet. Chem.* **1998**, *12*, 59–65.

52 L. J. Silverberg, J. L. Dillon, P. Vemishetti, *Tetrahedron Lett.* **1996**, *37*, 771–774.

53 P. D. Beer, C. Hazlewood, D. Hesek, J. Hodacova, S. E. Stokes, *J. Chem. Soc. Dalton Trans.* **1993**, 1327–1332.

54 O. Bagel, B. Limoges, B. Schöllhorn, C. Degrand, *Anal. Chem.* **1997**, *69*, 4688–4694.

55 A. L. Bordes, B. Limoges, P. Brossier, C. Degrand, *Anal. Chim. Acta* **1997**, *356*, 195–203.

56 X. Q. Guo, F. N. Castellano, L. Li, H. Szmacinski, J. R. Lakowicz, J. Sipior, *Anal. Biochem.* **1997**, *254*, 179–186.

57 M. Cais, *Methods Enzymol.* **1983**, *92*, 445–458.

58 M. Cais, N. Tirosh, *Bull. Soc. Chim. Belg.* **1980**, *90*, 27–35.

59 M. Cais, *L'Actualité Chimique* **1979**, 14–26.

60 F. Mariet, Ph D Thesis, Université de Bourgogne, Dijon, **1990**.

61 P. Cheret, Ph D thesis, Université de Bourgogne, Dijon, **1987**.

62 P. Cheret, P. Brossier, *Res. Commun. Chem. Pathol. Pharmacol.* **1986**, *54*, 237–253.

63 F. Mariet, P. Brossier, *Pharm. Acta Helv.* **1991**, *66*, 60–64.

64 F. Mariet, P. Brossier, *Res. Commun. Chem. Pathol. Pharmacol.* **1990**, *68*, 251–262.

65 I. Rémy, P. Brossier, *Analyst* **1993**, *118*, 1021–1025.

66 P. Brossier, I. Rémy, *Immunoanal. Biol. Spec.* **1992**, *34*, 17–24.

67 A. Vessières, M. Salmain, P. Brossier, G. Jaouen, *J. Pharm. Biomed. Anal.* **1999**, *21*, 625–633.

68 A. Vessières, G. Jaouen, M. Salmain, I. S. Butler, *Appl. Spectrosc.* **1990**, *44*, 1092–1094.

69 M. Salmain, A. Vessières, G. Jaouen, I. S. Butler, *Anal. Chem.* **1991**, *63*, 2323–2329.

70 V. Philomin, A. Vessières, G. Jaouen, *J. Immunol. Methods* **1994**, *171*, 201–210.

71 M. Salmain, A. Vessières, P. Brossier, I. S. Butler, G. Jaouen, *J. Immunol. Methods* **1992**, *148*, 65–75.

72 P. Brossier, G. Jaouen, B. Limoges, M. Salmain, A. Vessières-Jaouen, J. P. Yvert, *Ann. Biol. Clin.* **2001**, *59*, 677–691.

73 J. Leinonen, T. Lovgren, T. Vornanen, U. H. Stenman, *Clin. Chem.* **1993**, *39*, 2098–2103.

74 K. Pettersson, H. Alfthan,
U. H. Stenman, U. Turpeinen,
M. Suonpaa, J. Soderholm,
S. O. Larsen, B. Norgaard-Pedersen,
Clin. Chem. **1993**, *39*, 2084–2089.

75 Y. Y. Xu, K. Pettersson, K. Blomberg,
I. Hemmilä, H. Mikola, T. Lovgren,
Clin. Chem. **1992**, *38*, 2038–2043.

76 A. Varenne, A. Vessières, M. Salmain,
S. Durand, P. Brossier, G. Jaouen,
Anal. Biochem. **1996**, *242*, 172–179.

77 M. Salmain, A. Vessières, A. Varenne,
P. Brossier, G. Jaouen, *J. Organomet.
Chem.* **1999**, *589*, 92–97.

78 M. Salmain, A. Varenne, A. Vessières,
G. Jaouen, *Appl. Spectrosc.* **1998**, *52*,
1383–1390.

79 W. R. Heineman, H. B. Halsall,
Anal. Chem. **1985**, *57*, 1321 A–1331 A.

80 R. Lehmann, S. Kayrooz, H. Greuter,
G. A. Spinas, *Diabetes Res. Clin. Pract.*
2001, *53*, 121–128.

81 S. Weber, W. Purdy, *Anal. Lett.* **1979**,
12, 1–9.

82 A. E. G. Cass, G. Davis, G. D. Francis,
H. A. O. Hill, W. J. Aston,
I. J. Higgins, E. V. Plotkin,
L. D. L. Scott, A. P. F. Turner,
Anal. Chem. **1984**, *56*, 667–671.

83 B. Limoges, C. Degrand, *Anal. Chem.*
1996, *68*, 4141–4148.

84 T. K. Lim, Y. Komoda, N. Nakamura,
T. Matsunaga, *Anal. Chem.* **1999**, *71*,
1298–1302.

85 T. K. Lim, N. Nakamura,
T. Matsunaga, *Anal. Chim. Acta* **1997**,
354, 29–34.

86 T. K. Lim, N. Nakamura,
T. Matsunaga, *Anal. Chim. Acta* **1998**,
370, 207–214.

87 T. K. Lim, S. Imai, T. Matsunaga,
Biotechnol. Bioeng. **2002**, *77*, 758–763.

88 X. Q. Guo, F. N. Castellano, L. Li,
J. R. Lakowicz, *Anal. Chem.* **1998**, *70*,
632–637.

89 M. Dequaire, C. Degrand,
B. Limoges, *Anal. Chem.* **1999**, *71*,
2571–2577.

90 C. Valat, B. Limoges, D. Huet,
J. L. Romette, *Anal. Chim. Acta* **2000**,
404, 187–194.

91 D. J. Pritchard, H. Morgan,
J. M. Cooper, *Anal. Chim. Acta* **1995**,
310, 251–256.

92 SFRL, *Chiffres clefs de l'industrie du
diagnostic in vitro 2004*, www.sfrl.fr, **2004**.

93 C. A. K. Borrebaeck, *Immunol. Today*
2000, *21*, 379–382.

9
Genosensors Based on Metal Complexes

Shigeori Takenaka

9.1
Introduction

Biosensors are analytical devices used to monitor biomolecules on the basis of their specific interactions, such as antigen–antibody, enzyme–substrate, DNA–DNA or protein–protein interactions [1]. The information from these interactions may be converted to optical, electrochemical, or magnetic signals by specifically designed transducers. Recently DNA sensors or genosensors are attracting attention as biosensors, because of increasing demands for analysis of specific genes associated with diseases after elucidation of the human genome sequence [2–7]. Biosensors based on electrochemical methods are especially attractive because they can be easily miniaturized and automated while retaining high sensitivity. DNA is electrochemically inactive under ordinary conditions, though it shows some electrochemical response under limited conditions using some kinds of electrode and electrolyte [8–12]. Hence DNA needs to be labeled with an electrochemically active group, if one attempts to detect it electrochemically. Metal complexes showing a stable reversible redox response are effective as such labeling reagents for DNA. Water-soluble metal complexes showing a preference for double-stranded DNA (dsDNA) or electrochemically-active DNA probes covalently attached to a metal complex have been reported as the electrochemical genosensor. In this Chapter, electrochemical genosensors based on these metal complexes will be briefly summarized.

9.2
Metal Complexes as DNA Probes

9.2.1
Cationic Metal Complexes

DNA is electrochemically inactive under ordinary conditions. Many cationic metal complexes are known as DNA-binding ligands [13–46]. Typical examples of such

Bioorganometallics: Biomolecules, Labeling, Medicine. Edited by Gérard Jaouen
Copyright © 2006 Wiley-VCH Verlag GmbH & Co. KGaA, Weinheim
ISBN: 3-527-30990-X

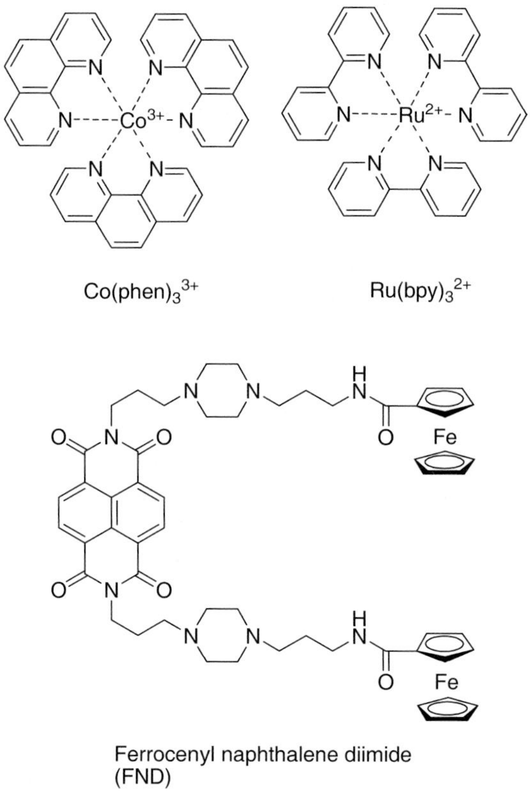

Co(phen)$_3^{3+}$ Ru(bpy)$_3^{2+}$

Ferrocenyl naphthalene diimide
(FND)

Fig. 9.1 Typical examples of DNA binding metal complexes.

cationic metal complexes that can bind to DNA are shown in Fig. 9.1. When the DNA-binding ligand is used for the detection of DNA, the amount of DNA can be estimated from the magnitude of electrochemical response of the ligand, as long as there is some correlation between the amounts of the bound ligand and DNA.

Bard and coworkers analyzed the interaction of the metal complex of CoL$_3^{3+}$ or FeL$_3^{2+}$ (L = phen or bpy, where phen stands for 1,10-phenanthrolinato and bpy for bipyridyl) with DNA by electrochemical techniques [13, 14]. They observed that both hydrophobic and electrostatic interactions of the DNA phosphate groups with the metal complex are involved in this binding, depending on the nature of the ligand and metal parts. For example, the electrostatic interaction of Fe(bpy)$_3^{2+}$ is the main force in the dsDNA binding, whereas the hydrophobic interaction of the ligand of Fe(phen)$_3^{2+}$ is the main one in the DNA groove binding. Furthermore, Barton and coworkers found that the Δ and Λ isomers of Ru(phen)$_3^{2+}$ form a more stable complex with the left- or right-handed DNA duplex, respectively, so that the configuration of the Ru complex has the best fitness for the DNA groove binding [44]. Mahardevan et al. confirmed this discriminating ability of the Ru complex by an electrochemical technique [45].

9.2.2

Metal Complexes Conjugated with a DNA Fragment or DNA-binding Ligand

A DNA probe method is used to detect the target DNA or gene. DNA can form a duplex structure based on the complementary hydrogen bonding between the nucleic acid bases such as adenine (A) and thymine (T). The two complementary single strands thus formed, can dissociate by heating (heat denaturation) and reform a duplex by cooling (annealing). This latter process of the DNA duplex formation is called "hybridization." For example, the DNA sequence of 5'-CGATGC-3' forms the best matched DNA duplex with its complementary sequence of 3'-GCTACG-5'. For detecting some genes, the DNA carrying a sequence complementary to that of the target DNA is prepared as a "DNA probe" attached with a labeling moiety such as a fluorophore and is allowed to hybridize with sample DNA. A target DNA fragment in the sample DNA mixture can be detected by the label introduced in the DNA probe. Fluorescent groups are commonly used as labeling agents, but when this group is replaced with an electrochemically active one, the target DNA can be detected electrochemically. For electrochemical detection, metal complexes have been conjugated with a short DNA fragment or oligonucleotide to obtain an electrochemically-active DNA probe [47–65]. Typical examples of such metal complexes attached to an oligonucleotide as an electrochemically-active DNA probe are shown in Fig. 9.2.

When a metal complex is formed with the nucleic acid base of single-stranded DNA (ssDNA), such ssDNA can be used as an alternative electrochemically-active DNA-binding ligand. For example, Palecek and coworkers reported the formation of a reversible redox-active metal complex by the reaction of osmium tetraoxide-pyridine with the thymine (T) base of ssDNA, and they developed an electro-chemical gene detection method based on this modified oligonucleotide as a DNA probe [3].

In another approach to electrochemical DNA detection, metal complexes conjugated with DNA-binding ligands such as intercalators can be used for that purpose instead of electrochemically-active DNA probes. A gene-detecting method is known that is based on electrochemically-active metal complexes that show preference for dsDNA [16, 21–24, 29, 32–43]. The principle of this technique is shown in Fig. 9.3.

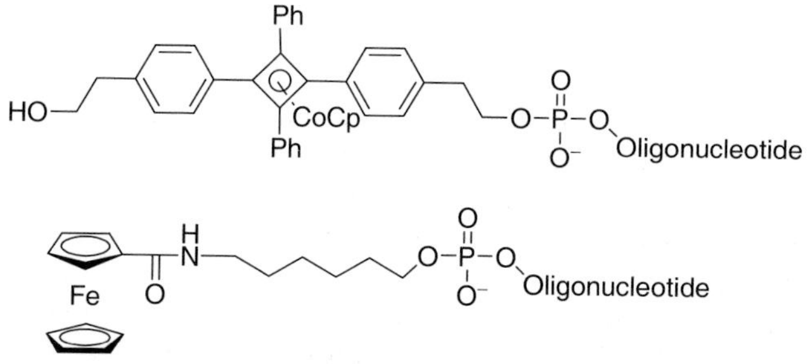

Fig. 9.2 Examples of oligonucleotides carrying an electrochemically active metal complex.

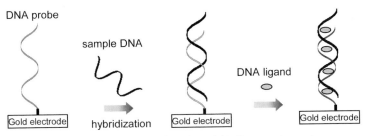

DNA probe

sample DNA

DNA ligand

Gold electrode hybridization Gold electrode Gold electrode

Fig. 9.3 Target DNA detection based on dsDNA-binding metal complexes.

A DNA probe is immobilized on the surface of an electrode and is allowed to hybridize with sample DNA. After hybridization, the electrode is dipped in an electrolyte containing a DNA-binding metal complex and electrochemical measurements are carried out. If the target DNA exists in the sample DNA solution, the metal complex is concentrated in the DNA duplex formed with the target DNA on the electrode. The target gene can be measured from the magnitude of this electrochemical signal of the metal complex concentrated on the electrode. In this respect, these complexes are called "electrochemical hybridization indicators". The key for the success of the analysis lies in the ability of these complexes to distinguish dsDNA from single-stranded DNA (ssDNA). Many kinds of organic ligands including an intercalator were studied for this purpose, yet they did not undergo a good reversible redox reaction, unlike the DNA-binding metal complexes [13–46]. The application of metal complexes, such as $Co(phen)_3^{3+}$ described above, was reported [19–24]. Recently, a metal complex of ferrocene attached with naphthalene diimide, which is the first example of the metal complex conjugated with a DNA-binding ligand, was reported as a high performance electrochemical hybridization indicator [32–43]. Naphthalene diimide derivatives are known to bind to dsDNA by the threading intercalation mode. In this complex, the two substituents of the naphthalene diimide project over the major and minor grooves to act as an anchor to prevent it from dissociating from the complex with dsDNA, resulting in its slow dissociation from the DNA duplex. In the case of ssDNA, there is no such interaction, and this ability of threading intercalators can be used to distinguish dsDNA from ssDNA.

9.3
Electrochemical Analysis of the Interaction of Metal Complexes with dsDNA

Bard and coworkers developed an electrochemical analysis method based on the interaction of a metal complex with dsDNA [13, 14]. For example, the cyclic voltammogram (CV) of $Co(phen)_3^{3+}$ on the glassy carbon electrode showed oxidation and reduction (redox) peaks with $E_{pc} = 0.107$ V and $E_{pa} = 0.137$ V (vs. Ag/AgCl), respectively. When DNA was added in the electrolyte, the peak potentials of this CV shifted to the anodic side of 0.120 and 0.182 V (vs. Ag/AgCl) with a decrease in the currents as shown in Fig. 9.4.

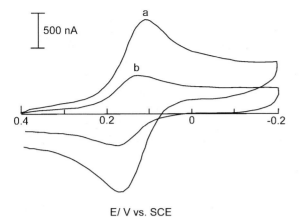

E/ V vs. SCE

Fig. 9.4 Cyclic voltammogram of 1.0×10^{-4} M $Co(phen)_3^{3+}$ in the absence (a) or presence of calf thymus DNA (b) in 5 mM Tris-HCl, 50 mM NaCl, pH 7.1. Scan rate 100 mV s^{-1}.

This decrease of the peak current may be due to an increase in the apparent size of $Co(phen)_3^{3+}$ bound to dsDNA. Since the peak currents (i_{pa} and i_{pc}) increased linearly with the root of the scan rate in this CV, the observed redox signal of $Co(phen)_3^{3+}$ seems to be due to the diffusion of $Co(phen)_3^{3+}$ from the bulk solution to the electrode. The slope of this plot for the free and bound forms of $Co(phen)_3^{3+}$ gave a diffusion coefficient of $D_f = 5.0$ ($\pm$ 0.6) $\times 10^{-6}$ cm^{-2} s^{-1} and $D_b = 3.2$ ($\pm$ 1.21) $\times 10^{-7}$ cm^{-2} s^{-1}. The diffusion coefficient of $Co(phen)_3^{3+}$ was decreased to 1/16 after binding to dsDNA. Bard and coworkers estimated the ratio of the binding constants for the oxidized and reduced species (K_{2+}/K_{3+}) as 1.74 for $Co(phen)_3^{3+}$ from the potential shift when bound to dsDNA, assuming a thermal cycle in this reaction. In other words, the divalent complex bound more strongly, suggesting that the hydrophobic rather than the electrostatic interaction with polyanionic DNA predominates in this complex formation. The electrostatic interaction was most prevalent with $Fe(bpy)_3^{2+}$. The binding constant K was estimated from a change in the peak potential of CV with the amounts of dsDNA on the basis of the following Scatchard-type equation.

$$C_b = b - \{b^2 - (2\ K^2 C_t\ [\text{DNA}]\ /\ s)^{1/2}\ /\ 2\ K,$$
$$b = 1 + KC_t + K\,[\text{DNA}]\ /\ 2\ s$$

Where s is the site size in base pairs excluded by the bound ligand complex and C_b and C_t are the concentration of the bound ligand and the total concentration of the ligand, respectively. The current obtained here is based on the species bound or not bound to dsDNA and therefore we must investigate whether these two species can interconvert in the measured timescale. In the case of no interconverting system (static, S), the square root of the diffusion coefficient is simply the sum of the free and bound forms of ligand weighted by the individual mole fractions as shown in the following equation:

$$i_{pc} = B \left(D_f^{1/2} C_f + D_b^{1/2} C_b \right)$$

where C_f represents the concentration of the free ligand. In the case of the interconverting system (mobile, M), the equation is as follows:

$$i_{pc} = B C_f \left(D_f C_f / C_t + D_b C_b / C_t \right)^{1/2}$$

For a Nernstian reaction in CV at 25 °C, B is expressed as $2.69 \times 10^5 \, n^{3/2} \, Av^{1/2}$, where n is the number of electrons transferred per metal complex and A is the surface area of the electrode. The binding constant of $Co(phen)_3^{3+}$ for calf thymus DNA was determined as $K = 1.6 \ (\pm 0.2) \times 10^4 \, M^{-1}$ (S, $s = 6$ bp) or $2.6 \ (\pm 0.4) \times 10^4 \, M^{-1}$ (M, $s = 5$ bp).

9.4
Gene Detection Based on a Cationic Metal Complex or Metal Complex Conjugated with DNA-binding Ligand

When a DNA-immobilized electrode is used in the CV measurement, the peak current of the cationic metal complex or metal complex conjugated with DNA-binding ligand on the dsDNA-immobilized electrode should be larger than that on the ssDNA-immobilized electrode. Mikkelsen's group carried out this experiment by using $Co(phen)_3^{3+}$ as a DNA-binding cationic metal complex [19–21]. The 18-meric oligonucleotide carrying ΔF sequence (5'-CAA ACA CCA ATG ATA TTT-3') attached to about dG_{25} sequence by enzymatic reaction was immobilized on the carbon paste electrode and was allowed to hybridize with sample DNA [21]. An increase in the current peak was observed upon hybridization with the complementary oligonucleotide as shown in Fig. 9.5.

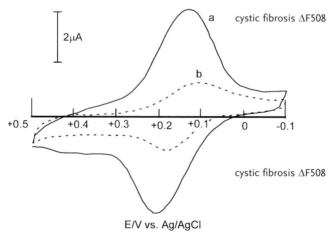

Fig. 9.5 Cyclic voltammogram of ΔF oligonucleotide (5'-CAA ACA CCA ATG ATA TTT-3') immobilized on the carbon paste electrode before (a) and after (b) hybridization with its complementary oligonucleotide in 0.12 mM $Co(phen)_3^{3+}$, 5 mM Tris-HCl (pH 7.0), and 20 mM NaCl [21].

This shows that the current obtained here is based on the $Co(phen)_3^{3+}$ complex concentrated on the electrode and the amount of this complex with dsDNA is larger than that with ssDNA. Metal complexes showing even higher preference for dsDNA are preferable for the electrochemical gene detection. Mikkelsen and coworkers prepared a DNA probe-carrying carbon paste electrode in the following way: a carboxyl moiety was generated by the oxidation of the surface of a carbon paste electrode and amino moieties of ssDNA were linked to it covalently [20]. They reported that this reaction was effective in the case of the guanine base of ssDNA and showed a detection limit of 2.5 ng for poly(dA)$_{4000}$ [20]. Wang and coworkers achieved highly sensitive gene detection by using $Co(phen)_3^{3+}$ coupled with a poten-tiometric stripping analysis (PSA) method, which is the technique where sample DNA is electrochemically concentrated on the DNA probe-immobilized electrode at a positive potential, and the amount of $Co(phen)_3^{3+}$ concentrated on this electrode is measured by chrono-potentiometry [22–24]. This complex has been widely used to detect oligonucleotides related to bacterial and viral pathogen sequences [22]. The PSA method based on this complex resulted in a detection limit of 0.05 µg ml^{-1} of oligonucleotide target sequence. Wang and coworkers also achieved the mis-matched DNA detection by the PSA method [23, 24]. Since the mismatched DNA duplex is less stable than the perfectly matched DNA duplex, the amount of the DNA hybrid on the electrode for the latter is larger than that for the former. The difference between the two may be expanded by the proper setting of the conditions of hybridization and subsequent washing. Wang et al. also tried to increase accuracy in the single-base mismatch detection by using a peptide nucleic acid (PNA) probe instead of the DNA probe [23, 24]. $Ru(bpy)_3^{3+}$ can mediate the oxidation reaction of guanine base electrochemically. Thorp and coworkers developed an electro-chemical gene detection method based on this phenomenon [25–30]. The DNA probe immobilized on the ITO electrode was allowed to hybridize with sample DNA. The mediating current for the guanine base of target DNA was observed in CV in an electrolyte containing the Ru complex as shown in Fig. 9.6. Since the mediating current was proportional to the amount of target DNA, one can measure the gene electrochemically. It should be noted, however, that the current signal changed with the nature of DNA samples with different GC content.

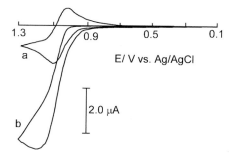

Fig. 9.6 Cyclic votammograms of $Ru(bpy)_3^{2+}$ in the absence (a) or presence (b) of calf thymus DNA at 25 mV s^{-1} in sodium phosphate buffer (pH 6.8) and 700 mM NaCl [26].

Takenaka and coworkers first synthesized ferrocenyl naphthalene diimide (FND) as a metal complex conjugated with DNA-binding ligand (a hybridization indicator) with a strong preference for dsDNA and reversible redox response at an anodic potential [32–43]. Thiolated oligonucleotide dT_{20} was immobilized on a gold electrode through a thiol–gold linkage as a DNA probe and the resulting electrode was allowed to hybridize with complementary dA_{20} or non-complementary dT_{20} [32, 33]. After dipping in FND solution, the electrode was transferred to an electrolyte without FND and a CV was measured. Fig. 9.7 shows a cyclic voltammogram of the electrode before and after hybridization with dA_{20}. A peak current was obtained only for the complementary oligonucleotide dA_{20} with this electrode, whereas the signal intensity for the non-complementary DNA, dT_{20}, was barely above background. This discrimination of dsDNA from ssDNA was based on the stable complex of FND with dsDNA through threading intercalation. They succeeded in the sequence specific DNA detection and gene expression analysis based on FND coupled with a DNA probe-immobilized electrode [32–34].

Mismatched DNA detection was also achieved with FND [35–43]. The affinity of FND for a mismatched DNA area decreases considerably from that for the matched one [38, 41]. Therefore, one can estimate the existence of mismatched DNA duplex from a decrease in the current compared with that of the perfectly matched one. Quartz crystal microbalance (QCM) experiments revealed that this current decrease is derived from a difference in the amount of FND concentrated per dsDNA on the electrode. Thus, a thiolated oligonucleotide was immobilized as DNA probe on the gold surface covered on the QCM chip and this chip was dipped in the same electrolyte as the electrochemical measurements. Complementary oligonucleotides with or without a single mismatched base were added in this system. A frequency decrease was observed upon their addition and a further decrease in the frequency was observed upon addition of FND. Double stranded DNA formation accounts for the first phase of frequency decrease and

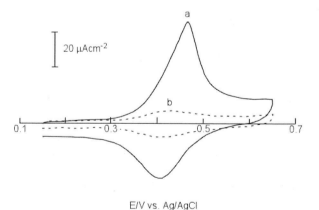

Fig. 9.7 Cyclic voltammograms of a dT_{20}-immobilized gold electrode in an electrolyte after hybridization with complementary dA_{20} (a) or non-complementary dT_{20}, (b) followed by immersion in FND solution.

FND binding to the dsDNA for the second phase. It was also demonstrated that nine molecules of FND were bound to 20-meric dsDNA on average, whereas seven molecules of FND bound to the same DNA but with a single base mismatch. This result shows that FND cannot bind to the mismatched base area on this dsDNA. Since this mismatched DNA detection based on the difference in the amount of FND bound per dsDNA does not depend on the stability of dsDNA, it can be extended to the simultaneous detection of DNA with a mismatch at many different sites. The dsDNA was stabilized by the bound FND and the mismatched DNA detection was achieved by a difference in the thermal stability of longer sample DNA fragments such as the PCR product. This mismatched DNA detection was extended to the single nucleotide polymorphism (SNP) analysis of the lipoprotein lipase (LPL) gene [42, 43]. Humans have two alleles, and where they are different from each other, the gene is called a heterozygote. Where one of the alleles is normal (wild type), while the other is a mutated form which may be associated with some disorder, the gene needs to be diagnosed early for treatment. (The heterozygote has the 1/2 gene when comparing with heterozygote as normal or mutated gene.) When two DNA probes for wild and mutant types were used in this DNA detection, the electrochemical signal from both DNA probes is observed for the heterozygote. In other words, one can detect the heterozygote. Fig. 9.8 shows an example of the detection of a heterozygote of the LPL gene [42].

When many different DNA probes immobilized on different electrodes are integrated, many different genes can be analyzed simultaneously. This is the same concept as the DNA chip. On the basis of a novel technology developed by Takenaka's group, TUM Gene Inc. has established an ECA (electrochemical array) system that consists of integrated multi electrodes and can achieve parallel analysis of many samples or many genes [35]. FND can be applied to the electrochemical

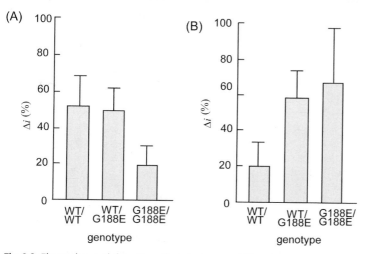

Fig. 9.8 Electrochemical detection of a heterozygous LPL gene. The heterozygote gave rise to a significant electrochemical signal from both electrodes for wild (A) and mutant type DNA probes (B) [42].

visualization of a DNA chip by using scanning electrochemical microscopy (SECM) [40]. A DNA chip was dipped in an electrolyte containing FND and measured by SECM. FND was concentrated on the dsDNA area on the DNA chip and gave a stronger signal. This technique can be extended to a hybrid with a DNA probe-immobilized on the insulating surface such as glass surface, offering a new interesting means to visualize the DNA chip electrochemically.

9.5
Gene Detection Based on Ferrocenyl Oligonucleotides as a Metal Complex Conjugated with DNA Fragments

Ferrocenyl oligonucleotide derivatives as metal complexes conjugated with DNA fragments are used as the DNA probe carrying an electrochemically-active moiety [47–65]. The electrochemical gene detection was reported by using a ferrocenyl oligonucleotide coupled with the high performance liquid chromatography (HPLC)-electrochemical detector (ECD) system [47]. The ferrocenyl oligonucleotide as a DNA probe was allowed to hybridize, and the unbound ferrocenyl oligonucleotide was removed by HPLC and the electrochemical response for the ferrocenyl moiety bound to the sample DNA was detected by ECD. For example, the hybrid of sample DNA having the yeast choline transporter gene with its complementary DNA probe sequence carrying a ferrocenyl moiety could be separated from the hybridized DNA probe by HPLC with the hydroxyapatite column and could be detected with ECD. The ECD signal was proportional to the amount of the target gene with a detection limit of several ten femtomoles. Eukaryotic mRNA has a polyA tail and therefore this can be detected with the ferrocenyl T_{20} oligonucleotide. The detection of mRNA was achieved for one taken from yeast total RNA. Ferrocenyl oligonucleotides could be used as a PCR primer, too. The PCR product carrying the ferrocene moiety could be measured electrochemically and the difference between the male (having one allele in this case) and female (two alleles) of the dystrophin gene could also be detected by quantitative PCR (Q-PCR) [49]. Kuhr and coworkers synthesized four different ferrocenyl oligonucleotides that show redox responses at four different potentials by connecting different electron-withdrawing or electron-donating moieties to the ferrocene part as shown in Fig. 9.9 [55]. They could be separated by capillary gel electrophoresis and detected by sinusoidal voltammetry. This result suggested that they can be used like the four "color" labels used in DNA sequencing.

Furthermore, Kuhr and coworkers succeeded in detecting the known SNP by using the enzymatic single base extension technique [62]. A ferrocenyl oligonucleotide carrying the sequence before the SNP site was allowed to hybridize with sample DNA as shown in Fig. 9.10. For example, T-type SNP could be detected by treating with ddATP and an extension enzyme. The extension product was analyzed by capillary gel electrophoresis by alternating current (AC) voltammetry. They improved this technique by using newly synthesized ferrocene-acyclo ATP [63].

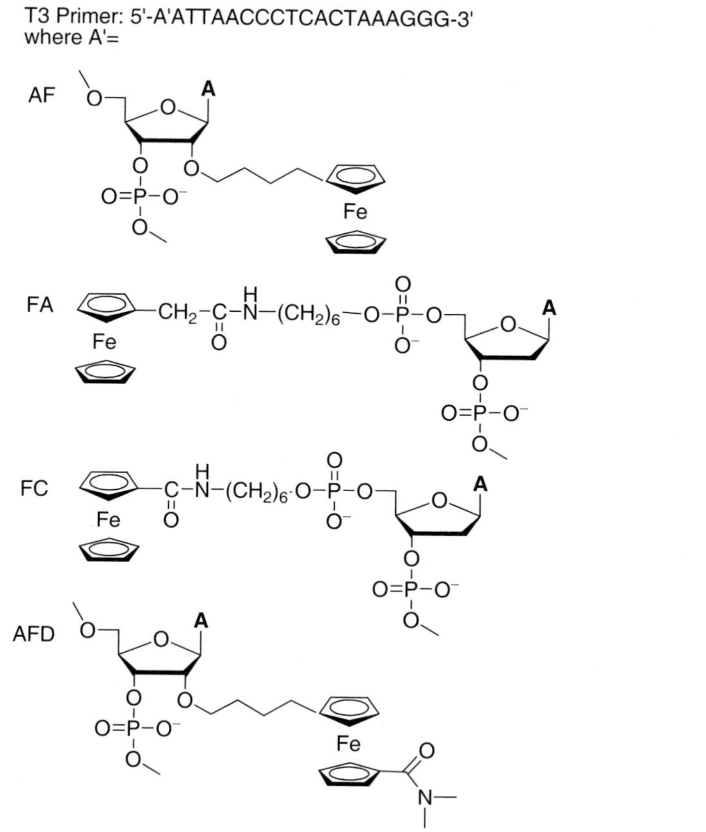

T3 Primer: 5'-A'ATTAACCCTCACTAAAGGG-3'
where A'=

AF

FA

FC

AFD

Fig. 9.9 Alkylferrocene-tagged T3 primer (AF), ferroceneacetate-tagged T3 primer (FA), ferrocenecarboxylate-tagged T3 primer (FC), and alkylferrocene dimethylcarboxamide-tagged T3 primer (AFD) were synthesized by attaching the corresponding redox-active tags to the 5'-adenosine moiety of a 20-meric oligonucleotide of T3 primer, 5'-AAT TAA CCC TCA CTA AAG GG-3' [55].

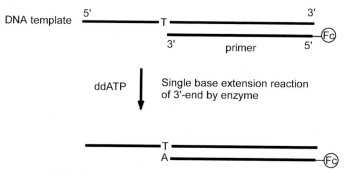

Extension reaction is occurred in the case of T SNP type.

Fig. 9.10 Model single base extension reaction system with a 5'-ferrocene labeled primer. The single base extended oligonucleotide was obtained for the T-type SNP [62].

In Fig. 9.11, the ferrocenyl oligonucleotide was replaced for an unlabeled oligo-nucleotide and ddATP was replaced for ferrocene-acycloATP. The oligonucleotide having the ferrocenyl moiety was obtained in the case of the corresponding SNP type (T-type). The former technique is necessary to separate the ferrocenyl oligonucleotides with one base difference. However, the ferrocenyl oligonucleotide was obtained for the target SNP only in this technique.

The principle of the electrochemical detection of M13 ϕ DNA as a cyclic ssDNA by using the ferrocene-dUTP as the substrate of polymerase is shown in Fig. 9.12 [65]. M13 ϕ DNA was allowed to hybridize with the DNA probe immobilized on the electrode. The polymerase chain reaction was then carried out on the electrode with dCTP, dATP, dGTP, and ferrocenyl dUTP instead of dTTP. The complementary strand of 2749-base M13 mp18 DNA was formed in the reaction and the ferrocenyl U instead of A was incorporated to this strand. They estimated the amount of the ferrocene incorporated by using the electron transfer reaction between the electrode and reduced glucose oxidase generated by its treatment with glucose. The detection of sub picomole of target M13 ϕ DNA was achieved by this method.

Ferrocenyl dUTP and ddUTP derivatives carrying another chemical linkage were also reported [60, 64]. Umek and coworkers developed the sequence specific

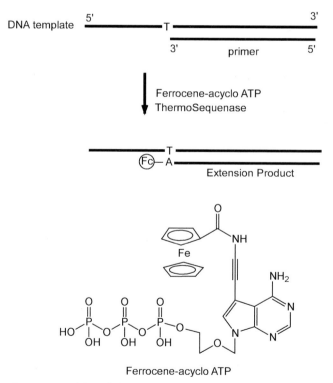

Fig. 9.11 A model single base extension reaction system utilizing a ferrocene-acycloATP terminator. Ferrocene-labeled acyclic ATP was incorporated in the case of T-type SNP [63].

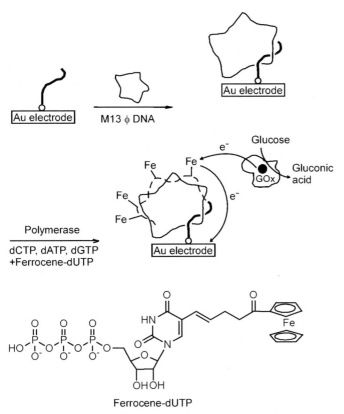

Fig. 9.12 Incorporation of ferrocene by the enzymatic complementary strand synthesis of M13φ DNA on the electrode and the electrochemical detection based on the mediation reaction of ferrocene [65].

DNA detection, mismatch recognition, and gene expression analysis based on sandwich-type assay of ferrocenyl oligonucleotides [58]. Two different DNA probes carrying complementary DNA sequences were prepared as shown in Fig. 9.13. One DNA probe was immobilized on the electrode and to the other was introduced a ferrocenyl moiety. Since sample DNA was allowed to hybridize with these DNA probes, the ferrocenyl moiety was immobilized on the electrode only in the presence of target DNA. The hybridized DNA was detected by the current corresponding to the redox of ferrocene. They adopted AC voltammetry in the target DNA detection. The insertion or deletion polymorphism in intron 16 of the angiotensin converting enzyme (ACE) gene was detected as a 192 bp DNA fragment. Amplicons, obtained by asymmetric PCR (A-PCR) from the human Hfe gene were also readily genotyped by virtue of the lower thermal stability of single base mismatched DNA. The sequences of fire apotosis-regulated gene (fas, p53, box, p21 and bcl-2) obtained by reverse transcription with total RNA taken from white blood cells, subsequently amplified by A-PCR, and finally detected on the electrode array. To increase rapidity and accuracy of single base mismatch

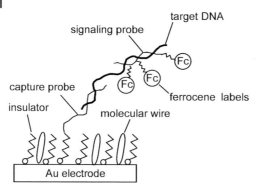

Fig. 9.13 Electrochemical detection of nucleic acid with the bioelectronic sensor based on a sandwich assay. A target nucleic acid is shown to anneal to a capture probe and a ferrocene-labeled signaling probe [58]. The thiol-terminated oligophenylethynyl molecules serve as molecular wires and provide a pathway for electron transfer between the ferrocenes and the gold in response to potential charges at the electrode. The alkanethiols terminated in ethylene glycol serve as insulators to block access of redox species in solution to the electrode, including free signaling probes.

detection, two different ferrocenyl signaling probes, one specific for the wild-type target and the other for the mutant, with two different redox potentials, were also developed. This technique is commercially available as a CMS e-sensor™ system of Motorola Inc. They prepared an amide reagent carrying the ferrocenyl adenine base and synthesized the multi ferrocene-labeled DNA probe on the DNA synthesizer.

Another method of grafting many ferrocene moieties to target DNA is based on the strong biotin–avidin interaction [67]. The biotinyl DNA probe was immobilized on the electrode through the target DNA hybridization as shown in Fig. 9.14. After treating this complex with avidin and then a biotinyl gold nanoparticle carrying many ferrocenyl moieties was immobilized on this complex. The amount of the target DNA can be detected by the redox current of these ferrocenyl moieties.

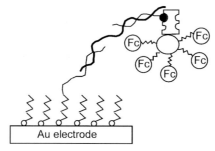

Fig. 9.14 Schematic representation of the amplified electrochemical detection of DNA hybridization via oxidation of the ferrocene caps on the gold nanoparticle/avidin conjugates [67].

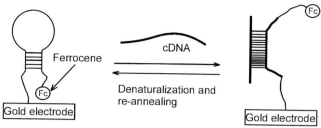

Fig. 9.15 Ferrocene was localized near the gold electrode by the hairpin structure of DNA. When the DNA probe region of the hairpin part was allowed to hybridize with target DNA, the stem duplex was destroyed and ferrocene falls apart from the electrode [66].

DNA detection with the oligonucleotide carrying ferrocene and thiol moieties at either ends was reported by Fan and coworkers [66]. The principle of their DNA detection is shown in Fig. 9.15. The oligonucleotide probe has a DNA sequence and the complementary sequence at either end. The hairpin structure is formed by attaching a thiol end on the gold electrode and the ferrocene moiety is located near the electrode. Here, the current based on the redox reaction of the ferrocene moiety flows because of its location near the electrode. When target DNA is allowed to hybridize, the distance between ferrocene and the electrode increases, resulting in a decrease in the current. This system does not rely on the ligand, yet it gives more convenient gene detection system.

9.6
Conclusions

Some of the metal complexes can be used as electrochemically-active DNA-binding ligands, as they undergo more stable reversible redox reactions compared with the other organic ligands, which enables electrochemical DNA detection. Furthermore, metal complexes with different redox potentials can be easily synthesized based on their molecular design. This feature gave the use of the metal complex for electrochemical DNA or gene sensor many advantages such as high performance, high sensitivity and low cost. Recently, DNA chip is applied for the gene diagnosis of a disease by using the fruit of the human genome sequencing project. Electrochemical DNA chip, which integrated many electrodes onto the small surface, is expected to be used for this diagnostic purpose and the metal complexes are useful elements for this electrochemical chip. The concept of the electrochemical DNA chip should be extended to other biologically important biochips such as protein chip or sugar chip.

References

1 B. Eggins, *Biosensors: An Introduction*, John Wiley & Sons Ltd, New York, **1997**.

2 S. Takenaka, *Bull. Chem. Soc, Jpn.* **2001**, *74*, 217–224.

3 E. Palecek, M. Fojta, *Anal. Chem.*, **2001**, *73*, 75A–83A.

4 E. Palecek, M. Fojta, F. Jelen, V. Vetterl, in G. S. Wilson (Ed.), *Encyclopedia of Electrochemistry*, Vol. 9, Bioelectrochemistry, Wiley-VCH, **2002**, pp. 365–429.

5 I. Willner, *Science* **2002**, *298*, 2407–2408.

6 F. Lucarelli, G. Marrazza, A. P. F. Turner, M. Mascini, *Biosens. Bioelectron.* **2004**, *19*, 515–530.

7 T. G. Drummond, M. G. Hill, J. K. Barton, *Nature Biotech.* **2003**, *21*, 1192–1199.

8 H. Berg, in S. Srinivasan, Y. J. Chizmadzhev, O. M. Bockris, B. E. Comway, E. Yeager (Eds.), *Comprehensive Treatise of Electrochemistry*, Plenum Press, New York, **1985**, pp. 189–229.

9 E. Palecek, *Bioelectrochem. Bioenerg.* **1986**, *15*, 275–295.

10 D.-W. Pang, Y. P. Qi, Z.-L. Wang, J.-K. Cheng, J.-W. Wang, *Electroanalysis* **1995**, *7*, 774–777.

11 E. Palecek, F. Jelen, C. Teijeiro, *Anal. Chim. Acta* **1993**, *273*, 175–186.

12 E. Palecek, *Electroanalysis* **1996**, *8*, 7–14.

13 M. T. Carter, A. J. Bard, *J. Am. Chem. Soc.* **1987**, *109*, 7528–7530.

14 M. T. Carter, M. Rodriguez, A. J. Bard, *J. Am. Chem. Soc.* **1989**, *111*, 8901–8911.

15 Y.-D. Zhao, D.-W. Pang, Z.-L. Wang, J.-K. Cheng, Y.-P. Qi, *J. Electroanal. Chem.* **1997**, *431*, 203–209.

16 Y. Mishima, J. Motonaka, S. Ikeda, *Anal. Chim. Acta* **1997**, *345*, 45–50.

17 M. Fojta, L. Havan, J. Fulneckova, T. Kubicarova, *Electroanalysis* **2000**, *12*, 926–934.

18 M. Aslanoglu, C. J. Isaac, A. Houlton, B. R. Horrocks, *Analyst* **2000**, *125*, 1791–1798.

19 K. M. Millan, A. J. Spurmanis, S. R. Mikkelsen, *Electroanalysis* **1992**, *4*, 929–932.

20 K. M. Millan, S. R. Mikkelsen, *Anal. Chem.* **1993**, *65*, 2317–2323.

21 K. M. Millan, A. Saraullo, S. R. Mikkelsen, *Anal. Chem.* **1994**, *66*, 2943–2948.

22 J. Wang, X. Cai, G. Rivas, H. Shiraishi, P. A. M. Farias, N. Dontha, *Anal. Chem.* **1996**, *68*, 2629–2634.

23 J. Wang, G. Rivas, X. Cai, M. Chicharro, C. Parrado, N. Dontha, A. Begleiter, M. Mowat, E. Palecek, P. E. Nielsen, *Anal. Chim. Acta* **1997**, *344*, 111–118.

24 J. Wang, E. Palecek, P. E. Nielsen, G. Rivas, X. Cai, H. Hiraishi, N. Dontha, D. Luo, P. A. Farias, *J. Am. Chem. Soc.* **1996**, *118*, 7667–7670.

25 T. W. Welch, A. H. Corbett, H. H. Thorp, *J. Phys. Chem.* **1995**, *99*, 11757–11763.

26 D. H. Johnston, K. C. Glasgow, H. H. Thorp, *J. Am. Chem. Soc.* **1995**, *117*, 8933–8938.

27 T. W. Welch, H. H. Thorp, *J. Phys. Chem.*, **1996**, *100*, 13829–13836.

28 M. E. Napier, C. R. Loomis, M. F. Sistare, J. Kim, A. E. Eckhardt, H. H. Thorp, *Bioconjugate Chem.* **1997**, *8*, 906–913.

29 M. E. Napier, H. H. Thorp, *Langumir* **1997**, *13*, 6342–6344.

30 P. M. Amistead, H. H. Thorp, *Anal. Chem.* **2001**, *73*, 558–564.

31 S. O. Kelly, N. M. Jackson, M. G. Hill, J. K. Barton, *Angew. Chem. Intern. Eng. Ed.* **1999**, *38*, 941–945.

32 S. Takenaka, Y. Uto, H. Saita, M. Yokoyama, H. Kondo, W. D. Wilson, *Chem. Comm.* **1998**, 1111–1112.

33 S. Takenaka, K. Yamashita, M. Takagi, Y. Uto, H. Kondo, *Anal. Chem.* **2000**, *72*, 1334–1341.

34 S. Sato, S. Fujii, K. Yamashita, M. Takagi, H. Kondo, S. Takenaka, *J. Organomet. Chem.* **2001**, *637–639*, 476–483.

35 H. Miyahara, K. Yamashita, M. Takagi, H. Kondo, S. Takenaka, *Trans. IEE of Jpn* **2001**, *121E*, 187–191.

36 S. Takenaka, H. Miyahara, K. Yamashita, M. Takagi, H. Kondo, *Nucleosides, Nucleotides & Nucleic Acids* **2001**, *20*, 1429–1432.

37 H. Miyahara, K. Yamashita, M. Kanai, K. Uchida, M. Takagi, H. Kondo, S. Takenaka, *Talanta* **2002**, *56*, 829–835.

38 K. Yamashita, M. Takagi, H. Kondo, S. Takenaka, *Chem. Lett.* **2000**, 1038–1039.

39 S. Takenaka, Y. Uto, M. Takagi, H. Kondo, *Chem. Lett.* **1988**, 989–990.

40 K. Yamashita, M. Takagi, K. Uchida, H. Kondo, S. Takenaka, *Analyst* **2001**, *126*, 1210–1211.

41 K. Yamashita, M. Takagi, H. Kondo, S. Takenaka, *Anal. Biochem.* **2002**, *306*, 188–196.

42 K. Yamashita, A. Takagi, M. Takagi, H. Kondo, Y. Ikeda, S. Takenaka, *Bioconjugate Chem.* **2002**, *13*, 1193–1199.

43 T. Nojima, K. Yamashita, M. Takagi, A. Takagi, Y. Ikeda, H. Kondo, S. Takenaka, *Anal. Sci.* **2003**, *19*, 79–83.

44 J. K. Barton, A. T. Danishefsky, J. M. Goldberg, *J. Am. Chem. Soc.* **1984**, *206*, 2172–2176.

45 S. Mahadevan, M. Palaniandavar, *Bioconjugate Chem.* **1996**, *7*, 138–143.

46 C. J. Waybright, C. P. Singleton, J. M. Tour, C. J. Murphy, U. H. F. Bunz, *Organometallics* **2000**, *19*, 368–370.

47 S. Takenaka, Y. Uto, H. Kondo, T. Ihara, M. Takagi. *Anal. Biochem.* **1994**, *218*, 436–443.

48 T. Ihara, Y. Maruo, S. Takenaka, M. Takagi, *Nucleic Acids Res.* **1996**, *24*, 4273–4280.

49 Y. Uto, H. Kondo, M. Abe, T. Suzuki, S. Takenaka, *Anal. Biochem.* **1997**, *250*, 122–124.

50 K. B. Jacobson, H. F. Arlinghaus, H. W. Schmitt, R. A. Sacleben, G. M. Brown, N. Thonnard, F. V. Sloop, R. S. Foote, F. W. Larimer, R. P. Woychik, M. W. England, K. L. Burchett, D. A. Jacobson, *Genomics* **1991**, *9*, 51–59.

51 F. V. Sloop, G. M. Brown, R. A. Sacleben, M. L. Garrity, J. E. Elert, *New J. Chem.* **1994**, *18*, 317–326.

52 R. C. Mucic, M. K. Herrlein, C. A. Mirkin, R. L. Letsinger, *Chem. Comm.* **1996**, 555–557.

53 C. Xu, P. He, Y. Fang, *Anal. Chim. Acta* **2000**, *411*, 31–36.

54 A. E. Beilstein, M. W. Grinstaff, *J. Organomet. Chem.* **2001**, *637–639*, 398–406.

55 S. A. Brazill, P. H. Kim, W. G. Kuhr, *Anal. Chem.* **2001**, *73*, 4882–4890.

56 C. J. Yu, H. Wang, Y. Wan, H. Yowanto, J. C. Kim, L. H. Donilon, C. Tao, M. Strong, Y. Chong, *J. Org. Chem.* **2001**, *66*, 2937–2942.

57 C. J. Yu, Y. Wan, H. Yowanto, J. Li, C. Tao, M. D. James, C. L. Tan, G. F. Blackburn, T. J. Meade, *J. Am. Chem. Soc.* **2001**, *123*, 11155–11161.

58 R. M. Umek, S. W. Lin, J. Vielmetter, R. H. Terbrueggen, B. Irvine, C. J. Yu, J. F. Kayyem, H. Yowanto, G. F. Blackburn, D. H. Farkas, Y.-P. Chen, *J. Mol. Diagn.* **2001**, *3*, 74–84.

59 A. Anne, A. Bouchardon, J. Moiroux, *J. Am. Chem. Soc.* **2003**, *125*, 1112–1113.

60 Y.-T. Long, C.-Z. Li, T. C. Sutherland, M. Chahma, J. S. Lee, H.-B. Kraatz, *J. Am. Chem. Soc.* **2003**, *125*, 8724–8725.

61 A. Anne, B. Blanc, J. Moiroux, *Bioconjugate Chem.*, **2001**, *12*, 396–405.

62 S. A. Brazill, W. G. Kuhr, *Anal. Chem.* **2002**, *74*, 3421–3428.

63 S. A. Brazill, N. E. Hebert, W. G. Kuhr, *Electrophoresis* **2003**, *24*, 2749–2757.

64 W. A. Wlassoff, G. C. King, *Nucleic Acids Res.* **2002**, *30*, e58 (1–7).

65 F. Patoisky, Y. Weizmann, I. Willner, *J. Am. Chem. Soc.* **2002**, *124*, 770–772.

66 C. Fan, K. W. Plaxco, A. J. Heeger, *PNAS* **2003**, *100*, 9134–9137.

67 J. Wang, J. Li, A. J. Baca, J. Hu, F. Zhou, W. Yan, D.-W. Pang, *Anal. Chem.* **2003**, *75*, 3941–3945.

10

Supramolecular Host Recognition Processes with Biological Compounds, Organometallic Pharmaceuticals, and Alkali-metal Ions as Guests

Richard H. Fish

10.1
Introduction

Supramolecular interactions that encompass recognition, reaction, transport, etc., are fundamental phenomena in biological systems that are involved in a number of processes between biologically important molecules, such as double and single strand DNA/RNA with proteins, drugs, and metal-ion containing probes, to name a few examples [1]. Organic chemists have exploited these interesting phenomena to a very significant extent and many supramolecular hosts have been synthesized to investigate the role of these interactions through the molecular recognition of guests, such as nucleosides, nucleotides, amino acids, peptides, and small organic molecules, by predominately non-covalent hydrogen bonding, π–π, and hydrophobic interactions [2].

Surprisingly, few molecular recognition studies have been attempted with organometallic hosts, while many more studies have occurred with inorganic hosts [3]. A pertinent organometallic example was the macrocyclic organopalladium hosts, synthesized by Loeb et al. [3a], that can recognize nucleobases via simultaneous first- and second-sphere coordination; i.e. σ-donation to Pd and hydrogen bonding to the macrocycle heteroatoms. It is important to note that these latter host–guest chemistry studies were performed in *non-aqueous* media, presumably because of the instability of their macrocyclic organometallic hosts in water or their lack of solubility in δaqueous solution. As well, chiral metalloporphyrin receptors that have been studied in organic solvents, show preferential binding to amino acids [3b], while Stang and coworker synthesized a series of platinium and palladium macrocyclic squares for host–guest complexation using a dihydroxynaphthalene compound as an example, also in organic solvents [3c]. An important review on inorganic and organometallic host–guest chemistry was published by Canary and Gibb [3h].

Among the limited inorganic or organometallic supramolecular hosts studied [3], none of them, however, were constructed by incorporating nucleobase, nucleoside, or nucleotide molecules as crucial components of the host framework.

Bioorganometallics: Biomolecules, Labeling, Medicine. Edited by Gérard Jaouen
Copyright © 2006 Wiley-VCH Verlag GmbH & Co. KGaA, Weinheim
ISBN: 3-527-30990-X

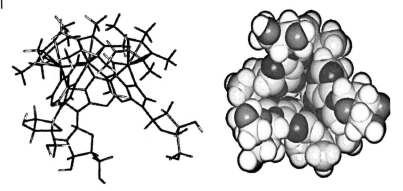

Fig. 10.1 [Cp*Rh(2'-deoxyadenosine)]$_3$(OTF)$_3$, **1**.

Recently, Fish et al. [4] reported on the molecular recognition of aromatic and aliphatic amino acid and aromatic and aliphatic carboxylic acid guests with supramolecular Cp*Rh-nucleotide cyclic trimer hosts, such as **1**, in aqueous solution at pH 7 [4a–d].

This type of host–guest chemistry in aqueous solution is important, since it could be considered as the simplest model for the interactions between DNA/RNA molecules and their binding proteins, such as those that regulate genes. Moreover, the non-covalent hydrophobic effect is more fully dramatized in water by solvophobic forces that enhance host–guest interactions [5].

In this chapter, I will discuss the scope of our molecular recognition studies with two organometallic hosts, by encompassing the guest examples from aromatic and aliphatic amino acids, to substituted aromatic and aliphatic carboxylic acids, and di- and tripeptides (Chart 10.1), while determining the importance of π–π, hydrophobic, and H-bonding effects as a function of steric, electronic, and conformational parameters, along with host–guest thermodynamic parameters, K_a (association constants) and ΔG^0 (free energies of complexation) values. Furthermore, I will also discuss the new supramolecular organometallic complexes that are models for ionophores, biological macrocylic ligands that selectively complex and transport alkali ions, such as Na$^+$ and Li$^+$ ions [6]. Finally, hormone receptor proteins that appear to play a key role in the proliferation of breast cancer tumors have been structurally characterized by X-ray crystallography, the estrogen receptor, and ERα, the estradiol binding site [4g]. Computer docking experiments with organometallic pharmaceuticals show dramatic changes in key helical proteins at the estrogen receptor site, ERα, in comparison to hydroxytamoxifen, the drug of universal use against breast cancer, and these conformational changes will be discussed as they pertain to antagonist and cytotoxic activity.

10.2
Host 1

10.2.1
Synthesis, Structure, and Aqueous Stability

Host **1** is shown in Fig. 10.1, the stick model (side view, left) and the CPK model (bottom view, right), while the self-assembly synthetic procedures have been described previously and includes reaction of $[Cp*Rh(H_2O)_3](OTf)_2$ with 2'-deoxy-adenosine at pH 10.5 [4a-d]. Trimer **1** is a diastereomeric mixture, while the single-crystal X-ray structure of an enantiomer of the 9-methyladenine cyclic trimer derivative was reported previously, and showed that it has a triangular dome-like supramolecular structure, with three Cp* groups stretching out from the top of the dome, three Me groups pointing to the bottom, three adenine planes forming the surrounding shell, and three Rh atoms embedded in the top of the dome (Fig. 10.2) [4a,d]. This molecule also possesses a C3 axis, which passes from the top of the dome to the bottom. The distance between the adjacent methyl groups at the bottom of the dome; i.e. at the opening of this molecular receptor, is about 7.5 Å, while the cavity depth is a consequence of the substituent on N9 of the nucleobase, nucleoside, or nucleotide and is in the range of 4 Å.

These Cp*Rh cyclic trimer derivatives are quite stable in aqueous solution; for example, complex **1** was observed by ^1H NMR spectroscopy, for two weeks, at pH 6–9, with no apparent decomposition [4a, b]. Therefore, all the critical parameters for host–guest chemistry, such as the supramolecular bowl shape, the large cavity size, and the aqueous stability of these Cp*Rh-nucleobase/nucleoside/nucleotide cyclic trimers provided the opportunity to utilize them as molecular receptors to recognize biologically relevant molecules in aqueous media at a physiological pH of 7 [4a, b].

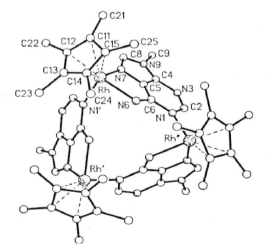

Fig. 10.2 X-ray structure of $[Cp*Rh(\mu-\eta^1(N1):\eta^2(N6, N7)-9-methyladenine]_3(OTf)_3$.

10.2.2
Molecular Recognition of Aromatic and Aliphatic Amino Acids

About one half of the twenty common amino acids were selected in this molecular recognition study, and several criteria were considered in the selection process: (1) solubility in H_2O; (2) representativeness; and (3) stability of the hosts in the presence of the amino acids. According to these criteria, tyrosine, cysteine, and methionine were excluded, since the first example is not soluble in H_2O, and the latter two apparently caused slight decomposition of the hosts. The structures of the key aromatic and aliphatic amino acids, aromatic and aliphatic carboxylic acids, and di- and tripeptides are shown in Chart 10.1, along with the relevant proton designations that were affected by host **1** [4a, b]. The pK_a values of these amino acids are indicative of the zwitterion forms being the predominant species at pH 7 [4a, b, 7].

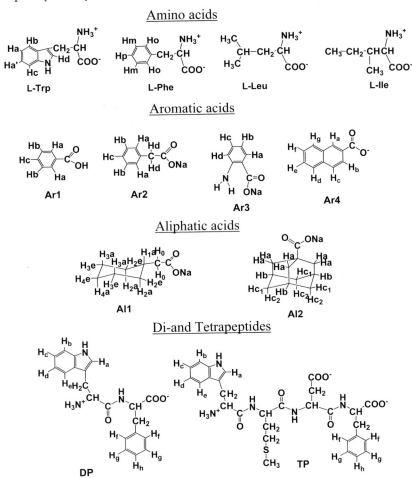

Chart 10.1 Guests used with host **1** in molecular recognition studies.

The molecular recognition process of these different amino acid guests with host **1** was studied using ^{1}H NMR spectroscopy at ambient temperature [4a,b]. The complexation induced ^{1}H NMR chemical shifts (CICS) of both guests and amino acid hosts were used to discern non-covalent interactions. The presence of upfield chemical shifts for any guest studied with host **1** was an indication of a possible host–guest interaction. We found that by varying the concentration of the host **1** from 0–1 equivalent in the presence of the appropriate amino acid guest, at a constant concentration of 1.0 equivalent, that the CICS values for the guests were maximized at ≈ 0.8–1.0 equivalents. Therefore, in all subsequent host–amino acid guest experiments, we utilized 1.0 equivalent of each host and 1.2 equivalents of each guest. Furthermore, the data show that cyclic trimer, **1**, can only recognize aromatic amino acids (L-Phe, L-Trp) and several aliphatic amino acids with relatively long hydrophobic side chains (L-Leu, L-Ile), pointing to the possibility of classical π–π and/or hydrophobic interactions. Other amino acids, such as L-valine, L-glycine, L-histidine, L-alanine, and L-proline (not shown in Chart 10.1), however, do not apparently interact with these hosts. It is important to note that no enantio- or diastereoselectivity was observed by ^{1}H NMR for host **1** in the molecular recognition reactions, and thus, it appears that all stereoisomers were affected in a similar manner.

These observations may be rationalized by the following three factors: (1) sterically, the cavity size of host **1** were large enough to fit L-Trp without any significant hindrance, and furthermore, L-Trp entered the cavities by using the most favored steric orientation. It is necessary to note that the major portion of L-Trp, which entered the receptor, was the benzene ring, which was very similar to the L-Phe case and; therefore, the steric influences of L-Trp and L-Phe during the recognition process were about the same; (2) electronically, the lone electron pair on the nitrogen atom of the five-member heterocyclic ring of L-Trp could donate electron density to the adjacent benzene ring to make this ring more electron-rich in comparison to L-Phe. Presumably, this electron enrichment is one reason that L-Trp has strong π–π interactions with the electron-deficient π system of **1** [2b, 5b], (3) the greater hydrophobicity of L-Trp (solubility = 0.01 g gH$_2$O^{-1}, Hansch partition coefficient, $\log P_{octanol} = -1.04$) appears to be another important reason, with the solvophobic effect of water as the driving force for this rather facile molecular recognition process. The strong interaction of L-Trp with **1** must be multi-component, encompassing π–π, hydrophobic, and solubility effects, since some amino acids, such as L-Ile and L-Leu, have relatively long hydrophobic side chains and were shown to be weakly associated with host **1**; the later mentioned aliphatic amino acid are also more soluble in water in comparison to L-Trp. As well, L-glycine apparently did not interact with **1**, and again dramatizes the solvophobic effect.

For nucleoside host **1**, the role of the hydroxyl groups on the ribose may be understood by comparing their interactions with guests, L-Phe and L-Trp. Host **1** has one OH group per each ribose unit; and therefore, at the opening of this host cavity, the hydrophobicity increases; a delicate balance of hydrophobicity and hydrophilicity exists. More importantly, the steric hindrance on the sugar of the nucleotide can decrease the non-covalent interactions with guests. For example, three ribose units with two OH groups and a monophosphate methyl ester at the 5'-position has the propensity for the surrounding H_2O molecules and to minimize this unfavorable interaction (desolvate) these three ribose units come close to each other, and presumably, this effect tends to increase the steric hindrance at the opening of the host cavity. Alternatively, the two OH groups on each ribose should have relatively favorable interactions with the surrounding H_2O molecules, and hence less steric demand at the opening of the host cavity. This rationale was supported by comparing the CICS values of L-Phe, which showed the largest value for the least hindered host and the smallest value for the most hindered host. In the case of L-Trp, the situation is more complicated, since the CICS values of L-Trp with **1** were the greatest. This sequence may be explained by considering the contribution of the hydrophobicity of the host, or hydrophobic effect of the host–guest complexation. As mentioned, L-Trp has greater hydrophobicity than L-Phe, and therefore, the hydrophobic interactions between L-Trp and **1** appears to play an important role during the recognition process, besides the π–π interactions and the steric effect at the opening of the host cavities. With the decreasing number of OH groups on the ribose units, the hydrophobicity increases, but at the same time, the steric hindrance increases as well. These two factors, which have the opposite influences on the recognition process of L-Trp, appear to be responsible for the interaction between **1** and L-Trp being optimal.

The ^1H NMR signals of Ha and Ha' on L-Trp were influenced to the greatest extent by host **1** [1a, 2, 5b], with a 0.45 ppm upfield shift, while those of the other two protons, Hb and Hc, had significantly smaller upfield-shifts (0.19 ppm). The ^1H NMR resonances of Hd on the five member ring, and the asymmetric CH_2 protons and the *C–H proton at the chiral center were only slightly affected with 0.01–0.02 ppm upfield shifts. It is apparent that the chemical shifts of host **1** do not show significant changes; only slight upfield shifts of 0.01 to 0.08 ppm were observed. Figure 10.3 illustrates the CICS values of guests L-Trp and L-Phe with host **1**.

Several points may be garnered from these results: (1) the Ha and Ha' side of L-Trp, which can be viewed as the "head" of this guest molecule, deeply penetrated the cavity of **1** and experienced the largest π–π influence; (2) the hydrophilic zwitterion end of L-Trp, which can be viewed as the "tail" of this molecule, was left outside the cavity in contact with H_2O; and (3) the cavity of **1** appears to be very shallow. These three points can be easily rationalized, since the "head" of has the

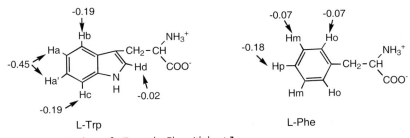

Fig. 10.3 CICS values of L-Trp and L-Phe with host **1**.

highest π-electron density available, and is very hydrophobic. In aqueous solution, this hydrophobic end wants to interact with other hydrophobic groups, such as the cavity of **1**, to minimize the thermodynamically unfavorable interactions with H_2O molecules. On the other hand, the hydrophilic zwitterion "tail" of L-Trp would likely be hydrogen-bonded to the surrounding H_2O molecules, which are mostly outside of the cavity of **1**.

This type of interaction in H_2O is reminiscent of protein molecules in which the majority of the hydrophilic, polar amino acid residues are located on the surface of these biopolymers in contact with H_2O, while the hydrophobic residues are mainly buried in the interior of these polymers to interact with each other. Finally, the depth of the cavity of **1** can easily be observed from the CPK model (Fig. 10.1), and this resulted in a significantly smaller upfield shift of Hb and Hc, which were not shielded as much as Ha and Ha' by the π-electron density of **1**. The above description of the molecular recognition process of L-Trp with **1** was shown in the energy minimized, space-filling model of **1** and the docking of L-Trp (Fig. 10.4).

These overall results suggest that the molecular recognition of L-Trp with **1** can be described in a way that places the L-Trp aromatic rings inside of the host cavity with the aromatic plane, or more specifically, the line which bisects the C–H(a) and C–H(a') bonds parallel to the C3 axis of host **1**.

The association constants (K_a) for the host–guest complexation were estimated by using a standard NMR method [4a, b, 8] to confirm the trends which were observed. The estimated K_a values, a value that encompasses the diastereomers of **1**, are in the range of 500–1000 M^{-1}, while free energies of host–guest complexation, ΔG^0 values are in the range of –4 to –6 kcal mol^{-1}. It is noteworthy to mention that L-Trp, with its optimized steric orientation, electron donating N atom, and hydrophobic effects, has similar K_a and ΔG^0 values with host **1** compared with similar interactions with L-Phe [4b].

The host–guest molecular recognition process was also further substantiated by an intermolecular NOE study between **1** and L-Trp. In that study, the line broadening parameter was set to 4 Hz to minimize the subtraction error. When H8 of **1** was irradiated, weak negative intermolecular NOE signals of L-Trp's Ha,

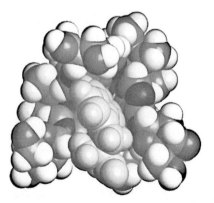

Fig. 10.4 Computer docking of host **1** with L-Trp.

Ha', Hb, and Hc aromatic protons were observed. It is important to note that no intermolecular NOE signal was found between **1** and the solvent, D_2O, which excludes the possibility that the NOE data was an artifact. The moderate association constant (K_a = 607) for **1** and L-Trp, in comparison to the range of literature reported values [2a, 2i, 3b] of 10 to 10^6 M^{-1}, was thought to be partially responsible for the somewhat weak intermolecular NOE signals that were observed. Negative intramolecular NOE signals of **1**'s H1', H2, H2', H2'', H3', and H4' protons were also observed when H8 was irradiated, and the intensities of these intramolecular NOE signals vary according to the distances between H8 and these protons. Similar results were seen when H2 of **1** was irradiated.

10.2.3
Molecular Recognition of Substituted Aromatic Carboxylic Acids

Three substituted aromatic carboxylic acids, *o*-, *m*-, and *p*-aminobenzoic acids were selected as guests to extend the scope of our molecular recognition studies by interaction with host **1**. All three guests, *o*-, *m*-, and *p*-aminobenzoic acids, have the same functional groups, i.e. an electron-donating NH_2 group and an electron-withdrawing COO^- group; however, the two groups are separated from each other at different distances (positional isomers). At pH 7, the anionic forms of *o*-, *m*-, and *p*-aminobenzoic acids are the predominant species in concert with their pK_a values [4a, b]. The CICS values of these three guests by host **1** are presented in Fig. 10.5, with the upfield shifts denoted on the structures, and the distances between the NH_2 and the COO^- groups designated in angstroms.

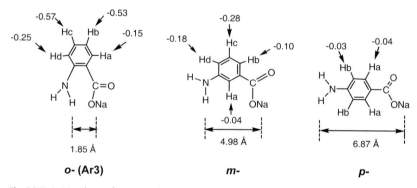

Fig. 10.5 CICS values of *o*-, *m*- and *p*-aminobenzoic acid with host **1**.

Interestingly, the minor substitution changes of these three positional isomers showed dramatically different CICS values and their interrelated K_a and ΔG^0 values, by both π–π and hydrophobic interactions with **1**. The largest CICS values for *o*-(Hc) and *m*- (Hc) are 14 and 7 times larger, respectively, than that of *p*- (Ha). This observation can be explained by the steric effects of the guests, since the two hydrophilic functional groups which form H-bonds with the bulk H_2O need to avoid unfavorable interactions with the hydrophobic cavity of **1**. Moreover, these

H-bonds with the bulk H_2O determine the orientations of o-, m-, and p-amino-benzoic acids as they approach host **1**; the hydrophobic end of guests o-, m-, and p-aminobenzoic acids must enter **1** more favorably. Therefore, guests with the most exposed hydrophobic portions, such as Ar3, should have the largest CICS values, as was observed. Molecular modeling studies confirm that the distances between the amino H atom and carboxylate O atom for o-, m-, and p-aminobenzoic acids are 1.85 Å, 4.98 Å, and 6.87 Å, respectively. With the increasing distances between these the two hydrophilic functional groups, from o-, m-, to p-amino-benzoic acids the steric hindrance also dramatically increases and, at the same time, the exposed hydrophobic portions decrease.

10.2.4
Molecular Recognition of Aromatic and Aliphatic Carboxylic Acids

Three aromatic carboxylic acids, benzoic acid (Ar1), phenylacetic acid (Ar2), and 4-methoxyphenylacetic acid were selected as guests to probe the depth of penetration in host **1**, and, as well, two aliphatic carboxylic acid guests, cyclo-hexylacetic acid (Al1), and 1-adamantanecarboxylic acid (Al2), were also used to further study the importance of hydrophobic effects in the molecular recognition process with host **1**. According to their pK_a values, all of these carboxylic acids existed in the anionic form at pH 7 [4a, b]. The CICS values of the three guests, Ar1, Ar2, and 4-methoxyphenylacetic acid, by host **1** are presented in a similar fashion as described above for the substituted aromatic carboxylic acids (Fig. 10.6).

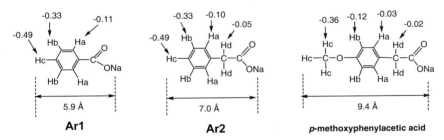

Fig. 10.6 CICS values of Ar1, Ar2 and 4-methoxyphenylacetic acid with host **1**.

It is apparent that the CICS values for Ar1 and Ar2 are almost identical, indicating that one more CH_2 group between the benzene ring and the carboxylate has very little or no influence on the π–π/hydrophobic recognition process; the CH_2 group is close to the hydrophilic group of the guest, and therefore, was not intimately involved in the molecular recognition process. The CICS values for p-methoxyphenylacetic acid, which has a CH_3O group on the 4-position, are quite different from those for Ar1 and Ar2. The CICS values for p-methoxyphenylacetic acid with **1** are –0.36, –0.12, –0.03, and –0.02 ppm for Hc (methyl), Hb (aromatic), Ha (aromatic), and the Hd (CH_2) protons, respectively. The small CICS values for aromatic protons of p-methoxyphenylacetic acid suggest that the hydrophobic interaction is the major recognition effect, while π–π stacking is the minor contributor in this molecular recognition example. This result also suggests that the cavity of host **1** is shallow, which agrees well with the estimated cavity depth of ≈ 4 Å.

Although the structure of guest Ar2 is very similar to that of L-Phe, their CICS values with **1** are quite different. This difference may be explained by inspecting the hydrophilic end of both guests. Guest L-Phe has two hydrophilic functional groups, NH_3^+, and COO^- while Ar2 only has only one, and therefore, the hydrophilic end of L-Phe forms stronger H-bonds with the bulk H_2O solution. The strong H-bonds between L-Phe and the surrounding H_2O should prevent L-Phe from entering the cavity of **1** too deeply; the desolvation energies appear to be higher in this case.

For aliphatic carboxylic acid guests, Al1 and Al2, the CICS values of the two guests with host **1** are shown in Fig. 10.7.

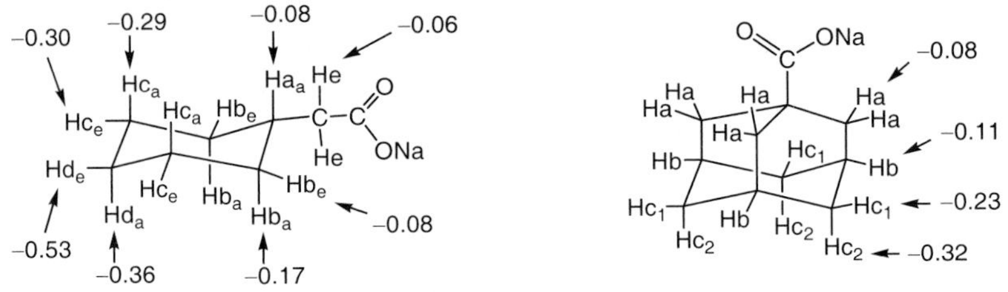

Fig. 10.7 CICS values of Al1 and Al2 with host **1**.

Surprisingly, the CICS values for these two aliphatic guests, especially Al1, are comparable, and in some cases, even greater than certain aromatic carboxylic guests. These results are in sharp contrast to the CICS values of several aliphatic amino acids, such as L-Leu and L-Ile, which have relatively long hydrophobic side chains, indicating that the conformation and the number of C atoms (hydrophobicity, solubility in H_2O) of the guest molecules are of significant importance in the molecular recognition process. The "chair" form of Al1 should be predominant during the host–guest interaction, since it should be locked in this conformation by the relatively large CH_2COO^- group, and this should be a less sterically demanding conformation than the alternative "boat" conformation. The molecular recognition process of Al1 with **1** ($K_a \approx 760$ M^{-1}, $\Delta G^0 = -3.9$) is shown in the energy minimized space-filling model of **1** and the docking of Al1 (Fig. 10.8).

The bulky and rigid aliphatic carboxylate guest, Al2, also showed relatively large CICS values, which further strengthens the argument that conformational parameters and the hydrophobic effect of the guest molecules are important in the overall molecular recognition process. The interactions between Al2 and **1** ($K_a \approx 15$, $\Delta G^0 = -1.6$; the $\Delta (\Delta G^0)$ between Al1 and Al2 with host **1** is ≈ 2.3 kcal mol^{-1}) may be also due to the flexibility of the three 2'-deoxyribose groups, which can possibly rotate away to make room for the bulky Al2; overall the apparent steric effect of Al2 limits this process.

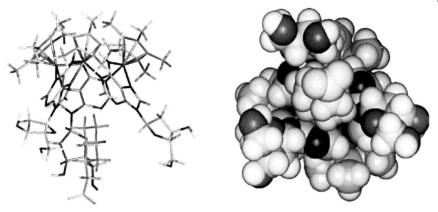

Fig. 10.8 Computer docking experiment of host **1** with guest, cyclohexylacetic acid.

In order to further determine the relative importance of π–π and hydrophobic effects in the molecular recognition process, a competition study between aromatic guest Ar2, and its closely related aliphatic guest, Al1, was undertaken. During this study, the concentrations of both competitors were kept constant, while that of **1** increased from 0 to 1 equivalent. Two control experiments, which held constant the concentration of *one* guest while varying the concentration of **1**, were also performed. The results show similar plots for the control and the competitive experiments for Ar2 and/or Al1, *which suggest that the π–π and hydrophobic effects may be of similar importance in aqueous solution.*

Two other guest molecules, *o*-methoxybenzoic acid and *o*-nitrobenzoic acid, were selected to study the steric and electronic influences of the different functional groups at the 2 position of benzoic acid with host **1**. Their CICS values, together with those of Ar3 and Ar1, were shown in Fig. 10.9 as before.

It is apparent that the electron-donating abilities of the four functional groups in the boxes (see above) follow the order of $NH_2 > OCH_3 > H > NO_2$, while the steric hindrance order is: $OCH_3 \approx NO_2 > NH_2 > H$. Favorable electronic and steric properties for Ar3 ($K_a \approx 810$, $\Delta G^0 = -4.0$) provide the largest CICS values, and hence K_a and ΔG^0 values, while the steric advantage of Ar1 ($K_a \approx 710$, $\Delta G^0 = -3.9$;

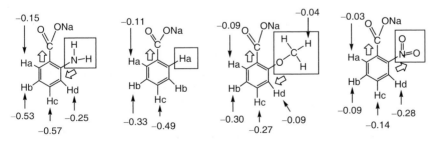

Fig. 10.9 CICS values of Ar1, Ar3, *o*-methoxybenzoic acid and *o*-nitrobenzoic acid with host **1**.

$\Delta (\Delta G^0)$ between Ar3 and Ar1 is ≈ 0.1 kcal mol^{-1}) makes its CICS values larger than those of *o*-methoxybenzoic acid ($K_a \approx 15$, $\Delta G^0 = -1.6$), even though OCH$_3$ is more electron-donating than an H atom. Therefore, sterically less demanding and more electron rich aromatic guests should have the largest CICS values in the molecular recognition process.

10.2.5
Discussion of the Molecular Recognition Process with Host 1

The driving force for the novel molecular recognition process that we presented in this chapter for host **1** (Fig. 10.1) and the designated guests (Chart 10.1) is directly related to the use of water as the solvent. As clearly pointed out by Breslow [5a], water maximizes the hydrophobic effect and the desolvation energies dictate the extent of the host–guest complexation. Therefore, water will solvate the hydrophilic end of the guests, while the desolvated hydrophobic substituent enters the host cavity and interact via non-covalent processes.

The unique structure of the cyclic trimer hosts, which can be modified by substituents on the N9 position of the adenine nucleus, represents a supramolecular bowl shaped molecule with a cavity opening of ≈ 7.5 Å that is sufficiently large enough to accommodate many guests. The adenine ligands form the inner shell of the bowl, with the rhodium atoms basically acting as an anchor for both the dome (Cp*) and for the inner shell (adenine). It is also apparent that the coordinatively-saturated Rh atoms are spectators during the molecular recognition process and do not directly take part in any of the non-covalent host–guest interactions.

The electron-deficient adenine inner shell, which forms intramolecular η^2 bonds to Cp*Rh via NH6 and N7 and an intermolecular η^1 bond to Rh via N1 of another adenine (a self-assembly mechanism that provides a cationic Cp*Rh cyclic trimer complex), was able to interact more favorably with the aromatic guests that contained electron-donating groups, such as L-Trp, Ar3, and *o*-methoxybenzoic acid. This result seemed to dictate that π–π interactions predominated in the molecular recognition process. However, by judiciously modifying the N9 substituent with the 2-deoxyribose group, host **1**, one could maximize the host–guest process and show that aliphatic carboxylic acids, such as Al1, interacted as favorably with host **1** as did aromatic guests, L-Trp, Ar3, and *p*-methoxyphenylacetic acid. Moreover, competition experiments with Ar2 and Al1, for host **1**, verified the equal importance of both non-covalent π–π and hydrophobic effects in the overall molecular recognition process.

Fish et al. also studied two other parameters, along with the above-mentioned electronic effect that appeared to affect the host–guest, molecular recognition process; namely, steric and conformational aspects. The importance of the steric effect can be seen with guests, *o*-, *m*-, and *p*-aminobenzoic acid, upon interaction with host **1**. As the positional isomers of aminobenzoic acids are changed from the *ortho*, to the *meta*, and then *para* positons, the extent of the host–guest interaction is dramatically decreased. The conformational effect is seen with a comparison of the aliphatic amino acid guests, such as L-Leu and L-Ile, that appeared not to interact with the host **1**, and guests Al1 and Al2, that were able to readily interact with the host **1**. Moreover, the solvophobic effects cannot be minimized, since the aliphatic amino acids are more soluble in water and, as well, their Hansch partition coefficients, log $P_{octanol}$, show them to have greater solubility in water than octanol [5b].

It appears that conformationally rigid guests, as epitomized by Al1 and Al2, were better able to interact with host **1**, presumably by a hydrophobic effect, although the apparent steric effect of Al2 somewhat limits this molecular recognition process. In contrast, the aliphatic amino acids, L-Leu, L-Ile, L-glycine, valine, alanine, and proline, suffer from conformational flexibility and, more importantly, their increased solubility in water, compared with L-Trp and L-Phe, significantly limits host–guest complexation. To reiterate, the electronic effect with electron-donating substituents, e.g. $-NH_2$ and N-heterocyclic rings, enhanced the host–guest interaction, as shown by guests, Ar3 and L-Trp, while a guest with an electron-withdrawing substituent, guest o-nitrobenzoic acid with a o-NO_2 group on a benzoic acid nucleus, appears to have a more complicated host–guest interaction. Interestingly, the proton ortho to the NO_2 group in o-nitrobenzoic acid is upfield shifted to as great of an extent as that ortho proton in guest Ar3 containing an electron-donating o-NH_2 group. Apparently, the o-NO_2 group positions itself in proximity to the host cavity opening via a possible non-covalent H-bonding effect with the 2′-deoxyribose group, but this reasoning is speculative as of now.

In contrast, guest o-methoxybenzoic acid with an o-CH_3O- group, that has the largest steric demand of the ortho substituted benzoic acid, appears to position itself quite differently, since the ortho proton is slightly affected by the host–guest interaction. This latter discussion clearly shows that the extent of these host–guest interactions is a consequence of subtle changes of a multiple of parameters that are far more complicated than our current understanding.

10.2.6
Conclusions for Host 1: Non-covalent Interactions with Various Guests

Fish et al. found that novel, bioorganometallic, supramolecular host, **1**, can readily recognize biologically important guests by a variety of non-covalent processes. The complexity of the interplay between these non-covalent processes; namely, π–π, hydrophobic, and subtle H-bonding effects, with further parameters of steric, electronic, conformational effects, and the ever present solvophobic effect in H_2O, clearly provides a driving force for future studies with these unique hosts. For example, Fish *et al.* were able to use host **1** to recognize certain protein sequences (Chart 10.1, DP, TP) [4b] with terminal amino acids that interact favorably via the non-covalent processes that we have attempted to elucidate in this chapter. Thus, conformational information that relates protein structure via the interaction of a supramolecular, bioorganometallic host is an exciting field that should be pursued further by organometallic chemists interested in this area of research.

10.3

A New Host, *trans*-[Cp*Rhη¹-(N3)-1-methylcytosine)(μ-OH)]₂(OTf)₂ 2

10.3.1
Molecular Recognition Studies with Aromatic Amino Acids:
H-bonding as the Critical Recognition Parameter

In the previous sections of this chapter, the molecular recognition process was described that involved π–π, hydrophobic, and possible subtle H-bonding effects. Now, I wish to discuss a new molecular recognition process based primarily on selective hydrogen bonding interactions between the host, *trans*-[Cp*Rh–η¹-N3)-1-methylcytosine)(μ-OH)]₂(OTf)₂, **2**, and several examples of aromatic amino acid guests, L-tryptophan and L-phenylalanine, L-Trp and L-Phe, in water at pH 7.0 [2].

The X-ray structure of host **2** (Fig. 10.10) [7, 8] clearly shows the unique intramolecular H-bonding aspects of the ligand, 1-methylcytosine, with the $Rh_2(\mu\text{-OH})_2$ center. Thus, the μ-OH groups act as both H-donor and acceptor with the 2-carbonyl (OH–O=C, 1.96 (1) Å) and NH_2 groups (HO–HNH, 1.93 (1) Å), respectively.

Moreover, it was thought that an intermolecular recognition process also based on H-bonding to the μ-OH groups and the cytosine NH_2 and C=O functionalities might be feasible with the aromatic amino acid NH_3^+ and COO^- groups, without seriously disrupting the intramolecular hydrogen bonding regime shown in Fig. 10.10.

¹H NMR techniques were again utilized to discern the complexation-induced ¹H NMR chemical shifts (CICS) for the host and the guests [8]. Table 10.1 shows the results with guest L-tryptophan (L-Trp) in the presence of host **2**.

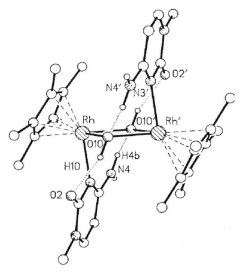

Fig. 10.10 X-ray structure of host **2** [7].

Table 10.1 CICS shifts upon host–guest recognition.[a]

Free tryptophan (δ)	With host 2 (δ)	$\Delta\delta$
a 7.13	7.02	−0.11
a' 7.05	6.89	−0.16
b 7.58	7.16	−0.42
c 7.39	7.14	−0.25
d 7.16	6.82	−0.34
e 3.15	3.00	−0.15
f 3.34	3.26	−0.07
g 3.90	3.77	−0.12

[a] ^{1}H NMR shifts at pH 7.0, 300 MHz, 1 : 1 host/guest ratio.

What is dramatically evident for L-Trp is the CICS values for H_d ($\Delta\delta = -0.34$); H_e ($\Delta\delta = -0.15$); H_f ($\Delta\delta = -0.07$); and H_g ($\Delta\delta = -0.12$), which were diametrically opposite to the previously reported Cp*Rh-2'-deoxyadenosine cyclic trimer molecular recognition studies with host **1**, where no upfield CICS for these designated protons were observed; in that process, the indole phenyl group was found inside the hydrophobic receptor, while the hydrophilic aromatic amino acid NH_3^+ and COO^- groups were outside in the water media, and the chiral C–H attached to these groups, as well as the adjacent asymmetric CH_2, were not affected by the magnetic anisotropy of the inner shell of the host adenosine ligands.

More importantly, two sets of signals for the host ligand, 1-methylcytosine, bound to Cp*Rh; the N-CH₃, H_5, and H_6 protons, were also observed. The CICS for one of the now apparently asymmetrical 1-methylcytosine ligands (Table 10.2) was similar to complex **2** alone, while the other had CICS upfield shifts for the N-CH₃ ($\Delta\delta = -0.41$); H_5 ($\Delta\delta = -0.31$); and H_6 ($\Delta\delta = -0.18$). Clearly, the CICS for one of the 1-methylcytosine ligands were affected by the non-covalent interactions with the indole ring of L-Trp and vice-versa. Thus, it appears plausible that the primary host–guest interaction of **2** with L-Trp was from a H-bonding process of the NH_3^+ and COO^- groups with **2**, enhancing non-covalent interactions of the 1-methylcytosine ligand with L-Trp.

In order to better understand these H-bonding and non-covalent interactions between host and guest, Fish et al. conducted computer docking experiments to provide the energy minimized, space-filling/ball and stick model of **2** with a ball and stick model of guest L-Trp, as shown in Fig. 10.11. The top view in Fig. 10.6 demonstrates the H-bonding of the NH_3^+ group to one μ-O and to the C=O group of one of the 1-methylcytosine ligands, while the COO^- group H-bonds to a NH₂

Table 10.2 Host **2** ^{1}H NMR data with guests L-Trp and L-Phe.[a]

2

Free host 2 (δ)		Guest L-Trp (δ)				Guest L-Phe (δ)			
			$\Delta\delta$		$\Delta\delta$		$\Delta\delta$		$\Delta\delta$
N-CH$_3$	3.24	2.83	−0.41	3.22	−0.02	3.23	−0.01	3.13	−0.11
H6	7.44	7.26	−0.18	7.41	−0.03	7.42	−0.02	7.35	−0.09
H5	5.83	5.52	−0.31	5.81	−0.02	5.81	−0.02	5.82	−0.01
Cp*	1.46	1.35	−0.12	1.35	−0.12	1.40	−0.06	1.40	−0.06

[a] ^{1}H NMR shifts at pH 7.0, 300 MHz, 1 : 1 host/guest ratio.

group of the 1-methylcytosine ligand. This H-bonding scheme of **2** with L-Trp then provides that the remaining structure of the guest is fixed in relation to the host, as shown in the top and middle views of Fig. 10.11.

Therefore, the indole group is positioned orthogonal to the plane of the 1-methyl-cytosine ligands in host **2**, while selectively affecting one of the two 1-methylcyto-sine groups, accounting for this ligand's asymmetry, and the upfield shifts observed in the NMR CICS values (Table 10.1). Figure 10.11 also shows the plausible reason that H$_b$ was appreciably shifted upfield due to its proximity (middle view) to the C=O group of 1-methylcytosine, while also noting the asymmetric CH$_2$ hydrogens, where H$_e$ is more affected by the CICS effects then H$_f$ (Table 10.1). It is also interesting to note the appreciable upfield shift for H$_d$ ($\Delta\delta$ = −0.34), shown in Fig. 10.11, middle, and which is attributed to the proximity to one of the Cp* ligands via a plausible CH-π non-covalent interaction. Moreover, the potentially asymmetric Cp*Rh groups are co-incident in the NMR (only one signal), even though the nitrogen ring of the indole nucleus appears (Fig. 10.11, middle) in the docking experiment to be somewhat orthogonal to one of the Cp* ligands.

Fish and coworkers then studied guest, L-phenylalanine (L-Phe), with host **2** (Table 10.3, see page 340), and found a striking difference in the CICS for the 1-methylcytosine ligands, as opposed to that with guest **2**, L-tryptophan. Relatively, smaller CICS values were observed for the 1-methylcytosine ligands in the presence of L-Phe; for example, one of the 1-methylcytosine ligands was not greatly affected

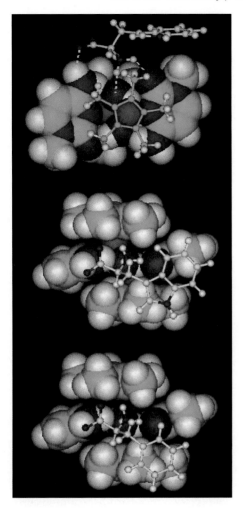

Fig. 10.11 The top view of **2-ʟ-Trp**: H-bonding of the NH_3^+ group to one μ-OH and to the C=O group of one of the 1-methylcytosine ligands, while the COO^- group H-bonds to a NH_2 group of the other 1-methylcytosine ligand. Middle view of **2-ʟ-Trp**: Bottom view of **2-ʟ-Phe**. N (blue); O (red); H (white) Rh (magenta).

by the host–guest interaction and showed the $N\text{-}CH_3$, H_5, and H_6 protons with average upfield shift values of $\Delta\delta \approx 0.017$ for the non-covalent interactions (Table 10.2). The other 1-methylcytosine ligand had $N\text{-}CH_3$, H_5, and H_6 proton upfield shifts of $\Delta\delta = -0.11$, -0.01, and -0.09, respectively.

Figure 10.11 (bottom) shows the docking experiment results with **2** and ʟ-Phe, and clearly a similar H-bonding process of the NH_3^+ group to one of the oxygen atoms of the Rh(μ-OH) assembly, and the C=O group of one of the 1-methyl-cytosine ligands, while that of the COO^- group to a NH_2 group, was deemed appropriate from the energy minimized structure found in Fig. 10.11, top. The critical aspect about this host–guest interaction is that, in analogy to the nitrogen ring proton, H_d, in ʟ-tryptophan, the H_p, H_m, and H_o protons at 7.273, 7.271, and 7.19 ΔD were upfield shifted, $\Delta\delta = -0.28$, -0.36 and -0.35, respectively (Table 10.3).

Table 10.3 CICS shifts upon host–guest recognition.[a]

Free L-Phe (δ)	Interaction with host 2 (δ)	$\Delta\delta$
a 3.85	3.79	−0.07
b 2.98	2.69	−0.29
c 3.15	2.73	−0.42
o,o' 7.19	6.84	−0.35
m,m' 7.271	6.91	−0.36
p 7.273	7.00	−0.28

[a] ^{1}H NMR shifts at pH 7.0, 300 MHz, 1 : 1 host/guest ratio.

Clearly, the aromatic protons of guest L-Phe were upfield shifted by the proximity to both the C=O of one of the 1-methylcytosine ligands, and one of the Cp* ligands. It is also important to notice that the asymmetric CH$_2$ protons are also substantially shifted upfield with values of $\Delta\delta$ = −0.29 and −0.42, respectively. We also observe only one Cp* signal that was shifted upfield with $\Delta\delta$ = −0.06, as was the case with the host–guest complex of **2** with L-Trp.

10.3.2
Conclusions for Molecular Recognition with Host 2

In this section, a new bioorganometallic recognition process with host **2** and several examples of aromatic amino acid guests has been demonstrated that depends on selective H-bonding of the NH$_3^+$ and COO$^-$ groups of the aromatic amino acids, L-Trp and L-Phe, to an oxygen of one of the Rh(μ-OH) assemblies, and C=O and NH$_2$ groups of the Rh bound 1-methylcytosine ligands, in water at pH 7.0. The significance of this new, highly selective, host–guest process is that it can be thought of as a model for H-bonding of biologically significant guests to metallo-enzymes and DNA/RNA [1a, 4a, b].

10.4
Computer Docking Experiments of Organometallic Pharmaceuticals at Estrogen Receptor Binding Sites: Selective, Non-Covalent Interactions with Hormone Proteins

Recently, Jaouen and coworkers discovered that an organometallic derivative of the known breast cancer drug, Tamoxifen (hydroxytamoxifen, a liver metabolite, is the active drug), Ferrocifen, **3**, and its derivatives, were potential candidates for breast cancer therapy, as well as other cancers [4g, 9]. This seminal finding has created a new paradigm; namely, the field of organometallic pharmaceuticals [9, 10]. Since the X-ray structure of the estrogen hormone receptor site (ERα) has been determined, and this is thought to be the major receptor protein implicated in hormone-dependant breast cancers, it was appropriate, using computer docking/energy minimization experiments at the receptor site, to discern the conformation and non-covalent interactions of Ferrocifen, **3**, and other organo-metallic drug derivatives, with the surrounding simplified protein structure [4g, 9, 10].

Moreover, the identification of novel targets of estrogen action provides an increasing degree of complexity to the understanding of mechanisms by which this hormone elicits many of its normal, as well as pathological effects. Estradiol, **4**, the archetype of estrogens, has been implicated in a number of problems from fertility questions to several types of cancer, including frequent diseases, such as osteoporosis, and cardiovascular and metabolic disorders. It is well known that the effect of estradiol is mediated through its ability to bind to the estrogen hormone receptor site.

A molecular view of the binding modes existing both with an agonist (**4**) as well as an antagonist; (Tamoxifen, 4-hydroxytamoxifen, blocks estradiol from the receptor site) with similar nanometer distances, based upon these X-ray determi-nations, can now be used to examine the consequences of the attachment of an organometallic moiety, for example, compound **3**, to a modified drug structure; i.e. hydroxytamoxifen modified to **3** (*note the optimized bioactivity for* **3** *was for a three-spacer methylene group on the ether oxygen*), with respect to the receptor binding site. Since we have two groups of organometallic drug derivatives based on an estrogenic, complex **5**, or an anti-estrogenic, complex **3**, structural effect, then we will illustrate the different non-covalent binding regimes with an example of each type of behavior.

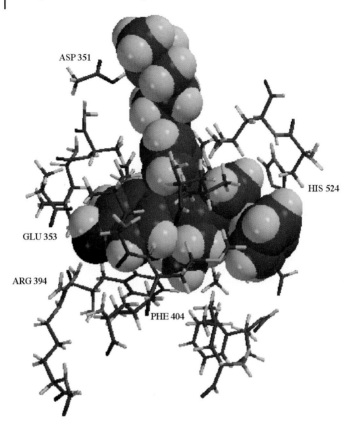

ASP 351

HIS 524

GLU 353

ARG 394

PHE 404

Fig. 10.12 Ferrocifen derivative (Z isomer), **3**, docked at the estrogen protein receptor site and shows the organometallic complex inside the antagonist binding site of the estrogen receptor.

Figure 10.12 defines the anti-estrogenic, organometallic complex, **3**, as to its conformation in computer docking/energy minimization experiments with the estrogen receptor site proteins, and demonstrates important non-covalent interactions with the amino acids depicted in the figure. Thus, several hydrogen bonding regimes are discernable, for example, between aspartic acid carboxylate 351 (1.868 Å) and one of the N-CH$_3$ groups of the ether side chain, O(CH$_2$)$_3$N(CH$_3$)$_2$, the carboxylate of glutamic acid 353 and the phenolic hydrogen (1.577 Å), and the arginine 394 NH with the phenolic oxygen (2.061 Å). Moreover, one of the Cp ligands of the ferrocene group has a non-covalent CH-π interaction with the histidine 524 imidazole ring.

In contrast to the anti-estrogenic, **3** (Z isomer), binding mode to the estrogen protein receptor site, the ligand binding domain for estrogenic **5** was similar to estradiol, **4**, with the exception of the ruthenocene Cp ligand, attached to a rigid acetylenic linkage. Clearly, Fig. 10.13 shows the dramatic conformational and non-covalent bonding differences with the estrogen protein binding site between the

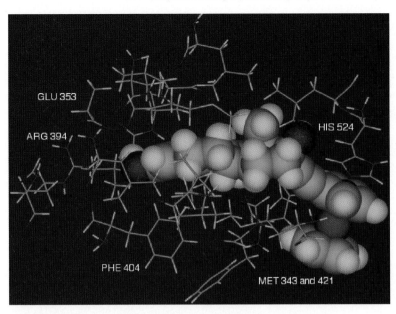

Fig. 10.13 17α-ruthenocenylethynylestradiol, **5**, docked at the estrogen protein receptor site. The ethynyruthenocenyl group is also bordered in its lower side by two hydrophobic amino acid residues Met 343 and Met 421. A shrinkage, which is well adapted to accommodate the rigid ethynyl group, can be clearly seen in front of the 17α-position of the hormone. This allows the ruthenocenyl group to avoid steric constraints inside the cavity.

two organometallic modified drugs, **3** and **5**. Significantly, one of the Cp rings of the ruthenocene group is now in a non-covalent π–π interaction (3.211 Å), with the histidine 524 imidazole group, while the imidazole ring NH group is hydrogen bonding (2.722 Å), to the 17 α OH group. Other pertinent hydrogen bonds are with the A ring phenolic OH with both the glutamic acid carboxylate 353 (2.722 Å), and the arginine 394 NH (3.101 Å).

Therefore, the exciting finding of possibly why organometallic pharmaceutial, **3**, is a potential anti-cancer agent from bioassay results, while the organometallic modified estradiol, **5**, is not, could be related to the conformational changes in the estrogen receptor protein upon binding of the drug. This can be depicted in the more complex receptor protein site with **3**, Fig. 10.14, where the apparent steric effects of the $O(CH_2)_3N(CH_3)_2$ side-chain appear to cause Helix 4 and Helix 12 to leave a gap between them. This factor is opposite to that of complex **5**, where there is no gap (similar to a mouse trap) between Helix 4 and Helix 12, and this plausible reason, among others, may explain why **3** is a potential drug for breast cancer, and **5** is not. Another important aspect is the fact that **3**, with a ferrocenyl ligand, can be readily oxidized to a ferrocenium ion, and in the process of degradation to Cp and Fe(III), can generate an oxygen radical species that can provide the cytotoxic effect, by possibly reacting with DNA in proximity to the binding domain at the estrogen receptor site [11].

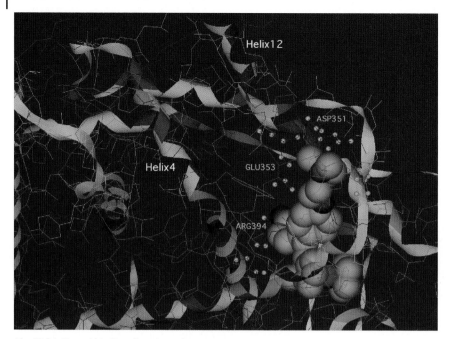

Fig. 10.14 Ligand binding domain at the estrogen receptor site of potential organometallic pharmaceutical, **3**.

10.5
Organometallic Ionophores

10.5.1
Structure and Selectivity to Alkali Metal Ions

Taking a similar synthetic approach to the self-assembled Cp*Rh cyclic trimer structures, such as **1**, that were used as hosts for biomolecules, Severin and coworkers have developed a novel, self-assembly approach to organometallic ionophores, biomolecular organometallamacrocycles that selectively sequester alkali metal ions [6]. Severin and coworkers have used this self-assembly approach to these synthetic targets for potential medical and analytical applications [12]. Scheme 10.1 demonstrates the versatility of both organic ligands and π-organometallic fragments as starting materials.

The self-assembly process begins with the chloro-bridged complexes [(π-ligand)MCl$_2$]$_2$ (M = Ru, Rh, Ir), which reacted with ligands L1-L3 in the presence of base. The 3-oxo-pyridonate ligand, L1, provided organometallamacrocyclic complexes of the general formula [(π-ligand)M(L1)]$_3$ with various arene ligands; for example, C$_6$H$_6$, C$_6$Me$_6$, 1,3,5-C$_6$H$_3$Et$_3$, cymene, C$_6$H$_5$CO$_2$Et, and with Cp* ligands (Fig. 10.15). Trinuclear complexes are likewise formed with the bridging ligands L2 and L3 in combination with (cymene)Ru^{2+} (L2, L3) and Cp*Ir^{2+} fragments (L2).

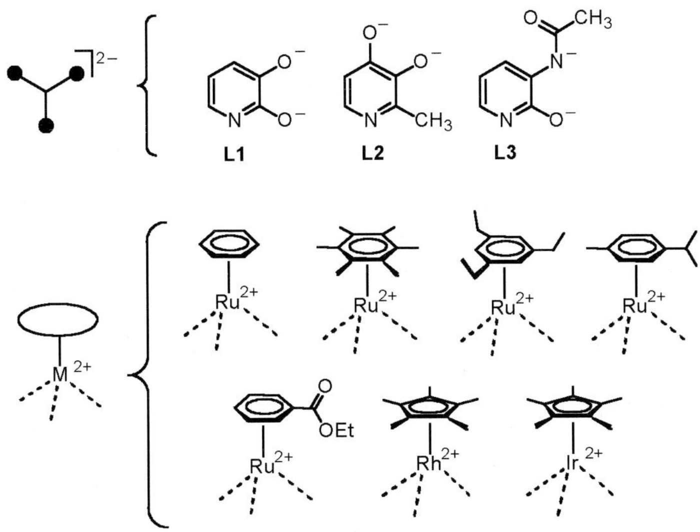

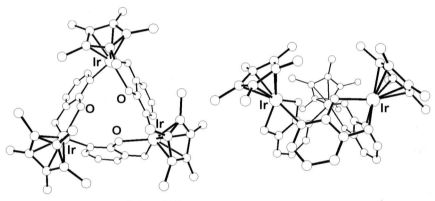

Scheme 10.1 Ligands, L1-L3, and organometallic fragments for the synthesis of potential organometallic ionophores.

Fig. 10.15 X-ray structure of [Cp*Ir(L1)]₃.
Left: view along the pseudo C₃ axis. Right: view from the side.

The tridentate ligand, L1, bridges the metals by coordination via the oxo groups and the pyridine nitrogen atom. The 12-membered organometallamacrocycles contain three oxygen atoms positioned in close proximity to each other, and can thus be regarded as organometallic analogs of 12-crown-3.

Similar to what was found for L1, complexes with L2 show a (pseudo) C_3 symmetric geometry with three tetrahedral (cymene)Ru or Cp*Ir corners (Fig. 10.16). Again, the ligand was coordinated via the oxygen atoms and the ring nitrogen atom to the metal atoms. The planes defined by the heterocyclic ligand; however, are almost perpendicular to the plane defined by the metal atoms, resulting in an expanded macrocycle with M···M distances of 7.24 Å.

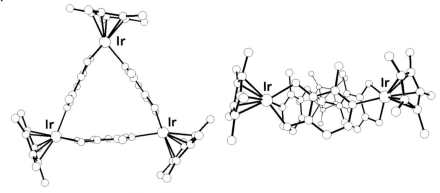

Fig. 10.16 X-ray structure of [Cp*Ir(L2)]₃.
Left: view along the C_3 axis. Right: view from the side.

The compounds [(π-ligand)M(L1)]₃ represent the first *organometallic crown complexes* to be synthesized [6]. Similar to their organic counterparts, organometallacrown complexes are able to bind cationic guests such as alkali metal ions. Thus far, Severin and coworkers have been able to isolate and structurally characterize adducts of [(π-ligand)M(L1)]₃ with LiCl, LiBF₄, LiF, LiFHF, NaCl, NaBr and NaI [12]. In all cases studied, the metal ion M′ was bound to the three adjacent oxygen atoms of the receptor. The fourth coordination site is generally occupied by the anion X (Scheme 10.2).

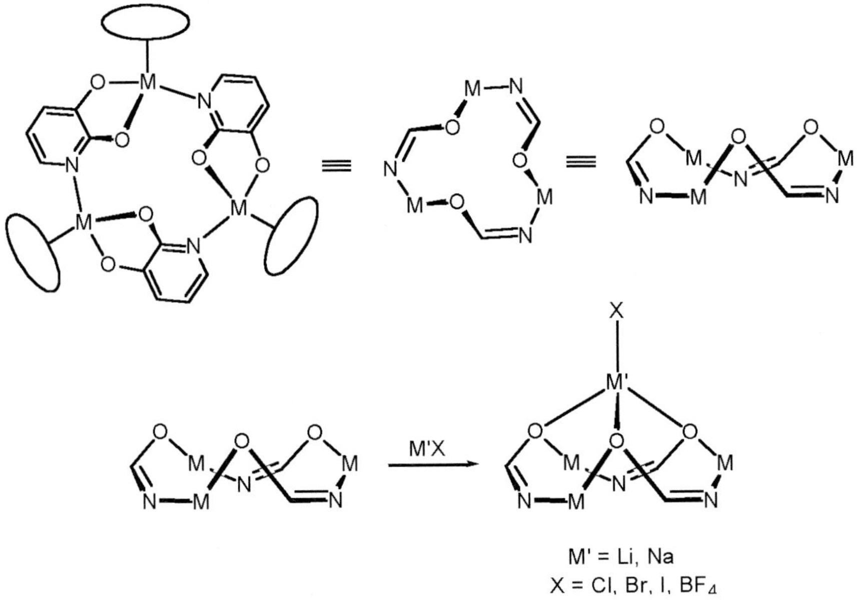

M′ = Li, Na
X = Cl, Br, I, BF₄

Scheme 10.2 Organometallic complexes of the general formula [(π-ligand)M(L1)]₃ as receptors for lithium and sodium salts.

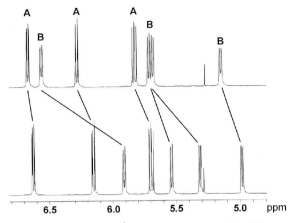

Fig. 10.17 Aspects of the ^{1}H NMR spectrum (CDCl$_3$) of the receptor [((cymene)Ru(L1)]$_3$ (bottom) and the corresponding NaCl adduct (top). The signals of the cymene π-ligand are labeled "B", the signals of the bridging pyridonate ligand L1 are labeled "A".

The coordination to guest molecules could be easily detected by ^{1}H NMR spectroscopy: upon binding to lithium or sodium salts, the signals for the pyridonate ligands, and the signals of the π-ligand, were shifted towards lower field (Fig. 10.17). If the guest M′X was added in sub-stoichiometric amounts, two sets of signals were observed, indicating that the exchange of M′X was slow on the NMR time scale. This makes the quantitation of adduct formation, under different conditions, very facile.

The stabilities of the host–guest complexes are remarkably high. If the adduct [(π-ligand)M(L1)]$_3$·M′X was dissolved in CDCl$_3$ (c = 10 mM), only the host–guest complex could be observed by ^{1}H NMR spectroscopy. The association constant must therefore be greater than 10^5 L mol^{-1}. To quantify the stability, competition experiments were performed with various crown ethers and cryptands. The experiments have revealed that the association constants of the organometalla-macrocyclic receptors [(π-ligand)M(L1)]$_3$ are significantly greater than those of crown ethers, while being similar to those of macrobicyclic ionophores, such as 2.2.1-cryptand (for NaCl adducts) and 2.1.1-cryptand (for LiCl adducts).

As usual, lower values were found when using more polar solvents. When the NaX adducts were dissolved in CD$_3$OD, signals for the free and the complexed receptors [(π-ligand)M(L1)]$_3$ could be observed by ^{1}H NMR spectroscopy. Thus, a direct calculation of K_a was then possible. For the halide salts, values between $K_a = 1.1 \pm 0.5 \cdot 10^2$ L mol^{-1} (X = Cl, (π-ligand)M = (C$_6$H$_6$)Ru) and $K_a = 3.5 \pm 0.5 \cdot 10^3$ L mol^{-1} (X = Cl, (π-ligand)M = (cymene)Ru) were obtained. Moreover, the stability of the Na$^+$ complexes depends on the nature of the anion. Thus, NaI adducts showed smaller K_a values than the NaCl adducts. The Li$^+$ adducts of the receptors [(π-ligand)M(L1)]$_3$ generally displayed a greater stability then the corresponding Na$^+$ adducts; even in polar solvents, such as methanol, the host–guest

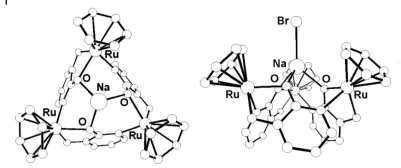

Fig. 10.18 X-ray structure of the receptor [(C$_6$H$_6$)Ru(L1)]$_3$ with a NaBr guest molecule. Left: view along the pseudo C$_3$ axis, the bromine atom is not shown. Right: view from the side.

complex was the only one that was detected by NMR with the [(π-ligand)M(L1)]$_3$. LiCl complexes that indicated a K_a value of > 10^5 L mol^{-1}. The X-ray structure of the NaBr adduct with [(π-C$_6$H$_6$)Ru(L1)]$_3$ is shown in Fig. 10.18.

The surprisingly high stability of the LiX and NaX complexes could be attributed to several facts: (1) The receptors are very rigid and clearly pre-organized to bind lithium or sodium ions. When binding the guest molecules, the bond length and angles change only slightly. (2) The salts are bound as an ion pair, which was energetically very favorable in organic solvents, such as chloroform. (3) The energetic costs for the desolvation of the receptors was very low, since a maximum of one water molecule could fit inside the binding cavity.

The receptors [(π-ligand)M(L1)]$_3$ not only showed a very high affinity for a certain alkali metal ion, but they also showed very high selectivity. The Li$^+$ adducts were formed with all receptors. The Na$^+$ adducts; however, were only formed with (π-ligand)M = (C$_6$H$_6$)Ru, (cymene)Ru and (C$_6$H$_5$CO$_2$Et)Ru, while the K$^+$ adducts have not been observed at all. The pronounced selectivity can be explained by the geometry of the organometallacrown complexes; the π-ligands formed the walls of a rather rigid binding cavity. For small π-ligands, such as benzene, sodium ions were able to easily enter. Larger π-ligands, such as hexamethyl-benzene, efficiently block the binding site and Li$^+$ specific receptors were obtained.

The host–guest chemistry of the ruthenium complex [(C$_6$H$_5$CO$_2$Et)Ru(L1)]$_3$ proved to be of special interest. Although this receptor is principally able to bind Na$^+$ ions, it showed an outstanding affinity and selectivity for Li$^+$ salts [12b]. This was evidenced by the following experiments: an aqueous solution containing LiCl (50 mM), together with a large excess of NaCl, KCl, CsCl, MgCl$_2$ and CaCl$_2$ (1 M each), was reacted with a chloroform solution of this receptor, where the exclusive and quantitative extraction of LiCl was observed (Scheme 10.3). This was very remarkable for two reasons. Firstly, the extraction of LiCl from water was in principle very difficult to accomplish, due to the high enthalpy of hydration of Li$^+$ (−521 kJ mol^{-1}) and Cl$^-$ (−363 kJ mol^{-1}). Secondly, the enthalpy of hydration of the other alkali metal ions is much smaller. Therefore, the exclusive formation of the LiCl adduct was indicative of extremely high selectivity for this alkali metal ion.

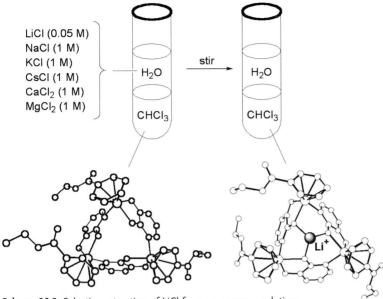

Scheme 10.3 Selective extraction of LiCl from an aqueous solution containing a large excess of alkali and earth alkaline metal salts using the receptor $[(C_6H_5CO_2Et)Ru(L1)]_3$.

As shown in Scheme 10.3, $[(C_6H_5CO_2Et)Ru(L1)]_3$ provides an organometalla-macrocycle that is highly selective to Li^+ ions over the more highly concentrated Na^+ ions [12]. The CPK models of this organometallic ionophore, and that of the Li complex (Cl omitted for clarity) are shown in Fig. 10.19.

In this context, it should be mentioned that lithium salts such as Li_2CO_3 are among the most frequently used drugs for the treatment of manic depression. Due to its narrow therapeutic range, the Li^+ concentration in the blood of the patients needs to be controlled on a regular basis. Since blood has a relative high concentration of Na^+, as compared with Li^+, the use of chemosensors has so far shown only very limited success. Thus, the use of these organometallamacrocycles, in conjunction with electrochemical techniques, for selective chemosensors, are presently being pursued by Severin and coworkers [12c].

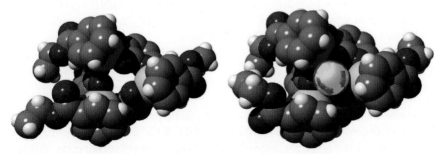

Fig. 10.19 Left: organometallic ionophore, $[(C_6H_5CO_2Et)Ru(L1)]_3$. Right: Li complex.

10.5.2
Specific Fluoride Receptors

Since the coordination of ion pairs to [(π-ligand)M(L1)]$_3$ receptors was energetically very favorable, the possibility of utilizing complexes of this nature as specific fluoride receptors was investigated. The basic idea is schematically shown in Scheme 10.4. A lithium ion, which serves as a binding site, is coordinated inside a [(π-ligand)M(L1)]$_3$ receptor. The accessibility of the Li$^+$ center is controlled by the steric requirements of the π-ligand. If large ligands were employed, only the small fluoride anion was able to enter the cavity, whereas larger anions were effectively blocked. Since the radius of the fluoride ion is significantly shorter than that of most other anions, a highly specific receptor was obtained.

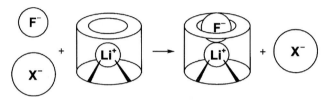

Scheme 10.4 Schematic representation of a specific fluoride receptor based on a Li$^+$ containing organometallamacrocycle.

To understand this concept, the complex [Cp*Ir(L1)]$_3$·LiBF$_4$ had appeared to be ideally suited as a highly specific receptor for a fluoride ion. NMR data showed that in solution (CDCl$_3$/CD$_3$CN, 2 : 1) the weekly bound BF$_4^-$ ion was not found to be coordinated to the lithium ion. If Bu$_4$NF was added to this solution, signals of the ion-paired complex [Cp*Ir(L1)]$_3$·LiF were immediately observed by ^{19}F and ^{7}Li NMR spectroscopy. The LiF complex was even formed in the presence of a large excess of other anions X (X$^-$ = Cl$^-$, Br$^-$, I$^-$, NO$_3^-$) indicative of a fluoride-X$^-$ selectivity of higher than 10^3.

The CPK model of an X-ray structure of [Cp*Ir(L1)]$_3$·LiF provides an explanation for this selectivity (Fig. 10.20). The fluoride ion was found to be positioned at the opening of the cavity, closely surrounded by the three Cp* ligands. As a result, the four very short CH···F contacts between the Cp* ligands and F$^-$ could be observed (CH···F = 2.15–2.28 Å). This very tight encapsulation of the fluoride ion was expected to contribute to the overall stability of the host–guest complex and prevented the coordination of larger anions.

When a solution of the complex [Cp*Ir(L1)]$_3$·LiBF$_4$ in CHCl$_3$/CH$_3$CN (2 : 1) was analyzed by differential pulse voltammetry, the peak potential for the first oxidation can be observed at 890 (± 3) mV (against Ag/AgCl). Upon addition of five equivalents of F$^-$ the complex is significantly easier to oxidize ($\Delta E = -203$ mV). A possible explanation for this shift is the reduced electron withdrawing character of the ion-paired LiF guest compared with that of the solvated lithium ion. In agreement with the NMR studies, only small changes were observed upon addition of Cl$^-$, Br$^-$, NO$_3^-$, HSO$_4^-$ or ClO$_4^-$ salts ($\Delta E < 24$ mV). Similar results were also

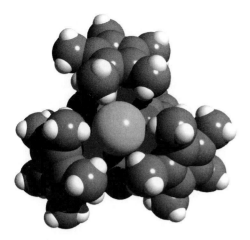

Fig. 10.20 CPK model of the X-ray structure of [Cp*Ir(L1)]₃ · LiF (view along the pseudo C₃ axis). The fluoride anion is tightly encapsulated by the Cp* ligands.

obtained in solutions containing methanol. The complex [Cp*Ir(L1)]₃·LiBF₄ was, therefore, a highly selective chemosensor, which allows the detection of fluoride anion by electrochemical means, even in protic solvents [6].

Several other LiF and LiFHF complexes were also characterized by single crystal X-ray analysis (Fig. 10.21). As expected, the lithium cation was coordinated to the three oxygen atoms of the receptors. The fourth coordination site was occupied either by the fluoride or the hydrogen difluoride anion. For the LiF complexes, the Li–F bond length was found to be between 1.77 and 1.81 Å. These values were found to amongst the smallest Li···F distances reported thus far, highlighting the unique situation of monomolecular LiF inside these organometallamacrocyclic hosts. For comparison, crystalline LiF has a Li···F distance of 2.009 Å. The difluoride anions were found to be coordinated in a bent fashion to the lithium ion (Li–F–F = 123–159°). The F···F distances observed (2.25–2.30 Å) were similar to what was found for crystals of hydrogen difluoride salts (2.24–2.28 Å).

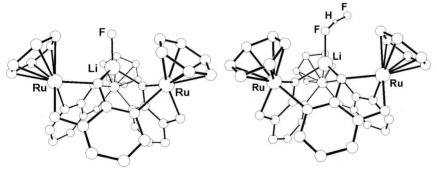

Fig. 10.21 X-ray structures of the LiF (left) and the LiF···HF (right) complexes using receptor [(cymene)Ru(L1)]₃. Cymene side groups have been omitted for clarity.

10.5.3
Conclusions for Organometallic Ionophores

There appears to be considerable interest in synthetic receptors with high affinity and selectivity for alkali metal ions. The current approaches towards this goal are often accompanied by substantial synthetic efforts that lack self-assembly techniques. Severin and coworkers have synthesized new receptors for small cations and anions by self-assembly of organometallamacrocyclic complexes [6, 12]. Compared with other synthetic ionophores, the Severin et al. approach offers major advantages: (1) these synthetic schemes could be accomplished in one step using simple starting materials; (2) the presence of guest molecules could be detected electrochemically; (3) due to a very high degree of pre-organization, excellent affinities and selectivities were observed.

The modular nature of these assemblies makes it possible to incorporate many structural variations of these organometallamacrocyclic complexes. Given the exceptional performance of these organometallamacrocyclic receptors when compared with classical ionophores, such as crown ethers and cryptands, various applications can be envisioned in the future, including studies in water [6, 12e].

10.6
Overall Conclusions for Molecular Recognition

Molecular recognition studies with organometallic supramolecular hosts, having structures based on biological ligands; for example, DNA/RNA bases, and Cp*Rh groups, critical for the superstructure, have been reviewed in their non-covalent, $\pi-\pi$, hydrophobic, and H-bonding interactions with various biological molecules, such as aromatic amino acids. Furthermore, self-assembled organometallamacro-cyclic ionophores, which selectively bind alkali metal ions, such as Li^+ ions, represent potential new applications in mental health analytical methods.

In addition, it was demonstrated that hormone receptors sites, that act as supramolecular hosts, non-covalently bind organometallic pharmaceuticals via specific H-bonding, $\pi-\pi$, and CH-π interactions, while modifying the conformation of helical proteins at the receptor site that must have a profound effect on biological activity for breast cancer therapy. This chapter represents new vistas in the bioorganometallic chemistry discipline, focused on supramolecular host structural diversity and molecular recognition studies, and we are hopeful that we have been able to enlighten the organometallic community to these new directions, and to envision the exciting possibilities for future directions. Clearly, as organo-metallic chemists, we have a vital role at the interface of chemistry and biology to create new paradigms for basic research and, for example, medical applications, for the betterment of the global society.

Acknowledgments

The studies at LBNL were generously supported by LBNL Laboratory Directed Research and Development funds and the Department of Energy under Contract No. DE AC03-76SF00098. The postdoctoral fellows and undergraduate students at LBNL, who conducted the molecular recognition NMR experiments and defined the structures of the organometallic hosts, are named in the references, while colleagues from Japan, Spain, and Israel, also named in the references, are also acknowledged for their contributions. RHF was a visiting professor at the Institute of Inorganic Chemistry, University of Zaragoza, Zaragoza, Spain, and the Weizmann Institute of Science, Rehovot, Israel during the course of aspects of our LBNL bioorganometallic chemistry program. RHF would like to thank colleagues, Gérard Jaouen and Kay Severin, for providing contributions to this review from their research programs.

References

1 (a) J.-M. LEHN, *Angew. Chem. Int. Ed. Engl.* **1988**, *27*, 90 and references therein; (b) D. J. CRAM, *Science* **1988**, *240*, 760 and references therein; (c) B. ALBERTS, D. BRAY, J. LEWIS, M. RAFF, K. ROBERTS, J. D. WATSON, *Molecular Biology of the Cell*, pp. 481–612, Garland Publishing, New York, **1989**; (d) T. D. TULLIUS, in T. D. TULLIUS (Ed.), *Metal-DNA Chemistry*, ACS Symposium Series, No. 402, American Chemical Society, Washington, D. C., **1989**, and references therein; (e) J. K. BARTON, *Comments Inorg. Chem.* **1985**, *3*, 321 and references therein.

2 (a) A. V. ELISEEV, H.-J. SCHNEIDER, *J. Am. Chem. Soc.* **1994**, *116*, 6081 and references therein; (b) C.-T. CHEN, J. S. SIEGEL, *J. Am. Chem. Soc.* **1994**, *116*, 5959; (c) J. KIKUCHI, K. EGAMI, K. SUEHIRO, Y. MURAKAMI, *Chem. Lett.* **1992**, 1685; (d) S. C. ZIMMERMAN, W. WU, Z. ZENG, *J. Am. Chem. Soc.* **1991**, *113*, 196 and references therein; (e) A. GALAN, D. ANDREU, A. M. ECHAVARREN, P. PRADOS, J. DE MENDOZA, *J. Am. Chem. Soc.* **1992**, *114*, 1511; (f) G. DESLONGCHAMPS, A. GALAN, J. DE MENDOZA, J. REBEK JR., *Angew. Chem. Int. Ed. Engl.* **1992**, *31*, 61 and references therein; (g) C. E. OSTERBERG, A. M. ARIF,

T. G. RICHMOND, *J. Am. Chem. Soc.* **1988**, *110*, 6903; (h) L. M. TORRES, L. G. MARZILLI, *J. Am. Chem. Soc.* **1991**, *113*, 4678; (i) C. S. WILCOX, J. C. ADRIAN JR., T. H. WEBB, F. J. ZAWACKI, *J. Am. Chem. Soc.* **1992**, *114*, 10189; (j) S. K. KURDISTANI, R. C. HELGESON, D. J. CRAM, *J. Am. Chem. Soc.* **1995**, *117*, 1659; (k) D. J. CRAM, M. T. BLANDA, K. PAEK, C. B. KNOBLER, *J. Am. Chem. Soc.* **1992**, *114*, 7765; (l) G. M. WHITESIDES, J. P. MATHIAS, C. T. SETO, *Science*, **1991**, *557*, 173; (m) H.-J. SCHNEIDER, *Angew. Chem. Int. Ed. Engl.* **1991**, *30*, 1417.

3 (a) J. E. KICKHAM, S. J. LOEB, S. L. MURPHY, *J. Am. Chem. Soc.* **1993**, *115*, 7031 and references therein; (b) T. MIZUTANI T. EMA, T. TOMITA, Y. KURODA, H. OGOSHI, *J. Am. Chem. Soc.* **1994**, *116*, 4240 and references therein; (c) P. J. STANG, D. H. CAO, S. SAITO, A. M. ARIF, *J. Am. Chem. Soc.* **1995**, *117*, 6273; (d) P. J. STANG, D. H. CAO, *J. Am. Chem. Soc.* **1994**, *116*, 4981 and references therein; (e) M. FUJITA, J. YAZAKI, K. OGURA, *J. Am. Chem. Soc.* **1990**, *112*, 5645; (f) M. FUJITA, S. NAGAO, K. OGURA, *J. Am. Chem. Soc.* **1995**, *117*, 1649; (g) J. E. KICKHAM, S. J. LOEB, *Inorg. Chem.* **1995**, *34*, 5656; (h) J. W. CANARY, B. GIBB, *Prog. Inorg. Chem.* **1997**, *45*, 1,

and references therein; (i) J. L. ATWOOD, J. E. DAVIES, D. D. MACNICOL, F. VÖGTLE, *Comprehensive supramolecular chemistry*, Pergamon, **1996**.

4 (a) H. CHEN, S. OGO, R. H. FISH, *J. Am. Chem. Soc.* **1996**, *118*, 4993; (b) S. OGO, S. NAKAMURA, H. CHEN K. ISOBE, Y. WATANABE, R. H. FISH, *J. Org. Chem.* **1998**, 63, 7151; (c) H. CHEN, M. F. MAESTRE, R. H. FISH, *J. Am. Chem. Soc.* **1995**, *117*, 3631; (d) D. P. SMITH, E. BARALT, B. MORALES, M. M. OLMSTEAD, M. F. MAESTRE, R. H. FISH, *J. Am. Chem. Soc.* **1992**, *114*, 10647; (e) D. P. SMITH, E. KOHEN, M. F. MAESTRE, R. H. FISH, *Inorg. Chem.* **1993**, 32, 4119; (f) R. H. FISH, *Coord. Chem. Rev.* **1999**, *185/186*, 569 and references therein; (g) R. H. FISH, G. JAOUEN, *Organometallics*, **2003**, 22, 2166 and references therein.

5 (a) R. BRESLOW, *Acc. Chem. Res.* **1991**, *24*, 159; (b) S. B. FERGUSON, E. M. SANFORD, E. M. SEWARD, F. DIEDERICH, *J. Am. Chem. Soc.* **1991**, *113*, 5410.

6 (a) K. SEVERIN, *Coord. Chem. Rev.* **2003**, *245*, 3 and references therein; (b) M. L. LEHAIRE, R. SCOPELLITI, L. HERDEIS, K. POLBORN, K. SEVERIN, *Inorg. Chem.* **2004**, *43*, 1609.

7 D. P. SMITH, M. M. OLMSTEAD, M. F. MAESTRE, R. H. FISH, *Organometallics* **1993**, *12*, 593.

8 A. ELDUQUE, D. CARMONA, L. A. ORO, M. EISENSTEIN, R. H. FISH, *J. Organomet. Chem.* **2003**, *668*, 123.

9 (a) G. JAOUEN, S. TOP, A. VESSIÈRES, R. ALBERTO, *J. Organomet. Chem.* **2000**, *600*, 25 and references therein; (b) G. JAOUEN, *Chemistry in Britain*, **2001**, 36.

10 S. TOP, H. EL HAFA, A. VESSIÈRES, M. HUCHÉ, J. VAISSERMANN, G. JAOUEN, G. *Chem. Eur. J.* **2002**, 8, 5241; (b) S. TOP, A. VESSIÈRES, C. CABESTAING, I. LAIOS, G. LECLERCQ, C. PROVOT, G. JAOUEN, *J. Organomet. Chem.* **2001**, *639*, 500.

11 D. OSELLA, M. FERRALI, P. ZANELLO, F. LASCHI, M. FONTANI, C. NERVI, G. CAVIGIOLIO, *Inorg. Chim. Acta.* **2000**, *306*, 42.

12 (a) H. PIOTROWSKI, K. POLBORN, G. HILT, K. SEVERIN, *J. Am. Chem. Soc.* **2001**, *123*, 2699; (b) H. PIOTROWSKI, K. SEVERIN, *Proc. Natl. Acad. Sci. USA* **2002**, *99*, 4997, and references therein; (c) H. PIOTROWSKI, G. HILT, A. SCHULZ, P. MAYER, K. POLBORN, K. SEVERIN, *Chem. Eur. J.* **2001**, 3196; (d) M.-L. LEHAIRE, R. SCOPELLITI, H. PIOTROWSKI, K. SEVERIN, *Angew. Chem. Int. Ed.* **2002**, *41*, 1419; (e) Z. GROTE, M.-L. LEHAIRE, R. SCOPELLITI, K. SEVERIN, *J. Am. Chem. Soc.* **2003**, *125*, 13638.

11
Structure and Mechanism of Metalloenzyme Active Sites

Juan C. Fontecilla-Camps

11.1
Introduction

Many fundamental reactions in biology are mediated by proteins containing metal cofactors. This is especially true for those processes that involve gases such as the reduction of inert N_2 to biologically essential ammonia, the oxidation and production of molecular hydrogen, various oxidative reactions that involve O_2 and the use of CO_2 and CO as a carbon source. Many reactions involving catalytic metal centers are typical of enzymes from anaerobic microorganisms and the respective active sites are very sensitive to inactivation by O_2. This effect has been interpreted as the consequence of the photosynthesis-derived increase in O_2 concentration over geological times. Thus, it is generally accepted that metal-catalyzed redox reactions involving CO, H_2, CO_2 and some other gases are at the very origin of life. In addition, because metals can behave as Lewis-type acids, a property that allow them to activate molecules such as water, they participate in many fundamental non-redox reactions that amino acids would not be able to catalyze.

In dealing with such a vast domain some decisions concerning the subjects addressed in this short chapter had to be made. Consequently, only selected enzymes containing the transition metals copper, iron, manganese, molybdenum/ tungsten, nickel and the related zinc, will be discussed; also, we will consider only X-ray structures of active sites published relatively recently and for which some discussion on the catalytic mechanism is included. Some reference is also made to Co in the context of the correnoid iron sulfur protein that interacts with acetyl Coenzyme A synthase in the synthesis or cleavage of acetyl CoA. With a few exceptions, the protein structure beyond the metal coordination sphere will not be described unless it impinges in the catalytic mechanism.

There are complicating factors when studying metalloenzymes. In many instances, metal sites seem to be rather indiscriminating, at least *in vitro*, as different ions can bind to them with various degrees of residual catalytic activity; this makes it difficult to establish which is the physiologically relevant metal.

Bioorganometallics: Biomolecules, Labeling, Medicine. Edited by Gérard Jaouen
Copyright © 2006 Wiley-VCH Verlag GmbH & Co. KGaA, Weinheim
ISBN: 3-527-30990-X

In other cases, it has been shown that the biologically relevant metal ion can be extensively replaced by (an)other(s) during purification or crystallization resulting in an inactive enzyme. (Inside the cell, where the concentration of free metal ions is negligible, such replacement is unlikely. However, once the protein is isolated, traces of metal ions present in the buffers will displace the original ion if their affinity for its site is high). Another problem arises when metalloenzyme crystals are exposed to X-rays, as photoelectrons can modify the initial redox states. Without having an effective way to monitor the redox state in the crystal, such as spectral changes due to reduction of a chromophore, it is difficult to be sure that even short X-ray exposures eliminate this problem.

The chapter is organized in sections according to the various metal ions and representative members are discussed in each case, in terms of active site structure and catalytic mechanisms.

11.2
Copper

This metal is both essential and toxic and, in humans, its improper metabolism leads to several pathological conditions such as Menkes and Wilson and Alzheimer diseases [1]. Copper-containing enzymes are known to participate in several oxygen-processing pathways. Biological accessible redox copper states are Cu(I) and Cu(II) with a remarkable flexibility in terms of coordination [2, 3]. However, Cu(I) prefers cysteine and methionine residues whereas Cu(II) is mostly coordinated by histidine, serine, threonine or tyrosine and water. Also, Cu(I) normally has tetrahedral coordination while Cu(II) is square planar [4]. Cu sites in enzymes can vary in their spectroscopic properties and three types have been defined. Type I sites display a strong absorption at 600 nm, responsible for the deep blue color of some oxidases. The coordination is typically either trigonal bipyramidal or distorted tetrahedral. A class of colorless Cu sites are found in both mono and dioxygenases, non-blue oxidases and CuZn superoxidase dismutase. The most common ligand for this type II copper centers is histidine. Type III binuclear Cu is found in both hemocyanines and tyrosinases where is associated with redox processes and displays a strong absorption at 300 nm [4]. Some examples of Cu-containing enzymes are discussed below.

11.2.1
Quercentine 2,3-dioxygenase

The recently published structure of this dimeric enzyme that contains one type II Cu(II) catalytic center per monomer has revealed that it belongs to the cupin superfamily, that includes Fe-homogentisate 1,2-dioxygenase and Mn-oxalate oxidase, enzymes which are evolutionary related, in spite of their different metal preferences [4]. The crystal structure shows that the Cu ion has two different coordinations. The first coordination environment, which represents approximately 70% of the active sites, is formed by three histidines and a water molecule, whereas in the second one, a glutamic acid is added, displacing the water to a different position. This is the first example of carboxylate coordination to a Cu(II). The carboxylate from the glutamate residue could be involved in catalysis by deprotonating an OH⁻ group from the substrate. Because neither tetrahedral nor trigonal planar coordination is expected for Cu(II) ions, the authors considered the possibility of reduction to Cu(I) in the crystal by photoelectrons produced by the X-ray beam during data collection [4]. No changes in the coordination were detected after collecting several data sets on the same crystal at the European Synchrotron Radiation Facility, in Grenoble, France. However, the possibility that the Cu ion was fully reduced during the collection of the first data set cannot be ruled out.

11.2.2
Amines Oxidases

These homodimeric enzymes, that are present in both prokaryotic and eukaryotic organisms, contain one Cu ion and one redox-active cofactor topaquinone (TPQ) per monomer [5, 6]. They catalyze the oxidative deamination of primary amines [7–9]. The Cu(II) ion is coordinated by three histidine residues and three water molecules (Fig. 11.1). The TPQ cofactor is not far from the Cu ion. The process can be divided into an initial reductive reaction followed by an oxidative step, based on the redox state of TPQ; the Cu ion is thought to be involved in the formation of the TPQ semiquinone through reduction of Cu(II) to Cu(I). An alternative hypothesis has been recently proposed where the copper ion stays as Cu(II) and the one-electron reduction of O_2 is carried out by a modified amino-resorcinol TPQ cofactor. The Cu(II) would provide electrostatic stabilization to the superoxide anion intermediate [10–12]. The reduction of molecular oxygen would result in weakly Cu-bound hydroperoxide which is subsequently displaced by a water molecule, gets protonated and it is eliminated as hydrogen peroxide. The rest of the reaction (the reductive step) is metal-independent and is mediated by different modified TPQ species [13].

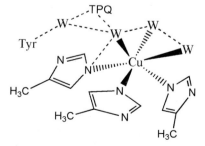

Fig. 11.1 The active site of amino oxidase. TPQ is a redox active topoquinone cofactor. Water molecules are depicted by W and X shows the position of the Cu ion. All the figures were prepared using ChemDraw.

11.2.3
CuZn Superoxide Dismutases (SOD)

These are critical components of the eukaryotic cell defences against oxygen-induced damage [14]. They are dimeric and contain one Cu(II) and one Zn(II) per monomer [15]. SODs catalyze the dismutation of the superoxide radical anion to hydrogen peroxide and oxygen through the oxidoreduction of Cu according to the following reactions [16]:

$$O_2^- + SOD\text{-}Cu(II) \rightarrow O_2 + SOD\text{-}Cu(I) \tag{11.1}$$

$$SOD\text{-}Cu(I) + O_2^- + 2\ H^+ \rightarrow SOD\text{-}Cu(II) + H_2O_2 \tag{11.2}$$

Several crystal structures are available for the Cu(II) form of SOD [17–20]. In all Cu(II)Zn SODs the metal ions are bridged by the side chain of a histidine residue. The copper ion is coordinated by three additional residues and a water molecule in a distorted square pyramidal geometry (Fig. 11.2). The Zn ion coordination is completed by two additional histidines and an aspartate side chain, forming a distorted tetrahedral geometry. In yeast and bovine reduced Cu(I)ZnSODs, the copper ion has been shown to be coordinated by only three histidine residues [17, 21]. Consequently, a key feature of the Cu(II) to Cu(I) transition is the change from penta- to tri-coordinated Cu [22]. This can be explained as follows: when Cu(II) is reduced by the superoxide ion (Eq. 11.1) the bridging histidine dissociates from copper, is protonated and the water molecule is displaced, which results in tri-coordinated Cu. Subsequently, Cu(I) is re-oxidized to Cu(II) by a second superoxide molecule (Eq. 11.2) which product binds two protons coming, respectively, from the previously bridging histidine and solvent, and subsequently leaves the active site as hydrogen peroxide.

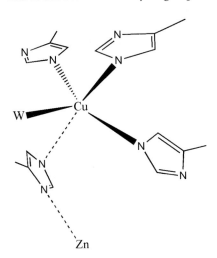

Fig. 11.2 The active site of CuZn superoxide dismutase.
Water molecules are depicted by W and Xs show the positions of the Cu and Zn ions.

A recent structural determination of the bovine enzyme at 1.7 Å resolution suggests that the Cu(I)/Cu(II) transition, along with their related coordination changes, is modulated by a protein loop that lines the active site cavity [15]. Again, possible photoreduction resulting from exposure to X-rays has been considered and (possibly) ruled out by examination of electron density maps and temperature factors corresponding to atoms at the active site.

11.2.4
Multicopper Oxidases

These proteins constitute a rather large and heterogeneous group of enzymes containing clusters of copper ions coordinated by both protein and non-protein ligands. Blue copper oxidases are often involved in iron metabolism and copper homeostasis. A well-studied example is mammal ceruloplasmin, that catalyzes Fe(II) to Fe(III) conversion and is the most abundant copper-containing protein in plasma [23–25]. In plants, both ascorbate oxidases and laccases are related enzymes, capable of oxidizing various phenolic substrates, the latter leading to productive polymers [26]. A brief description of some relevant enzymes follows.

11.2.4.1 Laccases

These enzymes couple the four-electron substrate oxidation of substrate to the four-electron reduction of molecular oxygen to water [23]. Molecular oxygen binds to the tri-nuclear active site made of one type II and two type III Cu ions and gets reduced to water through four one-electron transfer steps from the Cu type I site found about 13 Å away [27]. Laccases are involved in both lignin synthesis in plants and its degradation by fungi [28–30]. Although the three-dimensional structure of laccase has been known from some time, only recently have two groups reported the structure of the enzyme from the fungus *Trametes versicolor* which has a full complement of coppers [28, 29]. In one of these structures [29], laccase binds an arylamine, used as an enzyme inducer during culture. This rather poor substrate binds close but not directly to type I Cu. Although substrate binding to the related ceruloplasmin has been characterized using X-ray crystallography, these studies were carried out at medium resolution and there was no crystallographic refinement of the active site [25]. The fungal laccase structure also explains why the type II Cu is labile as it is only coordinated by two histidines and a water molecule, whereas the more stable two type III Cu ions have three histidine and a solvent ligand each (Fig. 11.3). The reaction catalyzed by laccases implies the capture of one electron from the substrate by a histidine residue that also binds the type I Cu ion. Subsequently, the electron is transferred to one of the two type III coppers at about 12 Å away, either through histidine side chains or by a through-space mechanism, and molecular oxygen is reduced. Laccases can operate at medium to very high potentials, ranging from 500 to 800 mV *versus* the normal hydrogen electrode. Conversely to other blue copper oxidases where type I copper is tetra-coordinated, in laccases this copper has trigonal coplanar coordination. The structural basis for redox potential modulation has been proposed to reside

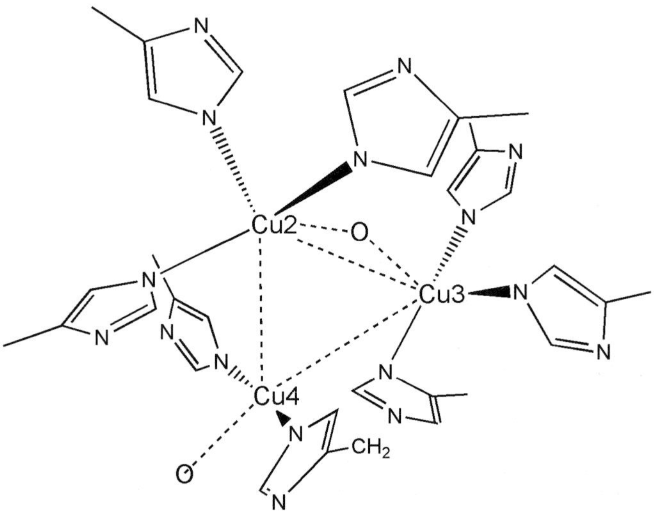

Fig. 11.3 The trinuclear active site of laccase. The Cu sites are depicted by X and oxo ligands by x.

in the nature of the Cu axial coordination [31, 32]. However, the highest resolution laccase structure obtained this far suggests that it is a lengthening in the Cu-ligand distance that increases redox potential through a reduction of electron density contribution at the metal center [28].

11.2.4.2 Nitrous Oxide Reductase (NOR)

Many organisms use oxidized nitrogen as a final electron acceptor in anaerobic respiration [33, 34]. This denitrification process can be described as a series of successive reducing steps: $NO_3^- \rightarrow NO_2^- \rightarrow NO \rightarrow N_2O \rightarrow N_2$.

The enzyme catalyzing the last reaction in this process has been recently shown to contain a novel multicopper center [35, 36]. It had been generally accepted for some time that NOR had four copper ions, forming a couple of dinuclear metal centers, one having properties similar to the Cu_A of cytochrome C oxidase and the other being a catalytic Cu_Z site. The crystal structure shows that in NOR, Cu_A is a symmetric site coordinated by two bridging cysteines, two terminal histidine ligands, a terminal methionine and a main chain carbonyl oxygen. Although this Cu site was well predicted using a series of approaches and is indeed similar to the Cu_A from cytochrome C oxidase, the NOR Cu_Z site turned out to be a complete surprise [35]. It is composed of four Cu ions disposed in a "butterfly" cluster with the metal ions protein ligands being all histidine side chains (Fig. 11.4). Three of the Cu ions are coordinated by two histidines each whereas the fourth Cu center has only one histidine bound to it. Three exogenous ligands are also present. One of these ligands caps all four Cu ions in a μ^4 fashion. Although initially it was modeled as OH^- [35], elementary analysis and Raman spectroscopy of isotopically labeled enzyme showed that this ligand is a sulfide ion [37]. Two additional water or hydroxide ligands complete the coordination of the Cu_Z cluster. The very

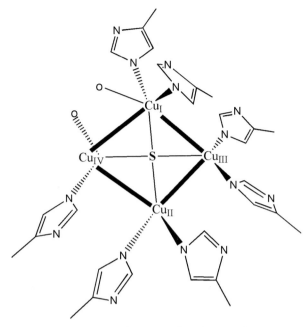

Fig. 11.4 The active site of nitrous oxide reductase. The Cu sites are depicted by X.
The central sulfido ion [37] was originally modelled as water [35].

important issue of substrate binding has not yet been resolved. Although the
fourth Cu ion of the Cu_Z cluster is an attractive candidate for binding N_2O, the
structure of either a substrate analog or an inhibitor bound to it is clearly needed
to clarify this point.

11.2.4.3 Copper Homeostasis

The recent high resolution structure of CueO, an enzyme required for Cu homeostasis in the gut bacterium *Escherichia coli* has provided the clearest view of the
four-copper active site so far [27]. For instance, the TI Cu site that has normally
been described as distorted tetrahedral [24] can now be characterized as trigonal
bipyramidal with a missing ligand as in the well-studied azurins. However, the
oxidation states of the Cu ions in the crystal structure have not been characterized.

11.3
Iron

Iron is the most common trace metal in most organisms. As heme iron, it is
involved in storage and transport of O_2 and respiration, it also mediates electron
transfer forming FeS clusters and catalyzes a great number of reactions, many of
them involving intermediates of the reduction of molecular oxygen to water and

the addition of oxygen atoms to substrate [38]. In the reducing environment of the cell, iron and other redox-active metal ions can catalyze the one-electron reduction of hydrogen peroxide to produce the highly reactive oxidant hydroxyl radical.

$$Fe(II) + H_2O_2 \rightarrow Fe(III) + OH^- + OH^\bullet \qquad (11.3)$$

Binding iron to proteins helps to prevent this reaction. Another way to avoid the generation of the aggressive hydroxyl radical is to lower the concentration of hydrogen peroxide by the action of catalases and peroxidases [38]. Other enzymes such as superoxide dismutases, which neutralize the reactive O_2^- ion, are also part of the detoxificating machinery of the cell. A few examples of iron enzymes are described below.

11.3.1
Cytochrome c Oxidase (COX)

This enzyme present in all animals and plants, aerobic yeast and some bacteria is involved in the next to the last step of oxidative phosphorylation, a process that couples electron transport to ATP synthesis; cytochrome c (cytc) delivers electrons to the final component of the respiratory chain, cytochrome c oxidase which reduces O_2 to water [39, 40]. Crystal structures for both bacterial and mitochondrial multisubunit membrane-bound COXs are available [41–45]. The enzyme contains two heme groups and three Cu ions. Of these, one heme and one Cu ion (called the heme a_3/Cu$_B$ site) constitute the catalytic center (Fig. 11.5), the remaining

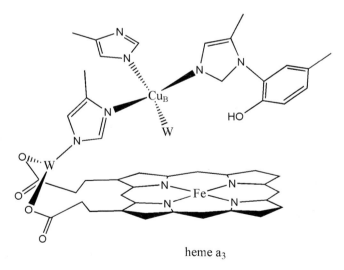

heme a$_3$

Fig. 11.5 The a_3/Cu$_B$ catalytic site of cytochrome c oxidase.
The heme group is depicted without internal double bonds.
W is a water molecule. Note the modified tyrosine-histidine couple.

heme *a* and the binuclear Cu_A center being involved in electron transfer. O_2 reduction takes place according to the following reaction:

$$4 \ cytc^{2+} + 2 \ H^+ \ (int) + O_2 \rightarrow 4 \ cytc^{3+} + 4 \ H^+ \ (out) + 2 \ H_2O \qquad (11.4)$$

where *int* stands for protons that are taken up from the inside of the mitochondrion or bacterium and out for protons released outside. There is a net proton transfer across the membrane in the reaction and the proton pumping mechanism has been extensively discussed [46]. Some postulated key intermediates of the reaction are discussed in [47].

11.3.2
Oxygenases

There are two classes of enzymes that catalyze reactions of O_2 with organic substrates: dioxygenases, that incorporate both oxygen atoms into the product, and monooxygenases where one oxygen atom is incorporated into the product and the other is eliminated as a water molecule [38]. Examples of the former are bacterial intradiol catechol dioxygenases. These enzymes cleave aromatic rings by inserting both atoms from O_2 to the ring [48–51]. They are important in the degradation of, for example, lignin and terpene and in tryptophane and tyrosine catabolism. From the structural standpoint, the best-studied enzyme of this group is protochatechuate 3,4-dioxygenase (3,4-PCD) which cleaves catecholic substrates to muconic acid derivatives [52]. The active site Fe ion of native 3,4-PCD has distorted trigonal bipyramidal coordination: the axial ligands are histidine and tyrosine and the equatorial plane is formed by histidine, tyrosine and a H_2O/OH^- ligand (Fig. 11.6). Crystallographic studies with both weak and strong monodentate inhibitors have shown that the former displaces the OH^- ligand whereas the latter

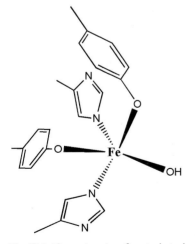

Fig. 11.6 The active site of protochatechuate 3,4-dioxygenase.

binds associatively in the equatorial plane forming an octahedral Fe coordination sphere [53]. The bidentate substrate displaces both the axial tyrosine and the OH⁻ ligand, forming a square pyramidal complex.

Another group of substrate-activating enzymes of known structures is that of lipooxygenases [54–57] which, along with cyclooxidases, react with fatty acids and generate species that are at the origin of the biosynthesis of eicosanoids. These molecules have been linked to pathological inflammatory processes and consequently have attracted significant attention [58]. Although it has been postulated that an Fe(III)-peroxide complex is an intermediate in the reaction, structural data of enzymes in any active Fe(III) state had been lacking. Recently it has been shown that, when treated with an excess of the lipid hydroperoxide product, solutions of soybean lipooxygenase-3 turned a distinct purple color, shown by EPR spectroscopy to originate from high-spin Fe(III). This metastable and photolabile material was crystallized and the structure of the complex was determined at 2.0 Å resolution [59]. From the substrate binding mode, it was concluded that purple lipohydrogenase contains an iron-peroxo complex (Fig. 11.7).

Monooxygenases can have either heme iron or nonheme iron but copper-containing enzymes, such as tyrosinase are also known. A representative and extensively studied group of heme iron-containing monooxygenases is the one consisting of cytochrome P-450 enzymes [60]. They catalyze a great variety of reactions such as hydroxylation of aliphatic compounds and aromatic rings, amine oxidation to amine oxides, sulfide oxidation to sulfoxides, dehalogenation and oxidative dealkylations [61]. The best studied member of the group is the P-450 enzyme from the bacterium *Pseudomonas putida* because, unlike other related enzymes, it is not membrane-bound but soluble. Proposed intermediates in several cytochrome P450 reactions are depicted in Fig. 11.8 [62].

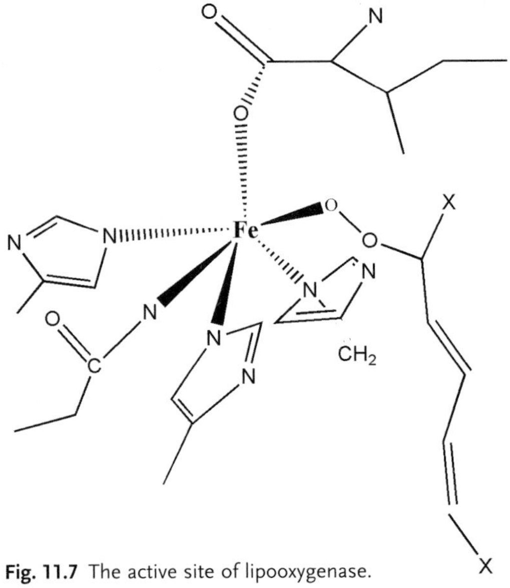

Fig. 11.7 The active site of lipooxygenase.

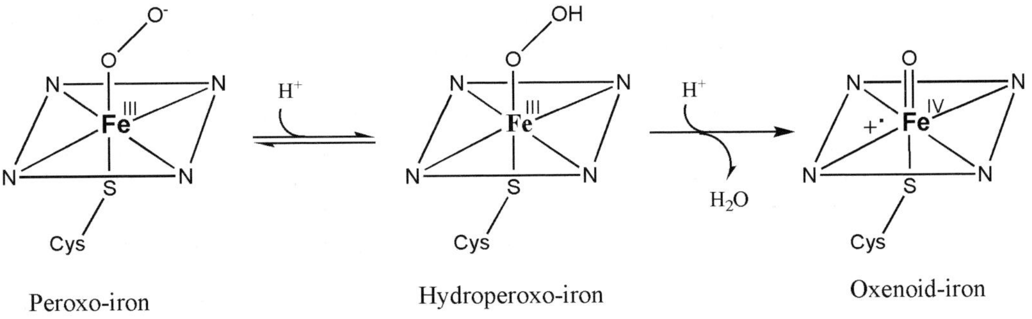

Fig. 11.8 Intermediates in cytochrome P450 reactions [62].

11.3.3
Ribonucleotide Reductase (RR)

Reduction of ribonucleotides to deoxyribonucleotides is an essential step in DNA synthesis. The reaction involves the substitution of the hydroxyl group on the 2' carbon of the ribose moiety by hydrogen in both nucleoside diphosphates and triphosphates. There are several classes of RR that can vary in cofactor requirements and amino acid sequences but they all contain metal and the reaction always involves radical chemistry [63]. Class I are tetrameric enzymes with α_2 and β_2 homodimers, called R1 and R2, respectively [64, 65]. R1 contains the binding site and R2 has a binuclear nonheme iron active center and a stable tyrosyl radical that is involved in the catalytic mechanism [66]. In the reduced Fe(II)–Fe(II) state each metal ion is tetra-coordinated by two bidentate glutamates and one monodentate carboxylate from either aspartate or glutamate, and a histidine (Fig. 11.9a). In the oxidized Fe(III)–Fe(III) state, one of the iron ions is hexa-coordinated by one histidine, two monodentate and one bridging glutamate, a solvent molecule and a μ-oxo ligand (Fig. 11.9b). The second iron center is also hexa-coordinate with one histidine, a bidentate aspartate, one bridging glutamate, a solvent molecule and a μ-oxo ligand. Tyrosyl radical generation requires O_2 binding to the binuclear Fe center. In most mechanisms a peroxo intermediate P has been proposed [67–69] and accumulation of a μ-1,2-peroxodiiron(III) has been demonstrated in the E. coli R2-D84E mutant. Residue 84 is a Fe ligand in the enzyme [70]. The reaction proceeds through an intermediate called X which has a Fe(III)–Fe(IV) pair that subsequently decays forming an oxo bridge and the stable unprotonated tyrosyl radical [71].

The crystal structure of the mouse R2 RR has also been solved but it contained only one iron ion, maybe because in this enzyme the iron/radical site is more exposed to the solvent [72]. A recent report on the crystal structure of this mammalian R2 with a dinuclear cobalt center replacing iron found a different orientation for one of the glutamate ligands when compared with the bacterial enzyme [73]. This could explain the increased lability of one of the iron centers. Because the substrate-binding site in R1 and the radical-generating site in R2 are far apart a

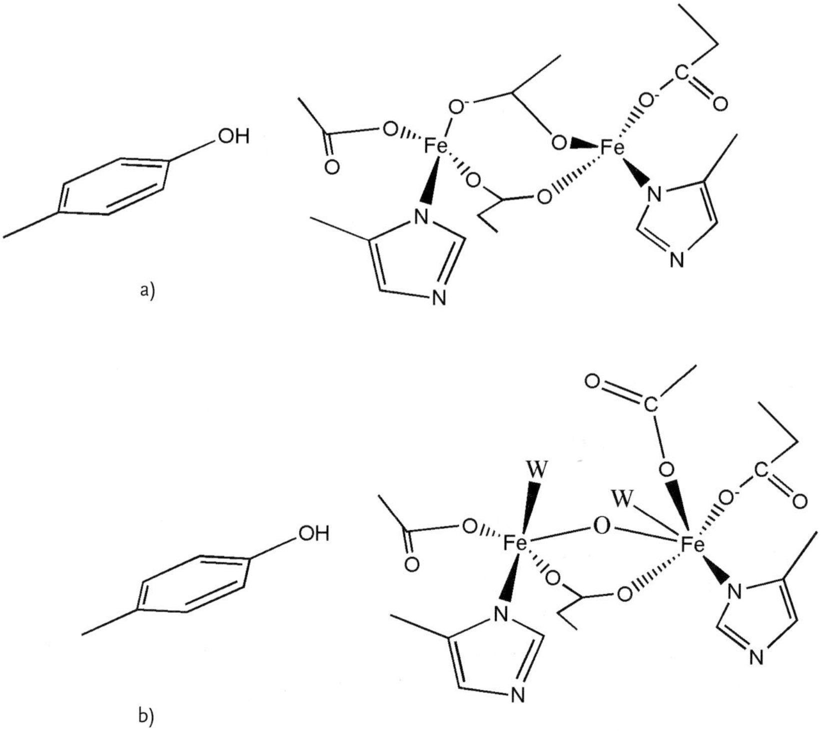

Fig. 11.9 Ribonucleotide reductase active site:
a) reduced Fe(II)–Fe(II) state; b) oxidized Fe(III)–Fe(III) state.

mechanism for radical transfer is required. It is likely that a H· radical is transferred through an array of hydrogen-bonded conserved amino acids [74–77].

In class II RRs, which are homodimeric enzymes, the radical species is generated by adenosyl cobalamine. They can function either aerobically or anaerobically and they are not found in higher organisms [78]. Anaerobic class III RRs have a $\alpha_2\beta_2$ quaternary structure, they contain an FeS cluster and a glycyl radical and belong to the radical enzyme family described below. No structure is available for class III RRs [79].

11.3.4
S-adenosyl Methionine (SAM)-dependent Radical Enzymes

SAM-dependent iron sulfur proteins can catalyze a great variety of radical-based reactions that have in common the production of 5'-deoxy-adenosine, methionine and an oxidized protein or substrate radical (Fig. 11.10) [80, 81]. All these enzymes contain an unusual [Fe$_4$S$_4$] cluster and the CxxxCxxC sequence motif but, otherwise, share very low amino acid sequence homology [82]. In most cases, the cleavage of SAM is irreversible and stoichiometric but some examples are known

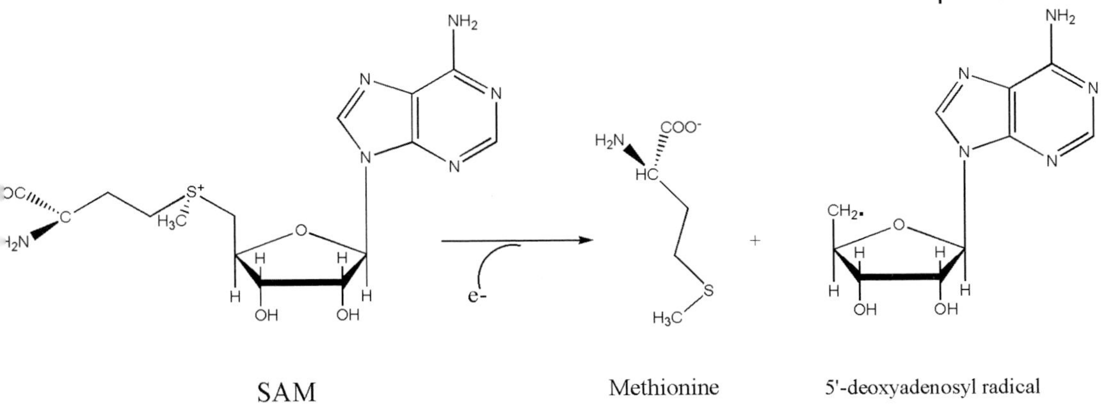

Fig. 11.10 S-adenosyl methionine cofactor.
Radical generation is mediated by a one-electron reduction.

where its role is catalytic [83]. Only very recently the first two crystal structures of radical SAM enzymes, coproporphyrinogen III oxidase (HemN) [84] and biotin synthase (BioB) [85], have been reported. The former catalyzes the oxidative decarboxylation of coproporphyrinogen III to protoporphyrinogen IX, by converting the propionate side chains into vinyl groups [84]. Unexpectedly, HemN binds two SAM molecules. One of them is directly coordinated to one of the iron ions of the $[Fe_4S_4]$ cluster through the amino and carboxylate groups and the sulfonium sulfur of the SAM methionine moiety. This suggests that electron transfer occurs through a favorable electronic interaction between the sulfonium sulfur and the $[Fe_4S_4]$ cluster. A second SAM molecule is found nearby but it is not clear whether it has been partially degraded or it is disordered. Model building indicates that the second SAM could occupy the substrate-binding site but a model with two SAMs and substrate bound to HemN is also plausible [84].

BioB catalyzes the conversion of dethiobiotin (DTB) to biotin by inserting a sulfur atom between two carbon atoms [85, 86]. Two different mechanisms have been proposed: the first involves the use of a $[Fe_2S_2]$ cluster as the sulfur source for biotin [87–88] whereas the second suggests that the sulfur atom is provided by an enzyme-bound persulfide obtained through an intrinsic pyridoxal-5'-phosphate(PLP)-dependent cysteine desulfurase activity [89, 90]. The crystal structure of BioB favors the first hypothesis because of the position of the two FeS clusters and the orientation of the bound SAM and DTB. Also, no typical PLP binding site was detected in the structure of BioB [85].

11.4
Manganese

The biochemistry of manganese is, in many respects, similar to that of iron and often, but not always [91], the two metals can be exchanged. One example of this is the cambialistic SOD from *Porphyromonas gingivalis* which exhibits maximal activity with either Fe or Mn [92, 93]. This is not surprising because both Fe-SODs and Mn-SODs belong to the same protein family and use equivalent amino acids and geometry to bind the active site metal. The Fe/Mn ratio in SOD from *P. gingivalis* seems to depend on the amount of air supplied to the cultures [94]. An extreme example of iron substitution by manganese occurs in the parasitic pathogen *Borrelia burgdorferi* that causes Lyme disease [95]. In this organism (which has no detectable iron) manganese, along with zinc and magnesium, are essential for cell growth. Replacement of iron-dependent enzymes by their manganese-containing counterparts is illustrated by the presence of a SOD with > 50% homology to the cambialistic enzyme from *P. gingivalis*. The reason for this adaptation is that host cells normally defend themselves against infection by restricting iron availability to pathogens and keeping levels of free Fe as low as 10^{-18} M. This is 10^{-11} M lower than the concentration required for microbial growth [96].

Like some other metal ions, Mn can act as a Lewis-type acid or be redox-active. Maybe the most studied and fascinating property of manganese is its ability to catalyze the water-splitting, oxygen-evolving reaction of photosystem II (PS II). A recent medium-resolution crystal structure of PS II [97] has revealed an unanticipated Mn_3Ca cluster that, along with some other representative Mn-containing active sites will be discussed below.

11.4.1
Superoxide Dismutases

These enzymes catalyze the following reactions:

$$M^{3+} + O_2^- \rightarrow M^{2+} + O_2 \tag{11.5}$$

$$M^{2+} + O_2^- + 2 H^+ \rightarrow M^{3+} + H_2O_2 \tag{11.6}$$

where M is a metal ion.

As discussed above, some SODs are cambialistic and they are equally active with either Fe or Mn bound at their catalytic centers [92]. On the other hand, the authentic Mn-depending SOD from *E. coli* can only be active when Mn is bound. However, this enzyme can also bind Fe and thus be rendered inactive. This is puzzling because the closely related *E. coli* Fe and Mn SODs have very similar first coordination spheres (Fig. 11.11) with three histidine and one aspartate side chain and a H_2O/OH^- (depending on the redox state) as ligands [98–102]. The coordination is almost exactly trigonal bipyramidal. The relationship between metal binding and activity seems to be controlled by outer-sphere ligands to the solvent

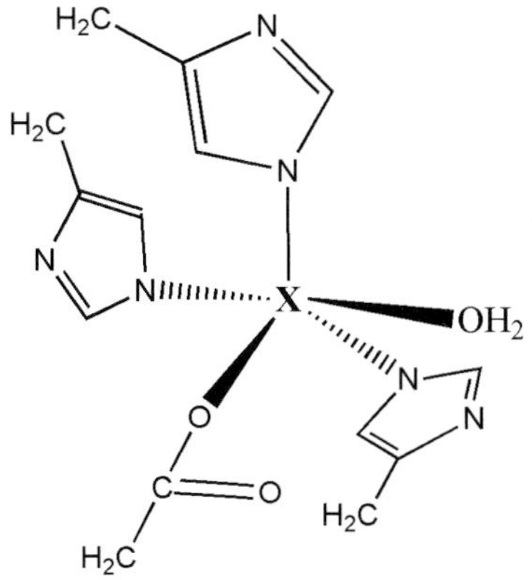

Fig. 11.11 The coordination sphere of Fe/Mn [X] ions in superoxide dismutase.

molecule. In MnSOD, the solvent is hydrogen-bonded to a glutamine side chain arising from the C-terminal domain whereas in Fe SODs, an equivalent glutamine is contributed by the N-terminal region of the enzyme. MnSOD is often purified as a mixture of Fe/Mn and apo species although the fully active (Mn_2)-enzyme can be selected by complementing the growth medium with manganese salts. It is interesting to note that the outer-sphere solvent ligand in cambialistic Mn/FeSOD is histidine [92]. This may represent an evolutionary solution to the lack of metal-binding specificity observed in MnSODs.

11.4.2
Phosphoprotein Phosphatases (PPPs)

Reversible phosphorylation is an essential element in the regulation of metabolism in all organisms [103]. Addition of the phosphoryl (PO_3^-) group to several amino acids is mediated by kinases and the reversed process is catalyzed by protein phosphatases. These enzymes have been classified according to their substrate specificities and metal requirements. A consensus "phosphodiesterase motif" with the sequence $DXH(X)_nGDXXD(X)_mGNHD/E$ has been identified in several members of this class of enzymes [104–106]; it corresponds to a $\beta\alpha\beta\alpha\beta$ secondary structure motif that defines the architecture of the catalytic dinuclear metal ion binding site. Residues shown in boldface are ligands to the M1 and M2 metal centers. M1 is coordinated by the first aspartate and histidine of the consensus sequence shown above and M2 binds the histidine contained in that sequence plus two histidine side chains found beyond the C-terminal end of the phospho-esterase motif. A conserved aspartate bridges the two metals in a μ-1,1 fashion. Previous studies of the Mn-containing PPP enzyme from bacteriophage λ (λPP) have shown that the two metal sites do not have the same affinity for Mn [107]. A recent crystallographic and EPR spectroscopic analysis of λPP has lead White and coworkers [103] to propose a mechanism where the Mn at the high affinity M2 site stays in the active center throughout catalysis whereas the low-affinity

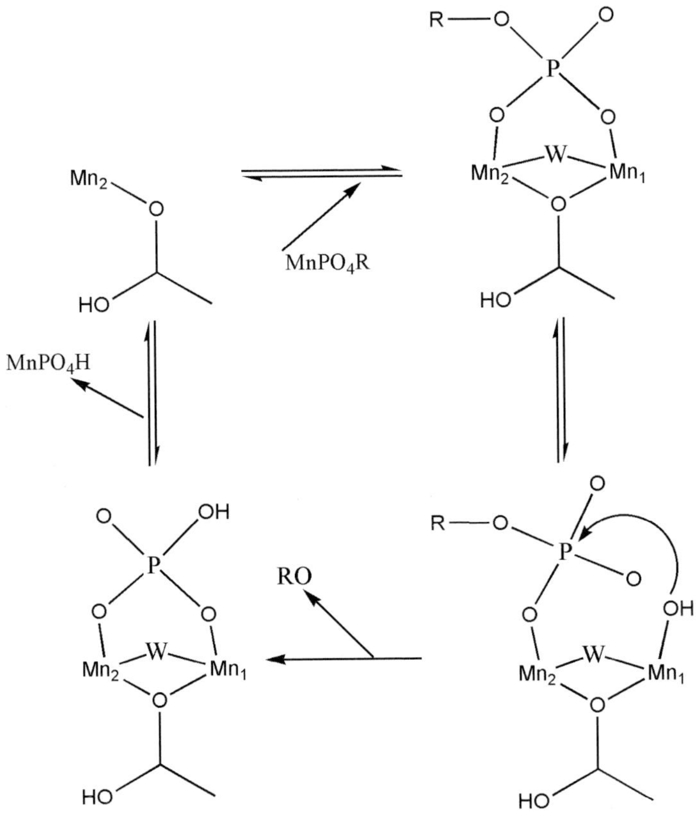

Fig. 11.12 Proposed catalytic cycle for phosphoprotein phosphatase from bacteriophage λ [103].

Mn ion binds the active site complexed to the substrate and subsequently leaves as $MnPO_4H$ (Fig. 11.12). An important question is how the ligand field generated for the two Mn(II) ions favors Lewis acid-type hydrolytic chemistry over redox reactions such as the ones observed in Mn catalases that have otherwise similar active sites.

11.4.3
Oxalate Decarboxylases (Pyruvate Carboxylase)

Oxalic acid poses a problem to both leafy plants and vertebrates because these organisms cannot catabolize it [108]. Although accumulation of oxalate leads to stress in plants, in vertebrates this molecule can be metabolized by bacteria present in the intestinal tract [109]. Oxalate can be catabolized in different ways: by oxidation, by decarboxylation of oxalyl-coenzyme A or by direct decarboxylation. Both oxidation and decarboxylation of oxalate are catalyzed by Mn-containing enzymes. Here we will discuss the oxalate decarboxylate reaction that produces formate and CO_2. The crystal structure of oxalate oxidase from *Bacillus subtilis*

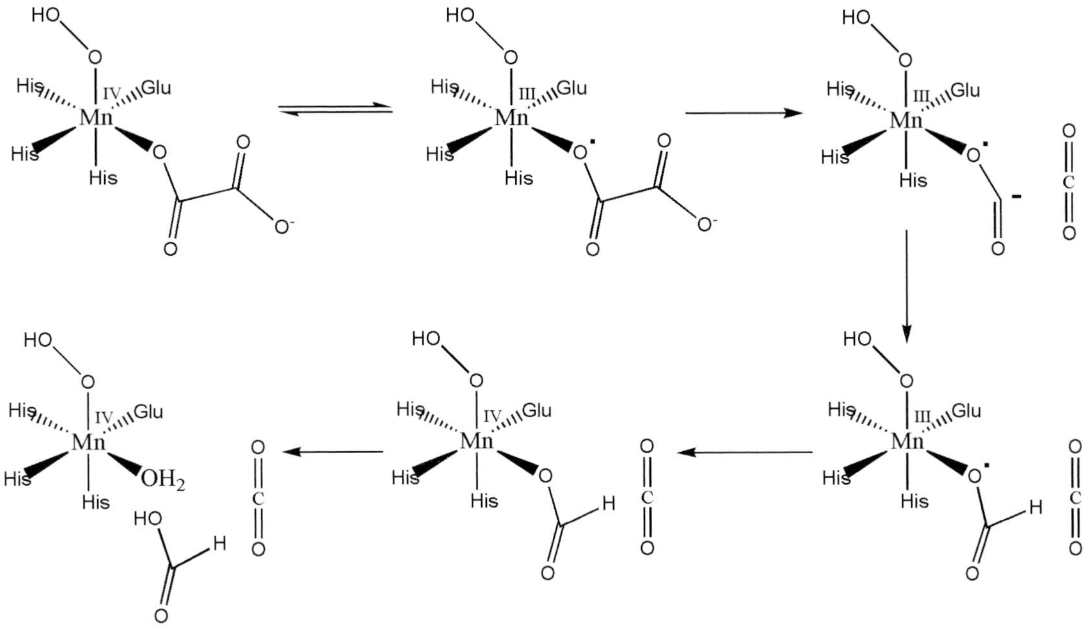

Fig. 11.13 Possible mechanism for oxalate decarboxylase.

complexed to formate has been recently solved to 1.75 Å resolution. The enzyme
has two homologous cuspin domains, each binding one Mn ion. The two Mn
sites are separated by about 26 Å and both have octahedral geometry with three
histidines and one glutamate highly conserved ligands being contributed by the
protein [108]. The Mn center in domain 1 is further coordinated by a formate ion
and a water molecule whereas its counterpart in domain 2 binds two water
molecules. A superposition of the active sites of oxalate decarboxylase and oxalate
oxidase shows that they are very similar and raises the question again, as to which
elements in the coordination sphere determine the difference between the oxidative
decarboxylation and the non-redox decarboxylation. Several amino acid replace-
ments which remove proton donors from oxalate oxidase may prevent it from
catalyzing the former reaction to give formate. A possible catalytic mechanism is
outlined in Fig. 11.13.

11.4.4
Arginases

These enzymes catalyze the hydrolysis of L-arginine to form L-ornithine and urea.
Mammalian arginases have been divided into classes I and II as a function of
their location: arginase I is predominantly found in the liver whereas arginase II
is located in non-hepatic tissues and functions primarily in L-arginine homeostasis
[110]. The crystal structure of rat arginase I has shown the active site to be
composed of a Mn binuclear center. The $Mn(II)_A$ and $Mn(II)_B$ ions which are

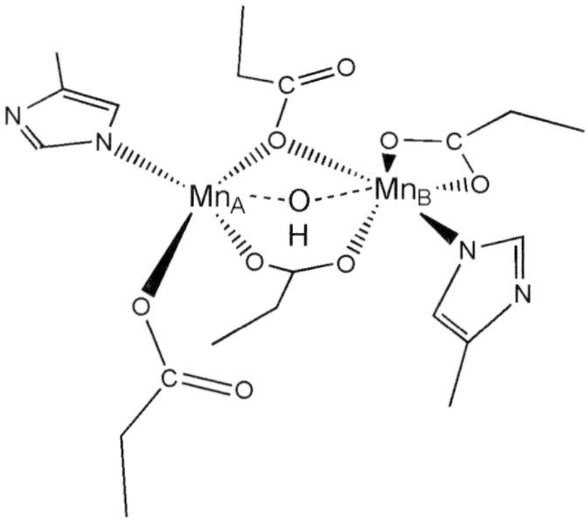

Fig. 11.14 The active site of arginase.

separated by 3.3 Å are respectively coordinated by one histidine and three aspartate residues and a solvent molecule that bridges the two metals and donates a hydrogen bond to one of the aspartate ligands of Mn_A (Fig. 11.14). It has been determined that an intact active site is essential for catalysis [111]. In the most widely accepted mechanism, both Mn ions serve to polarize and orient a bridging hydroxide ion for nucleophilic attack at the guanidinium carbon of the substrate.

Arginase II is found in kidney, small intestine, mammary gland and penile corpus cavernosus. It can play a very important role in L-arginine bioavailability to NO synthase because both enzymes compete for the same substrate. Indeed, it has been shown that NO biosynthesis is enhanced by inhibiting arginase II with boronic acid derivatives providing an alternative treatment to penile dysfunction [112]. The binuclear Mn active site of arginase II is almost identical to that of arginase I so their different activities are solely determined by their dissimilar locations.

11.4.5
The Photosynthetic Oxygen-evolving Center (OEC) of Photosystem II (PSII)

It is impossible to overemphasize the determining role that oxygenic photosynthesis has played in determining the atmosphere composition, and biological evolution, of this planet. The central reaction of photosynthesis is the photoinduced oxidation of water, one of the most thermodynamically demanding processes in Nature. It has been said that "if you can split water you can split anything" [113]. It has been known for some time that the metal center responsible for this crucial reaction contains both manganese and calcium. In a real *tour de force*, Barber, Iwate and coworkers [97] have very recently published the structure of the PSII from the cyanobacterium *Thermosynechococcus elongatus* at 3.5 Å resolution. Although lower resolution structures of PSII have already been determined, the structure of the Mn-containing EOC has remained elusive. The Barber–Iwata model, based on X-ray anomalous electron density maps calculated from data collected both at the Mn absorption edge and at a wavelength of 2.25 Å, where

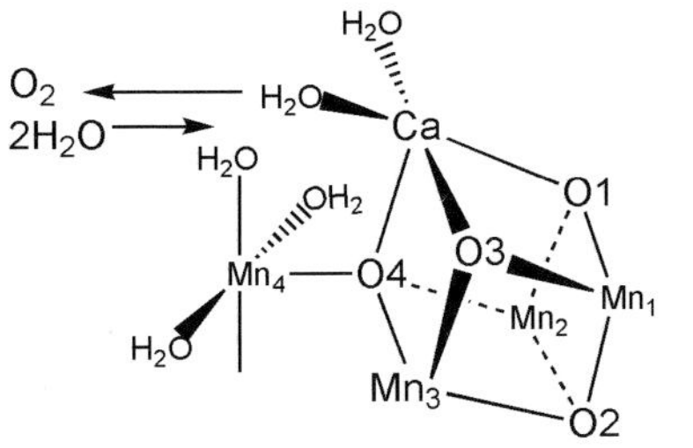

Fig. 11.15 The postulated Mn_3CaO_4 cubane-like cluster at the water-splitting centre of Photosystem II [97].

Ca(II) has a stronger f" anomalous signal than Mn, suggest that the active site is composed of a cubane-like cluster Mn(3)CaO(4) and a Mn ion connected to the cluster through one of the oxo ligands (Fig.11.15). Several mechanisms have been proposed where a Ca–Mn pair is a prerequisite for water oxidation. They include the binding of a water molecule to one of the Mn ions that, prior to oxygen formation, produces a highly electrophilic intermediate, either a Mn(IV) oxyl radical or a Mn(V) oxo species. The O=O bond formation would occur by the nucleophilic attack from a second water molecule ligated to the Ca ion. In the reaction, Ca(II) could act as a weak Lewis-type acid helped by the Mn_3 cluster and a Cl$^-$ ion [97].

11.5
Molybdenum and Tungsten

Both molybdenum and tungsten are elements of group VI in the periodic table. They have equal atomic and ionic radii and oxidation states and are relatively rare, ranking 53rd and 54th in natural abundance [114]. Mo and W are the only second and third row transition metals that are essential for growth at least for some organisms. Mo is found at the catalytic center of nitrogenase, where is part of a cluster that also contains iron and sulfur (FeMo-co), and as a protein-bound molybdenum-cofactor (Mo-co) in many other enzymes. Mo (or W)-Co contains a mononuclear metal ion and an organic molecule called molybdopterin (MPT) (Fig. 11.16). Although the structure of FeMo-co has been known for over 10 years [115], the reaction mechanism of nitrogenase has not been completely elucidated. FeMo-co has been extensively studied and will not be discussed here any further.

In terms of the reaction catalyzed, molybdopterin-containing enzymes can be divided in two groups: those that mediate oxygen atom transfer, such as dimethyl sulfoxide (DMSO) reductase and sulfite oxidase (SO), and those that catalyze hydroxylation reactions of aromatic heterocyclic compounds and aldehydes [116], for instance xanthine oxidoreductase (XOR) and aldehyde oxidoreductase (AOR). However, this functional classification does not coincide with structural properties that suggest that the enzymes should be grouped into five families, whose most representative members are: (1) DMSO reductase; (2) XOR; (3) SO; (4) aldehyde-

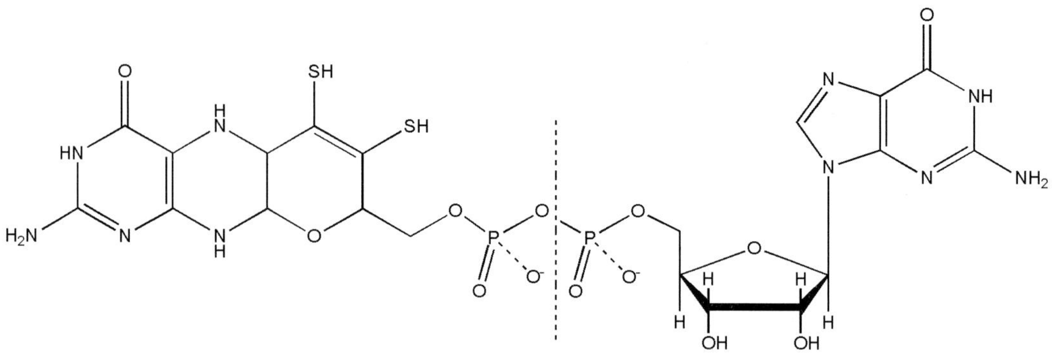

Fig. 11.16 The molydopterin cofactor. In prokaryotes the cofactor contains a nucleotide (right side of the dotted line) whereas this moiety is absent in eukaryotes.

ferredoxin oxidoreductase and (5) carbon monoxide dehydrogenase (CODH). Eukaryotic Mo-enzymes are characterized by binding the Mo-MPT unit, whereas in bacteria the Mo-co can incorporate a second nucleotide, GMP, AMP, IMP or CMP [117]. The functional role of this modification is not clear. All Mo-co containing enzymes, even if they can be more or less phylogenetically related, are characterized by having the cofactor buried deeply in the protein matrix and a funnel-shaped cavity that connects the active site with the molecular surface. It is the size, shape and overall charge of this region that determines substrate specificity. The catalytic mechanism is otherwise very similar in all the Mo-co enzymes and involves the 4+, 5+ and 6+ oxidation states of Mo or W. The different families are discussed below.

11.5.1
DMSO Reductase Family

The prototypic DMSO reductase catalyzes the following reaction:

$$(CH_3)_2S{=}O + 2\ H^+ + 2\ e^- \rightarrow (CH_3)_2S + H_2O \qquad (11.7)$$

The transfer of an oxygen atom from the substrate to water has been shown by ^{17}O-labeling experiments [118]. Similar oxygen atom transfer reactions are catalyzed by the related trimethylamine-N-oxide reductase (TMAO), biotin sulfoxide reductase (BSO), monomeric and heterodimeric dissimilatory nitrate reductases (NR) [119, 120], both Mo and W-containing formate dehydrogenases and W-containing N-formylmethanofuran dehydrogenase. All these enzymes contain a bis(MGD)-molybdenum form of the cofactor. The four thiolates, along with the Oγ from a serine in DMSO and BSO (Fig. 11.17a), a Sγ from a cysteine in NR, or a selenocysteine in formate dehydrogenase H from *E. coli*, coordinate the metal ion [116, 121, 122]. In addition, crystals structures have suggested additional Mo coordination by oxo ligand(s) [122–125].

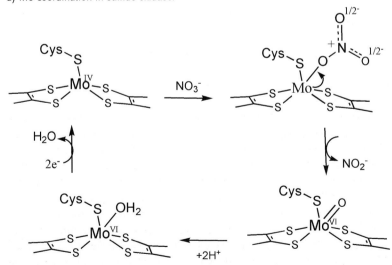

Fig. 11.17 a) bis[MGD]-molybdopterin; b) W-coordination (no protein ligands observed); c) sulfido form of the AOR active site; d) Mo coordination in sulfide oxidase.

Fig. 11.18 Catalytic cycle of nitrate reductase [119].

However, some of these structures show unusual features and a crowded active site. EXAFS studies of *Rhodobacter capsulatus* DMSO reductase show the expected four Mo–S ligands and one Mo–O bond arising from a serine residue plus one Mo=O in the reduced Mo(IV) state of the active site. This oxo ligand is removed upon oxidation of the metal ion to Mo(VI). In the oxidized structure, an aquo ligand is postulated to coordinate the molybdenum ion (Fig. 11.17c) [126–128]. An equivalent result has been reported for BSO that catalyzes the reduction of D-biotin D-sulfoxide to D-biotin [129]. A similar overall catalytic mechanism is expected for nitrate reductases (Fig. 11.18), which catalyze the following reaction:

$$NO_3^- + 2\,H^+ + 2\,e^- \rightleftharpoons NO_2^- + H_2O \tag{11.8}$$

These enzymes are present in both eukaryotes and prokaryotes and play an important role in nitrogen assimilation and dissimilation [130, 131]. Crystal structures have been reported for the monomeric NR from *Desulfovibrio desulfuricans* [119] and the heterodimeric enzyme from *R. sphaeroides* [120]. The former contains Mo bound to the bis(MGD) cofactor and a $[Fe_4S_4]$ cluster separated by about 12 Å [119]. The large subunit of the dimeric enzyme is homologous to the *D. sulfuricans* enzyme whereas its small subunit contains two c-type hemes that act as electron transfer units [120]. As in oxidized DMSO reductase, the Mo ion is coordinated by the four thiolates from the bis(MGD) cofactor, a protein side chain (cysteine in this case) and a hydroxo/water ligand. Consequently, the proposed catalytic mechanism is similar to the one advanced for DMSO reductase [119].

11.5.2
Xanthine Oxidoreductase (XOR) Family

XOR accelerates the hydroxylation of purines, pyrimidines, pterins and aldehydes [132]. In humans, the enzyme catalyzes the last two steps of purine catabolism: the oxidation of hypoxanthine to xanthine and of the latter to uric acid. An unusual property of this, but not all XOR enzymes [133], is its interconversion between xanthine dehydrogenase and xanthine oxidase activities which implies a switch between NAD^+ and molecular oxygen being used as the final electron acceptor [134]. Structural studies suggest that this switch, that can be irreversibly induced by proteolysis [135], results from conformational changes that lead to both restricted access to the NAD cofactor to its binding site and changes in the redox potential of the FAD cofactor [136].

Other members of the XOR family are aldehyde oxidase and *D. gigas* aldehyde oxidoreductase (AOR). All these enzymes are homodimeric and, with the exception of AOR, are only found in eukaryotes. They generally contain four domains, starting at the N terminus with two $[Fe_2S_2]$ cluster-containing regions, followed by a flavin-binding domain (absent from the bacterial enzyme) [122] and completed by the Mo-co containing stretch. The mammalian proteins are collectively known as MFE (mammalian Molybdo-Flavo Enzymes) [134]. The first crystal structure determined for this family was that from the *D. gigas* AOR. The active site Mo

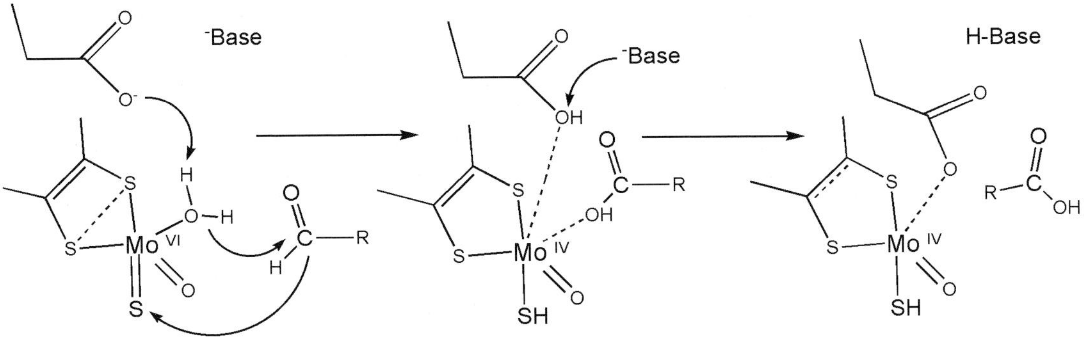

Fig. 11.19 Proposed mechanism for *D. gigas* AOR.

was coordinated by the dithiolene sulfurs of a single molybdopterin cytosine dinucleotide, or MCD, and three nonprotein oxygen ligands. Because EXAFS studies of the active enzyme have indicated the presence of a Mo sulfido ligand, this crystal structure is thought to correspond to the desulfo, inactive form. However, it does have the credit of being the first structural example of Mo-co covalently attached to a nucleotide and bound to a protein. Another important observation coming from the AOR structure is that MPT adopts the tricyclic structure already seen in the W-containing AOR from *Pyrococcus furiosus*, underscoring the basic similarities between W- and Mo-containing enzymes [116], already mentioned above. A proposed mechanism for *D. gigas* AOR, involving the Mo sulfido ligand is shown in Fig. 11.19. It differs from the mechanism proposed for other Mo-co containing enzymes, such as DMSO reductase, because it is the oxygen atom of a water molecule which is transferred to the substrate, rather than a Mo-bound oxo group.

11.5.3
Sulfite Oxidase Family

Assimilatory nitrase reductases (ANR), which catalyze the reduction of nitrate to nitrite, subsequently converted to NH_4^+ by nitrite reductase, are also members of this group [137, 138]. The prototypic SO, an enzyme located in the mitochondrial intermembrane, catalyzes the biologically essential oxidation of sulfite to sulfate, the terminal reaction in the oxidative degradation of sulfur-containing methionine and cysteine amino acids.

$$SO_3^{2-} + H_2O + 2 \text{ (cytc)}_{ox} \rightleftharpoons SO_4^{2-} + 2 \text{ (cytc)}_{red} + 2 H^+ \qquad (11.9)$$

Amino acid sequence comparisons have shown that ANR and SO are structurally unrelated to the MFE group mentioned above. Although they are both homo-dimeric enzymes of similar molecular weights, they differ in domain organization and the number of bound cofactors. ANR is composed of three domains: the N-terminal region binds the Mo-co, the central domain binds a b-type cytochrome and the C-terminal stretch contains FAD and a $NAD(P)^+$ binding site [139]. Although the crystal structure of SO from chicken liver [140] shows that this protein is also clearly divided into three domains; the cofactor composition is different: in SO, the heme-containing N-terminal region, which is 31% identical to the central domain of spinach ANR, is structurally similar to bovine cytochrome b_5; the second domain, with 38% identity to the N-terminal domain of ANR, binds the Mo-co in a protein fold that had not been previously observed, and the C-terminal region, that does not bind any cofactor, closely resembles the fold of cell adhesion molecules [141]. The active site of SO, as determined by the crystallographic analysis, can be described as follows. The Mo-co consists of only one MPT that, as in other structures, forms a tricyclic ring system with a pyran ring fused to the pyrazine ring of the pterin [142]. Consistent with its eukaryotic origin, the MPT of SO does not contain an associated nucleotide. The Mo is coordinated by five ligands with approximate square pyramidal coordination geometry. The terminal oxo occupies the axial position, the equatorial plane being composed of two sulfur ligands which are contributed by the dithiolene sulfurs from the molybdopterin, a thiolate ligand provided by a protein cysteine side chain and a water/hydroxo ligand [140]. The cysteine ligand is typical of this family and its mutation to serine results in inactive enzyme [143]. This metal coordination was expected from EXAFS experiments to correspond to the reduced Mo(IV), and not to the fully oxidized crystallized enzyme which should have had a Mo(VI) ion [144]. It is likely that photoreduction of the Mo ion took place during the X-ray data collection. X-ray-induced photoreduction has been documented in other cases [145].

In humans, genetic deficiency of SO, that can arise from either mutations in the synthetic machinery of MPT or from changes in the gene encoding SO, has dramatic consequences. These include severe neurological abnormalities, dislocated ocular lenses and mental retardation [146]. Of the four identified human mutations, two affect the active site and the others may disturb the dimer interface [140].

11.5.4
Aldehyde Ferredoxin Oxidoreductase Family

As indicated by its name, this group's members catalyze the interconversion of aldehydes to carboxylates. Other enzymes of the AFOR family are formate dehydrogenase, formaldehyde ferredoxin oxidoreductase, acetylene hydratase and carboxylic acid reductase. With the exception of the latter, these W-containing enzymes have been identified in hyperthermophilic archaea. As opposed to most Mo-containing enzymes, where this metal ion can be replaced by W with more or less significant loss in catalytic activity, the proteins mentioned above are strictly W-enzymes [147]. The absolute requirement for W in these supposedly old enzymes may be related to the higher water solubility of WS_2, as compared with MoS_2 making W more readily available to living organisms in the anoxic conditions of ancient Earth.

The prototypic AFOR from *Pyrococcus furiosus* was the first Mo-co-type enzyme to be characterized by X-ray crystallography [148]. It is a homodimer with three different types of metal centers: two W-co, two [Fe$_4$S$_4$] clusters and a single metal ion (most likely Fe) found at the dimer interface. AFOR has a unique fold unrelated to either DMSO reductase or AO. The W ion is coordinated by a bis(MPT) unit; there are no coordinating protein side chains and no oxo ligands were detected by the crystallographic analysis, possibly due to the presence of multiple enzyme forms in the crystal [148]. Thus the W coordination geometry appears to be distorted square planar (Fig. 11.17b). Little is known about the catalytic mechanism of AOR although the potentially relevant W(IV), W(V) and W(VI) redox states have been identified in the enzyme [149].

11.5.5
Carbon Monooxide Dehydrogenase (CODH$_{Mo}$)

Although structurally CODH$_{Mo}$, from the aerobic organism *Oligotropha carboxy-dovorans*, is a member of the XO family, its unusual dinuclear metal center suggests a unique catalytic mechanism. Like its anaerobic nickel-containing counterpart (see the section on nickel) the enzyme catalyzes the following reaction:

$$CO + H_2O \rightarrow CO_2 + 2\ H^+ + 2\ e^- \qquad (11.10)$$

The protons produced by the oxidation of CO are used to generate a gradient across the cytoplasmic membrane and channel electrons to a CO-insensitive respiratory chain [150]. The cell then uses the Calvin cycle for the fixation of the CO$_2$ product. CODH$_{Mo}$ is a dimer of heterotrimers, each trimer being composed of one large (L), one medium-size (M) and one small (S) subunit. The L subunit carries a mononuclear Mo complexed to MCD [151] whereas the M subunit binds FAD and the S subunit harbors two [Fe$_2$S$_2$] clusters. The four cofactors are roughly aligned in the sequence FAD–[Fe$_2$S$_2$]–[Fe$_2$S$_2$]–MCD–Mo (see Fig. 2 in [152]). The L and S subunits are significantly homologous to C-terminal and N-terminal domains of AOR from *D. gigas*, respectively [153, 154]. The M subunit belongs to the flavin-binding alcohol oxidase family. Like in all other Mo-co enzymes, in CODH$_{Mo}$ the cofactor is deeply buried. However, the tunnel is relatively narrow and hydrophobic which probably prevents access to the active site to substrates much bulkier than CO [152].

The structure of the active site was initially modeled with a modified S-selenyl-cysteine [152]. However, subsequent work by the same authors using anomalous scattering effects at higher resolution, showed that an active site Cu(I) ion had been mistakenly interpreted as a SeH moiety. Thus the CODH$_{Mo}$ active site is in fact a dinuclear (CuSMo(=O)OH) cluster [155]. In the revised active site structure, the Mo ion is equatorially coordinated by the two thiolates from the pterin cofactor, a μ-sulfido and an hydroxo ligand. The Mo coordination sphere is completed by an apical oxo ligand, the whole arrangement resulting in distorted square pyramidal geometry. Fig. 11.20 shows the recently proposed revised catalytic mechanism for CODH$_{Mo}$ [155].

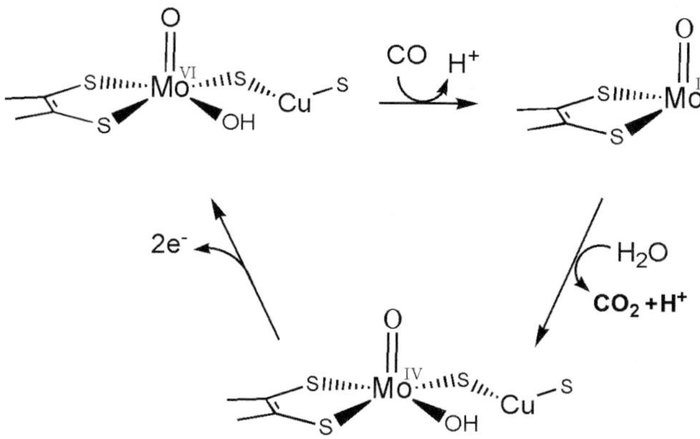

Fig. 11.20 The revised mechanism of CODH$_{Mb}$ [155].

11.6
Nickel

Nickel is a rare metal in biology. Only a handful of enzymes are known to contain this ion. The most common redox state in Nature is Ni(II). In ureases, Ni acts as a Lewis acid but in other enzymes it has been considered to be involved in redox processes. Because Ni is both toxic and essential to many microorganisms its transport is highly regulated by the cell. Reflecting its likely role in primordial biological processes, Ni is present in the active site of enzymes dealing with H_2, CO, CO_2 and CH_4 metabolism, gases that were abundant at early stages of Earth's atmospheric evolution. Special attention will be given here to the recently determined structures of the Ni-containing Acetyl-Coenzyme Synthase (ACS) and Carbon Monooxide Dehydrogenase [156, 157] and to the controversy generated by the binding of three different ions to the catalytic center of the former.

11.6.1
Carbon Monoxide Dehydrogenase

CODHs which form homodimers, can be associated to other catalytic proteins; in methanogens, where it is called α, it is associated to four other proteins βγδε generating a pentameric arrangement that oligomerizes to form αβγδε$_g$ clusters [158]. In acetogens, CODH, called β, forms heterotetramers $α_2β_2$ with ACS. Isolated homodimeric CODHs have been characterized in *Rhodospirillum rubrum* and *Carboxidotropha hydrogenoformans*. The crystal structures of these two proteins have indicated that the CODH dimer has two B-clusters, two C-clusters and one D-cluster [159, 160]. The active site responsible for the reversible oxidation of CO to CO_2 is the C-cluster that was postulated to be composed of one Ni and several Fe ions, maybe as a Ni–X–[Fe$_4$S$_4$] center similar to the A-cluster of ACS (see below).

The C-cluster of ACS/CODH displays two EPR signals with g_{av} of 1.82 (also called Cred1) and 1.86 (Cred2). It has been suggested that Cred1 is 2 e$^-$ more oxidized than Cred2 [161, 162] and that both states are involved in catalysis.

$$Cred1 + CO + OH^- \rightleftharpoons Cred2 + CO_2 + H^+ \qquad (11.11)$$

Mössbauer spectroscopy has shown that there is an unusual Fe species in the C-cluster, called Ferrous Component II (FeCII) that is present in both Cred1 and Cred2, but not in the Ni-depleted C-cluster [163]. Also, there are no changes in the UV/visible spectrum in going from Cred1 to Cred2 suggesting that the 2 e$^-$ are not located in the Fe-containing part of the C-cluster. This leads to the conclusion that Ni, possibly associated to a non-metal redox species at the active site, is likely to be responsible for the oxidative addition of CO_2. So far, four different C-cluster structures have been reported: (1) the one from *C. hydrogenoformans* [159], (2) the low-resolution one from *R. rubrum* [160], and (3) two different conformations in the C-cluster of *M. thermoacetica* [156, 157]. 1) is a high resolution structure that shows a square planar Ni with 3 labile sulfides and one thiolate ligand (Fig. 11.21a). As such, this structure does not provide the two empty Ni sites thought to be required to bind both CO and OH$^-$. The structures in 3) provide a possible answer to this problem: one of the labile sulfides (the one bridging Ni and the unique Fe, designated as Fe$_u$ (thought to be FCII, see below) is missing in these structures (Fig. 11.21b). Since COS is also a substrate of CODH, SH$^-$ should be a product of this reaction.

Thus the *C. hydrogenoformans* structure strongly suggests that, as the observed SH$^-$, OH$^-$ should bridge Ni and Fe$_u$. Since the Ni ion has square planar coordina-

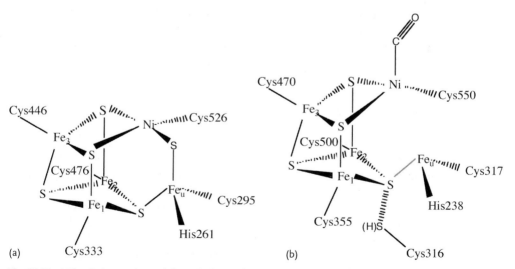

Fig. 11.21 a) The C-cluster reported for *C. hydrogenoformans* CODH (note that the Ni ion is four-coordinated). b) The cluster of *Moorella thermoacetica* [157].

tion it is most likely in the 2+ valence state. In both *M. thermoacetica* structures the Ni is distorted tetrahedral. In the CO-treated enzyme [157], the Ni (partially) binds CO (or CO_2?) and moves away from the plane in the direction of the gas tunnel. In the structure determined by Drennan et al. [156], the Ni moves towards Fe_u forming a formal Ni–Fe bond. Thus the presence of CO/CO_2 determines the geometry of the Ni coordination sphere. In the CO-treated enzyme Ni is likely to be reduced because of the excess of electrons coming from CO oxidation and the absence of electron acceptors. Very recent work suggests that treating the enzyme with excess CO results in the dissociation of the Ni–Fe bridging sulfide, maybe by forming COS [164].

The substrate/product CO_2 binds to clusters in both Cred1 and Cred2 states. However, the binding must be different in the two cases because CO_2 does not affect the g_{av} = 1.86 signal but it does modify g_{av} = 1.82 [162]. The simplest interpretation is that CO_2 binds terminally to tetrahedral Ni(0) and it bridges Ni and Fe_u in Cred1, when Ni is square planar Ni(II).

11.6.2
Acetyl Coenzyme A synthase (ACS)

ACS are present in methanogens, acetogens and carboxydotrophic micro-organisms where they catalyze the synthesis and/or the decarbonylation of acetylCoA according to the following reaction, where CoFeS correspond to a correnoid/cobalt containing iron–sulfur protein [158, 165].

$$CH_3CoS\text{-}CoA + Co^I FeSP + H^+ \rightleftharpoons CH_3Co^{III} FeSP + CO + CoA\text{-}H \qquad (11.12)$$

The oligomeric association of ACS subunits varies according to the function. In ACDS from methanogens they are part of a $\alpha\beta\gamma\delta\varepsilon$ pentamer [166]. In acetogens, ACS is associated to CODH as a $\alpha_2\beta_2$ heterotetramer [167]. In carboxydotrophic bacteria, which can grow with CO as the only carbon source, ACS can be purified as a monomer (very recently, however, a CODH that is presumably its partner has been identified [168]). Thus although CODH can function as a homodimer when CO is used as a source of carbon and reducing power (see section on Mo), it seems that ACS is always associated with CODH.

The A-cluster has been shown by several techniques to contain Ni and Fe. Mössbauer spectroscopy showed that Fe is found in a classic cubane, i.e. [Fe_4S_4] and that Ni is probably bridged to that cluster through an unidentified ligand [169]. A fraction of the Ni can be extracted using *ortho*-phenanthroline and the Ni-deficient enzyme is completely inactive. This fraction has been called labile Ni and could be reinserted in the enzyme by treating ACS with $NiCl_2$ [170–172]. Both methylation and reduction prevented labile Ni extraction. The remaining Ni was reported to have square planar coordination and be diamagnetic, and should correspond to Ni(II) [173]. It was puzzling to note that the Ni site seemed to exist under two very different states and that different preparations contained variable amounts of labile, active Ni.

There has been significant controversy concerning the order of substrate binding; in acetyl-CoA synthesis CO can bind to the A-cluster to form an EPR-active species [156, 174, 175]. The binding requires the presence of reductants in *M. thermoacetica* but apparently this is not the case for the NiCl$_2$-reconstituted recombinant β* fragment of *Methanosarcina thermophila* where the EPR species forms upon the addition of CO alone [175]. Isotopic substitutions using ^{61}Ni, ^{57}Fe and ^{13}C have induced changes in the EPR spectrum of the "NiFeC" signal suggesting that the unpaired electron resides on one of these atoms and that they are close to each other. This signal disappears when a methylating species is added to the reaction medium. This has been interpreted in two different ways. For Ragsdale and coworkers [176] the reaction is the following:

$$\text{Ni(I)-CO} + \text{CH}_3^+ \rightleftharpoons \text{Ni(III)-(CO)CH}_3 + 1\ e^- \text{ (from CODH)}$$
$$\rightarrow \text{Ni(II)-COCH}_3 \tag{11.13}$$

For Lindahl et al. [177] Ni(I)–CO is not a catalytically competent intermediate; there is no example of Ni(I) that supports methylation and forms Ni(III)–CH$_3$. Lindahl et al. did not observe reduction of either the B or C-cluster of CODH during methylation, so instead of invoking CODH/ACS inter-electron transfer, they proposed a "D-site", possibly a pair of cysteine thiolates, to provide the two electrons required for the oxidative methylation [177].

$$\text{Ni(II)} + \text{D}_{red} + \text{CH}_3^+ \rightleftharpoons \text{Ni(II)-CH}_3 + \text{D}_{ox}$$

$$\text{Ni(II)-CH}_3 + \text{CO} \rightarrow \text{Ni(II)-COCH}_3 \tag{11.14}$$

In this case, if the Ni(I)–CO complex is already present, it must dissociate and Ni has to be oxidized to Ni(II) prior to methylation. Grahame et al. have also studied the NiFeC signal using the enzyme from *M. barkeri* and came to the same conclusion, namely, that the CO complex is not part of the catalytic cycle [175]. Clearly, the major objection to the EPR active CO complex being an intermediate is the fact that after methylation, a two-electron reaction, the EPR signal disappears. Very recently, Grahame, Cramer and coworkers have reported XAS studies of reconstituted β* and concluded that the NiFeC signal does not arise from a Ni(I) species but from a low-spin Ni(II)–CO complex, with a CO carbon-based radical [178, 179]. In addition to acetyl synthesis (and acetyl–CoA formation), the A-cluster is capable of carrying out *in vitro* CO/acetyl–CoA exchange and transacetylation. It has been reported that added CO inhibits the transacetylation reaction in a non-competitive manner, probably by binding to the A-cluster at a site other than the one involved in transferring the acetyl group. Although CO does not inhibit the CO/acetyl–CoA exchange, it is also an acetylation inhibitor at relatively high concentrations by preventing methylation of the A-cluster.

A major breakthrough in the field was the publication of the three-dimensional structure of *M. thermoacetica* CODH/ACS by Drennan et al. [156]. As it is often the case with crystal structures, the active site was surprising in more than one way. Although, as predicted from spectroscopic studies, the A-cluster contained both a [Fe$_4$S$_4$] cubane and a Ni ion, it also had an additional metal site occupied by Cu(I). Thus, the cluster had the following arrangement: [Fe$_4$S$_4$]–S–Cu–S$_2$–Ni, the S atoms coming from cysteine residues. The Cu(I) ion had tetrahedral coordination whereas the Ni coordination was square planar. This unexpected Cu–Ni binuclear center required an unusual catalytic mechanism involving the carbonylation of the Cu(I) ion followed by methylation of the Ni ion. The two electrons required for the oxidative addition of the methyl cation were proposed to come from the oxidation of the [Fe$_4$S$_4$] cluster from +1 to +2 and of Ni from Ni(I) to Ni(II). The next step in the reaction implied the migration of the methyl group to the tetrahedral Cu(I) center followed by acetyl synthesis.

It is remarkable that the Cu proposition would turn out to be so short-lived. A first element of concern was raised by Gencic and Grahame using recombinant β* fragment. [175]. Reconstitution of this *E. coli*-expressed molecule could only yield active protein when Fe and Ni were supplemented, binding β* with a 2 : 1 ratio. Other ions, such as Mn^{2+}, Cu^{1+}, or Co^{2+} were ineffectual in conferring activity to the recombinant fragment. It should be noted, however, that its acetyl–CoA synthesis activity was only one three-thousandth (1/3000) of that of the native enzyme [175]. Subsequently, the second crystal structure determination of the *M. thermoacetica* CODH/ACS [157] showed that the A-clusters could contain Zn or Ni, instead of Cu, at the proximal (relative to the [Fe$_4$S$_4$] cluster) site (Fig. 11.22). Using monomeric α_{M_r} subunit, Lindahl et al. showed that incubation of the apo enzyme with Cu actually prevented subsequent activation by NiCl$_2$ [180]. Thus Cu is not only not a normal active site component but it is in fact an inhibitor of Ni binding to the A-cluster. The promiscuity of the proximal metal site can be explained in terms of the enzyme adopting either open or closed configurations [157]. In the latter, the proximal Ni ion (Ni$_p$) has square planar coordination (Fig. 11.23) and it is very exposed to the solvent medium. Consequently, it is easy to imagine that it could readily exchange with Cu(I) or Zn(II), trapping the proximal site in a tetrahedral coordination that so far has only been observed in the closed ACS and results in inactivation of the enzyme. On the other hand, square planar Ni$_p$ has also been observed in the very recently determined open structure of the monomeric ACS from *C. hydrogenoformans* [168]. We have proposed a mechanism

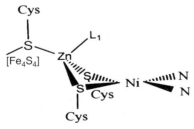

Fig. 11.22 The Zn/Ni-containing A-cluster in the closed form of the ACS from *M. thermoacetica*.

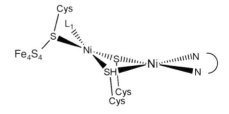

Fig. 11.23 The Ni/Ni-containing A-cluster in the open form of the ACS from *M. thermoacetica*.

where tetrahedral Ni_p coordination correlates with closed ACS and square planar Ni_p correlates with the open form of the enzyme [157].

The crystal structure of ACS_{Mt} has revealed the presence of an extensive tunnel network that connects all four active sites (the two A and two C-clusters of the heterotetramer). Their existence was predicted from experiments where CO synthesized from CO_2 and electrons at the C-cluster did not escape into the solvent but got incorporated as the carbonyl group of acetyl–CoA [181, 182]. Even in the presence of a saturating concentration of externally supplied CO, only C-cluster derived CO was used at the A-cluster [182, 183]. Conversely, CO_2 from acetyl–CoA undergoes exchange more productively than CO_2 from solution. This can be explained by the fact that the tunnel connecting the C-clusters to A-clusters points directly at the proximal metal site of the latter. Thus newly synthesized CO will migrate through the tunnel and then bind axially to Ni_p, either completing a tetrahedral or square pyramidal coordination of the metal (this depends on whether CO or the methyl group binds first during catalysis). External CO, on the other hand, is likely to bind equatorially to square planar Ni_p, at least initially, based on the open ACS_{Mt} structure [157]. CO_{ext} can inhibit acetyl–CoA synthesis, most likely by interfering with the methylation of Ni_p. This makes sense because methylation should also occur at the exposed equatorial open site of this metal center. Examination of the A-cluster crystal structure indicates that CO could indeed bind at three different sites: one axial and one equatorial site at Ni_p and the available axial site at the distal Ni site (Fig. 11.23).

Studies resorting to exogenous CO binding to ACS have considerably complicated the overall picture because it seems clear now that the source of CO (external vs. internal) influences the course of the reaction. In this respect, we have suggested that the tunnel connecting the A and C-clusters is subjected to a gating mechanism of substrate diffusion that is controlled by the opening and closing of ACS [157]. Neither the β* fragment nor the monomeric α subunit from *M. thermoacetica* are likely to bind CO the way the native enzymes would when being associated to their corresponding CODH subunits.

The crystallographic results, that support the existence of a gating mechanism for CO diffusion from the C-cluster, where it is synthesized, to the A-cluster, can be used to re-interpret several previously reported hypotheses. For instance, Lindahl et al. proposed that the C and A-clusters, known to be magnetically isolated, were nevertheless "coupled" through conformational changes [184]. This can now be understood in terms of C-cluster-synthesized CO accessibility to the A-cluster. The proposed "reductase" (CO synthesis) and "synthase" (acetyl-CoA synthesis) modes for CODH/ACS were based on the observation that when the CO_2 reduction was carried out in the absence of substrates for acetyl-CoA synthesis, CO production from CO_2 and reductant had lower k_{cat} and higher K_m than when those substrates were provided. This can now be rationalized as an accumulation of CO within the tunnels in the absence of acetyl-CoA synthesis because ACS would then be mostly in the "closed" configuration. When suitable substrates are provided, however, internal CO, along with the methyl group provided by FeSCP, are used to make acetyl-CoA, and CO synthesis should proceed at higher rates, as observed [184].

11.6.3
NiFe Hydrogenases

Many microorganisms are capable of using hydrogen as a source of reducing power or protons as final acceptors according to the following reaction.

$$H_2 \rightleftharpoons H^+ + H^- \rightleftharpoons 2\,H^+ + 2\,e^-$$ (11.15)

This reaction is catalyzed by two major classes of enzymes, NiFe and Fe-only hydrogenases [185, 186]. Only the latter will be discussed in this chapter (the active site structure of Fe-only hydrogenases is addressed in chapter 12). As indicated in Eq. (11.15), the uptake reaction is heterolytic generating a proton and a hydride. The crystal structure of the NiFe hydrogenase from *D. gigas* [187] and simultaneous FTIR studies, have established that the active site of these enzymes is composed of a $Fe(CN)_2CO$ unit bridged to a Ni ion through two thiolate sulfurs [188, 189]. The Ni center is further coordinated by two terminal cysteine thiolates (see Chapter 12). The active site is buried deeply within the structure and cavity calculations performed with the refined crystal structure indicated that several tunnels connect it to the molecular surface [190]. High pressure Xe-binding experiments performed on hydrogenase crystals show that the noble gas binds within these tunnels and theoretical molecular dynamics calculations using hydrogen placed at the experimentally determined Xe sites strongly suggested that the tunnels are the only access pathway for substrate to bind the catalytic center [190]. Furthermore, the tunnel points at a vacant Ni coordination site suggesting that the protein environment controls hydrogen binding to the active site. EPR experiments have identified three major paramagnetic species, called Ni-A, Ni-B and Ni-C because they have been shown to be Ni-based using isotopic (^{61}Ni) substitution [191–194]. Of these three paramagnets, only Ni-C is part of the catalytic cycle. Ni-A and N-B species are obtained when the enzyme is exposed to oxidizing conditions and have been called "unready" and "ready", respectively [186, 193, 194]. Ni-B reacts very fast in the presence of hydrogen whereas Ni-A requires a long reductive activation process that can be accelerated by heat. Also, Ni-B can interact directly with hydrogen and Ni-A cannot. Crystallographic analyses have shown that the enzyme in the Ni-A state has a NiFe bridging ligand that was initially postulated to be an oxo/hydroxo species [188, 195]. Recent results from our laboratory, however, have indicated that in the Ni-B state the enzyme also has a bridging ligand, making the structural distinction between Ni-A and Ni-B difficult. This problem may have been solved very recently with our re-interpretation of the N-A state structure which shows that the bridging ligand should be diatomic, possible a (hydro)peroxide species [195]. The different outcome in the oxidation of Ni-B and Ni-A states seems to depend on the number of electrons initially available [196]. With four electrons per hydrogenase molecule, the oxidation of O_2 can be carried out all the way to the formation of one water molecule and the bridging putative hydroxide ligand, generating the Ni-B species. However, if only two electrons per enzyme molecule are available, the reaction stops after the formation of (hydro)peroxide. It is thought now that it is either the release of this ligand or the reduction of peroxide to water that is responsible for the sluggish activation kinetics of the Ni-A state [195].

FTIR spectroscopy has been very useful in monitoring redox states of NiFe hydrogenases, including those that are diamagnetic through active site CO and CN^- band shifts due to redox changes. Thus several EPR-silent states have been identified [189, 197]: (1) from the Ni-A state, a one-electron more reduced Ni-SU is produced; (2) after activation two EPR-silent reduced species, $NiSI_I$ and $NiSI_{II}$ are observed. Work by Fernandez and coworkers has shown that these two latter species differ by one H^+ [189]. Finally, (3) an additional one-electron reduction generates the Ni-C state which can be further reduced to the silent Ni-R. Although the subject of some initial debate, it seems now that the Ni-A, Ni-B and Ni-C EPR signals arise from (formal) Ni(III) species. This is easy to understand in the case of the oxidized Ni-A and Ni-B states. In the Ni-C species, however, the formal Ni(III)-H^- assignment may be thought of as the isoelectronic Ni(I)-H^+ species. *Bone fide* Ni(I) may be obtained when the Ni-C state is photolyzed and gets deprotonated. For a description of the catalytic cycle, see Fig. 12.11 [QQQ].

The presence of CO and CN^- at the active site of NiFe hydrogenases requires a synthetic machinery to generate these diatomic ligands. Considerable light has been shed on this problem by the excellent work of Böck and coworkers who have identified carbamoyl phosphate as a precursor [198].

11.7
Zinc

Zn is essential for the growth of all types of life. In humans, it is the second most abundant trace metal after iron [199, 200]. There are over 300 characterized enzymes that use Zn(II) for catalysis divided into 6 groups [201, 202]. Zn(II) because it is a d^{10} ion, is not redox active at the potentials normally found in biology [203] which prevents the study of its reactivity by, for instance, EPR or optical spectroscopic approaches [204]. On the other hand, the lack of redox activity makes Zn(II) stable in a biological medium of fluctuating potential. In proteins, zinc most frequently binds to histidine, aspartate, glutamate and cysteine [205]. In all cases the catalytic Zn ion acts as a Lewis acid [206], often binding a water molecule that upon deprotonation results in a reactive nucleophilic OH^- species. Zn(II) has a similar effect on cysteine thiolates and enzymes containing Zn–S bonds have attracted considerable attention in recent years [207]. The ligand-field stabilizing energy of Zn(II) is zero in all liganding geometries so this ion has no preferred coordination geometry [208]. However, in zinc metalloenzymes of known three-dimensional structure, the ion often adopts a distorted tetrahedral coordination (although, as indicated below, distorted trigonal bipyramidal geometry is also observed). Changes in coordination are central to the catalytic properties of Zn(II). Representative and widely studied Zn-enzymes are carbonic anhydrases, β-lactamases, alcohol dehydrogenases, alkaline phosphatases, nucleases and peptidases.

In addition, Zn can be part of heterodinuclear active sites such as CuZn superoxide dismutase, an enzyme that has been discussed in the Cu section. Here, we will describe the active site structure and, whenever possible, the postulated catalytic mechanism of representative members of the enzyme classes mentioned above. We will also address some very recent structural results concerning other less well-studied Zn-enzymes.

11.7.1
Carbonic Anhydrases (CA)

These proteins, which were the first Zn-containing enzymes to be characterized, are ubiquitous in nature where they are involved in processes such as respiration, photosynthesis, ion transport, and pH homeostasis [209, 210]. The reaction catalyzed by CA is the following where proton transfer to the bulk solvent is the rate-limiting step in the reaction.

$$CO_2 + H_2O \rightleftharpoons HCO_3^- + H^+ \tag{11.16}$$

There are three different classes of CAs called α, β and γ. In α CAs, found in mammals, algae and prokaryotes, and in γCAs, isolated from methanogens, the Zn ion is tetrahedrally coordinated by three His residues and a H_2O/OH^- species [211, 212]. In both cases the Zn(II) acts as an electrostatic catalyst that lowers the pK_a of the bound solvent molecule and stabilizes the negative charge of the transition state. It is noteworthy that, although the active sites are very similar in these two classes of enzymes, their respective polypeptide foldings are different. According to recent results, in the β enzymes the Zn ion is tetrahedrally coordinated by two Cys, one His and one Asp residue (Fig. 11.24) [213]. Unexpectedly, there is no solvent molecule in the first coordination sphere of the metal, although the electron density map indicates the presence of a putative water molecule nearby [213]. It has been proposed that during catalysis this water molecule displaces the Asp carboxylate, binds the Zn ion and gets activated by deprotonation, as in other Zn-enzymes. This suggests that although the catalytic mechanism of the β class of CAs is, at least in terms of Zn coordination, different

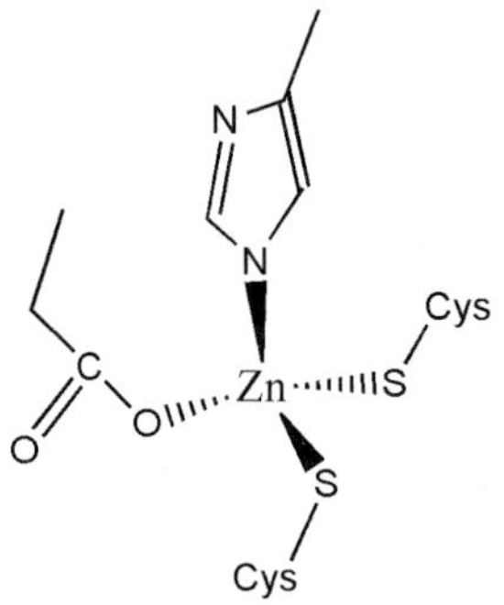

Fig. 11.24 The active site of β-carbonic anhydrase.

from those already well-established for the α and γ types, the underlying chemistry is very similar.

11.7.2
Alcohol Dehydrogenases (ADH)

These widely distributed oxidoreductases require either NAD(H) or NADP(H) as a coenzyme to catalyze the reversible interconversion of alcohols to aldehydes/ketones. They can display different substrate specificities and have been divided into three classes as a function of their molecular weights [214–216]. Short and long-chain alcohol dehydrogenases, about 250 and 385 residues long, respectively, are devoid of metal and will not be discussed any further. Medium-chain ADHs, about 350 residues long, are, for the most part, Zn-containing enzymes. So far, they have been found in bacteria, fungi, plants, cephalopods and vertebrates. The best studied enzyme from the structural viewpoint, is the one purified from horse liver (HLADH) [217–220] which is dimeric, as are all the other characterized ADHs from plants and animals. In the crystals, HLADH adopts an open conformation in the absence of the NAD cofactor and a closed conformation in its presence [218, 221, 222]. In the latter, the active site region becomes more hydrophobic, as required for hydride transfer from the substrate to the cofactor during catalysis. In the apo (no NAD) form of ADH from the archeon *Sulfolobus solfataricus*, the Zn ion is coordinated by two Cys and a His residue, the distorted tetrahedral coordination being completed by a water molecule [223]. The second coordination sphere of the metal ion contains the carboxylate group from a Glu residue located opposite to the substrate binding site. This carboxylate has been assigned the role of coordinating the Zn ion when either water or alcohol dissociates from Zn during catalysis (although this idea has been contested). Accordingly, two mechanisms have been put forward to explain the removal of the two hydrogen atoms from the substrate as required for alcohol oxidation. The first mechanism proposes that the Zn-bound water is displaced by substrate and the active site metal ion retains its tetrahedral coordination throughout catalysis [224, 225]. In the second mechanism, the substrate binds to tetrahedral Zn(II) without ligand displacement, forming a penta-coordinated species [226, 227]. It has been difficult to distinguish between these two possibilities because no kinetic studies were carried out until very recently. New time-dependent XAS studies of the (evolutionary unrelated) tetrameric NADP-dependent ADH from the hyperthermophilic bacterium *Thermoanaerobacter brockii* (TbADH) have favored the second hypothesis as they suggested the presence of two penta-coordinated transient Zn species during turnover [204]. In the first species, the Zn ion first coordination sphere would be composed of one water molecule along with Glu, Asp, His and Cys side chains from TbADH (Fig. 11.25). A similar transient intermediate could be present in HLADH where, as indicated above, a Glu carboxylate group is at binding distance from the catalytic Zn(II) ion and could form a penta-coordinated sphere around the metal ion composed of one water molecule, along with two cysteines, one histidine and one glutamate from the enzyme [223]. The second putative

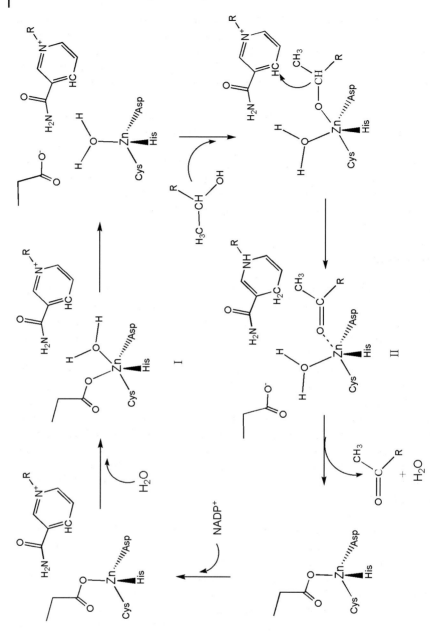

Fig. 11.25 The catalytic cycle of alcohol dehydrogenase from *T. brockii* [204].

penta-coordinated adduct of TbADH corresponds to Zn(II) complexed to product, a water molecule and Cys, His, and Asp protein side chains. Subsequent dissociation of water and product restores the tetrahedral Glu-coordinated Zn ion and the cycle can start again (Fig. 11.25). TbADH has been used in several chiral syntheses due to its thermostability and resistance to organic co-solvents [228].

The enzyme from the hyperthermophilic archaeon *Sulfolobus sulfataricus* (SsADH) appears to be similar to the bacterial TbADH although, as the vertebrate enzyme it uses NAD(H) instead of NADP(H) as cofactor. Overall, SsADH shares 24–26% similarities with both HLADH and TbADH. The three enzymes belong to the A-specific dehydrogenase family for, when operation in the alcohol synthesis mode, they transfer the *pro*-R hydrogen atom of the cofactor onto the *re*-face of the substrate ketone thus forming (S)-alcohol [223].

11.7.3
Metallo-β-lactamases

Because over half of all the commercially available antibiotics are β-lactams, and in recent years bacterial resistance to these molecules has become so prevalent (to the point that some infections can no longer be treated), metallo-β-lactamases are not only important from the organometallic standpoint but they are enzymes central to human health [229]. Most of the β-lactamases identified to date belong to the A, C and D serine hydrolase classes [230–233]. Class B enzymes, on the other hand, require divalent (Zn, Co, Cd or Mn) metal ions for catalytic activity [234, 235]. Because only 10% of the 200-odd β-lactamases identified so far are metalloenzymes little attention has been given to these proteins as potential threats to antibiotic therapy [236, 237]. However, metallo-β-lactamases are able to hydrolyze compounds that are resistant to serine-β-lactamases and they show little or no susceptibility to traditional β-lactamase inhibitors [238, 239]. This has resulted in ineffective use of the most available antibiotics in bacteria capable of producing metallo-β-lactamases [240, 241].

In general these enzymes catalyze the hydrolytic opening of the β-lactam ring according to the reaction depicted in Fig. 11.26. The hydrolytic product, unlikely the intact β-lactam, is not able to interfere with bacterial cell wall synthesis, the key effect of the antibiotic [229]. The crystal structures of several metallo-β-lactamases have shown that the active site can have one or two Zn(II) ions (Zn1 and Zn2) then separated by 3.5 Å [242–247]. In these structures Zn1 is tetrahedrally coordinated to three His and one water molecule. When present, Zn2 is either penta-coordinated to Cys, His and Asp residues and two water molecules, or to two His, one Asp and two water molecules, in a trigonal bipyramid. In the structure where the active site is composed of a binuclear Zn center, a H_2O/OH^- bridges the two metals. As in other binuclear centers with bridging solvent molecules, its activation provides the OH^- nucleophile for the reaction [248]. It should be pointed out that not all β-lactam antibiotics bind to the active site in the same way so mechanisms may differ from molecule to molecule.

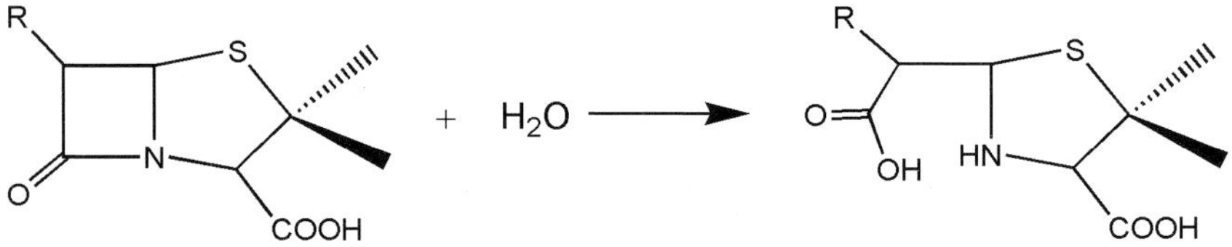

Fig. 11.26 Hydrolysis of the β-lactam ring by β-lactamases.

11.7.4
Alkaline Phosphatases

These homodimeric enzymes catalyze the hydrolysis of a great variety of phosphate monoesters. When a phosphate acceptor is present, they carry out transphosphorylation reactions [249]. There is a covalent phosphoseryl intermediate in the reaction that either produces inorganic phosphate or transfer the phosphoryl group to alcohols. Alkaline phosphatases are present in both prokaryotes and eukaryotes. In the former, their function may be related to the acquisition of phosphate whereas they may be involved in mineralization in humans and other higher organisms. The best studied enzyme is the one from *E. coli* [250] which was shown long ago to have three metal binding sites at the catalytic center. These are normally occupied by two Zn ions and a Mg ion. However, the latter can be substituted by Zn resulting in reduced activity [249]. Overall, the catalytic metal triad is similar to those found in phospholipase C and P1 nuclease [251, 252]. The Zn1 ion is coordinated by an aspartate, two histidines and two oxygen atoms from an active site-bound phosphate ion. Zn2 is tetrahedrally bound by two aspartates, a histidine and either the nucleophilic serine residue (SAP) or an oxygen atom from the phosphate inhibitor (ECAP, PLAP). Finally, the metal at the third site can be coordinated either by an aspartate, a threonine, a glutamate, a histidine and a water molecule (Zn) or by aspartate, a glutamine, a threonine/serine and three water molecules in an octahedral geometry. A possible catalytic mechanism is described in [253].

11.7.5
Metallopeptidases

These enzymes are characterized by the presence of short conserved signature Zn-binding amino acid sequences containing histidine and glutamic residues. The most widespread motif is HexxH assigned to "zincins", best represented by the well-studied thermolysin, but other sequences such as HxxEH (inverzincins), HxxE (carboxypeptidases) and HxH have also been described [254]. In zincins, the two histidine residues contained in the signature motif are always Zn(II) ligands that together with either His or Glu and a water molecule form the trigonal-pyramidal coordination sphere of the catalytically active metal center [254, 255]. The catalytic mechanism in this class of enzymes is believed to rely on the polarization of both the carbonyl group of the scissile amide bond (by the Zn ion) and a water molecule (by both Zn(II) and a conserved glutamate) to generate a

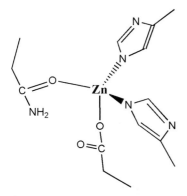

Fig. 11.27 The postulated "non-active" site of LytM autolysin.

reactive nucleophilic OH⁻ species [255]. Very recent results concerning the crystal structure of the HxH-type LytM autolysin from the human pathogen *Staphylococcus aureus* indicates that the first histidine in that motif is not a Zn ligand but the second one does bind to the metal center [256]. This has been interpreted as a "latent", non-active, state of the enzyme: it lacks an ordered water that could ligand the Zn ion (Fig. 11.27). Furthermore, the tetrahedral Zn coordination by four protein ligands would not leave room for substrate or water to contact the Zn ion. Indeed, site-directed mutagenesis of the first histidine of the HxH motif completely abolishes the enzyme activity confirming that the crystal structure does not correspond to a fully functional enzyme.

The recently published crystal structure of the thermolysin-like angiotensin I-converting enzyme (AnCE) from *Drosophila melagonaster* has shown that this enzyme belongs to the gluzincin family, which contain the HExxH motif [257]. In AnCE, the Zn ion was penta-coordinated by conserved Glu, and His protein ligands and a couple of water molecules [258]. By comparison, in thermolysin, another member of this family, a key glutamate residue plays a central role by polarizing a nucleophilic water molecule and by transferring protons to the substrate. In addition, a tyrosine and a histidine residue promote formation of the tetrahedral intermediate by stabilizing negative charges.

11.7.6
Other Zn-dependent Enzymes

11.7.6.1 D-aminoacylase
D-aminoacylase is an unusual member of the α/β barrel amidohydrolase super-family (a group also including urease, phosphotriesterase and other enzymes containing dinuclear metal active sites as well as nucleoside diaminases which have mononuclear active centers [259]). The enzyme from *Alcaligenes fecaelis* has been defined as a mononuclear metalloprotein with a binuclear active site [260]. Of the two Zn ion sites detected by anomalous X-ray scattering data from the crystals, one is fully occupied, tightly bound and tetrahedrally coordinated by two

cysteines, one histidine and an acetate ion from the crystallization solution. The second Zn ion site, separated from the first Zn by 3.1 Å, has 25% occupancy, it is tetrahedrally coordinated by two histidines and one cysteine residue and, again, by an acetate ion from the crystallization medium. Only the tightly bound Zn center is required for activity. The presence of the binuclear Zn site, with widely different metal-binding affinities, is reminiscent of the situation in the β-lactamase from *Bacillus cereus* which is also a monozinc metalloenzyme having two Zn sites [261, 262].

11.7.6.2 **GTP Cyclohydrolase I**
In plants and several microorganisms the conversion of GTP to dihydroneopterin triphosphate provides the first committed intermediate in the biosynthesis of tetrahydrofolate [263]. In animals, dihydroneopterin triphosphate is converted to tetrahydrobiopterin which functions as an essential cofactor to NO synthases, and monooxygenases [264]. The active site Zn(II) is postulated, as in other Zn-containing enzymes, to polarize its bound water ligand by increasing its acidity, subsequently generating an hydroxyl nucleophile in close proximity of the imidazole ring of the bound substrate.

11.7.6.3 **Enzymes Containing Zn–S Bonds**
When acting as a Lewis acid-type catalyst, Zn(II) can transform the thiol ligand into an effective nucleophile capable of participating in alkylating reactions [265–267]. On the other hand, zinc can confer redox properties to active sites by activating thiolate ligands so that it is the ligands, and not the inert metal ion that get oxidized and reduced [268]. When oxidized, the modified cysteinesulfenic acid does not bind Zn(II), so redox processes result in a zinc switch that can induced major conformational changes. An example of redox zinc switch is the conformational change experienced upon thiolate ligand oxidation by the heat-shock protein Hsp-33 [269, 270]. This change induces chaperon activity. Another fundamental role of Zn–S bonds is to control the cellular distribution of the metal ion [207]. Several ZnS clusters are depicted in Fig. 11.28.

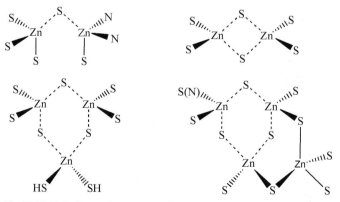

Fig. 11.28 ZnS clusters in proteins [207].

11.8
Conclusions

In this chapter an attempt has been made to describe representative active sites of metalloenzymes containing a variety of metal ions. Metals can perform catalytic functions that are difficult or impossible to carry out by amino acid functional groups or organic cofactors. These include activation of water molecules, reactions involving gases and, to some extent, redox processes. Although recent results have shed considerable light on this field, some caution should be exerted when interpreting active site structures. On the one hand, many metal sites seem to be rather promiscuous, being able to bind different ions, sometimes even keeping the catalytic activity. On the other hand, photoreduction by X-ray photons may be a very common phenomenon. Without some spectroscopic technique to monitor possible redox changes during or after X-ray data collection, only tentative conclusions should be drawn.

References

1 C. Rensing, B. Fan, R. Sharma, B. Mitra, B. P. Rosen, *Proc. Natl. Acad. Sci. USA* **2000**, 652–656.

2 B. Abolmaali, H.V Taylor, U. Weser, *Struct. Bond.* **1998**, *91*, 91–190.

3 R.H Holm, P. Kennepohl, E. I. Salomon, *Chem. Rev.* **1996**, *96*, 2239–2314.

4 F. Fusetti, K. H. Schroter, R. A. Steiner, P. I. Van Noort, T. Pijning, H. J. Rozeboom, K. H. Kalk, M. R. Egmond, B. W. Dijkstra, *Structure* **2002**, *10*, 259–268.

5 S. Kishishita, T. Okajima, M. Kim, H. Yamaguchi, S. Hirota, S. Suzuki, S. Kuroda, K. Tanizawa, M. Mure, *J. Am. Chem. Soc.* **2003**, *125*, 1041–1055.

6 S. M. Janes, D. Mu, D. Wemmer, A. J. Smith, S. Kaur, D. Maltby, A. L. Burlingame, J. P. Klinman, *Science* **1990**, *248*, 981–987.

7 W. S. McIntire, C. Hartman, V. L. Davidson, *Principles and Applications of Quinoproteins*, New York, **1993**, pp. 97–171.

8 J. P. Klinman, D. Mu, *Annu. Rev. Biochem.* **1994**, *63*, 299–344.

9 P. F. Knowles, D. M. Dooley, G. Sigel, A. Sigel, in H. Sigel, A. Sigel (Eds.), *Metalloenzymes involving amino acid residue and related radicals*, Metal Ions in Biological Systems, Vol. 30, Marcel Dekker, New York, **1994**, pp. 361–403.

10 Q. Su, J. P. Klinman, *Biochemistry* **1998**, *37*, 12513–12525.

11 S. A. Mills, J. P. Klinman, *J. Am. Chem. Soc.* **2000**, *122*, 9897–9904.

12 B. Schwartz, A. Olgin, J. P. Klinman, *Biochemistry* **2001**, *40*, 2954–2963.

13 A. Rinaldi, A. Gialtosio, G. Floris, A. Finazzi-Agro, *Biochem. Biophys. Res. Commun.* **1984**, *120*, 242–249.

14 I. Fridovich, *Annu. Rev. Bioch.* **1975**, *44*, 147–159.

15 M. A. Hough, R. Strange, W. H. S. Samar, *J. Mol. Biol.* **2000**, *304*, 231–241.

16 R. W. Strange, S. Antonyuk, M. A. Hough, P. A. Doucette J. A. Rodriguez, P. J. Hart, L. J. Hayward, J. S. Valentine, S. S. Hasnain, *J. Mol. Biol.* **2003**, *328*, 877–891.

17 J. A. Tainer, E. D. Getzoff, J. S. Richardson, D. C. Richardson, *Nature* **1983**, *306*, 284–287.

18 K. Djinovic, A. Coda, L. Antolini, G. Pelosi, A. Desideri, M. Falconi, G. Rotilio, M. Bolognesi, *J. Mol. Biol.* **1992**, *226*, 227–238.

19 K. Djinovic, G. Gatti, A. Coda, L. Antolini, G. Pelosi, A. Desideri,

M. Falconi, F. Marmocchi, G. Rotilio, M. Bolognesi, *J. Mol. Biol.* **1992**, *225*, 791–809.

20 K. Djinovic-Carugo, A. Battistoni, M. T. Carri, F. Poticelli, A. Desideri, G. Rotilio, A. Coda, K. S. Wilson, M. Bolognesi, *Acta Crystallog. Sect. D* **1996**, *52*, 176–188.

21 N. J. Blackburn, S. S. Hasnain, N. Binsted, *Biochem J.* **1984**, *219*, 985–990.

22 M. E. McAdam, R. A. Fox, F. Lavelle, M. F. Fielden, *Biochem. J.* **1977**, *165*, 71–79.

23 R. R. Crichton, J. L. Pierre, *BioMetals* **2001**, *14*, 90–112.

24 E. I. Solomon, U. M. Sundaram, T. E. Manchonkin, *Chem. Rev.* **1996**, *96*, 2563–2605.

25 I. Zaitseva, V. Zaitsev, G. Card, K. Moshkov, B. Bax, A. Ralph, P. Lindley, *J. Biol. Inorg. Chem.* **1996**, *1*, 15–23.

26 J. M. Henson, M. J. Butler, A. W. Day, *Annu. Rev. Phytopathol.* **1999**, *37*, 447–471.

27 S. A. Roberts, A. Weichsel, G. Grass, K. Thakali, J. T. Hazzard, G. Tollin, C. Rensing, W. R. Montfort, *Proc. Natl. Acad. Sci. U. S.A* **2002**, *99*, 2766–2771.

28 K. Pionteck, M. Antorini, T. Choinowski, *J. Biol. Chem.* **2002**, *277*, 37663–37669.

29 T. Bertrand, C. Jolivalt, P. Briozzo, E. Caminade, N. Joly, C. Madzak, C. Mougin, *Biochemistry* **2002**, *41*, 7325–7333.

30 M. A. McGuirl, D. Dooley, *Curr. Opin. Chem. Biol.* **1999**, *3*, 138–144.

31 T. E. Machonkin, L. Quintanar, A. E. Palmer, R. Hassett, K. Severance, E. I. Salomon, *J. Am. Chem. Soc.* **2001**, *123*, 5507–5517.

32 B. G. Malmström, B. Reinhammar, T. Vanngard, *Biochim. Biophys. Acta* **1970**, *205*, 48–57.

33 B. C. Berks, S. J. Fergusson, J. W. B. Moir, D. J. Richardson, *Biochim. Biophys. Acta* **1997**, *1232*, 97–173.

34 W. G. Zumft, *Microbiol. Mol. Biol. Rev.* **1997**, *61*, 533–616.

35 K. Brown, M. Tegoni, M. Prudencio, A. S. Pereira, S. Besson, J. J. Moura, I. Moura, C. Cambillau, *Nature Struct. Biol.* **2000**, *7*, 191–195.

36 A. C. Rosenzweig, *Nature Struct. Biol.* **2000**, *7*, 169–171.

37 T. Rasmussen, B. C. Berks, J. Sanders-Loehr, D. M. Dooley, W. G. Zumft, A. J. Thomson, *Biochemistry* **2000**, *39*, 12753–12756.

38 J. S. Valentine, in I. Bertini, H. B. Gray, S. J. Lippard, J. S. Valentine (Eds.), *Bioinorganic Chemistry*, University Science Books, New York, **1994**, pp. 253–267.

39 M. Wikström, *Nature* **1977**, *266*, 271–273.

40 P. Mitchell, *Nature* **1961**, *191*, 144–148.

41 S. Iwata, C. Ostermeier, B. Ludwig, H. Michel, *Nature* **1995**, *376*, 660–669.

42 C. Ostermeier, A. Harrenga, U. Ermler, H. Michel, *Proc. Natl. Acad. Sci. U. S. A.* **1997**, *94*, 10547–10553.

43 T. Tsukihara, H. Aoyama, E. Yamashita, T. Tomizaki, H. Yamaguchi, K. Shinzawa-Itoh, R. Nakashima, R. Yaono, S. Yoshikawa, *Science* **1995**, *269*, 1069–1074.

44 T. Tsukihara, H. Aoyama, E. Yamashita, T. Tomizaki, H. Ymaguchi, K. Shinzawa-Itoh, R. Nakasshima, R. Yaono, S. Yoshikawa, *Science* **1996**, *272*, 1136–1144.

45 S. Yoshikawa, K. Shinzawa-Itoh, R. Nakashima, R. Yaono, E. Yamashita, N. Inoue, M. Yao, M. J. Fei, C. P. Libei, T. Muzyshima, H. Yamagucchi, T. Tomizaki, T. Tsukihara, *Science* **1998**, *280*, 1723–1729.

46 D. M. Popovic, A. A. Stuchebrukhov, *J. Am. Chem. Soc,* **2004**, *126*, 1858–1871.

47 H. Michel, *Biochemistry* **1999**, *38*, 15129–15140.

48 J. D. Lipscomb, A. M. Orville, in H. Sigel, A. Sigel (Eds.), *Degradation of environmental pollutants by micro-organisms and their metalloenzymes,* Metal Ions in Biological Systems, Vol. 28, Marcel Dekker, New York, **1992**, pp. 243–298.

49 T. D.H Bugg, C.J Winfield, *Nat. Prod. Rep.* **1998**, 513–530.

50 S. Dagley, *Am. Sci.* **1975**, *63*, 681–689.

51 B.R Glick, J. J. Pasternak (Eds.), *Molecular Biotechnology: Principles and Applications of Recombinant DNA,* 2nd

edition, American Society for
Microbiology, Washington DC, **1998**.

52 E. I. Solomon, T. C. Brunold,
M. I. David, J. N. Kemsley, S. K. Lee,
N. Lehnert, F. Neese, A. J. Slulan,
Y. S. Yang, J. Zhou, *Chem. Rev.* **2000**,
100, 235–349.

53 J. D. Lipscomb, A. M. Orville,
R. W. Frazee, K. B. Dolbeare,
N. Elango, D. H. Ohlendorf,
A. C. S. *Symp. Ser.* **1998**, *692*, 387–402.

54 J. C. Boyington, B. J. Gaffney,
L. M. Amzel, *Science* **1993**, *260*,
1482–1486.

55 W. Minor, J. Steczko, B. Stec,
Z. Otwinowski, J. T. Bolin, R. Walter,
B. Axelrod, *Chemistry* **1996**, *35*,
10687–10701.

56 E. Skrzypczak-Jankun, L. M. Amzel,
B. Kroa, M. O. Funk, *Proteins Struct.
Funct. Genet.* **1997**, *29*, 15–31.

57 S. A. Gillmor, A. Villasenor,
R. Fletterick, E. Sigal, M. F. Browner,
Nature Struct. Biol. **1997**, *4*, 1003–1009.

58 G. Dannhardt, W. Kiefer, H. Eur,
J. Med. Chem. **2001**, *36*, 109–126.

59 E. Skrzypczak-Jankun, R. A. Bross,
R. T. Carroll, W. R. Dunham,
M. O. Funk, *J. Am. Chem. Soc.* **2001**,
123, 10814–10820.

60 M. Katagiri., B.N Ganguli,
I. C. Gunsalus, *J. Biol. Chem.* **1968**,
3543–3546.

61 F. M. Guengerich, *Chem. Res. Toxicol.*
2001, 611–650.

62 M. J. Coon, A. D. N. Vaz,
D. F. McGinnity, H. M. Peng, *Drug
Metabol. Dispos.* **1998**, *26*, 1190–1193.

63 P. Reichard, *Science* **1993**, *260*,
1773–1777.

64 B. M. Sjöberg, *Struct. Bond.* **1997**, *88*,
139–173.

65 P. Nordlund, B.M Sjöberg,
H.Eklund, *Nature* **1990**, *345*, 593–598.

66 J. Stubbe, *Curr. Opin. Chem. Biol.* **2003**,
7, 183–188.

67 K. E. Liu, D. Wang, B. H. Huynh,
D. E. Edmondson, A Salifoglou,
S. J. Lippard, *J. Am. Chem. Soc.* **1994**,
116, 7465–7466.

68 K. E. Liu, A. M. Valentine., D. Wang,
B. H. Huynh, D. E. Edmondson,
A. Salifoglou, S. J. Lippard, *J. Am.
Chem. Soc.* **1995**, *117*, 10174–10185.

69 A. M. Valentine, S. S. Stahl,
S. J. Lippard, *J. Am. Chem. Soc.* **1999**,
121, 3876–3887.

70 W. C. Voegtli, N. Khidekel,
J. Baldwin, B. A. Ley, J. M.Bollinger,
A. C. Rosenzweig, *J. Am. Chem. Soc.*
2000, *122*, 3255–3261.

71 J. M. Bollinger Jr., W. H. Tong.,
N. Ravi, B. H. Huynh,
D. E. Edmondson, J. Stubbe, *J. Am.
Chem. Soc.* **1994**, *116*, 8015–8023.

72 B. Kauppi, B. B. Nielsen, S. Ramaswamy,
I. Kjoller-Larsen, M. Thelander,
L. Thelander, J. Eklund, *J. Mol. Biol.*
1996, *262*, 706–720.

73 K. R. Strand, S.Karlsen,
K. K. Andersson, *J. Biol. Chem.* **2002**,
277, 34229–34238.

74 U. Rova, K Goodtzova,
R. Ingemarsson, G. Behravan,
A. Gräslund, L. Thelander,
Biochemistry, **1995**, 4267–4275.

75 M. Ekberg, M. Sahlin, M. Eriksson,
B. M. Sjöbert, *J. Biol. Chem.* **1996**, *271*,
20655–20659.

76 P. E. M. Siegbahn, M. R. A. Blomberg,
R. H. Crabtree, *Theor. Chem. Acc.* **1997**,
97, 289–300.

77 P. P. Schmidt, U. Rova, B. Katterle,
L. Thelander, A. Gräslund, *J. Biol.
Chem.* **1998**, *273*, 21463–21472.

78 M. D. Sintchak, G. Arjara,
B. A. Kellogg, J. Stubbe,
C. L. Drennan. *Nat. Struct. Biol.* **2002**,
9, 293–300.

79 M. Fontecave, E. Mulliez,
S. Ollagnier-des-Choudens, *Curr.
Opin. Chem. Biol.* **2001**, *5*, 506–511.

80 J. T. Jarrett, *Curr. Opin. Chem. Biol.*
2003, *7*, 174–172.

81 C. Rödenbeck, S. Houweling,
M. Gloor, M. Heimann, *Atm. Chem.
Phys. Discuss.* **2003**, *3*, 2575.

82 H. J. Sofia, G. Chen, B. G. Hetzler,
J. F. Reyes-Spindola, N. E. Miller,
Nucleic Acids Res. **2001**, *29*, 1097–1106.

83 J. Cheek, J. B. Broderick, *J. Am.
Chem. Soc.* **2002**, *124*, 2860–2861.

84 G. Layer, J. Moser, D. W. Heinz,
D. Jahn, W. D. Schubert, *EMBO J.*
2003, *22*, 6214–6224.

85 F. Berkovitch, Y. Nicolet, J. T. Wan,
J. T. Jarrett, C. L. Drennan, *Science*,
2004, *303*, 76–79.

86 F. Escalettes, D. Florentin,
B. Tse Sum Bui, D. Lesage,
A. Marquet, *J. Am. Chem. Soc.* **1999**,
121, 3571.

87 B. Tse Sum Bui, D. Florentin,
F. Fournier, O. Ploux, A. Mejean,
A. Marquet, *FEBS Lett.* **1998**, *440*, 226.

88 B. Tse Sum Bui, R. Benda,
V. Schunemann, D. Florentin,
A. X. Trautwein, A. Marquet,
Biochemistry **2003**, *42*, 8791–8798.

89 S. Ollagnier-de-Choudens,
E. Mulliez, S. Hewiston,
M. Fontecave, *Biochemistry* **2002**, *41*,
9145–9152.

90 S. Ollagnier-de-Choudens,
E. Mulliez, M. Fontecave, *FEBS Lett.*
2002, *532*, 465–468.

91 J. W. Whittaker, *Biochem. Soc. Trans.*
2003, *31*, 1318–1321.

92 S. Sugio, B. Y. Hiraoka, F. Yamakura,
Eur. J. Biochem. **2000**, *267*, 3487–3495.

93 Y. Yakamura, R. L. Rardin,
G. A. Petsko, D. Ringe, B. Y. Hiraoka,
K. Nakayama, T. Fujimuta, H. Taka,
K. Murayama, *Eur. J. Biochem.* **1998**,
253, 49–56.

94 A. Amano, S. Shizukuishi,
H. Tamagawa, K. Iwakura,
S. Tsunasawa, A. Tsunemitsu,
J. Bacteriol. **1990**, *172*, 1457–1463.

95 J. E. Posey, F. C. Gherardini, *Science*
2000, *288*, 1651–1653.

96 J. J. Bullen, *Rev. Infect. Dis.* **1981**, *3*,
1127–1138.

97 K. N. Ferreira, T. M. Iverson,
K. Maghlaoui, J. Barber, S. Iwata,
Science **2004**, *303*, 1831–1838.

98 M. S. Lah, M. M. Dixon,
K. A. Pattridge, W. C. Stallings,
J. A. Fee, M. L. Ludwig, *Biochemistry*
1995, *34*, 1646–1660.

99 B. L. Stoddard, P. L. Howell,
D. Ringe, G. A. Petski, *Biochemistry*
1990, *29*, 8885–8893.

100 M. L. Ludwig, A. L. Metzger,
K. A. Pattridge, W. C. Stallings,
J. Mol. Biol, **1991**, *219*, 335–358.

101 J. B. Cooper, K. McIntyre,
M. O. Badasso, S. P. Wood, Y. Zhang,
T. R. Garbe, D. Young, *J. Mol. Biol.*
1995, *246*, 531–544.

102 M. Schmidt, B. Meier, F. Parak,
J. Bioinorg. Chem. **1996**, *1*, 532–541.

103 D. J. White, N. J. Reiter, R. A. Sikkink,
L. Yu, F. Rusnak, *Biochemistry* **2001**, *40*,
8918–8929.

104 E. V. Koonin, *Protein. Sci.* **1994**, *3*,
256–358.

105 G. J. Barton, P. T. W. Cohen,
D. Barford, *Eur. J. Biochem.* **1994**, *220*,
225–237.

106 S. Zhuo, J. C. Clemens, D. J. Hakes,
D. Barford, *J. Biol. Chem.* **1993**, *268*,
17754–17761.

107 F. Rusnak, L.Yu, S. Todorovic,
P. Mertz, *Biochemistry* **1999**, *38*,
6943–5952.

108 R. Anand, P. C. Dorrestein,
C. Kinsland, T. P. Begley,
S. E. Ealick, *Biochemistry* **2002**, *41*,
7659–7669.

109 P. C. Maloney, *Curr. Opin. Cell Biol.*
1994, *6*, 571–582.

110 E. Cama, F. A. Emig, D. E. Ash,
D. W. Christianson, *Biochemistry*
2003, *42*, 7748–7758.

111 R. S. Reczkowski, D. E. Ash, *J. Am.
Chem. Soc.* **1992**, *114*, 10992–10994.

112 T. J. Bivalaqua, W. J. Hellstrom,
P. J. Kadowitz, H. C. Champion,
Biochem. Biophys. Res. Commun. **2001**,
283, 923–927.

113 N. Lane, *Oxygen*, Oxford University
Press, Oxford, **2002**, p. 132.

114 N. P. L'vov, A. N. Nosikov, *Biochemistry
(Moscow)* **2002**, *67*, 196–200.

115 J. Kim, D. C. Rees, *Nature* **1992**, *360*,
553–560.

116 C. Kisker, H. Schindelin, D. C. Rees,
Annu. Rev. Biochem. **1997**, *66*, 233–67.

117 K. V. Rajagopalan, J. L. Johnson,
J. Biol. Chem. **1992**, *267*, 10199–202.

118 B. E. Schultz, R. Hille, R. H. Holm,
J. Am. Chem. Soc. **1995**, *117*, 827–828.

119 J. M Dias., M. E. Than, A. Humm,
R. Huber, B. G. P. Bourenkov,
H. D.Bartunik, S. Bursakov,
J. Calvete, J. Caldeira, C. Carneiro,
J. J. G. Moura, I. Moura, M. Romao.
Structure **1999**, *7*, 65–79.

120 P. Arnoux, M. Sabaty, J. Alric,
B. Frangioni, B. Guigliarelli,
J. M. Adriano, D. Pignol, *Nature
Struct. Biol.* **2003**, *10*, 929–934.

121 J. C. Hilton, K. V. Rajagopalan,
Arch. Biochem. Biophys. **1996**, *325*,
139–143.

122 H. Schindelin, C. Kisker, J. Hilton, K. V. Rajogopalan, D. C. Rees, *Science* **1996**, *272*, 1615–1621.

123 F. Schneider, J. Löwe, R. Huber, H. Schindelin, C. Kisker, J. Knäblein, *J. Mol. Biol.* **1996**, *263*, 53–69.

124 A.S McAlpine, A. G. McEwan, A. L. Shaw, S. Bailey, *J. Biol. Inorg. Chem.* **1997**, *2*, 690–701.

125 A. S. McAlpine, A. G. McEwan, A. L. Shaw, S. Bailey, *J. Mol. Biol.* **1998**, *275*, 613–623.

126 G. N. George, J. Hilton, C. Temple, R. Prince, K. V. Rajagopalan, *J. Am. Chem. Soc.* **1999**, *121*, 1256–1266.

127 M. Czjzek, J. P. Dos Santos, J. Pommier, G. Giordano, V. Mejean, R. Haser *J. Mol. Biol.* **1998**, *284*, 435–447.

128 G. N. George, J. Hilton, K. V. Rajagopalan, *J. Am. Chem. Soc.* **1996**, *118*, 1113–1117.

129 C. A. Temple, G. N. George, J. C. Hilton, M. J. George, R. C. Prince, M. J. Barber, K. V. Rajagopalan, *Biochemistry*, **2000**, *39*, 4046–4052.

130 J. T. Lin, V. Steward, *Adv. Microbiol. Physiol.* **1998**, *39*, 1–30.

131 W. H. Campbell, *Plant Physiol.* **1996**, *111*, 355–361.

132 K. Okamoto, T. Eger Bryan, T. Nishino, S. Kondo, F. Pai Emil, T. Nishino. *J. Biol. Chem.* **2003**, *278*, 17, 1848–1855.

133 R. Hille, T. Nishino, *FASEB J.* **1995**, 995–1003.

134 E. Garattini, R. Mendel, M. J. Romao, R. Wright, M. Terao, *Biochem. J.* **2003**, *372*, 15–32.

135 E. Della Corte, F. Stirpe, *FEBS Lett.* **1968**, *2*, 83–94.

136 C. Enroch, B. T. Eger, K. Okamoto, T. Nishino, E. F. Pai, *Proc. Natl. Acad. Sci. U. S. A.* **2000**, *97*, 10723–10728.

137 J. C. Boyington, V. N. Gladyshev, S. V. Khangulov, T. C. Stadtman, P. D. Sun, *Science* **1997**, *275*, 1305–1308.

138 J. C. Wooton, R. E. Nicholson, J. M. Cock, D. E. Walters, J. F. Burke, W. A. Doyle, R. C. Bray, *Biochim. Biolphys. Acta.* **1991**, *1057*, 157–185.

139 Y. Kubo, N. Ogura, H. Nakagawa, *J. Biol. Chem.* **1998**, *263*, 19684–19689.

140 C. Kisker, H. Schindelin, A. Pacheco, W. A. Wehbi, R. M. Garrett, K. V.Rajagopalan, J. H. Enemark D. C. Rees, *Cell* **1997**, Vol. 91, 973–983.

141 E. Y. Jones, K. Harlos, M. J. Bottomley, R. C. Robinson, P. C. Driscoll, R. M. Edwards, J. M. Clements, T. J. Dudgeon, D. J. Stuart, *Nature* **1995**, *373*, 539–544.

142 J. C. Boyington, V. N. Gladyshev, S. V. Khangulow, T. C. Stadtman, P. D. Sun, *Science* **1997**, *275*, 1305–1308.

143 R. M. Garrett, K.V Rajagopalan, *J. Biol. Chem.* **1996**, *271*, 7387–7391.

144 G. N. George, R. M. Garrett, R. C. Prince, K. V. Rajagopalan, *J. Am. Chem. Soc.* **1996**, *118*, 8588–8592.

145 E. Garman, *Current Op. Struct. Biol.* **2003**, *13*, 545–561.

146 J. L. Johnson, S. K. Wadman, in C. R. Scriver (Ed.), *The Metabolic Basis of Inherited Disease*, 6th edition, McGraw-Hill Health Professions Division, New York, **1989**.

147 S. Mukund, M. W. W. Adams, *J. Bacteriol.* **1996**, *178*, 163–167.

148 M. K. Chan, S. Mukund, A. Kletzin, M. W. W. Adams, D. C. Rees, *Science* **1995**, *267*, 1463–1469.

149 M. K. Johnson, D. C. Rees, M. W. W. Adams, *Chem. Rev.* **1996**, *96*, 2817–2839.

150 V. Svelitchnji, C. Peschel, G. Acker, O. Meyer, *J. Bacteriol.* **2001**, *183*, 5134–5144.

151 O. Meyer, K. Frunzke, J. Tachil, J. M. Wolk, in E. I. Stiefel, D. Coucouvanis, W. E. Newton (Eds.), *Molybdenum Enzymes, Cofactors and Model Systems*, ACS Symposium Series No. 535, Washington DC, **1993**, pp. 433–459.

152 H. Dobbek, L. Gremer, O. Meyer, R. Huber, *Proc. Natl. Acad. Sci. U.S.A.* **1999**, *96*, 8884–8889.

153 U. Schübel, M. Kraut, G. Mörsdorf, *J. Bacteriol.* **1995**, *177*, 2197–2203.

154 M. J. Romao, M. Archer, I. Moura, J. J. G. Moura, J. LeGall, R. Engh, M. Schneider, P. Hof, R. Huber, *Science* **1995**, *270*, 1170–1176.

155 H. DOBBEK, L. GREMER,
R. KIEFERSAUER, R. HUBER, O. MEYER,
Proc. Natl. Acad. Sci. U. S. A. **2002**, *99*,
15971–15976.

156 T. I. DOUKOV, T. M. IVERSON,
J. SERAVALLI, S. W. RAGSDALE,
C. L. DRENNAN, *Science* **2000**, *298*,
567–572.

157 C. DARNAULT, A. VOLBEDA, E. J. KIM,
P. LEGRAND, X. VERNEDE,
P. A. LINDAHL, J. C. FONTECILLA-CAMPS,
Nat. Struct. Biol. **2003**, *10*, 271–279.

158 B. BHASKAR, E. DEMOLL, D. A. GRAHAME,
Biochemistry **1998**, *37*, 14491–14499.

159 H. DOBBEK, V. SVETLITCHNYI,
L. GREMER, R. HUBER, O. MEYER,
Science **2001**, *293*, 1281–1285.

160 C. L. DRENNAN, J. HEO,
M. D. SINTCHAK, E. SCHREITER,
P. W. LUDDEN, *Proc. Natl. Acad. Sci.*
USA **2001**, *98*, 11973–11978.

161 M. E. ANDERSON, P. A. LINDAHL,
Biochemistry **1994**, *33*, 8702–8711.

162 M. E. ANDERSON, P. A. LINDAHL,
Biochemistry **1996**, *35*, 8371–8380.

163 Z. G. HU, N. J. SPANGLER,
M. E. ANDERSON, J. Q. XIA,
P. W. LUDDEN, P. A. LINDAHL,
E. MÜNCK, *J. Am. Chem. Soc.* **1996**, *118*,
830–845.

164 H. DOBBEK, V. SVETLITCHNYI, J. LISS,
O. MEYER, *J. Am. Chem. Soc.* **2004**, *126*,
5382–5387.

165 J. CHEN, S. HUANG, D. A. GRAHAME,
J. Biol Chem. **1991**, *266*, 22227–22233.

166 J. G. FERRY, *Biofactors* **1997**, *6*, 25–35.

167 J. XIA, J. F. SINCLAIR, T. O. BALDWIN,
P. A. LINDAHL, *Biochemistry* **1996**, *35*,
1965–1971.

168 V. SVETLITCHNYI, H. DOBBEK,
W. MEYER-KLAUCKE, T. MEINS,
B. THIELE, P. ROMER, R. HUBER,
O. MEYER. *Proc. Natl. Acad. Sci. USA*
2004, *101*, 446–451.

169 P. A. LINDAHL, E. MUNCK,
S. W. RAGSDALE, *J. Biol. Chem.* **1990**,
265, 3873–3879.

170 W. SHIN, P. A. LINDAHL, *Biochemistry*
1992, *31*, 12870–12875.

171 W. SHIN, M. E. ANDERSON,
P. A. LINDAHL, *J. Am. Chem. Soc.*, **1993**,
115, 5522–5526.

172 W. SHIN, P. A. LINDAHL, *Biochem.*
Biophys. Acta **1993**, *1161*, 317–322.

173 W. K. RUSSEL, C. M. B. STAALHANSKE,
J. XIA, R. A. SCOTT, P. A. LINDAHL,
J. Am. Chem. Soc. **1998**, 120,7502–7510.

174 X. S. TAN, C. SEWELL, Q. YANG,
P. A. LINDAHL, *J. Am. Chem. Soc.* **2003**,
125, 318–319.

175 S. GENCIC, D. A. GRAHAME, *J. Biol.*
Chem. **2003**, *278*, 6101–6110.

176 J. CHEN, S. HUANG, J. SERAVALLI,
H. GUTZMAN JR., D. J. SWARTZ,
S. W. RAGSDALE, K. A. BAGLEY,
Biochemistry, **2003**, *42*, 14822–14830.

177 D. P. BARONDEAU, P. A. LINDHAL,
J. Am. Chem. Soc., **1997**, *119*, 3959–3970.

178 W. GU, S. GENCIC, S. P. CRAMER,
D. A. GRAHAME, *J. Am. Chem. Soc.*
2003, *125*, 15343–15351.

179 T. FUNK, W. GU, S. FRIEDRICH,
H.WANG, S. GENCIC, D. A. GRAHAME,
S. P. CRAMER, *J. Am. Chem. Soc.* **2004**,
126, 88–95.

180 M. R. BRAMLETT, X. TAN, P. A.LINDAHL
J. Am. Chem. Soc. **2003**, *125*,
9316–9317.

181 E. L. MAYNARD, P. A. LINDAHL, *J. Am.*
Chem. Soc. **1999**, *121*, 9221–9222.

182 J. SERAVALLI, S. W. RAGSDALE,
Biochemistry **2000**, *39*, 1274–1277.

183 D. A. GRAHAME, E. DEMOLL,
Biochemistry **1995**, *34*, 4617–4624.

184 E. L. MAYNARD, P. A. LINDAHL,
Biochemistry **2001**, *40*, 13262–13267.

185 M. W. W. ADAMS, *Biochim. Biophys. Acta*
1990, *1020*, 115–145.

186 S. P. J. ALBRACHT, *Biochim. Biophys.*
Acta **1994**, *1188*, 167–204.

187 A. VOLBEDA, M. H. CHARON,
M. C. PIRAS, E. C. HATCHIKIAN,
M. FREY, J. C. FONTECILLA-CAMPS,
Nature **1995**, *373*, 580–587.

188 A. VOLBEDA, E. GARCIN, C. PIRAS,
A. L. DE LACEY, V. M. FERNANDEZ,
E. C. HATCHIKIAN, M. FREY,
J. C. FONTECILLA-CAMPS, *J. Am. Chem.*
Soc. **1996**, *118*, 12989–12996.

189 A. L. DE LACEY, E. C. HATCHIKIAN,
A. VOLBEDA, M. FREY, J. C. FONTECILLA-
CAMPS, V. M. FERNANDEZ, *J. Am. Chem.*
Soc. **1997**, *119*, 7181–7189.

190 Y. MONTET, P. AMARA, A. VOLBEDA,
X. VERNEDE, E. C. HATCHIKIAN,
M. FIELD, M. FREY, J. C. FONTECILLA-
CAMPS, C. *Nat. Struct. Biol.* **1997**, *4*,
523–526.

191 R. Cammack, D. S. Patil,
E. C. Hatchikian, V. M. Fernandez,
Biochim. Biophys. Acta **1987**, *912*, 98–109.

192 M. Teixeira, I. Moura, A. V. Xavier,
J. J. G. Moura, J. LeGall,
D. V. Der Vartanian, H. D. Peck,
B. H. Huynh, *J. Biol. Chem.* **1989**, *264*,
16435–16450.

193 V. M. Fernandez, E. C. Hatchikian,
R. Cammack, *Biochim. Biophys. Acta*
1985, *832*, 69–79.

194 V. M. Fernandez, E. C. Hatchikian,
D. S. Patil, R. Cammack, *Biochim.
Biophys. Acta* **1986**, *883*, 145–154.

195 A. Volbeda, L. Martin, C. Cavazza,
M. Matho, B. W. Faber, W. Roseboom,
S. P. J. Albracht, E. Garcin,
M. Rousset, J. C. Fontecilla-Camps,
J. Biol. Inorg. Chem. **2005**, *10*, 239–249.

196 S. E. Lamle, S. P. J. Albrach,
F. A. Armstrong, *J. Am. Chem. Soc.*
2005, *127*, 6595–6604.

197 K. A. Bagley, E. C. Duin, W. Roseboom,
S. P. J. Albracht, W. H. Woodruff,
Biochemistry **1995**, *34*, 5527–5535.

198 S. Reissmann, E. Hochleitner,
H. Wang, A. Paschos, F. Lottspeich,
R. S. Glass, A. Böck, *Science* **2003**, *299*,
1067–1070.

199 B. L. Vallee, in I. Bertini, H. B. Gray
(Eds.), *Zinc enzymes*, Birkhauser,
Boston, **1986**.

200 R. A. McCance, C. W. Widdowson,
Biochem. J. **1942**, *36*, 692–693.

201 D. W. Christianson, *Adv. Prot. Chem.*
1991, *42*, 281–335.

202 J. E. Coleman, *Annu. Rev. Biochem.*
1992, 897–946.

203 R. J. P. Williams, *Polyhedron* **1987**, *6*,
61–69.

204 O. Kleifeld, A. Frenkel,
J. M. L. Martin, I. Sagi, *Nat. Struct.
Biol.* **2003** *10*, 98–103.

205 D. S. Gregory, A. C. R. Martin,
J. C. Cheetham, A. R. Rees, *Prot. Eng.*
1993, *6*, 29–35.

206 A. Butler, *Science* **1998**, *281*, 207–209.

207 W. Maret, *Biochemistry* **2004**, *43*,
3301–3309.

208 J. E. Huheey, E. A. Keiter, R. L. Keiter,
*Inorganic chemistry: Principles of structure
and reactivity*, 4th edition, Harper
Collins College Publishers, New York,
1993.

209 R. E. Tashian, *Bioessays* **1990**, *10*,
186–192.

210 M. R. Badger, G. D. Price, *Annu. Rev.
Plant Physiol. Plant Mol. Biol.* **1994**, *45*,
369–392.

211 A. Liljas, K. K. Kannan,
P. C. Bergstén, I. Waara,
K. Fridborg, B. Strandberg,
U. Carlbom, L. Järup, S. Lövgren,
M. Petef, *Nat. New. Biol.* **1972**, *235*,
131–137.

212 K. Hakansson, M. Carlsson,
L. A. Svensson, A. Liljas, *J. Mol. Biol.*
1992, *227*, 1192–1204.

213 S. Mitsuhashi, T. Mizushima,
E. Yamashita, M. Yamamoto,
T. Kumasaka, H. Moriyama, T. Ueki,
S. Miyachi, T. Tsukihara, *J. Biol.
Chem.* **2000**, 275,5521–5526.

214 B. Persson, M. Krook, H. Jörnvall,
Eur. J. Biochem. **1991**, *200*, 537–543.

215 H. Jörnvall, J. O. Höög, *Alcohol*, **1995**,
30, 153–161.

216 T. Inoue, M. Sunagawa, A. Mori,
C. Imai, M. Fukuda, M. Takagi,
A. Mori, *J. Bacteriol.* **1989**, *171*,
3115–3122.

217 H. Eklund, B. Norström,
E. Zeppezauer, G. Söderlund,
I. Ohlsson, T. Boiwe, *J. Mol. Biol.* **1976**,
102, 27–59.

218 H. Eklund, J. P. Samma, L. Wallen,
C. I. Brändén, A. Akeson,
T. A. Jones, *J. Mol. Biol.* **1981**, *146*,
561–587.

219 H. Eklund, C. I. Brändén, in
F. A. Jurnak, A. McPherson (Eds.),
*Biological Macromolecules and
Assemblies*, Vol. 3, John Wiley & Sons,
New York, **1987**, pp. 73–142.

220 R. Meijers, R. J. Morris,
H. W. Adolph, A. Merli, V. S. Lamzin,
E. S. Cedergren-Zeppezauer, *J. Biol.
Chem.* **2001**, *276*, 9316–9321.

221 F. Colona-Cesari, D. Perahia,
M. Karplus, H. Eklund,
C.-I. Brändén, O. Tapia, *J. Biol. Chem.*
1986, *261*, 15273–15280.

222 S. Ramaswany, *Adv. Exp. Med. Biol.*
1999, *463*, 275–284.

223 L. Esposito, F. Sica, C. A. Raia,
A. Girdano, M. Rossi, L. Mazzarella,
A. Zagari, *J. Mol. Biol.* **2002**, *318*,
463–477.

224 M. F. Dunn, J. F. Biellmann,
G. Branlant, *Biochemistry* **1975**, *14*,
3176–3182.

225 H. Eklund, B. V. Plapp, J. P. Samama,
C. I. Brändén, *J. Biol. Chem.* **1982**, *257*,
14349–14358.

226 R. T. Dworschack, B. V. Plapp,
Biochemistry **1977**, *16*, 111–116.

227 M. W. Makinen, W. Maret, M. B. Yim,
Proc. Natl. Acad. Sci. USA **1983**, *80*,
2584–2588.

228 C. A. Raia, S. D'Auria, M. Rossi,
Biocatalysis **1994**, *11*, 143–150.

229 N. V. Kaminskaia, B. Spingler,
S. J. Lippard, *J. Am Chem. Soc.* **2000**,
122, 6411–6422.

230 K. Bush, G. A. Jacoby, A. A. Medeiros,
Antimicrob. Agents Chemother. **1995**, *39*,
1211–1233.

231 S. J. Cartwright, A. F. W. Coulson,
Philos. Trans. R. Soc. London, Ser. B
1980, *289*, 370–372.

232 S. J. Cartwright, A. K. Tan,
A. L. Fink, *Biochem. J.* **1989**, *263*,
905–912.

233 N. Strynadka, M. N. G. James, *Nature*
1992, *359*, 700–705.

234 M. I. Page, A. P. Laws, *Chem. Commun.*
1998, 1609–1617.

235 R. B. Davies, E. P. Abraham, *Biochem.
J.* **1974**, *143*, 129–135.

236 B. A. Rasmussen, K. Bush, *Antimicrob.
Agents Chemother.* **1997**, *41*, 223–232.

237 T. R. Walsh, S. Gamblin, D. C. Emery,
A. P. MacGowan, P. M. Bennett,
J. Antimicrob. Chemother. **1996**, *37*,
423–431.

238 K. Bush, *Clin. Infect. Diseases*, **1998**, *27*,
S48–S53.

239 D. M. Livermore, *J. Antimicrob.
Chemother, Suppl. D*, **1998**, *41*, 25–41.

240 G. G. Hammond,
J. L. Huber, M. L. Greenlee, J. B. Laub,
K. Young, L. L. Silver, J. Balkovec,
K. D. Proyer, J. K. Wu, B. Leiting,
D. L. Pompliano, J. H. Toney, *FEMS
Microbiol. Lett.* **1999**, *179*, 289–296.

241 J. H. Toney, P. M. D. Fitzgerald,
N. Grover-Sharma, S. H. Olson,
W. J. May, J. G. Sundelof,
D. E. Vanderwall, K. A. Cleary,
S. K. Grant, J. Wu, J. W. Kozatich,
D. L. Pompliano, G. G. Hammond,
Chem. Biol. **1998**, *5*, 185–196.

242 P. M. D. Fitzgerald, J. K. Wu,
J. H. Toney, *Biochemistry* **1998**, *37*,
6791–6800.

243 J. H. Toney, P. M. D. Fitzgerald,
N. Groversharma, S. H. Olson,
W. J. May, *Chem. Biol.* **1998**, *5*,
185–196.

244 A. Carfi, S. Parès, E. Duee, M. Galleni,
C. Duez, J. M. Frère, O. Dideberg,
EMBO J. **1995**, *14*, 4914–4821.

245 A. Carfi, E. Duée, M. Galleni,
J. M. Frère, O. Dideberg, *Acta
Crystallogr.* **1998**, D54, 313–323.

246 N. O. Concha, B. A. Rasmussen,
K. Bush, O. Herzberg, *Structure* **1996**,
4, 823–836.

247 A. Carfi, E. Duée, R. Paulsoto,
M. Galleni, J. M. Frère, O. Dideberg,
Acta Crystallogr. **1998**, D54, 47–57.

248 D. E. Wilcox, *Chem. Rev.* **1996**, *96*,
2435–2458.

249 M. Backer, S. McSweeney,
H. B. Rasmussen B. W. Riise,
P. Lindley, E. Hough, *J. Mol. Biol.*
2002, *318*, 1265–1274.

250 J. R. Knox, H. W. Wyckoff, *J. Mol.
Biol.* **1973**, *74*, 533–545.

251 E. Hough, L. K. Hansen, B. Birknes,
K. Jynge, S. Hansen, A. Hordvik,
C. Little, E. Dodson, Z. Derewenda,
Nature **1989**, *338*, 357–360.

252 A. Volbeda, A. Lahm, F. Sakiyama,
D. Suck, *EMBO J.* **1991**, *10*, 1607–1618.

253 B. Stec, K. M. Holtz,
E. R. Kantrowitz, *J. Mol. Biol.* **2000**,
299, 1303–1311.

254 N. M. Hooper, *FEBS Lett.* **1994**, *354*,
1–6.

255 W. Bode, F. X. Gomis-Ruth,
W. Stockler, *FEBS Lett.* **1993**, *331*,
134–140.

256 S. G. Odintsov, I. Sabala,
M. Marcyjaniak, M. Bochtler,
J. Mol. Biol. **2004**, *335*, 775–785.

257 W. N. Lipscomb, N. Strater, *Chem.
Rev.* **1996**, *96*, 2375–2434.

258 H. M. Kim, D. R. Shin, O. J. Yoo,
H. Lee, J-O Lee, *FEBS Lett.* **2003**, *538*,
65–70.

259 E. Jabri, M. B. Carr, R. P. Hausinger,
P. A. Karplus, *Science* **1995**, *268*,
998–1004.

260 S.-H. Liaw, S.-J. Chen, T.-P. Ki,
C.-S. Hsu, C.-J. Chen, A. H.-J. Wang,

Y.-C Tsai, *J. Biol. Chem.* **2003**, *278*,
4957–4962.

261 S. M. Fabiane, M. K. Sohi, T. Wan,
D. J. Payne, J. H. Bateson,
T. Mitchell, B. J. Sutton, *Biochemistry*
1998, *37*, 12404–12411.

262 L. Chantalat, E. Duee, M. Galleni,
J. M. Frere, O. Dideberg, *Protein Sci.*
2000, *9*, 1402–1406.

263 G. M. Brown, J. M. Williamson, in
F. C. Neidhardt, J. L. Ingreham,
K. B. Low (Eds.), *Escherichia Coli and
Salmonella Typhimurium*, American
Society for Microbiology, Washington
DC, **1987**, pp. 521–538.

264 S. Kaufman, *Annu. Rev. Nutr.* **1993**, *13*,
261–286.

265 R. G. Matthews, C. W. Goulding,
Curr. Opin. Chem. Biol. **1997**, *1*, 332–339.

266 K. E. Hightower, C. A. Fierke, *Curr.
Opin. Chem. Biol.* **1999**, *3*, 176–181.

267 J. J. Wilker, S. J. Lippard, *J. Am. Chem.
Soc.* **1995**, *117*, 8682–8683.

268 W. Maret, B. L. Vallee, *Proc. Natl.
Acad. Sci. USA* **1998**, *95*, 3478–3482.

269 U. Jacob, W. Muse, M. Eser,
J. C. A. Bardwell, *Cell* **1999**, *96*,
341–352.

270 P. C. F. Graf, U. Jacob, *Cell. Mol. Life
Sci.* **2002**, *59*, 1624–1631.

12
Synthetic Models for Bioorganometallic Reaction Centers

Rachel C. Linck and Thomas B. Rauchfuss

12.1
Introduction

Interest in bioorganometallic reaction centers arises from the novelty of the active sites and the remarkable reaction mechanisms of organometallic enzymes. The field of synthetic bioorganometallic modeling may inspire new approaches to old reactions, e.g. C–H activation, carbonylation, and hydrogenation. Furthermore, the biomimetic approach may show how platinum metal catalysts can be supplanted by base metals, such as Ni and Fe [1].

It appears that transition metals are central to all biological processes involving H_2, CO, N_2, and CH_4, which are extremely important feedstocks in industry as well. The one exception, "metal-free" hydrogenases, was in fact, after several years of research, shown to contain Fe and to be inhibited by CO [2].

The interface between organometallic chemistry and enzymology is quite broad. In this review we have restricted our attention to naturally occurring processes and have, for example, excluded the interaction of xeno-organometallics (metallo-drugs, $HgMe_2$, etc.) with biological systems [3]. We do not discuss vitamin B_{12} and related cobalamin cofactors either, because this huge research theme is more mature than the systems we discuss [4]. The underlying chemistry of vitamin B_{12} is spectacular mechanistically and profound in its health implications. We also choose not to discuss related bioorganometallic themes, including ethylene sensor proteins (e.g. ETR1) [5, 6], CO-sensor proteins (e.g. CooA) [7], and nitrogenase [8, 9].

Bioorganometallics: Biomolecules, Labeling, Medicine. Edited by Gérard Jaouen
Copyright © 2006 Wiley-VCH Verlag GmbH & Co. KGaA, Weinheim
ISBN: 3-527-30990-X

12.2
Fe-only Hydrogenase

12.2.1
Overview

Hydrogenases catalyze the interconversion of protons and H_2 (Eq. 12.1).

$$2 H^+ + 2 e \rightleftharpoons H_2 \qquad\qquad (12.1)$$

They have been found in bacteria and some fungi and algae, but the Fe H_2-ase gene is known to occur widely in eukaryotes, including humans [10]. The usual biological role for the Fe-only hydrogenases (Fe H_2-ases) is the reduction of protons to H_2, although the enzyme is bidirectional. Catalytic rates are extraordinary: H_2 is produced and oxidized at 6000–9000 and 28 000 molecules per second per hydrogenase molecule (30 °C), respectively [11].

The structures of all Fe-only hydrogenases appear to be similar [12, 13]. Illustrative is the periplasmic enzyme isolated from the sulfate reducing microorganism, *Desulfovibrio desulfuricans*, a 53 kDa αβ heterodimeric protein; the larger subunit (42 kDa) contains the active site [11, 14].

12.2.2
Structure of the Active Site

The active site consists of a hexairon aggregate, called the H cluster. The H cluster features an unusual diiron center linked to an otherwise typical $Fe_4S_4(SR)_4$ cluster (Fig. 12.1). The single Fe–S bond between the diiron and tetrairon sites is the only covalent linkage between the binuclear active site and the protein, which is consistent with the binuclear active site being evolutionarily primitive. The binuclear center features an astonishing ligand set containing three CO ligands, two CN^- ligands, and a thiolate bridging from the Fe_4S_4 cluster. IR spectroscopy corroborates the presence of both the carbonyl and cyanide ligands [15, 16].

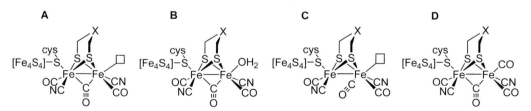

Fig. 12.1 Crystallographically characterized structures of Fe-only hydrogenases partially reduced hydrogenase of *Desulfovibrio desulfuricans* (DdH) (A), oxidized form of *Clostridium pasteurianum* hydrogenase I (CpI) (B), fully reduced form from DdH (C), and CO-inhibited form of CpI (D). Vacant or unresolved ligand sites are represented by a box. X is assumed to be NH_x, but could also be CH_2 or O [15].

The two iron atoms are bridged by a novel dithiolate cofactor wherein the two sulfur atoms are connected via a three-atom chain. This three-atom chain has been postulated to be $-CH_2NHCH_2-$ [15]. Heteroatom X (especially an amine nitrogen) is within hydrogen-bonding distance to a nearby cysteine-SH. For example, X and the S of cys 331 are separated in DdH by only 3.1 Å, a distance that is appropriate for a hydrogen bond [15]. An azadithiolate cofactor could participate in H_2 formation (Eq. 12.2).

$$(12.2)$$

It is agreed that the diiron center exists in two functional oxidation states, H_{red} and H_{ox}, as well as a more oxidized, inactive "as isolated" state, H_{ox}^{air}. The presence of multiple CO and cyanide ligands virtually ensures low-spin configurations. Mössbauer analysis has confirmed that the $[Fe_4S_4]_H$ cluster remains in the conventional 2+ oxidation state [17], thus the redox-induced changes in the EPR and Mössbauer spectra reflect on the diiron site. Mössbauer analysis of the diamagnetic H_{red} state of the CpI enzyme indicates either a $[Fe(I)]_2$ or $[Fe(II)]_2$ state [17]. Oxidation of H_{red} to the paramagnetic H_{ox} state requires one electron, implying that H_{ox} should adopt the mixed-valence $S = 1/2$ state, consistent with the EPR spectrum [18]. $H_{ox\text{-}CO}$ is also an $S = 1/2$ system. Although Mössbauer and IR data were originally interpreted as supporting the assignment $H_{red} = [Fe(II)]_2$, IR data are more consistent with the presence of $[Fe(I)]_2$ [19, 20]. The Mössbauer data indicate that in H_{ox}, the Fe(II) is proximal to the [4Fe–4S] cluster. This conclusion is consistent with IR measurements that indicate that the CO inhibition occurs at a labile site on the distal Fe(I) center [21]. If H_{red} is $[Fe(I)]_2$ and H_{ox} is $\{[Fe(I)][Fe(II)]\}$, then H_{ox}^{air} is $[Fe(II)]_2$, consistent with its diamagnetism.

The main mechanistic clue obtained from the crystallographic characterization of the diiron center in Fe H_2-ase is the variable coordination at one site on the distal Fe. The H_{red} form of the DdH enzyme features an apparently vacant coordination site on the distal Fe center. This site is widely assumed to be the location of H_2 formation and activation. In the structural study on a crystal of a more oxidized sample of the CpI enzyme [15, 22], this site is occupied by a water molecule, which can be displaced by CO, as confirmed crystallographically [23]. The enzymes feature a hydrophobic tunnel extending from the vicinity of this binding site to the surface of the protein; this tunnel would permit transport of H_2 (and CO) [14,24]. ^{13}CO-labeling experiments in conjunction with FT-IR measurements show that the exogenous and endogenous CO ligands do not exchange [19].

12.2.3
Proposed Mechanism

Several lines of evidence point to substrate (H^+ and H_2) binding to a terminal (vs. bridging) site on the Fe center distal from the Fe_4 cluster: (1) CO inhibition involves binding at this site, (2) in the H_{red} and H_{ox} states, this site is either vacant or occupied by exchangeable ligands, (3) the central atom of the dithiolate cofactor, which is thought to have a functional role, is tilted toward this site, and (4) the hydrophobic, H_2-conveying tunnel terminates at this Fe site.

A catalytic cycle begins with 1e reduction of H_{ox}, in which the proximal iron becomes Fe(I) (Eq. 12.3). Reduction shifts the bridging CO ligand to a terminal position on the distal Fe.

(12.3)

$$\mathbf{H_{ox}} \qquad\qquad\qquad \mathbf{H_{red}}$$

The H_{red} form is then protonated, and subsequently or concomitantly with protonation, an electron is delivered, probably conveyed via the Fe_4S_4 center. This second reducing equivalent enables a second protonation, again at the distal Fe, thereby forming a H_2 ligand. Some metal hydrido complexes are known to protonate to give H_2 complexes [25]. It is suggested that the azadithiolato cofactor participates in the delivery of protons to, or the removal of protons from, the metal in the case of H_2 production and H_2 oxidation, respectively [26]. Mechanistic proposals, however, must be viewed with skepticism pending further spectroscopic characterization of enzyme-H_x complexes ($x = 1, 2$), or similar studies on models bearing terminal hydrido ligands.

12.2.4
Fundamental Chemistry of $Fe_2(SR)_2(CO)_6$, Including Cyano Derivatives

The dinuclear active site of the Fe-H_2ase is reminiscent of $Fe_2(SR)_2(CO)_6$ species, which were discovered in the 1920s [27, 28] and well developed in the 1960–1980s [29–31]. Attesting to their considerable stability, these hexacarbonyls arise via many pathways, but commonly synthesis entails the reactions of iron carbonyls with the corresponding thiol or disulfide.

Addition of two equivalents of CN^- to $Fe_2(SR)_2(CO)_6$ efficiently yields $[Fe_2(SR)_2(CN)_2(CO)_4]^{2-}$ for a wide range of substituents R, including both chelating and simple thiolates (Fig. 12.2) [32–34]. The *di*cyanide forms particularly

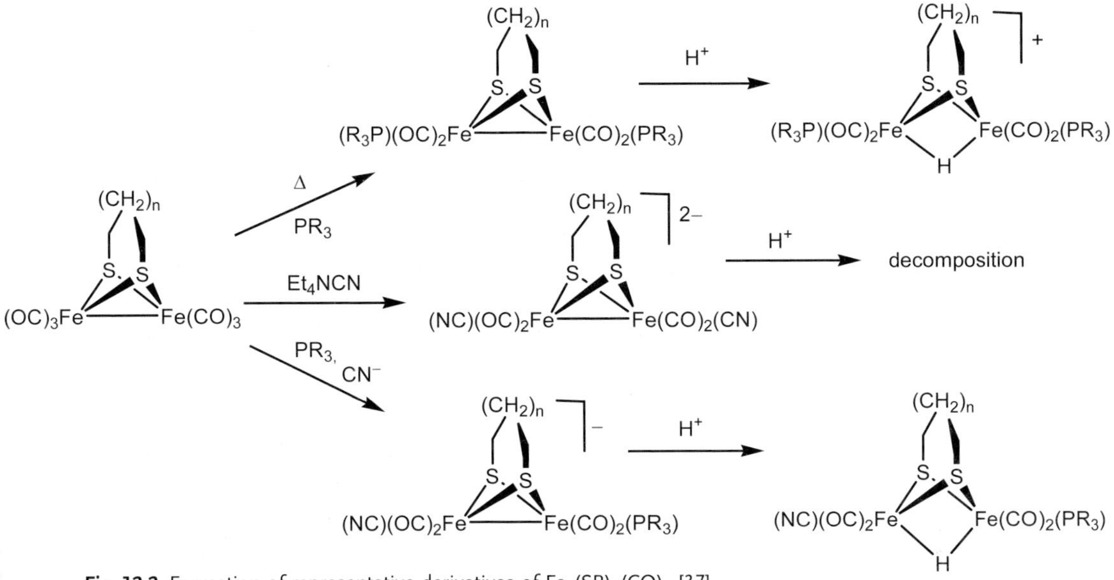

Fig. 12.2 Formation of representative derivatives of $Fe_2(SR)_2(CO)_6$ [37].

rapidly and cleanly, even when the initial $[CN^-]/[Fe_2(SR)_2(CO)_6]$ ratio is less than two [35]. The CN^--for-CO substitution reaction follows associative kinetics, as for other CO-substitution reactions of $Fe_2(SR)_2(CO)_6$ [36]. The monocyanide $[Fe_2(SR)_2(CN)(CO)_5]^-$ reacts more slowly with Et_4NCN than does the corresponding hexacarbonyl starting material, as expected (Fig. 12.2) [20].

The ability of cyanide to induce unusual substitution pathways was demonstrated in studies on $Fe_2[(SCH_2)_2C(Me)CH_2SMe](CO)_5$, which bears a pendant thioether ligand. Treatment of this pentacarbonyl with 2 equiv of CN^- yields the *dicyano-tetra*carbonyl-thioether complex. This 36e dicyanide is noteworthy because it contains a bridging CO ligand (Eq. 12.4), and decomposes at temperatures above $0\,°C$ via dissociation of the thioether concomitant with reformation of the Fe–Fe bond [38, 39]. The novel reactivity of these diiron compounds is further manifested by the finding that Et_4NCN induces a reaction of $Fe_2(S_2C_3H_6)(CO)_6$ and PMe_3 to give $Et_4N[Fe_2(S_2C_3H_6)(CN)(CO)_4(PMe_3)]$ [20]. Diverse phosphine-substituted derivatives, e.g. $Fe_2(S_2C_3H_6)(CO)_4(PMe_3)_2$, are also well known [40].

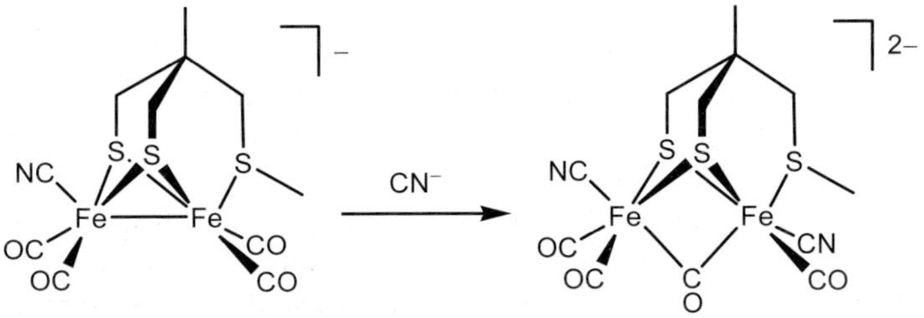

(12.4)

12.2.5
Modeling the Azadithiolate Cofactor

The dithiolate cofactor has been proposed to be SCH_2NHCH_2S [15], although protein crystallography cannot distinguish between N, O, and C. 1,3-Propane-dithiol is unknown biologically, so despite its routine chemical character, this species is no more plausible as the cofactor than the hetero-substituted derivatives.

Azadithiolato diiron species can be synthesized by treatment of $Li_2[Fe_2(S)_2(CO)_6]$ with bis(chloromethyl)amines. A range of alkyl and aryl substituents can be introduced by this method, including thioethers and alkenes that can coordinate to the Fe center [41–43]. A milder route to the azadithiolato diiron complexes entails the reaction of CH_2O, RHN_2, and $Fe_2(SH)_2(CO)_6$, a reaction that is proposed to proceed via the intermediacy of $Fe_2(SCH_2OH)_2(CO)_6$. Other iron reagents including $Fe_3S_2(CO)_9$ and even freshly generated FeS (under an atmosphere of CO) are competent precursors, although the latter is far less so than the preformed metal carbonyls (Fig. 12.3) [37].

Structural and acid-base studies on the diiron azadithiolato complexes show that the basicity of the amine is diminished versus Et_3N in the same solvent [37, 42]. Theoretical calculations suggest that the reduced basicity is arises from an electronic interaction between the nitrogen lone pair and the C–S σ* orbital (Fig. 12.4). The conjugate acid of this weakly basic amine would be capable of protonating a weakly basic iron hydride to generate hydrogen.

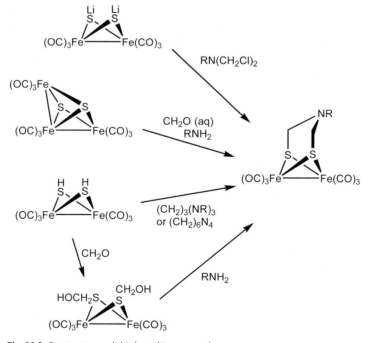

Fig. 12.3 Routes to azadithiolato diiron complexes.

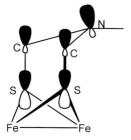

Fig. 12.4 Orbital interactions proposed for the azadithiolate cofactor [42].

It is well known from classical organometallic chemistry that non-coordinating amine bases affect the behavior of H_2 complexes [44]. The azadithiolate motif has been exploited in the design of new catalysts for hydrogen oxidation, as described in the next section on [NiFe] hydrogenases [45].

12.2.6
Reactions of Model Complexes with H_2

Disubstituted complexes of the type $Fe_2(SR)_2(CO)_4L_2$ readily protonate at the Fe–Fe bond (Fig. 12.3) [46, 47]. In contrast, however, to the prevalent models $[Fe_2(\mu\text{-}H)(SR)_2(CO)_4L_2]^+$, the hydrido ligand in the functioning enzyme is assumed to be terminal (see Eq. 12.2). Nonetheless, the photochemically induced H/D exchange reaction of $[HFe_2(SR)_2(CO)_4L_2]^+$ with D_2 is intriguing because, for the first time, $Fe_2(SR)_2\text{-}H_2$ intermediates are implicated, although none were observed. The 36e species $[Fe_2(SMe)_3(CO)_4(PMe_3)_2]^+$, also photocatalyzes exchange between D_2 and H_2O, but, in contrast to $[HFe_2(SR)_2(CO)_4L_2]^+$, not between D_2 and H_2 [48]. Hydrogenase enzymes are characteristically assayed by monitoring H/D exchange between solvent water and D_2 [13].

12.2.7
H_2 Production Catalysts

The overall process of generating H_2 from protons effectively entails the intermediate reduction of one proton into a hydridic entity, which then sustains protonation, i.e. $H^+ + H^- \rightarrow H_2$. The conversion of a proton to such a hydridic entity begins with protonation of an electron-rich metal center concomitant with transfer of charge from the metal. Often the resulting metal hydride reacts spontaneously with a further equivalent of H^+ to give H_2, but more commonly the metal hydrido intermediate requires further reduction to enable protonation. Protonation of $[Fe_2(S_2C_3H_6)(CN)_2(CO)_4]^{2-}$ produces modest amounts of H_2 [32], which suggests that an initially formed hydride is in fact sufficiently reducing to undergo further protonation. Unfortunately, the oxidized iron complex generated by protonation of $[Fe_2(S_2C_3H_6)(CN)_2(CO)_4]^{2-}$ is insoluble and cannot be re-reduced [32]. The production of H_2, however, demonstrates the highly electron-rich nature of these di-

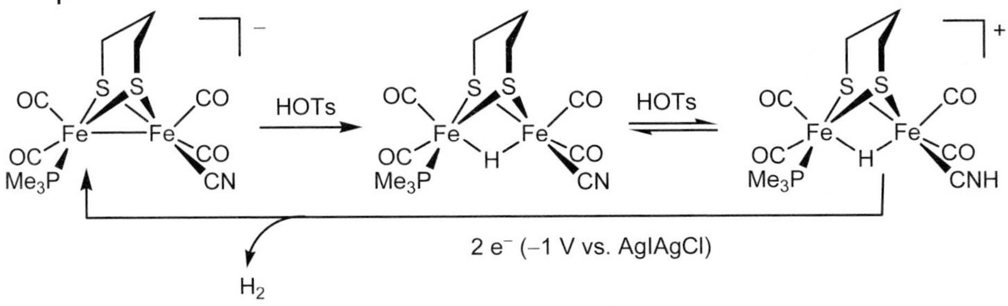

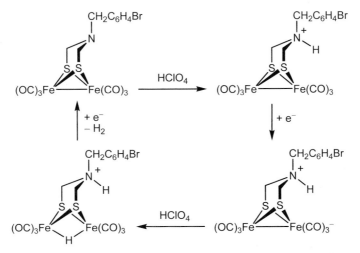

Fig. 12.5 Steps proposed for the biomimetic proton reduction pathway [49].

anionic dicyanides. The mixed ligand complex $[Fe_2(S_2C_3H_6)(CN)(CO)_4(PMe_3)]^-$ has, gratifyingly, proven to be a robust catalyst for the sought-after biomimetic reduction of protons to H_2 [46, 49]. These reactions are typically conducted in MeCN solution using HOTs ($pK_a = 8.0$ in MeCN soln). The cycle begins with protonation at the Fe–Fe bond to produce the hydride $HFe_2(S_2C_3H_6)(CN)(CO)_4(PMe_3)$ (Fig. 12.5), which has been fully characterized. Further equivalents of acid protonate at the cyanide, giving the cation $[HFe_2(S_2C_3H_6)(CNH)(CO)_4(PMe_3)]^+$ [46, 49].

At ≈ -1 V (vs. Ag|AgCl), this cation reduces protons to H_2 with perfect Faradaic efficiency. The corresponding bis-phosphine complex, $Fe_2(SR)_2(PMe_3)_2(CO)_4$ also exhibits good catalytic activity in the presence of excess protons [46]. Cyclic voltammetric experiments indicate that the diphosphine derivatives are less catalytically active than the phosphine–cyanide complex [46, 50]. This difference is possibly attributable to the ability of the CN^- to convey a proton to the Fe–H–Fe group.

The azadithiolate-containing hexacarbonyl $Fe_2[(SCH_2)_2NCH_2C_6H_4Br)](CO)_6$ also catalyzes hydrogen evolution [43]. Due to the presence of six CO ligands, this complex is more easily reduced than the substituted derivatives, operating at 0.9 V vs. Ag|AgCl (Fig. 12.6).

Fig. 12.6 Proposed cycle for H_2 production catalyzed by azadithiolate-modified diiron hexacarbonyl.

After protonation of the pendant amine, catalysis involves reduction of the diiron unit, which then sustains a second protonation to give an iron hydride, which couples, presumably intramolecularly, with the ammonium center to liberate H_2.

12.2.8
Modeling Diferrous Dithiolates

High-spin binuclear ferrous thiolates, e.g. $[Fe_2(SEt)_6]^{2-}$, are unlikely to display reactivity relevant to hydrogenase, i.e. coordination of CO, H_2, H^+ [51, 52]. The hydrides $[HFe_2(S_2C_3H_6)(CO)_4(PR_3)_2]^+$ are formally diferrous, but as mentioned above, the hydride occupies a bridging rather than the biologically relevant terminal site [53]. These considerations have led to new approaches to diferrous dithiolates. One-electron oxidation of the pendant thioether species $[Fe_2[(SCH_2)_2C(Me)-CH_2SMe](CN)_2(CO)_4]^{2-}$ gives an Fe(I)–Fe(II) derivative that has EPR and IR signatures similar to those of the CO-inhibited, oxidized enzyme (Eq. 12.5) [54]. This mixed-valence species, however, decomposes rapidly at ambient temperatures.

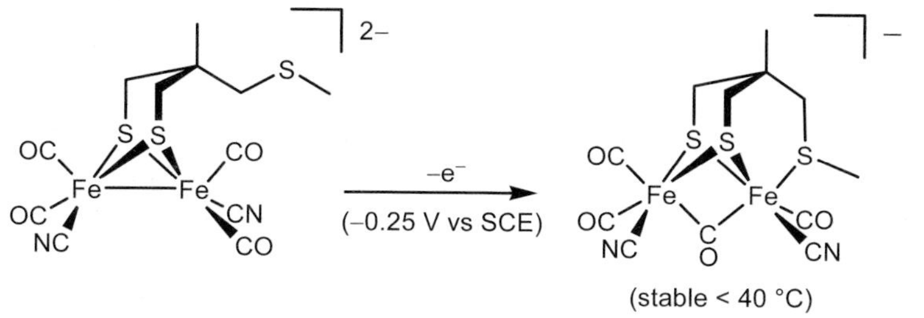

$$(12.5)$$

(stable < 40 °C)

As previously stated, oxidation of $[Fe_2(SR)_2(CN)_2(CO)_4]^{2-}$ is irreversible, producing insoluble solids thought to be CN bridged polymers. Stable diferrous dithiolates, however, result upon oxidation of the polysubstituted derivatives $Fe_2(S_2C_nH_{2n})(CO)_{6-x}L_x$, where $x \geq 2$ and L = RNC, PR_3, and CN^-. These oxidations are conducted in the presence of additional trapping ligands such as RNC, PR_3, and CN^-. Illustrative diferrous carbonyls prepared in this way are $[Fe_2(S_2C_nH_{2n})(\mu-CO)(CO)_2(CN)_4]^{2-}$ and $[Fe_2(S_2C_2H_4)(\mu-CO)(CNMe)_6]^{2+}$. The presence of the μ-CO ligand is characteristic of these diferrous species as well as the H_{ox}^{air} state (Fig. 12.7) [55, 56].

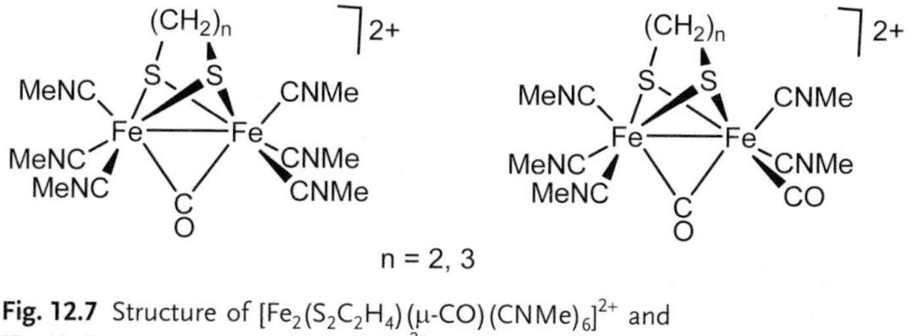

n = 2, 3

Fig. 12.7 Structure of $[Fe_2(S_2C_2H_4)(\mu-CO)(CNMe)_6]^{2+}$ and $[Fe_2(S_2C_nH_{2n})(\mu-CO)(CO)(CNMe)_5]^{2+}$ [56].

12.3
NiFe Hydrogenases

12.3.1
Overview

The NiFe hydrogenases ([NiFe] H_2-ases) are the best known hydrogenases, occurring in many bacteria and archaea. For example, the *E. coli* genome codes for four separate [NiFe] H_2-ases [11]. Most often, these enzymes are involved in the oxidation of H_2 to give protons and reducing equivalents. The organism responsible for stomach ulcers, *H. pylori*, relies on an [NiFe] H_2-ase for energy production using H_2 available in the gut [57]. The rates for both H_2 oxidation and proton reduction are quoted at 700 per enzyme per second (30°C), about 10–100 less active than the Fe H_2-ases. The affinities of these enzymes for H_2 is higher than the Fe H_2-ases [11]. [NiFe]-H_2ase absorbed on electrodes catalyzes the oxidation of H_2 at rates at least comparable to Pt metal but is less sensitive to poisoning by CO [58]. [NiFe]-based proteins also serve as H_2-sensors, whereby they induce transcription of [NiFe]-H_2-ases.

12.3.2
Structure of the Active Site

A decisive development in bioorganometallic chemistry was the determination of the structure of *D. gigas* [NiFe] H_2-ase at 2.54 Å in conjunction with IR studies that indicated the presence of CN⁻ and CO ligands [59, 60]. The protein, which has been subsequently heavily studied [61], is a 88 kDa heterodimer, with the larger 60 kDa subunit bearing the [NiFe] active site. Most crystallographic studies have examined the air-deactivated forms, probably a mixture of the A- and B-state (*vide infra*) wherein nickel, assumed to be Ni(III), resides in an irregular coordination environment bound to four cysteine thiolates as well as a hydroxide. Two of these thiolates and the hydroxide bridge to an octahedrally-coordinated iron atom, which is further ligated by two cyanide ligands and one carbonyl ligand [60, 62]. Crystallographic characterization of a [NiFeSe] H_2-ase, wherein one of the two terminal cysteine residues is replaced by selenocysteine, revealed the absence of the bridging oxygen ligand and a corresponding contraction of the Ni $\cdots$ Fe distance from 2.9 Å in the aforementioned denaturized form, to 2.5 Å [63, 64]. Like other enzymes that process gaseous substrates, the [NiFe] H_2-ases feature tunnels that convey gaseous substrates from the exterior of the protein to the active site [65].

The biologically unusual CN⁻ ligands are derived from cysteinyl thiocyanate, which arises from the reaction of the thiol with carbamoylphosphate $(H_2NC(O)OPO_3^{2-})$ followed by dehydration of the S-cysteinylthiocarbamate [66]. Alkylthiocyanates react with anionic metal centers to give cyanides. The carbamoylphosphate pathway is conceivably applicable to the formation of CO ligands via the deamination of Fe-C(O)NH$_2$ species [67]. It is unknown if this biosynthetic pathway is relevant to the formation of cyanide and CO ligands in the Fe H_2ases.

Mutagenesis has been used to explore the importance of hydrogen-bonding of nearby serine and arginine residues to the CN^- ligands. Replacement of the serine residue by alanine has been accomplished, but substitution of the arginine, which would be expected to more strongly hydrogen-bond to cyanide, prevents expression, indicating that it critically stabilizes the active site [68].

12.3.3
Proposed Mechanism

The [NiFe] H_2-ases can exist in as many as seven states, distinguished by their EPR and IR spectra as well as their relative reactivity towards H_2 (Fig. 12.8) [61].

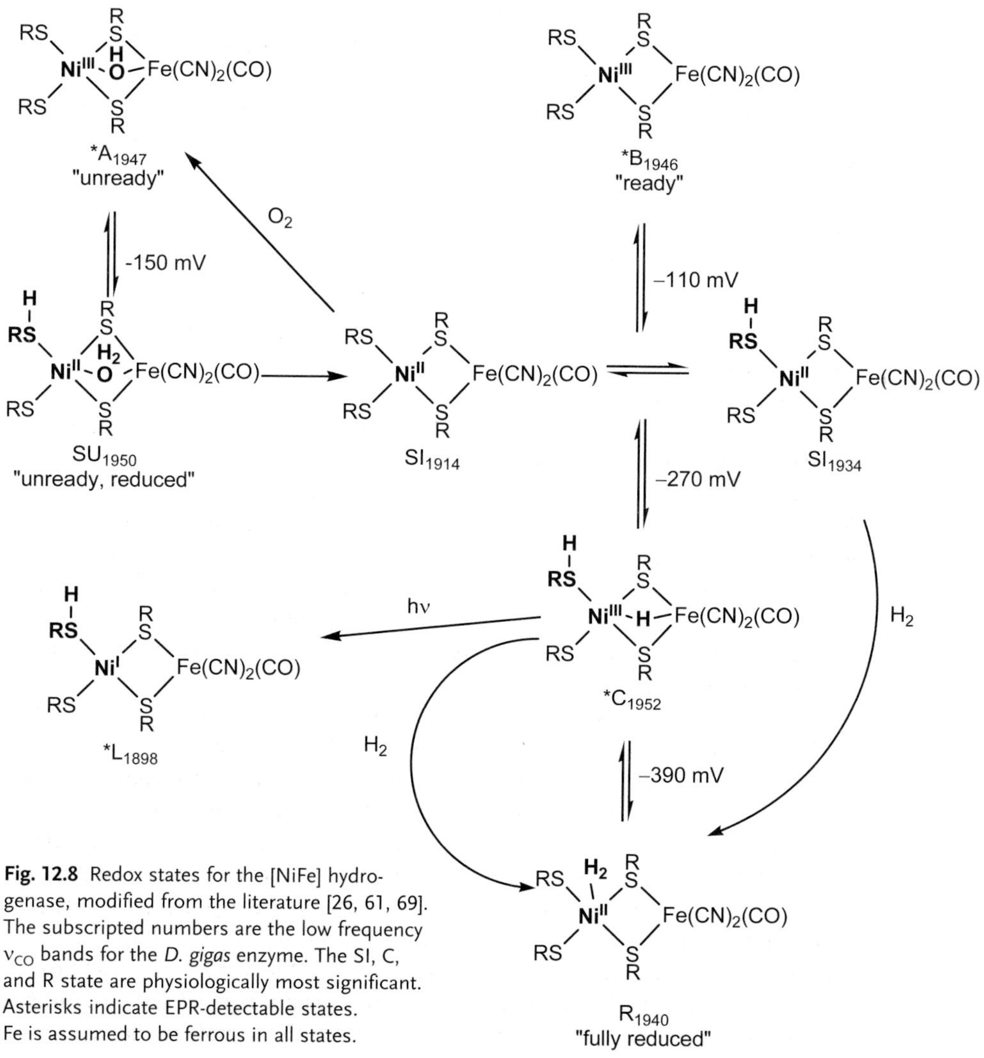

Fig. 12.8 Redox states for the [NiFe] hydrogenase, modified from the literature [26, 61, 69]. The subscripted numbers are the low frequency v_{CO} bands for the *D. gigas* enzyme. The SI, C, and R state are physiologically most significant. Asterisks indicate EPR-detectable states. Fe is assumed to be ferrous in all states.

These states differ with regards to the presence of aerobically-derived μ-OH ligands, protonation at a terminal thiolato ligand, the redox state of the Ni center, and the presence of hydride ligands bridging Ni and Fe. In their IR spectra, v_{CO} varies from 1914 to 1952 cm^{-1}, the modest range of which indicates that most of the changes occur at Ni since the CO is bound to Fe. The most dramatic spectroscopic changes are caused by photolysis of the Ni–C state to generate the Ni–L state, which is proposed to involve deprotonation of Ni(III) hydride to generate a Ni(I) intermediate [69].

Enzyme activation begins with reduction of the Ni(III), the defining feature of the Ni–A and Ni–B states [70]. Crystallographic analysis of the CO-inhibited H$_2$-reduced form of [NiFe] H$_2$-ase reveals that CO has added to Ni, consistent with EPR measurements [71]. This result points to Ni as a direct participant in the hydrogen activation. Terminal thiolato ligands are typically basic [72], thus a transiently formed H$_2$ ligand could reasonably be expected to transfer a proton to a coordinated cysteinyl thiolate, as has been demonstrated in model systems [73]. It has, however, proven difficult to probe protonation at thiolate in both models and especially in the protein. It is clear that H$_2$ activation to give the Ni–C state produces a Ni–H–Fe substructure (vs. Ni–OH$_x$–Fe substructures in inactive states). ENDOR–ESEEM (HYSCORE) studies of the [NiFe]. H$_2$-ase from *Ralstonia eutropha*, which conveniently features Fe–S clusters that do not spectroscopically interfere with the detection of NiFe center, identified the Ni–*H*–Fe signal, which disappears after exposure to ^{2}H$_2$O, and upon activation with ^{2}H$_2$, a corresponding signal appears in the ^{2}H–M region [74]. The signal also disappears reversibly upon photolysis to give the Ni–L form [69]. The iron site, [Fe(SR)$_2$(CN)$_2$(CO)]^z, is expected to be electronically unsaturated (< 18e configuration) [26], and this aspect stabilizes the formation of bridging hydride and hydroxide ligands.

12.3.4
Models for the Fe Site

A sophisticated model of the Fe site of the [NiFe] H$_2$-ase enzyme is provided by the anion ["S3"Fe(CN)$_2$(CO)]$^{2-}$, where "S3" is a facially-chelating thioether-dithiolate [75]. The Fe(CN)$_2$(CO) fragment can also be stabilized with facial triamine ligand and cyclopentadienyl ligands to produce even more viable spectroscopic models, e.g. [*fac*-L$_3$Fe(CN)$_2$(CO)]^z [76, 77]. The IR spectrum of the Cp derivative in the 1900–2100 cm^{-1} region closely resembles that for the aerobically deactivated enzyme. The authors conclude "that the Ni(μ-SCys)$_2$(μ-OH) donor set of Ni–A or Ni–B has the same electronic effect on the IR spectroscopic properties of the pyramidal Fe(CO)(CN)$_2$ fragment as ...C$_5$H$_5^-$" [77]. Five-coordinate Fe(II)(CN)$_2$(CO) centers can be stabilized with π-donor ligands (Eq. 12.6) [78–80].

$$\text{(12.6)}$$

Coordinative unsaturation of such species is indicated by their ability to bind CO, although H_2 activation was not observed.

12.3.5
Structural Models for the Bimetallic Site

Despite the availability of high-resolution crystal structures of the [NiFe] H_2-ase for several years, progress toward accurate bimetallic models has proven difficult. The nickel subsite presents a distinct challenge as *tetrahedral* $Ni(SR)_4$ species are rare [81, 82]. Particular emphasis has been placed on species bearing the minimal coordination set $(XY)Fe(\mu\text{-}SR)_2Ni$ where XY is NO, CO, or CN (Fig. 12.9). A remarkable model arises via the reaction of a nickel dithiolate and a dithiolate-iron carbonyl, resulting in an $Ni(SR)_4$ site bound to the $Fe(SR_2)(CO)(PMe_3)_2$ fragment. Although the extra thioether is undesired and the Ni site is square planar, the $(CO)(PMe_3)_2$ set reasonably approximates the $(CO)(CN)_2^{2-}$ set (both CN^- and PMe_3 are strong sigma donor ligands). The position of ν_{CO} for this Ni(II)–Fe(II) model resembles that for the most reduced "Ni–R" state of the enzyme, implicating an Ni(II)–Fe(II) center [83]. The reaction of $NiCl_2(Ph_2PC_2H_4PPh_2)$ with the anion $[Fe(NS_3)CO]^-$ $(NS_3 = N(CH_2CH_2S^-)_3)$ under a CO atmosphere has also afforded a sophisticated structural model featuring a $(amine)(RS)(OC)_2Fe(\mu\text{-}SR)_2NiCl(PR_3)_2$ site [84].

The aforementioned work highlights the challenges associated with controlling the coordination behavior at Fe, both with regard to its penta-coordination and the presence of both CO and CN^- ligands. Numerous $[Fe(CN)_{6-x}(CO)_x]^{n-}$ complexes have been prepared in recent years [85–87]; this line of research could provide sources of the $\{Fe(CO)(CN)_2\}$ fragment. Furthermore, the spontaneous assembly of such Fe–CO–CN species is potentially relevant to the biosynthesis of the [NiFe]

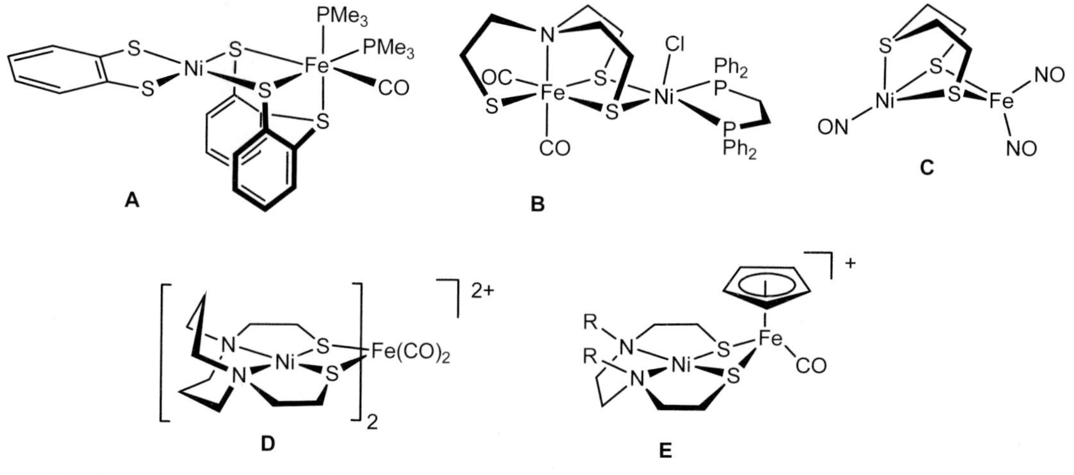

Fig. 12.9 Synthetic analogs for the [NiFe] H_2-ase active site, focusing on $(OC)Fe(\mu\text{-}SR)_2Ni$ systems. Compounds C, D, E are not discussed in the text [88–90].

H$_2$-ases. From the perspective of preparing functional models, it is unclear if these diatomic ligands play any direct catalytic role, thus their electronic influence might be adequately simulated with a facial tridentate ligand.

Arguably the most advanced structural model for the active site is the species [Fe(NS$_3$)CO]$_2$Ni (Fig. 12.10), despite its trimetallic nature. Each Fe center is five-coordinate, bearing CO (v_{CO} = 1933 cm^{-1}) and bridged via a pair of thiolato ligands to a tetrahedral Ni(SR)$_4$ center [91].

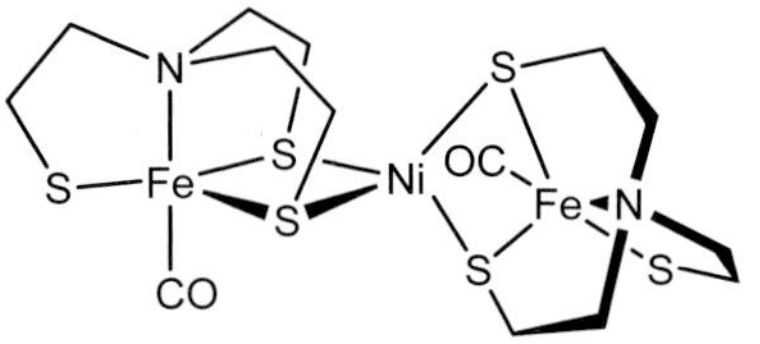

Fig. 12.10 Synthetic analog for the [NiFe] H$_2$-ase active site [91].

Numerous models are based on the use of Ni(SR)$_2$L$_2$ (L = amine, phosphine, thioether) as metalloligands, which can be attached to iron components (Fig. 12.9, D and E). In this context, the Ni(SR)$_2$L$_2$ motif is utilized as a mimic for the [Ni(SR)$_4$]$^{2-}$ site observed in the enzyme. The main deficiency with this approach is that the nickel center in the enzyme adopts a distinctly non-planar geometry, whereas these Ni(SR)$_2$L$_2$ building blocks are exclusively square planar. A rare family of NiS$_2$N$_2$-derived bimetallic complexes that react with small molecules are of the type [(C$_5$Me$_5$)Ru(NiS$_2$N$_2$)]$^+$, where S$_2$N$_2$ is a tetradentate diamino-dithiolate ligand. These complexes form adducts with CO and MeCN and undergo aerobic "deactivation" via the irreversible binding of O$_2$ and S$_2$ (Fig. 12.11) [92].

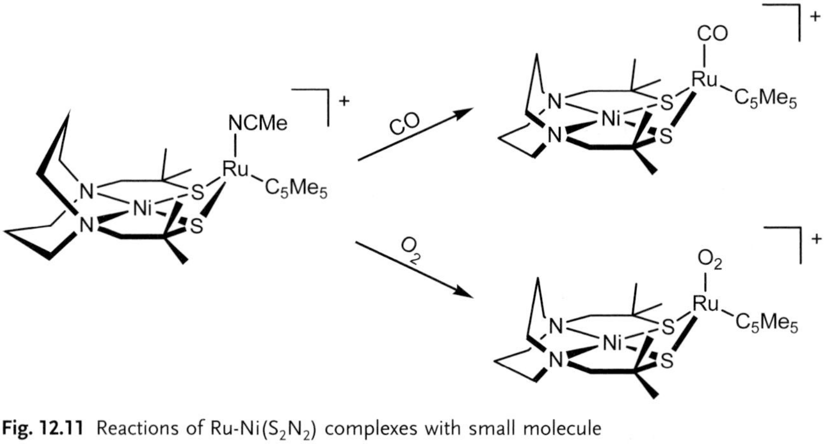

Fig. 12.11 Reactions of Ru-Ni(S$_2$N$_2$) complexes with small molecule poisons of NiFe hydrogenase.

12.3.6
Functional Models

No Ni–Fe complex has been reported to catalyze the redox interconversion of H_2 and protons. Relevant reactivity, however, has been achieved with mononuclear Ni species. For example, a nickel(II) dithiolate iminophosphine complex undergoes H–D exchange upon exposure to D_2 [93]. Particularly impressive is the ability of nickel(II) complexes of aminodiphosphines to catalyze the oxidation of H_2. Catalysis requires the presence of an intramolecular amine base that deprotonates an incipiently formed H_2 complex (Eq. 12.7). Thus catalysis is observed for $[Ni(Et_2PCH_2NMeCH_2PEt_2)_2]^{2+}$ but not the related $[Ni(Et_2PCH_2CH_2CH_2PEt_2)_2]^{2+}$, which lacks the internal amine base [45].

$$(12.7)$$

Treatment of the amino-phosphine complex with H_2 generates a hydrido nickel complex with a pendant ammonium substituent. Catalysis, which was established electrochemically, is proposed to involve oxidation of the Ni(II) hydride to a Ni(III) hydride, a process which enhances the acidity of the hydrido ligand sufficiently to allow its deprotonation by the pendant amine (Fig. 12.12). This proton transfer gives an easily oxidized Ni(I) intermediate. Catalysis of the H_2 oxidation by mononuclear nickel complexes foreshadows the preparation of related bimetallic species exhibiting hydrogenase reactivity.

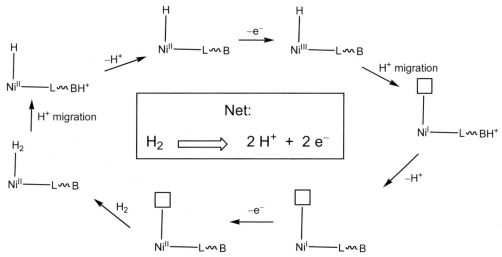

Fig. 12.12 Schematic representation for the H_2 oxidation cycle catalyzed by $[Ni(PNP)_2]^{2+}$ complexes, see also Eq. (12.7) [45].

12.4
Acetyl CoA Synthase

12.4.1
Overview

Central to acetogenesis [94], acetyl CoA synthase (ACS) effectively catalyzes the formation of acetyl CoA from CO and a cobalamin-like methyl carrier called the corrinoid iron sulfur protein (CoFeSP) (Eq. 12.8).

$$CO + CH_3 + CoFeSP + CoSH \xrightarrow{ACS} CH_3C(O)\text{-}SCoA + CoFeSP + H^+ \quad (12.8)$$

CoFeSP obtains its methyl groups from an N-methylated heterocycle, the identity of which differs for bacteria and archaea [95].

ACS-CODH usually exists as an $\alpha_2\beta_2$ heterotetramer of 310 kDa. The A clusters, the site of the ACS activity, occupy the α subunits (82 kDa each) and the C clusters, site of the CODH activity, occupy the β subunits (73 kDa) [96]. Three [4Fe–4S] clusters also are located in the β subunits. The α and β subunits are interconnected by a hydrophobic channel that allows transport of CO between all four catalytic clusters. While ACS is usually found as part of the bifunctional enzyme ACS-CODH, it can be obtained as a monomer [96].

Related to ACS-CODH, acetyl CoA decarbonylase/synthase (ACDS) occurs in methanogenic archaea (e.g. *Methanosarcina sp.*), which extract energy from the decomposition of acetate into CH_4 and CO_2. The A and C clusters in ACDS closely resemble those in ACS-CODH, but the ACDS protein is structurally more complex with an $(\alpha_2\epsilon_2)_4\beta_8(\gamma\delta)_8$ organization, the A cluster being located in the β subunits [97, 98]. The importance of the metabolic pathway supported by ACDS is indicated by the fact that in some methanogens, ACDS accounts for 25% of the soluble protein content.

12.4.2
Structure of the Active Site

The active site (Fig. 12.13) consists of a square planar (distal) nickel center linked via a pair of bridging thiolates to proximal nickel that in turn is bound to one thiolate bridged to an otherwise typical $Fe_4S_4(SR)_4$ cubane. The distal Ni is accommodated in a tetra-anionic diamido-dithiolato site provided by a cys–gly–cys sequence. The complexation of metals through deprotonated (peptidyl) amide is unusual but precedented in the nitrile hydratases [99]. Although crystallographic analysis originally suggested that the proximal metal was Cu [100], it has been demonstrated that this site is occupied by Ni in the active form [101]. The proximal metal is labile (e.g. removable by 1,10-phenanthroline), thus incorporation of adventitious Cu^+ and Zn^{2+} occurs readily [102]. The protein superstructure and the geometry at this metal center may depend on the metal that occupies the proximal site [101].

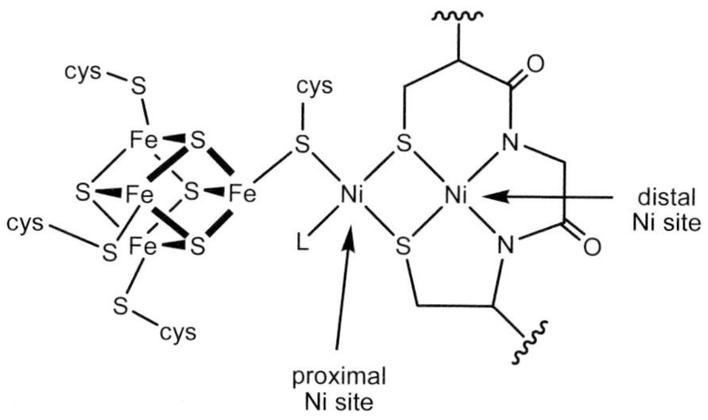

Fig. 12.13 Structure of the active site (A cluster) of ACS. The geometry and non-peptide ligation (L) of the proximal nickel remain uncertain.

The EPR spectrum of the reduced enzyme under ^{13}CO features the so-called NiFeC signal, which displays hyperfine coupling to ^{61}Ni, ^{57}Fe, and ^{13}CO [103, 104]. The Mössbauer spectrum indicates that this species contains a diamagnetic $[Fe_4S_4]^{2+}$ cluster exchange-coupled to a Ni(I)CO center, which is proposed to be the proximal site [105]. It remains unclear if the species responsible for the NiFeC signal is relevant to catalysis [106, 107].

12.4.3
Proposed Mechanism

ACS effects many transformations: methylation, complexation of CO, coupling of methyl to CO to give acetyl, transfer of the acetyl group to the thiol of CoA, as well as associated redox chemistry. Catalysis is assumed to be localized on the proximal Ni. The functional role of the distal Ni remains unclear – it may simply serve a structural role, providing a pair of highly basic thiolate ligands and thereby modulating the electronic environment of the proximal Ni center. Theoretical calculations indicate that the distal Ni center is unlikely to bind CO or accept the equivalent of CH_3^+ [108]. Although the sequence of carbonylation and methylation *in vivo* are unclear, it has been established that the reduced (S = 1/2) A cluster can be methylated by MeCoFeSP to give a non S= 1/2 species. Treatment of this species with CO and CoA quantitatively affords acetyl CoA [109]. The NiFeC signal is not observed during this cycle. One of the more perplexing mechanistic issues is that the reaction of MeCoFeSP, a source of a methyl cation equivalent, with Ni(I) would produce a Me–Ni(III) product. In contrast a Me–Ni(II) species is implied by the biochemical studies.

12.4.4
Spectroscopic Models for Proximal Nickel

An informative spectroscopic model for the CO-bound, reduced proximal Ni site is derived from the anionic tripodal ligand PhTt^{tBu}, or phenyl-tris(*tert*-butylthio-

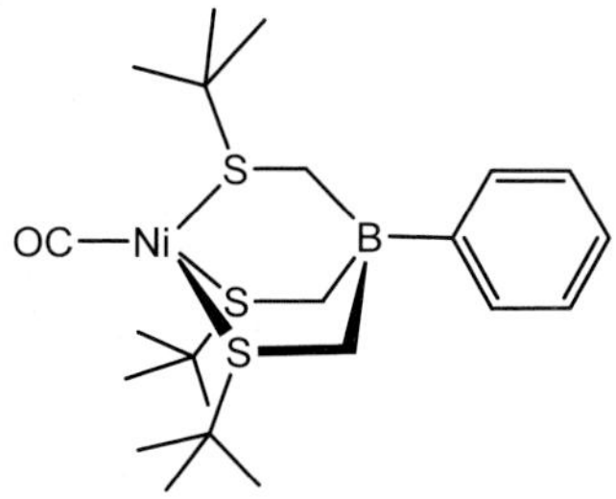

Fig. 12.14 Structure of Ni(PhTt^{tBu})CO, a spectroscopic model for the reduced CO-bound proximal Ni site in ACS.

methyl)borate (Fig. 12.14). The EPR spectrum of [PhTt^{tBu}]Ni(CO) is related to the "NiFeC" EPR signal, but lacks the ^{13}C hyperfine coupling, indicating that the structure of the model is imperfect, albeit close [104]. The v_{CO} band for the CO-reduced enzyme and this model complex are identical at $\approx$ 1995 cm^{-1}.

12.4.5
Structural Models for the Dinickel Site

The unusual structure of the ACS active site spurred active investigations into the bimetallic complexes containing square-planar nickel bound through thiolate to tetrahedral metal centers. Initial modeling work was motivated by the incorrect proposal that the active enzyme contained Cu in the proximal site. NiS$_2$N$_2$ metalloligands, which were mentioned above, react with [Cu(MeCN)$_4$]$^+$ to give a variety of derivatives [110, 111]. No Cu–Ni species binds CO, however. Several di- and tri-nickel models for the distal-proximal assembly in ACS arise from the use of NiS$_2$N$_2$ building blocks illustrated in Fig. 12.15 [110–112]. For example, treatment of a diamidodithiolato nickel complex with Ni(cyclooctadiene)$_2$ in the presence of CO gives a mixed valence Ni^{II}S$_2$N$_2$–Ni^0(CO)$_2$ derivative.

This air-sensitive species degrades to a trimetallic Ni–Ni–Ni dianion upon oxidation in air. The bimetallic dicarbonyl species also reacts with two equivalents of PPh$_3$ to yield Ni(CO)$_2$(PPh$_3$)$_2$ and the original metalloligand. The facility of this demetallation is consistent with the lability of the proximal nickel center, which can be removed using chelators such as 1,10-phenanthroline [114]. Model studies support the lability of the proximal metal [115]. Nickel binds to the tripeptide cys–gly–cys (protected at the N and O termini) to give the expected diamidodithiolates that also bind nickel via the thiolato bridges [113].

A dinickel(II) system has been described that undergoes reversible reduction at the "proximal" nickel center to a Ni(II)–Ni(I) state and further irreversible reduction to a Ni(II)–Ni(0) state (Fig. 12.15) [112]. Both of the nickel atoms in the beginning complex exhibit square planar geometry, but the Ni(SR)$_2$(PR$_3$)$_2$ center becomes tetrahedral upon reduction to the Ni(II)–Ni(I) state, as evidenced by EPR spectroscopy. A related geometric change during the catalytic cycle of the ACS enzyme has been suggested.

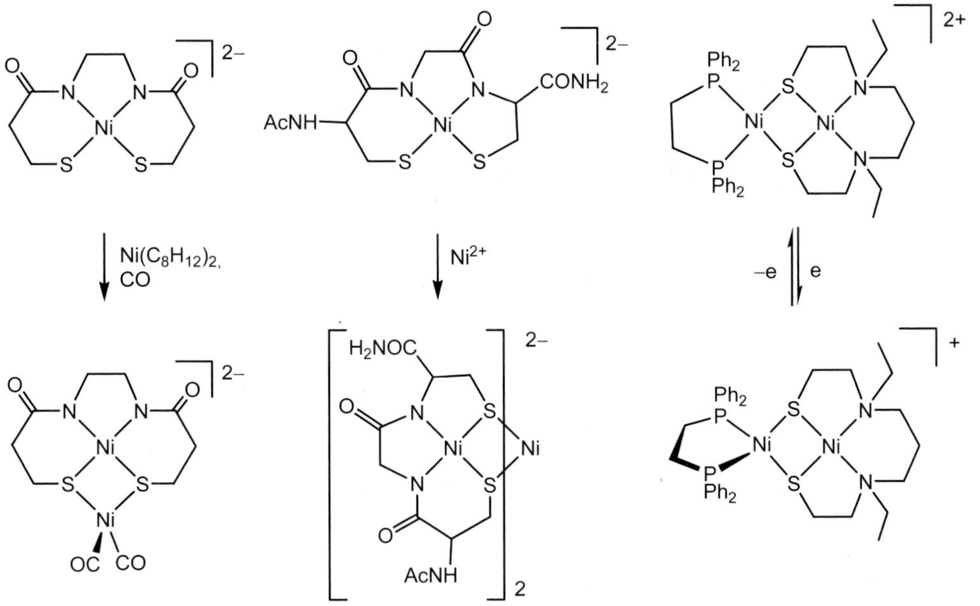

Fig. 12.15 Dinickel complexes illustrating different features of the ACS enzyme: CO binding [110], peptidomimetic ligation [113], and redox-activity [112].

12.4.6
Functional Models

Obviously modeling all of these functions represents a formidable challenge. But several of the individual steps have been simulated with model complexes in solution.

12.4.6.1 Transmetalation of Methyl

An elegant model had been developed to describe the transfer of a methyl cation from the cobalt-corrinoid Fe–S protein to a proximal nickel center [116]. Methyl transfer from a methyl-cobalt species to a Ni(I) complex yields a stable Ni(II)-Me species concomitant with oxidation of a second equivalent of the Ni(I) complex (Eq. 12.9). In this system, the methyl cobalt species is equivalent to a methyl cation source, although mechanistic studies revealed the involvement of methyl radicals formed via homolysis of an initially generated Co(II)-methyl species [117].

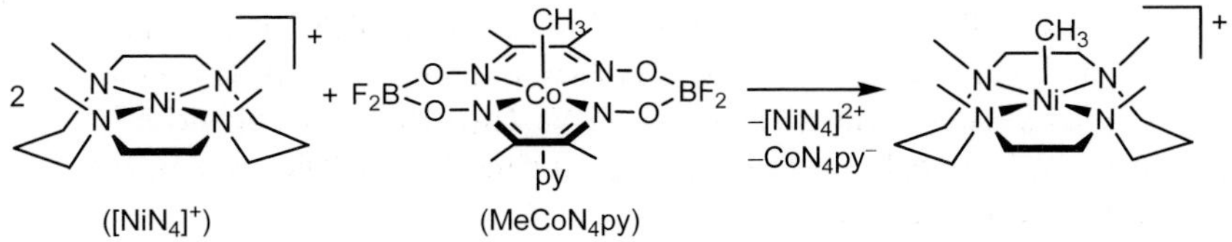

$$(12.9)$$

12.4.6.2 Thioester-Forming Steps

Metal complexes are known to insert CO into carbon–sulfur bonds [118], even catalytically [119]. Stoichiometric precedents exist for the formation of thioesters from nickel–alkyls, CO, and thiols [120]. For example, $NiMe_2$(bipy) reacts with thiols to afford methylnickel(II) thiolates, which carbonylate to afford acetylnickel(II) thiolates. These acetylnickel(II) thiolates reductively eliminate thioester in the presence of CO [121]. More biologically relevant is the reactivity of nickel acyls toward thiolates, which gives the thioester concomitant with reduction to Ni(0) (Eq. 12.10) [122]. Thiolates are known to reduce Ni(II) to Ni(0) under an atmosphere of CO [123].

$$\text{(12.10)}$$

12.5
Anaerobic Carbon Monoxide Dehydrogenase

12.5.1
Overview

In anaerobic environments, the biological redox-processing of carbon monoxide and carbon dioxide is accomplished by carbon monoxide dehydrogenases (CODH), which are often found in combination with a second enzyme, acetyl CoA synthase (ACS, see above) [124]. In the CODH/ACS bifunctional enzymes, the two substrate-binding prosthetic groups, the A cluster (site of acetyl CoA synthesis and decarbonylation, see previous section) and the C cluster (site of CO/CO_2 interconversion) are located in the α and β subunits, respectively. CODH from *R. rubrum* is not associated with ACS. The CO produced at the C-cluster travels via a tunnel through the enzyme to the α subunit where it is processed by acetyl CoA synthase (the A-cluster) [125]. CODH catalyzes the interconversion of CO_2 and CO (Eq. 12.11).

$$CO + H_2O \rightleftharpoons CO_2 + 2\,e + 2\,H^+ \tag{12.11}$$

This equilibrium is central to acetogenesis, i.e. the mildly exothermic reaction $H_2 + CO_2 \rightarrow$ acetic acid. Complementarily, CODH is utilized by some methanogenic archeons to extract energy by the disproportionation of acetic acid into CO_2 and CH_4 [126]. Purified CODH from *Moorella thermoacetica* is an exceptional electrocatalyst for the reduction of CO_2 (700 turnovers/hour, 100% conversion, –0.57 V vs. NHE) [127].

12.5.2
Structure of the Active Site

Crystallographic studies, while generally in agreement, differ in the details of the coordination environment for Ni and Fe_{ext}. These differences may be due to denaturation, reflecting the sensitivity of the enzyme. The active site features an exceptional [4Fe–Ni–5S] cluster (Fig. 12.16), referred to as the C cluster. This cluster can be viewed as being composed of two subunits, an external bimetallic pair Ni–Fe_{ext} and a Fe_3S_4 cubane fragment. All five metals are bound to cysteinyl thiolates. Fe_{ext} and Ni are bound to the Fe_3S_4 portion by one and two sulfides, respectively. Fe_{ext}, which adopts a distorted tetrahedral geometry, with one histidine ligand, is linked to Ni via an inorganic sulfur atom [128]. The Fe_{ext}–S–Ni linkage is apparently labile, and some crystallographic studies have found Fe–S(cys)–Ni linkages [129].

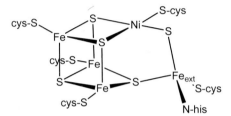

Fig. 12.16 Structure for the proposed active form of the C cluster in anaerobic CODH, based on crystallographic analysis.
Other structures have been reported that do not feature the Ni-S-Fe linkage [128].

12.5.3
Proposed Mechanisms

Numerous precedents exist for the oxidation of CO by water, a well known example being the reaction of OH^- and $Fe(CO)_5$, which cogenerates CO_2 and $[HFe(CO)_4]^-$. Enzyme catalysis is assumed but not proven to involve an analogous reaction involving attack of FeOH at Ni–CO [130, 131]. Inhibition by cyanide presumably occurs by blocking coordination of substrate to a metal, although it is neither clear which substrate nor which metal. Attack of OH^- at coordinated CO would generate a metallacarboxylic acid ($M–CO_2H$), which would undergo deprotonation prior to loss of CO_2 and formal 2e reduction of the cluster (Fig. 12.17).

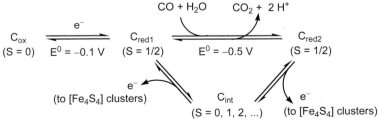

Fig. 12.17 Proposed states and steps for anaerobic CODH [132].

Redox states for the C cluster relevant to catalytic CO oxidation are the C_{red1}, proposed to be $\{(Ni^{2+}Fe^{2+}):[Fe_3S_4]^{1-}\}$ and C_{red2}, which is 2e more reduced than C_{red1} [125, 132]. These redox states are both S = 1/2, thus 2e reduction could convert C_{red2} from C_{red1} although 1e steps are also conceivable.

12.5.4
Functional Studies

The basic reaction catalyzed by CODH is related to the water gas shift reaction employed industrially to generate H_2 by the reaction of CO and H_2O. Industrially the process is iron-catalyzed, and numerous organometallic complexes effect the same reaction [133]. Additionally, square-planar Ni(II) complexes electrocatalyze the reduction of CO_2 to CO, including $[Ni^{II}(cyclam)]^{2+}$ (cyclam = 1,4,8,11-tetraazacyclotetradecane), nickel–porphyrins and phthalocyanines. The cluster $[Fe_4S_4(SR)_4]^{2-}$ and certain Fe–porphyrins also catalyze this reaction [124].

The prevailing hypothesis that Ni is the CO receptor has focused research on the binding and activation of CO by Ni^{II} complexes. On the one hand, Ni(II) tends not to bind CO, but, once bound, the CO ligand would be electrophilic and hence susceptible to addition of hydroxide at carbon. Ni(II) complexes of CO tend to be neutral or anionic [93, 134].

In terms of functional models, N–O–S complexes of Ni(II) catalyze the oxidation of CO by H_2O to give CO_2, protons, and reducing equivalents (Eq. 12.8) [135]. The complex [Ni(tmtss)(MeOH)] (tmtss = 2'-hydroxy-4'5'-dimethylacetophenone-4-methyl-thiosemicarbazone) features a four-coordinate Ni(II) center that is stable in MeOH (Eq. 12.12). Catalysis proceeds at 1.04 turnovers per hour (25 °C) and is inhibited by CN^- and MeI, although these reagents inhibit many forms of catalysis.

$$CO + H_2O + 2\ MV^{2+} \xrightarrow{\text{catalyst}} CO_2 + 2\ MV^+ + 2\ H^+ \qquad (12.12)$$

The reaction is assumed to proceed via formation of a CO adduct, which undergoes nucleophilic attack by water followed by loss of CO_2 to produce a nickel hydride, which reduces MV^{2+} (MV = methyl viologen). The catalytic cycle is closed by reoxidation of the resulting Ni(I) complex with MV^{2+}.

12.5.5
Structural Models

Cuboidal [3Fe–Ni–4S] species related to the C cluster have been prepared using a specially designed tris(mercaptide) ligand, which allows the stabilization of the three Fe centers bound to a fourth, reactive metal site [136]. Alternatively, and without the need for the specialized ligands, the linear triiron species $[Fe_3S_4(SR)_4]^{3-}$ reacts with $Ni(PPh_3)_4$ to give $[(Ph_3P)NiFe_3S_4(SR)_3]^{2-}$ (Eq. 12.13). The resulting $NiFe_3$ cubanes, which are described as $\{[Fe(III)]_3Ni(II)\}$, undergo selective reactions at the Ni site, although treatment with CO degrades the cluster to give $[Fe_4S_4(SR)_4]^{2-}$ [137, 138].

$$(12.13)$$

for z = 3–, L = CN, SR
for x = 2–, L = PR$_3$, RNC

12.6
Aerobic Carbon Monoxide Dehydrogenase

12.6.1
Overview

The aerobic family of CODHs are derived from aerobic bacteria. In toto, CODHs are estimated to process $\approx 10^{11}$ kg y^{-1} of CO. The electrons from CO oxidation are transferred through an electron transport chain that ultimately oxidizes O_2. *In vitro* the CODH also displays some H_2 oxidation catalysis.

12.6.2
Structure of the Active Sites and Proposed Mechanisms

The enzyme from *O. carboxidovorans* has been crystallographically characterized to 1.1 Å resolution as a 277 kDa dimer of heterotrimers. Each heterotrimer is composed of a molybdoprotein, a flavoprotein, and an iron–sulfur protein. In the molybdoprotein, Mo is bound to one molybdopterin-based dithiolene, one sulfide, and two oxo/hydroxo ligands. Copper, the site of CO oxidation, is bound linearly between the Mo=S center and cysteinyl thiolate (Fig. 12.18, early reports on an Mo–S–Se–S site should be disregarded) [139, 140]. The Mo–molybdopterin portion of CO dehydrogenase is very similar to xanthine oxidase/dehydrogenase. Another Cu–Mo–S-containing protein has been shown to contain $[S_2MoS_2CuS_2MoS_2]^{3-}$, although the function of this cofactor remains unclear [141].

A proposed mechanism of action entails initial binding of CO at the two-coordinate cuprous center. The resulting coordinated CO undergoes attack by an Mo-bound oxo or hydroxo ligand, with participation by the sulfido group. Excellent mechanistic information is provided by crystallographic characterization of the *t*-BuNC derivative of the enzyme [139]. The isocyanide is chemically related to CO by replacement of O by N-*t*-Bu, but it is also a far better donor ligand that tends to form more robust Cu$^+$ complexes than does CO. In the *t*-BuNC adduct the carbon is bound to the sulfido and oxo groups on Mo, with the nitrogen attached to Cu. This structure indicates that reduction of Mo occurs upon insertion of CO to give the thiocarbonato complex. Release of CO_2 occurs concomitantly with reformation of the Mo=S$\cdots$Cu bond (Fig. 12.18) [139].

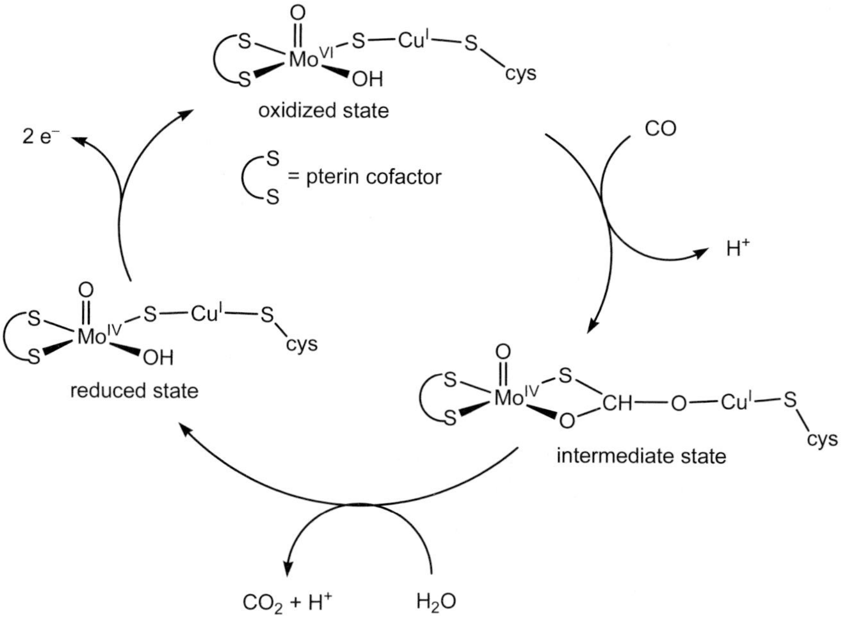

Fig. 12.18 Proposed mechanism of action for Cu-Mo CODH (adapted from [139]).

12.6.3
Structural and Functional Models

In terms of precedent, copper(I) is well known to bind CO and such adducts typically exhibit high values of v_{CO}, which suggest susceptibility to attack by nucleophiles. Nucleophilic attack at Cu–CO by this enzyme, however, is proposed to lead to an O-bound thiocarbonate, which entails scission of an initially formed Cu–CO bond. Precedent for this reactivity is provided by the reaction of Pd(II) complexes of CNMe with W–SH species (Eq. 12.14) [142].

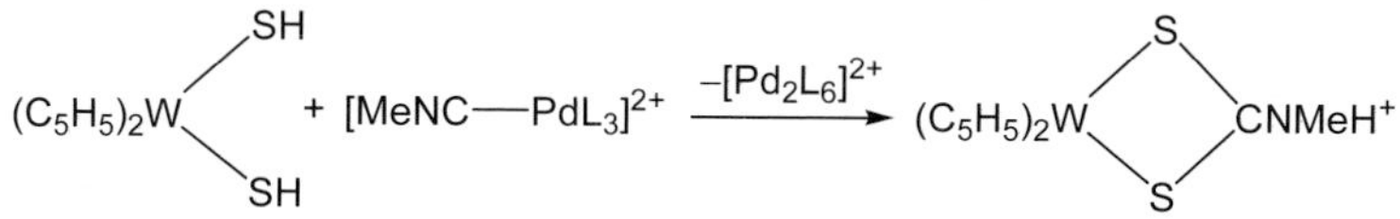

Models for monodithiolene molybdenum species related to xanthine oxidase and aerobic CODH have been described, e.g. $[Mo(S_2C_6H_4)O_2(SR)]^-$ [143]. The synthesis of related species with terminal sulfido ligands would enable assembly of the indicated (dithiolene)MoO_2S–CuSR ensembles. The interaction of MoS_4^{2-} and Cu^+ has received considerable study, motivated by interest in the ability of dietary Mo to inhibit copper uptake in ruminants [144]. Bi-, tri-, tetra-, or higher nuclearity Cu–Mo clusters result from the reaction of MoS_4^{2-} and Cu^+ in the presence of tertiary phosphines, amines, or cyanide [144].

12.7
Methyl Coenzyme M Reductase

12.7.1
Overview

Methyl Coenzyme M Reductase (MCR) is a central enzyme in methogenesis [126], the process that is probably responsible for the earth's vast natural gas reserves (ca. 10^{17} liters). Methanogenesis remains shrouded in mystery, at least from the molecular perspective, but the biochemical facts are well established. Occurring exclusively in archaea, methanogenesis utilizes seven unique cofactors, making it one of the most exciting and underestimated areas in biochemistry. Some archaeal genera also contain a separated coded isozyme. It is also suspected that the same set of enzymes participate in methane consumption – or reverse methanogenesis – when these organisms have access to both methane and reducible substrates such as sulfate [145].

12.7.2
Structure of the Active Site

Inactive enzymes from *Methanobacter marburgensis* have been crystallographically characterized at 1.16 Å resolution [146]. At 300 kDa, MCR has an $\alpha_2\beta_2\gamma_2$ organization and bears two identical nickel-containing F_{430} coenzymes. Enzyme preparations typically also contain both CoM (2-mercaptoethylsulfonic acid) and CoB (7-thioheptanoylthreonine-phosphate) located near the F_{430} cofactor. The F_{430} cofactor is a nickel complex of an elaborately substituted corphin macrocycle, a reduced porphyrinoid macrocycle (Fig. 12.19). Ni is bound to the four pyrollic

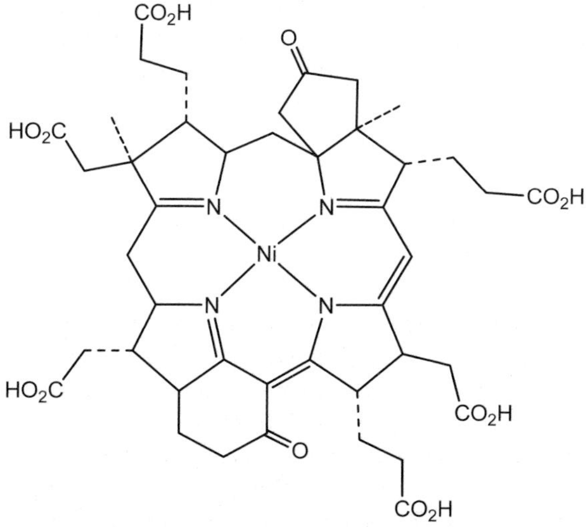

Fig. 12.19 Structure of cofactor F_{430} of MCR [126].

nitrogen atoms. An O-bound glycine is bound to Ni on the substrate-inaccessible side of the macrocycle, representing the only covalent attachment of the cofactor to the protein. Crystallographic analysis shows that the occupancy of the sixth position depends on the particular preparation, varying from the thiol (or thiolate) of HS–CoM, with an Ni–S distance of 2.45 Å, the sulfonate from CoB–S–S–CoM, or nothing. In the structure containing HSCoM, the two organic cofactors occupy adjacent compartments of the 42 Å cavity above the sixth position of Ni. EXAFS examination of the reduced active form of the enzyme/HS–CoM complex indicates the absence of Ni$\cdots$S bonding [147]. Four amino-acids that decorate the active site cavity are methylated. In this same region one glycine residue, in place of its carbonyl features a thiocarbonyl, which could participate in catalysis [148].

12.7.3
Proposed Mechanism

MCR catalyzes the reaction of methyl CoM and HSCoB (Eq. 12.15).

HS coenzyme M is a competitive inhibitor for the reaction of methyl-CoM and HS-coenzyme B. ^{33}S labeling studies show that when both HS-cofactors are bound, HS-coenzyme M is bound to Ni [149]. It is striking that *both* organic cofactors are required for MCR to function. The requirement for both organic cofactors points to an unprecedented reaction mechanism, one which likely couples formation of the heterodisulfide linkage with methylation of nickel. The substrate specificity is very high although the enzyme will accept *ethyl* CoM. Using optically active CH_3CDT-CoM (where D and T refer to deuterium and tritium, respectively), it has been shown that the formation of CH_3CHDT occurs with inversion at carbon [150], consistent with attack by a nickel nucleophile (Fig. 12.20).

The active enzyme contains Ni(I), as evidenced by EPR studies as well as the requirement for reducing agents for enzyme activation [126]. The reduction potential (≈ -650 mV) is comparable to that for cobalamin (II/I) couple. The methylated F_{430} cofactor, which is a CH_3–Ni(II) derivative, can be prepared by treatment of the reduced cofactor with CH_3I or alkylation of the Ni(II) form with $(CH_3)_2Mg$. Protonolysis of the CD_3–Ni(II) derivative produces CD_3H [151].

Mechanisms have also been proposed where no methyl–Ni bond is formed, rather methane is generated initially concomitant with the formation of side-chain stabilized thiyl radicals that subsequently couple [152].

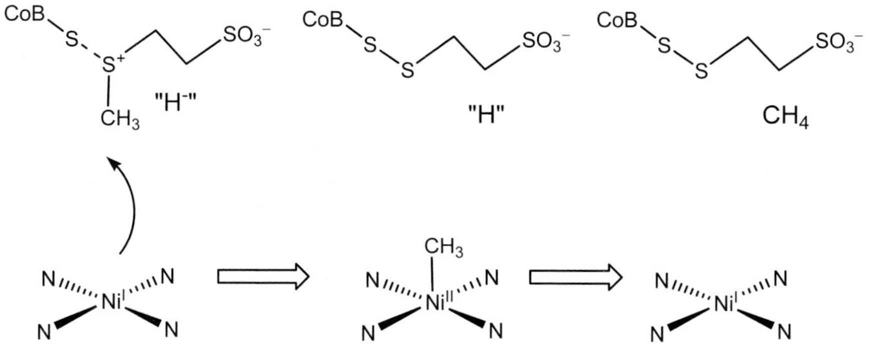

Fig. 12.20 Conceptual depiction of the MCR-catalyzed reaction of Me–CoM and HS–CoB to give methane. The nature of "H⁻" and "H" is unknown.

12.7.4
Structural/Functional Models

In terms of its chemical properties, the isolated cofactor F_{430} resembles other nickel (II) complexes of tetraamine macrocycles with respect to redox and the binding of axial ligands. Me–Ni and RSNi derivatives of these macrocyclic complexes represent possible model intermediates for the C–S bond scission process [134].

Model studies are guided by three main assumptions: (1) Ni(I) is important, (2) both organic cofactors are required for catalysis, and (3) the sulfonate group of CoM is not involved in metal coordination. The required presence of *both* CoB and CoM is particularly curious and points to a novel mechanism. One hypothesis related to the implied cooperativity of the cofactors is that sulfonium (R_3S) cations or radicals are involved. Cofactor F_{430} indeed catalyzes the reduction of Me_3S^+, $(CH_2)_4SMe^+$, and methylated cyclic disulfides to give methane (Eq. 12.16) [153, 154].

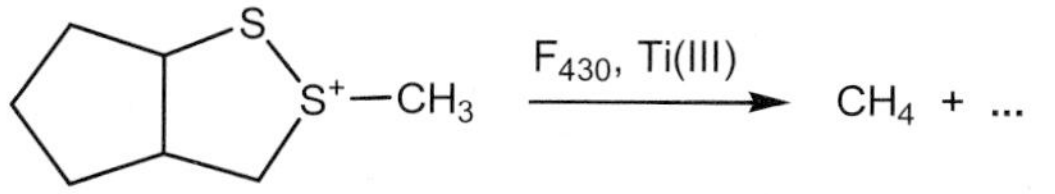

$$(12.16)$$

The reduction of related sulfonium ions, i.e. R_3S^+ to $R_2S + RH$ can be effected electrochemically and chemically using $[Fe_4S_4(SEt)_4]^{2-}$ [155].

Complexes of 3-methylthioether-propanethiolato ligands release modest yields of methane upon photolysis, concomitant with formation of the cyclic disulfide (Fig. 12.21) [156].

The efficiency of the photoreaction is greater still in the presence of additional thioether-thiol. Even though these experiments relied on photoreactions (the biological process is thermal) to initiate the radical chemistry, it is interesting to observe the formation of methane concomitant with the disulfide.

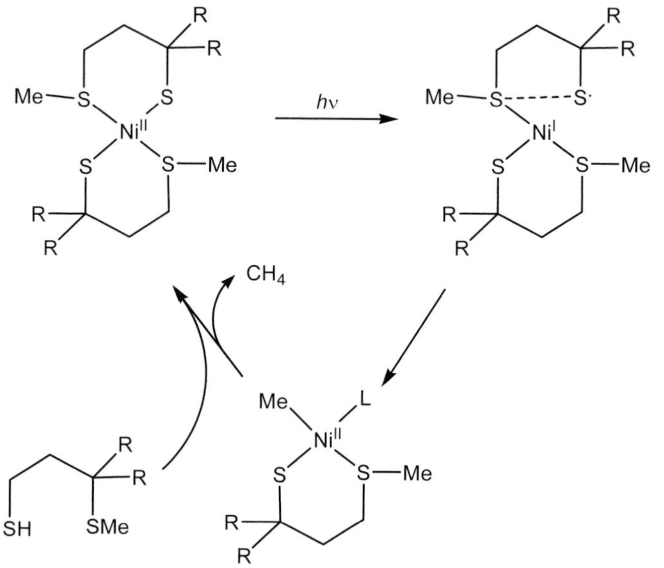

Fig. 12.21 Model methanogenesis system illustrating a proposed interaction between a methyl thioether and a thiyl radical.

12.8
Conclusions

It is clear that enzymes from extremophiles and primitive organisms often rely on organometallic chemistry to accomplish their metabolic needs. The area of bioorganometallic enzymology has opened new opportunities for organometallic chemistry to contribute to important themes in enzymology. Organometallic methodology is required for the synthesis and characterization of low molecular weight active site models. It will be interesting to see the extent to which the lessons from enzymology contribute to the development of of organometallic catalysts relevant to needs in industry and organic synthesis.

Note added in proof

Tatsumi et al. have recently described the synthesis of a sophisticated active site model for [NiFe]-hydrogenases via the attachment of an NiS_4 site to an $Fe(CO)_2(CN)_2$ center: Z. Li, Y. Okhi, K. Tatsumi, *J. Am. Chem. Soc.* **2005**, *127*, 8950–8951. Pickett et al. have recently described the synthesis of a sophisticated active site model for the H-cluster of the Fe H$_2$-ases via the attachment of an [4Fe-4S] cluster to a $Fe_2(SR)_2(CO)_5$ center: C. Tard, X. Liu, S. K. Ibrahim, M. Bruschi, L. De Goiia, S. C. Davies, X. Yang, L.-S. Wang, G. Sawers, C. J. Pickett, *Nature* **2005**, *433*, 610–614.

References

1 R. Cammack, M. Frey, R. Robson, *Hydrogen as a Fuel: Learning from Nature*, Taylor & Francis, London, 2001.

2 E. J. Lyon, S. Shima, G. Buurman, S. Chowdhuri, A. Batschauer, K. Steinbach, R. K. Thauer, *Eur. J. Biochem.* 2004, 271, 195–204.

3 W. Beck, K. Severin, *Chem. Unserer Zeit* 2002, 36, 356–365.

4 R. Banerjee, S. W. Ragsdale, *Ann. Rev. Biochem.* 2003, 72, 209–247.

5 J. Hirsch, S. D. George, E. I. Solomon, B. Hedman, K. O. Hodgson, J. N. Burstyn, *Inorg. Chem.* 2001, 40, 2439–2441.

6 A. B. Bleecker, H. Kende, *Ann. Rev. Cell Dev. Biol.* 2000, 16, 1–18.

7 S. Aono, *Acc. Chem. Res.* 2003, 36, 825–831.

8 S. C. Lee, R. H. Holm, *Chem. Rev.* 2004, 104, 1135–1157.

9 P. C. Dos Santos, D. R. Dean, Y. Hu, M. W. Ribbe, *Chem. Rev.* 2004, 104, 1159–1173.

10 D. S. Horner, B. Heil, T. Happe, T. M. Embley, *Trends Biochem. Sci.* 2002, 27, 148–153.

11 M. Frey, *ChemBioChem* 2002, 3, 153–160.

12 J. W. Peters, *Curr. Opin. Struct. Biol.* 1999, 9, 670–676.

13 M. W. W. Adams, *Biochem. Biophys. Acta* 1990, 1020, 115–145.

14 Y. Nicolet, B. J. Lemon, J. C. Fontecilla-Camps, J. W. Peters, *Trends Biochem. Sci.* 2000, 25, 138–143.

15 Y. Nicolet, A. L. de Lacey, X. Vernede, V. M. Fernandez, E. C. Hatchikian, J. C. Fontecilla-Camps, *J. Am. Chem. Soc.* 2001, 123, 1596–1601.

16 Z. Chen, B. J. Lemon, S. Huang, D. J. Swartz, J. W. Peters, K. A. Bagley, *Biochemistry* 2002, 41, 2036–2043.

17 C. V. Popescu, E. Münck, *J. Am. Chem. Soc.* 1999, 121, 7877–7884.

18 G. Wang, M. J. Benecky, B. H. Huynh, J. F. Cline, M. W. Adams, L. E. Mortenson, B. M. Hoffman, E. Munck, *J. Biol. Chem.* 1984, 259, 14328–14331.

19 A. L. De Lacey, C. Stadler, C. Cavazza, E. C. Hatchikian, V. M. Fernandez, *J. Am. Chem. Soc.* 2000, 122, 11232–11233.

20 F. Gloaguen, J. D. Lawrence, M. Schmidt, S. R. Wilson, T. B. Rauchfuss, *J. Am. Chem. Soc.* 2001, 123, 12518–12527.

21 A. L. De Lacey, V. M. Fernandez, M. Rousset, C. Cavazza, E. C. Hatchikian, *J. Biol. Inorg. Chem.* 2003, 8, 129–134.

22 J. W. Peters, W. N. Lanzilotta, B. J. Lemon, L. C. Seefeldt, *Science* 1998, 282, 1853–1858.

23 B. J. Lemon, J. W. Peters, *Biochemistry* 1999, 38, 12969–12973.

24 Y. Nicolet, C. Piras, P. Legrand, C. E. Hatchikian, J. C. Fontecilla-Camps, *Structure* 1999, 7, 13–23.

25 G. J. Kubas, *Metal Dihydrogen and σ-Bond Complexes*, Kluwer Academic/Plenum Publishers, New York, 2001.

26 D. J. Evans, C. J. Pickett, *Chem. Soc. Rev.* 2003, 32, 268–275.

27 H. Reihlen, A. Gruhl, G. V. Hessling, *Justus Liebigs Ann. Chem.* 1929, 472, 268–287.

28 W. Hieber, P. Spacu, *Z. Anorg. Allgem. Chem.* 1937, 233, 852–864.

29 N. S. Nametkin, B. I. Kolobkov, V. D. Tyurin, A. N. Muratov, A. I. Nekhaev, M. Mavlonov, A. Y. Sideridu, G. G. Aleksandrov, A. V. Lebedev, *J. Organomet. Chem.* 1984, 276, 393–7.

30 R. B. King, *J. Am. Chem. Soc.* 1962, 84, 2460.

31 D. Seyferth, R. S. Henderson, L. C. Song, *Organometallics* 1982, 1, 125–33.

32 M. Schmidt, S. M. Contakes, T. B. Rauchfuss, *J. Am. Chem. Soc.* 1999, 121, 9736–7.

33 E. J. Lyon, I. P. Georgakaki, J. H. Reibenspies, M. Y. Darensbourg, *Angew. Chem., Int. Ed. Engl.* 1999, 38, 3178–3180.

34 A. Le Cloirec, S. C. Davies, D. J. Evans, D. L. Hughes, C. J. Pickett, S. P. Best, S. Borg, *Chem. Commun.* 1999, 2285–2286.

35 E. J. Lyon, I. P. Georgakaki,
J. H. Reibenspies, M. Y. Darensbourg,
J. Am. Chem. Soc. **2001**, *123*,
3268–3278.

36 P. C. Ellgen, J. N. Gerlach, *Inorg.
Chem.* **1973**, *12*, 2526–32.

37 H. Li, T. B. Rauchfuss, *J. Am. Chem.
Soc.* **2002**, *124*, 726–727.

38 S. J. George, Z. Cui, M. Razavet,
C. J. Pickett, *Chem. Eur. J.* **2002**, *8*,
4037–4046.

39 M. Razavet, S. C. Davies,
D. L. Hughes, J. E. Barclay,
D. J. Evans, S. A. Fairhurst, X. Liu,
C. J. Pickett, *Dalton Trans.* **2003**,
586–595.

40 M. S. Arabi, R. Mathieu,
R. Poilblanc, *J. Organomet. Chem.*
1979, *177*, 199–209.

41 J. D. Lawrence, H. Li,
T. B. Rauchfuss, *Chem. Commun.*
2001, 1482–1483.

42 J. D. Lawrence, H. Li,
T. B. Rauchfuss, M. Bénard,
M.-M. Rohmer, *Angew. Chem., Int. Ed.
Engl.* **2001**, *40*, 1768–1771.

43 S. Ott, M. Kritikos, B. Akermark,
L. Sun, R. Lomoth, *Angew. Chem.,
Int. Ed.* **2004**, *43*, 1006–1009.

44 J. C. Lee Jr., E. Peris, A. L. Rheingold,
R. H. Crabtree, *J. Am. Chem. Soc.*
1994, *116*, 11014–11019.

45 C. J. Curtis, A. Miedaner,
R. Ciancanelli, W. W. Ellis,
B. C. Noll, M. Rakowski DuBois,
D. L. DuBois, *Inorg. Chem.* **2003**, *42*,
216–227.

46 F. Gloaguen, J. D. Lawrence,
T. B. Rauchfuss, M. Bénard,
M.-M. Rohmer, *Inorg. Chem.* **2002**, *41*,
6573–6582.

47 X. Zhao, I. P. Georgakaki,
M. L. Miller, J. C. Yarbrough,
M. Y. Darensbourg, *J. Am. Chem. Soc.*
2001, *123*, 9710–9711.

48 I. P. Georgakaki, M. L. Miller,
M. Y. Darensbourg, *Inorg. Chem.* **2003**,
42, 2489–2494.

49 F. Gloaguen, J. D. Lawrence,
T. B. Rauchfuss, *J. Am. Chem. Soc.*
2001, *123*, 9476–9477.

50 D. Chong, I. P. Georgakaki, R. Mejia-
Rodriguez, J. Sanabria-Chinchilla,
M. P. Soriaga, M. Y. Darensbourg,

J. Chem. Soc., Dalton Trans. **2003**,
4158–4163.

51 K. S. Hagen, R. H. Holm, *Inorg. Chem.*
1984, *23*, 418–427.

52 V. E. Kaasjager, R. K. Henderson,
E. Bouwman, M. Lutz, A. L. Spek,
J. Reedijk, *Angew. Chem. Int. Ed.* **1998**,
37, 1668–1670.

53 K. Fauvel, R. Mathieu, R. Poilblanc,
Inorg. Chem. **1976**, *15*, 976–978.

54 M. Razavet, S. J. Borg, S. J. George,
S. P. Best, S. A. Fairhurst,
C. J. Pickett, *Chem. Commun.* **2002**,
700–701.

55 J. D. Lawrence, T. B. Rauchfuss,
S. R. Wilson, *Inorg. Chem.* **2002**, *41*,
6193–6195.

56 C. D. Boyke, T. B. Rauchfuss,
S. R. Wilson, *J. Am. Chem. Soc.* **2004**,
126, 15151–15160.

57 J. W. Olson, R. J. Maier, *Science* **2002**,
298, 1788–1790.

58 A. K. Jones, E. Sillery,
S. P. J. Albracht, F. A. Armstrong,
Chem. Commun. **2002**, 866–867.

59 A. Volbeda, E. Garcin, C. Piras,
A. L. de Lacey, V. M. Fernandez,
E. C. Hatchikian, M. Frey,
J. C. Fontecilla-Camps, *J. Am. Chem.
Soc.* **1996**, *118*, 12989–12996.

60 R. P. Happe, W. Roseboom,
A. J. Pierik, S. P. J. Albracht,
K. A. Bagley, *Nature* **1997**, *385*, 126.

61 A. Volbeda, J. C. Fontecilla-Camps,
Dalton Trans. **2003**, 4030–4038.

62 A. J. Pierik, W. Roseboom,
R. P. Happe, K. A. Bagley,
S. P. J. Albracht, *J. Biol. Chem.* **1999**,
274, 3331–3337.

63 E. Garcin, X. Vernede,
E. C. Hatchikian, A. Volbeda,
M. Frey, J. C. Fontecilla-Camps,
Structure **1999**, *7*, 557–566.

64 Y. Higuchi, H. Ogata, K. Miki,
N. Yasuoka, T. Yagi, *Structure* **1999**, *7*,
549–556.

65 Y. Montet, P. Amara, A. Volbeda,
X. Vernede, E. C. Hatchikian,
M. J. Field, M. Frey, J. C. Fontecilla-
Camps, *Nat. Struct. Biol.* **1997**, *4*, 523–526.

66 S. Reissmann, E. Hochleitner,
H. Wang, A. Paschos, F. Lottspeich,
R. S. Glass, A. Böck, *Science* **2003**, *299*,
1067–1070.

67 R. J. Angelici, *Acc. Chem. Res.* **1972**, *5*, 335–341.

68 A. Volbeda, Y. Montet, X. Vernede, E. C. Hatchikian, J. C. Fontecilla-Camps, *Int. J. Hydrogen Energy* **2002**, *27*, 1449–1461.

69 M. Stein, W. Lubitz, *J. Inorg. Biochem.* **2004**, *98*, 862–877.

70 M. Carepo, D. L. Tierney, C. D. Brondino, T. C. Yang, A. Pamplona, J. Telser, I. Moura, J. J. G. Moura, B. M. Hoffman, *J. Am. Chem. Soc.* **2002**, *124*, 281–286.

71 H. Ogata, Y. Mizoguchi, N. Mizuno, K. Miki, S.-I. Adachi, N. Yasuoka, T. Yagi, O. Yamauchi, S. Hirota, Y. Higuchi, *J. Am. Chem. Soc.* **2002**, *124*, 11628–11635.

72 V. Autissier, W. Clegg, R. W. Harrington, R. A. Henderson, *Inorg. Chem.* **2004**, *43*, 3098–3105.

73 M. Schlaf, A. J. Lough, R. H. Morris, *Organometallics* **1996**, *15*, 4423–4436.

74 M. Brecht, M. van Gastel, T. Buhrke, B. Friedrich, W. Lubitz, *J. Am. Chem. Soc.* **2003**, *125*, 13075–13083.

75 D. Sellmann, F. Geipel, F. W. Heinemann, *Chem. Eur. J.* **2002**, *8*, 958–966.

76 A. C. Moreland, T. B. Rauchfuss, *Inorg. Chem.* **2000**, *39*, 3029–3036.

77 C.-H. Lai, W.-Z. Lee, M. L. Miller, J. H. Reibenspies, D. J. Darensbourg, M. Y. Darensbourg, *J. Am. Chem. Soc.* **1998**, *120*, 10103–10114.

78 T. B. Rauchfuss, S. M. Contakes, S. C. N. Hsu, M. A. Reynolds, S. R. Wilson, *J. Am. Chem. Soc.* **2001**, *123*, 6933–6934.

79 W.-F. Liaw, J.-H. Lee, H.-B. Gau, C.-H. Chen, S.-J. Jung, C.-H. Hung, W.-Y. Chen, C.-H. Hu, G.-H. Lee, *J. Am. Chem. Soc.* **2002**, *124*, 1680–1688.

80 C.-H. Chen, Y.-S. Chang, C.-Y. Yang, T.-N. Chen, C.-M. Lee, W.-F. Liaw, *Dalton Trans.* **2004**, 137–143.

81 D. Swenson, N. C. Baenziger, D. Coucouvanis, *J. Am. Chem. Soc.* **1978**, *100*, 1932–1934.

82 A. Mueller, G. Henkel, *Z. Naturforsch, Teil B* **1995**, *50*, 1464–1468.

83 D. Sellmann, F. Geipel, F. Lauderbach, F. W. Heinemann, *Angew. Chem., Int. Ed. Engl.* **2002**, *41*, 632–634.

84 S. D. Davies, D. J. Evans, D. L. Hughes, S. Longhurst, J. R. Sanders, *Chem. Commun.* **1999**, 1935–1936.

85 S. M. Contakes, S. C. N. Hsu, T. B. Rauchfuss, S. R. Wilson, *Inorg. Chem.* **2002**, *41*, 1670–1678.

86 A. Kayal, T. B. Rauchfuss, *Inorg. Chem.* **2003**, *42*, 5046–5048.

87 J. Jiang, S. A. Koch, *Inorg. Chem.* **2002**, *41*, 158–160.

88 W.-F. Liaw, C.-Y. Chiang, G.-H. Lee, S.-M. Peng, C.-H. Lai, M. Y. Darensbourg, *Inorg. Chem.* **2000**, *39*, 480–484.

89 C.-H. Lai, J. H. Reibenspies, M. Y. Darensbourg, *Angew. Chem., Int. Ed. Engl.* **1996**, *35*, 2390–2393.

90 A. L. Eckermann, *Ph.D. Dissertation*, University of Illinois at Urbana – Champaign **2003**.

91 M. C. Smith, S. Longhurst, J. E. Barclay, S. P. Cramer, S. P. Davies, D. L. Hughes, W. W. Gu, D. J. Evans, *J. Chem. Soc., Dalton Trans.* **2001**, 1387–1388.

92 M. A. Reynolds, T. B. Rauchfuss, S. R. Wilson, *Organometallics* **2003**, *22*, 1619–1625.

93 D. Sellmann, P. Prakash, F. W. Heinemann, *Eur. J. Inorg. Chem.* **2004**, *9*, 1847–1858.

94 H. L. Drake, *Acetogenesis*, Chapman & Hall, New York, **1994**.

95 B. E. H. Maden, *Biochem. J.* **2000**, *350*, 609–629.

96 V. Svetlitchnyi, H. Dobbek, W. Meyer-Klaucke, T. Meins, B. Thiele, P. Römer, R. Huber, O. Meyer, O. *Proc. Nat. Acad. Sci., USA* **2004**, *101*, 446–451.

97 S. Gencic, D. A. Grahame, *J. Biol. Chem.* **2003**, *278*, 6101–6110.

98 W. Gu, S. Gencic, S. P. Cramer, D. A. Grahame, *J. Am. Chem. Soc.* **2003**, *125*, 15343–15351.

99 W. Huang, J. Jia, J. Cummings, M. Nelson, G. Schneider, Y. Lindqvist, *Structure* **1997**, *5*, 691–699.

100 T. I. Doukov, T. M. Iverson, J. Seravalli, S. W. Ragsdale, C. L. Drennan, *Science* **2002**, *298*, 567–572.

101 C. Darnault, A. Volbeda, E. J. Kim, P. Legrand, X. Vernede,

P. A. LINDAHL, J. C. FONTECILLA-CAMPS, *Nat. Struct. Biol.* **2003**, *10*, 271–279.

102 M. R. BRAMLETT, X. TAN, P. A. LINDAHL, *J. Am. Chem. Soc.* **2003**, *125*, 9316–9317.

103 S. W. RAGSDALE, H. G. WOOD, W. E. ANTHOLINE, *Proc. Nat. Acad. Sci., USA* **1985**, *82*, 6811–6814.

104 J. L. CRAFT, B. S. MANDIMUTSIRA, K. FUJITA, C. G. RIORDAN, T. C. BRUNOLD, *Inorg. Chem.* **2003**, *42*, 859–867.

105 J. XIA, Z. HU, C. V. POPESCU, P. A. LINDAHL, E. MÜNCK, *J. Am. Chem. Soc.* **1997**, *119*, 8301–8312.

106 J. SERAVALLI, M. KUMAR, S. W. RAGSDALE, *Biochemistry* **2002**, *41*, 1807–1819.

107 X. S. TAN, C. SEWELL, P. A. LINDAHL, *J. Am. Chem. Soc.* **2002**, *124*, 6277–6284.

108 C. E. WEBSTER, M. Y. DARENSBOURG, P. A. LINDAHL, M. B. HALL, *J. Am. Chem. Soc.* **2004**, *126*, 3410–3411.

109 D. P. BARONDEAU, P. A. LINDAHL, *J. Am. Chem. Soc.* **1997**, *119*, 3959–3970.

110 R. C. LINCK, C. W. SPAHN, T. B. RAUCHFUSS, S. R. WILSON, *J. Am. Chem. Soc.* **2003**, *125*, 8700–8701.

111 R. KRISHNAN, J. K. VOO, C. G. RIORDAN, L. ZAHKAROV, A. L. RHEINGOLD, *J. Am. Chem. Soc.* **2003**, *125*, 4422–4423.

112 Q. WANG, A. J. BLAKE, E. S. DAVIES, E. J. L. McINNES, C. WILSON, M. SCHROEDER, *Chem. Commun.* **2003**, 3012–3013.

113 R. KRISHNAN, C. G. RIORDAN, *J. Am. Chem. Soc.* **2004**, *126*, 4484–4485.

114 W. SHIN, P. A. LINDAHL, *Biochemistry* **1992**, *31*, 12870–12875.

115 M. L. GOLDEN, M. V. RAMPERSAD, J. H. REIBENSPIES, M. Y. DARENSBOURG, *Chem. Comm.* **2003**, 1824–1825.

116 M. S. RAM, C. G. RIORDAN, *J. Am. Chem. Soc.* **1995**, *117*, 2365–2366.

117 C. G. RIORDAN, M. S. RAM, G. P. A. YAP, L. LIABLE-SANDS, A. L. RHEINGOLD, A. MARCHAJ, J. R. NORTON, *J. Am. Chem. Soc.* **1997**, *119*, 1648–1655.

118 S. ANTEBI, H. ALPER, *Organometallics* **1986**, *5*, 596–598.

119 M. FURUYA, S. TSUTSUMINAI, H. NAGASAWA, N. KOMINE, M. HIRANO, S. KOMIYA, S. *Chem. Commun.* **2003**, 2046–2047.

120 D. SELLMANN, D. HAEUSSINGER, F. KNOCH, M. MOLL, *J. Am. Chem. Soc.* **1996**, *118*, 5368–5374.

121 G. C. TUCCI, R. H. HOLM, *J. Am. Chem. Soc.* **1995**, *117*, 6489–6496.

122 P. STAVROPOULOS, M. C. MUETTERTIES, M. CARRIE, R. H. HOLM, *J. Am. Chem. Soc.* **1991**, *113*, 8485–8492.

123 K. TANAKA, Y. KAWATA, T. TANAKA, *Chem. Lett.* **1974**, 831–832.

124 S. W. RAGSDALE, M. KUMAR, *Chem. Rev.* **1996**, *96*, 2515–2539.

125 P. A. LINDAHL, *Biochemistry* **2002**, *41*, 2097–2105.

126 R. K. THAUER, *Microbiology* **1998**, *144*, 2377–2406.

127 W. SHIN, S. H. LEE, J. W. SHIN, S. P. LEE, Y. KIM, *J. Am. Chem. Soc.* **2003**, *125*, 14688–14689.

128 H. DOBBEK, V. SVETLITCHNYI, J. LISS, O. MEYER, *J. Am. Chem. Soc.* **2004**, *126*, 5382–5387.

129 C. L. DRENNAN, J. W. PETERS, *Curr. Op. Struct. Biol.* **2003**, *13*, 220–226.

130 C. L. DRENNAN, J. HEO, M. D. SINTCHAK, E. SCHREITER, P. W. LUDDEN, *Proc. Natl. Acad. Sci.* **2001**, *98*, 11973–11978.

131 H. DOBBEK, V. SVETLITCHNYI, L. GREMER, R. HUBER, O. MEYER, *Science* **2001**, *293*, 1281–1285.

132 J. FENG, P. A. LINDAHL, *Biochemistry* **2004**, *43*, 1552–1559.

133 H. ARAKAWA, M. ARESTA, J. N. ARMOR, M. A. BARTEAU, E. J. BECKMAN, A. T. BELL, J. E. BERCAW, C. CREUTZ, E. DINJUS, D. A. DIXON, K. DOMEN, D. L. DuBOIS, J. ECKERT, E. FUJITA, D. H. GIBSON, W. A. GODDARD, D. W. GOODMAN, J. KELLER, G. J. KUBAS, H. H. KUNG, J. E. LYONS, L. E. MANZER, T. J. MARKS, K. MOROKUMA, K. M. NICHOLAS, R. A. PERIANA, L. QUE, J. ROSTRUP-NIELSON, W. M. H. SACHTLER, L. D. SCHMIDT, A. SEN, G. A. SOMORJAI, P. C. STAIR, B. R. STULTS, W. TUMAS, W. *Chem. Rev.* **2001**, *101*, 953–996.

134 C. G. RIORDAN, in T. J. MEYER, J. A. McCLEVERTY (Eds.), *Comprehensive Coordination Chemistry II*, Pergamon Press, London, **2004**, Vol. 8.

135 Z. LU, R. H. CRABTREE, *J. Am. Chem. Soc.* **1995**, *117*, 3994–3998.

136 J. Zhou, J. W. Raebiger,
C. A. Crawford, R. H. Holm, *J. Am.
Chem. Soc.* **1997**, *119*, 6242–6250.

137 S. Ciurli, P. K. Ross, M. J. Scott,
S. B. Yu, R. H. Holm, *J. Am. Chem. Soc.*
1992, *114*, 5415–5423.

138 R. Panda, Y. Zhang,
C. R. McLauchlan, P. V. Rao, F. A. T.
de Oliveira, E. Münck, R. H. Holm,
J. Am. Chem. Soc. **2004**, *126*, 6448–6459.

139 H. Dobbek, L. Gremer, R. Kiefersauer,
R. Huber, O. Meyer, *Proc. Nat. Acad.
Sci.* **2002**, *99*, 15971–15976.

140 M. Gnida, R. Ferner, L. Gremer,
O. Meyer, W. Meyer-Klaucke,
Biochemistry **2003**, *42*, 222–230.

141 G. N. George, I. J. Pickering, E. Y. Yu,
R. C. Prince, S. A. Bursakov,
O. Y. Gavel, I. Moura, J. J. G. Moura,
J. Am. Chem. Soc. **2000**, *122*, 8321–8322.

142 C. J. Ruffing, T. B. Rauchfuss,
Organometallics **1985**, *4*, 524–528.

143 B. S. Lim, M. W. Willer, M. Miao,
R. H. Holm, *J. Am. Chem. Soc.* **2001**,
123, 8343–8349.

144 S. Sarkar, S. B. S. Mishra, *Coord.
Chem. Rev.* **1984**, *59*, 239–264.

145 M. Krüger, A. Meyerdierks,
F. O. Glöckner, R. Amann, F. Widdel,
M. Kube, R. Reinhardt, J. Kahnt,
R. Böcher, R. K. Thauer, S. Shima,
Nature **2003**, *426*, 878–881.

146 W. Grabarse, F. Mahlert, E. C. Duin,
M. Goubeaud, S. Shima,
R. K. Thauer, V. Lamzin, U. Ermler,
J. Mol. Biol. **2001**, *309*, 315–330.

147 E. C. Duin, N. J. Cosper, F. Mahlert,
R. K. Thauer, R. A. Scott, *J. Biol.
Inorg. Chem.* **2003**, *8*, 141–148.

148 U. Ermler, W. Grabarse, S. Shima,
M. Goubeaud, R. K. Thauer, *Cur. Op.
Struct. Biol.* **1998**, *8*, 749–758.

149 C. Finazzo, J. Harmer, C. Bauer,
B. Jaun, E. C. Duin, F. Mahlert,
M. Goenrich, R. K. Thauer,
S. Van Doorslaer, A. Schweiger,
J. Am. Chem. Soc. **2003**, *125*, 4988–4989.

150 Y. Ahn, J. A. Krzycki, H. G. Floss,
J. Am. Chem. Soc. **1991**, *113*, 4700–4701.

151 S. K. Lin, B. Jaun, *Helv. Chim. Acta*
1991, *74*, 1725–1738.

152 V. Pelmenschikov, M. R. A. Blomberg,
P. E. M. Siegbahn, R. H. Crabtree,
J. Am. Chem. Soc. **2002**, *124*,
4039–4049.

153 B. Jaun, A. Pfaltz, *J. Chem. Soc.,
Chem. Commun.* **1988**, 293–294.

154 A. Berkessel, *Bioinorg. Chem.* **1997**,
431–445.

155 C. J. A. Daley, R. H. Holm,
Inorg. Chem. **2001**, *40*, 2785–2793.

156 L. Signor, C. Knuppe, R. Hug,
B. Schweizer, A. Pfaltz, B. Jaun,
Chem. Eur. J. **2000**, *6*, 3508–3516.

Subject Index

Bioorganometallics: Biomolecules, Labeling, Medicine. Edited by Gérard Jaouen
Copyright © 2006 Wiley-VCH Verlag GmbH & Co. KGaA, Weinheim
ISBN: 3-527-30990-X